Vermitechnology: Economic, Environmental and Agricultural Sustainability

Sesan Abiodun Aransiola · Naga Raju Maddela
Editors

Vermitechnology: Economic, Environmental and Agricultural Sustainability

Springer

Editors
Sesan Abiodun Aransiola
Department of Microbiology, Faculty of Science
University of Abuja
Abuja, Nigeria

Naga Raju Maddela
Departamento de Ciencias Biológicas, Facultad de Ciencias de la Salud
Universidad Técnica de Manabí
Portoviejo, Ecuador

ISBN 978-3-031-96211-0 ISBN 978-3-031-96212-7 (eBook)
https://doi.org/10.1007/978-3-031-96212-7

This Springer imprint is published by the registered company Springer Nature Switzerland AG
The registered company address is: Gewerbestrasse 11, 6330 Cham, Switzerland

Foreword

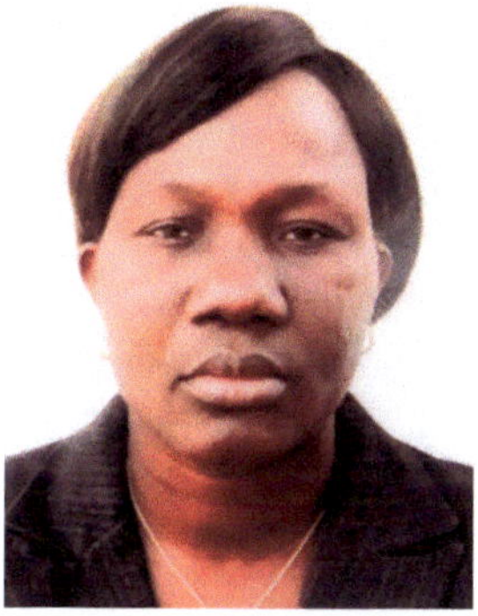

The idea of this edited volume *Vermitechnology: Economic, Environmental and Agricultural Sustainability* is highly instructive with current scientific information about an aspect of biotechnology called vermitechnology, which involves the use of earthworms with some influencing factors, such as water and microorganisms, in converting organic wastes to useful quality manure for sustainable agricultural practices. This extensively interesting acceptable technology of this topic includes vermi-protection, vermi-remediation, vermi-production, vermi-filtration, vermicomposting, vermi-agro production, which presents an economically viable, eco-friendly, and socially acceptable. The economic implications of vermitechnology, however, generate quality by-products and sustainable organic farming, which have a place in mitigating global warming. In the management of global safe food, vermitechnology is employed in waste management to replace inorganic or chemical farming with the organic system of farming and its safety precursors by contributions to renewing the soil fertility and plant growth for healthy living and effective symbiotic relationship of microorganisms in the soil. This technology has a direct relationship with humans in bioconversion and biofiltration of wastewater, the agro-industrial waste treatments, vermiremediations of polluted soil with heavy metals and (polycyclic aromatic hydrocarbon) PAHs, and the role in plastics conversions. The book, *Vermitechnology: Economic, Environmental and Agricultural Sustainability*, is a good collection of independent chapters that present full insights into the study of vermitechnology for economic, environmental, and agricultural sustainability. In an expansive form, this book focuses on factors influencing vermitechnology, waste to wealth and economic impact of vermitechnology, usage of this technology to remediate different environmental pollutants, and the symbiotic relationship of microbial diversity in the soil, among others. I therefore have no doubt that the above focused areas provide adequate information on vermitechnology and do fill the scientific knowledge gaps.

March 2023

Agarry Oluwabunmi Olaitan, Ph.D., Professor
Professor and Dean, Faculty of Science
University of Abuja
Abuja, Nigeria

Preface

This book, titled *Vermitechnology: Economic, Environmental and Agricultural Sustainability*, has been designed to give up-to-date scientific information on the economic, environmental, and agricultural sustainability of vermitechnology. Already, it has been made clear that one of the prominent problems which are being faced by mankind globally is solid waste management. Due to the successive increase in the amount of waste being generated from various developments worldwide, there is an encroachment of fertile areas and a population explosion. Also, due to fast urbanization, a massive amount of waste is being generated. For the conversion of solid organic waste to compost, vermitechnology is employed. Vermiculture as a discipline of biotechnology involves the breeding and propagation of earthworms. Vermiculture has become a significant tool of waste recycling worldwide and has also been considered a separate and fruitful discipline of biology. "Vermitechnology" involves a low-cost and environmentally sound waste management practice involving the use of earthworms in the form of natural bioreactors. Vermitechnology is a system harnessing earthworms for bio-conservation of waste into vermicompost, which has extensive application in waste management and sustainable organic farming, and has proved to be one of the efficient methods of managing waste with least complexity and economic viability. Vermicomposting is a viable option to handle solid waste in an environmentally friendly way.

This book reviews the history and economic implications of vermitechnology. This technology has utilized domestic market as well as industrial marketing of it is being done in several counties like USA, Canada, Australia, Italy and Japan. The process of vermicomposting, which started in Ontario (Canada) in the year 1970, has enormous economic profits. Vermitechnology, being a biological process, involves interactions between earthworms and microorganisms. Due to this, there is an efficient conversion of different types of organic wastes into nutrient-rich manure. There is also the use of worms during vermicomposting process, due to which worms are able to transfer organic waste into a nutrient-rich fertilizer. Mutual actions exist between earthworms and microorganisms, which alter the physical, chemical, and biological properties of waste material and convert it into vermicompost. Vermicompost and vermiculture associated with other biological inputs have actually been used to grow vegetables and other crops successfully and have been found to be economical and productive. Therefore, organic farming helps to provide many advantages such as eliminating the use of chemicals in the form of fertilizers/pesticides; recycling and regenerating waste into wealth; improving soil, plant, animal, and human health; and creating an eco-friendly, sustainable, and economical bio-system models.

With these interesting areas in vermitechnology, this book has been structured to accommodate the economic, environmental, and agricultural sustainability. To justify this, 3 parts have been designed with 18 chapters: *First Part: Vermitechnology—Preparation, Distribution and Economical Impacts. Second Part: Vermitechnology—Application in Environmental Sustainability and Third Part: Vermitechnology—Application in Agricultural Sustainability.* First Part consists of five chapters. Chapter "General Perspectives of Vermitechnology—An Overview" was general perspectives of vermitechnology. Chapters "Vermitechnology and the Influencing Factors", "Vermitechnology as a Sustainable Solution: Transforming Organic Waste into Economic and Environmental Wealth", "Vermicomposting Process—Optimisation", and "Molecules of Therapeutic and Prophylactic Value in the Earthworm" revealed the vermitecnology and influencing factors, waste to wealth-strength of vermitechnology and economic impact, role of various factors influencing vermicomposting process, and molecules of therapeutic and prophylactic value in the earthworm, respectively. Second Part consists of 6 chapters, assisted phytoremediation of trace toxic elements (TTE): the role of vermicompost, vermiconversion and vermifiltration in waste water treatment, vermistabilization of earthworms, Agro-industrial sludge and vermitechnology, biorestoration of polycyclic aromatic hydrocarbons and heavy metals contaminated soil: the role of vermitechnology and sustainable management of agro-industrial waste using vermitechnology. Third Part consists of 7 chapters, which focused on phytoremediation and vermicomposting, symbiotic relationship of microbial communities and vermicompost, vermicast, vermiwash: a substitute to chemical fertilizer, role of vermitechnology in plant growth and nutrient, organic farming- the role of vermitechnology, vermicompost by-products for healthy agriculture and the future directions guidelines of research in vermitechnology was adequately discussed.

The chapters were contributed by 73 Academicians/Scientists/Researchers from 10 different countries (China, Ecuador, India, Nigeria, Canada, Ghana, South Africa, the United States, the United Kingdom, and Hong Kong) across the globe.

Abuja, Nigeria Sesan Abiodun Aransiola, Ph.D.
Portoviejo, Ecuador Naga Raju Maddela, Ph.D.

Acknowledgments

With great honor, the authors acknowledge the support of the chapter's contributors for their valuable contributions and timely responses for the success of this project. All contributors are immensely appreciated for their eagerness and inordinate support for this volume to be ready in the scheduled time; the authors therefore appreciate their teamwork and partnership. They also thank anonymous reviewers for their constructive criticism, which has helped us in improving the quality of this book by inviting the experts to contribute the additional chapters. The authors greatly acknowledge the Springer Editorial and Production team for their support; without their guidelines, this project would not have been finished in such a very short time. It is an honest honor and privilege to work with them all. Finally, yet importantly, we are very much thankful to the management and colleagues at the University of Abuja (Nigeria) and Universidad Técnica de Manabí (Ecuador) for their unrestricted backing and for the establishment of treasured propositions at the time of book proposal and final book preparation.

Abuja, Nigeria — Sesan Abiodun Aransiola, Ph.D.
Portoviejo, Ecuador — Naga Raju Maddela, Ph.D.

Contents

About the Editors

Sesan Abiodun Aransiola is a notable Ph.D. holder in Environmental Microbiology in the Department of Microbiology, Federal University of Technology, Minna, Nigeria. He obtained his first degree (B.Tech.) and Master's Degree (M.Tech.) from the same Department in 2009 and 2014, respectively. He has demonstrated his research expertise in the production of vermicasts from vermicoposting (Vermitechnology) of organic wastes to assist plants in the remediation of polluted soil with heavy metals. Currently, he is a Lecturer at the Department of Microbiology, University of Abuja, Nigeria. His expertise spans environmental health, phytoremediation, vermicomposting, marine resources and bioremediation of pollutants, and soil, including plastics. Aransiola's research is deeply rooted in innovative solutions for soil and water remediation, with significant contributions to microbial biotechnology and waste management. He is an award-winning researcher and has over 100 publications, including book chapters, research, and review articles of good international repute with a high impact factor. He has coedited scientific books of global interest, which include *Microbial Biotechnology for Bioenergy* (Elsevier). *Ecological Interplays in Microbial Enzymology, Prospects for Soil Regeneration and Its Impact on Environmental Protection, Soil Microbiome in Green Technology Sustainability, Marine Bioprospecting for Sustainable Blue-Bioeconomy* (Springer). *Microbial Biofilms—Applications and Control, Marine Greens:*

Environmental, Agricultural, Industrial and Biomedical Applications, Emerging Contaminants in Food and Food Products, Phytoremediation in Food Safety: Risks and Prospects, Biotoxins in Food: Threats and Detection (CRC Press), with many more in the process. Also, has worked to the rank of an assistant chief scientific officer at Bioresources Development Centre, National Biotechnology Research and Development Agency, Nigeria, where he was involved in the use of vinasse (a by-product of ethanol) as bio-fertilizer to reclaim polluted soil for agricultural purposes. Dr. Aransiola's academic journey is complemented by a robust teaching career and active involvement in community services, where he leverages his knowledge to foster sustainability. His contributions continue to impact both academia and the broader environmental sciences community. Abiodun is a member of the Nigerian Society for Microbiology and the American Society for Microbiology, among others.

Naga Raju Maddela received his M.Sc. (1996–1998) and Ph.D. (2012) in Microbiology from Sri Krishnadevaraya University, Anantapuramu, India. During his doctoral program in the area of Environmental Microbiology, he investigated the effects of industrial effluents/insecticides on soil microorganisms and their biological activities, and he has been working as a Faculty member in Microbiology since 1998, teaching undergraduate and postgraduate students. He worked on Prometeo Investigator (fellowship received from SENESCYT) at Universidad Estatal Amazónica, Ecuador, during 2013–2015, and received "Postdoctoral Fellowship" (2016–2018) from Sun Yat-sen University, China. He also received external funding from the "China Postdoctoral Science Foundation" in 2017, internal funding from the "Universidad Técnica de Manabí" in 2020. He participated in national/international conferences and presented research data in China, Cuba, Ecuador, India, and Singapore. Currently, he is working as a Full Professor at the Facultad de Ciencias de la Salud, Universidad Técnica de Manabí, Portoviejo, Ecuador. He has been actively publishing scientific articles, books (authored and edited), and chapters since 2007. As of now, he has published 150 articles, 30 books, and 50 book chapters.

Vermitechnology—Preparation, Distribution and Economical Impacts

General Perspectives of Vermitechnology—An Overview

S. A. Aransiola, A. E. Oyewumi, U. R. Attah-Olottah, O. P. Abioye, and Naga Raju Maddela

Abstract Vermicomposting is a biotechnological method of composting in which specific species of earthworms are employed to speed up waste conversion and provide a higher-quality end product. It is a nourishing plant diet that is high in nitrogen and carbon, also necessary for soil microbiomes including nitrogen-fixing bacteria and mycorrhizal fungi, which are good growth boosters. It can function as a good nutritional biofertilizer that is very effective at promoting growth. Earthworms are important detritus feeders that are critical to the decomposition of organic materials and the metabolism of soil. They are known as soil health indicators. Enhancing soil fertility is the outcome of a complex process that involves the partial breakdown of organic waste and mixing it with mucus and gut microbial flora in the form of earthworm cast. Vermitechnology is an important technology in solid waste management, especially organic wastes. For plant growth and productivity, vermicomposting through vermifiltration plays a very vital and useful role. Due to the importance of earthworms, vermitechnology, which makes use of both surface- and subsurface-dwelling indigenous earthworm varieties, has been developed. This technology is used for composting and soil management. As a result, the vermicomposting process may effectively recycle organic waste, producing vermiwash and vermicompost, both of which have been shown to be crucial to the health and productivity of plants. Agriculture has traditionally stood for the sustainable production of food with the least possible disruption to the natural environment. This chapter focuses on the general perspectives of vermitechnology, its role in plant growth and

S. A. Aransiola (✉)
Department of Microbiology, University of Abuja, Abuja, Nigeria
e-mail: blessedabiodun@gmail.com

Bioresources Development Centre, National Biotechnology Development Agency, Ogbomoso, Nigeria

A. E. Oyewumi · U. R. Attah-Olottah · O. P. Abioye
Department of Microbiology, Federal University of Technology, Minna, Nigeria

N. R. Maddela
Department of Biological Sciences, Faculty of Health Sciences, Universidad Técnica de Manabí, Portoviejo, Ecuador

S. A. Aransiola and N. R. Maddela (eds.), *Vermitechnology: Economic, Environmental and Agricultural Sustainability*, https://doi.org/10.1007/978-3-031-96212-7_1

how phytoremediation processes could be enhanced by the use of vermicomposting by-products.

Keywords Vermitechnology · Earthworm · Biofertilizer · Vermiculture · Vermicompost

1 Introduction

Multiple crops can be grown on the same piece of land each year, thanks to chemical herbicides and fertilizers, but this puts stress on the soil and depletes its nutrient supply (Ansari et al., 2016). This cycle of chemical inputs is required continuously to maintain the high levels of production required in many developing countries such as Nigeria, Bangladesh, and so on (Ansari et al., 2016; Ojuolape et al., 2015). The soil structure is no longer optimal for sustainable farming, and soil fertility is diminishing (Ansari et al., 2016). Historically, agricultural practices have been a strong advocate for the sustainable production of food with minimal disruption to the natural environment. Chemical fertilizers were introduced during the first Green Revolution, which greatly increased production and helped underdeveloped nations feed their expanding populations. However, over time, the soil's productivity has declined due to a shortage of organic nutrients. Although it has not gained much traction, this is quickly emerging as the best way to counteract the negative effects of chemical pesticides. A subfield of biotechnology known as vermitechnology can neutralize the effects of these chemical fertilizers (Ansari et al., 2016; Aransiola et al., 2024).

Vermitechnology is an important aspect of biotechnology that deals with the involvement of earthworms to process the organic matter and transform it into high-quality manure. This term has been defined as the method of converting wastes into useful agricultural resources through the processing done by earthworms. It is a highly eco-friendly, economically viable, and socially acceptable technology having several categories: vermicomposting, vermifiltration, vermiremediation, vermi-agro production, vermiprotection, and vermiproduction (Sinha et al., 2010).

Vermitechnology uses surface- and subsurface-dwelling local earthworm species for soil management and composting. It is widely acknowledged that earthworms play a crucial role in agriculture. Along with other microbes, earthworms have been crucial in controlling soil processes, preserving soil fertility, and promoting nutrient cycling. With their ability to create aggregates and enhance the physical conditions for plant growth and nutrient uptake, earthworms have a significant impact on soil structure. They also increase soil fertility by accelerating the breakdown of organic debris and plant litter, which releases nutrients in a form that plants can absorb. Vermiculture and vermicomposting, however, are solid cornerstones of vermitechnology. The major protagonists, earthworms, are regarded as natural bioreactors because they multiply with other microbes and create the ideal environment for the biodegradation of trash (Sinha et al., 2009). This chapter will, therefore, review the general perspective of vermitechnology.

2 Vermiculture

Vermiculture is the practice of raising large populations of earthworms on organic material that can be composted or broken down. The term "vermiculture" (Latin for "worm farming") has been used for at least a century. Studies on vermiculture are related to sustainable agricultural methods, soil detoxification, and waste management. There are just two regions where commercial vermiculture is practiced. The first is the processing of vermicompost, and the second is the creation of worm biomass. Worm biomass is produced for use as a protein source in fish and poultry farms. The vermicomposting of sewage, sewage sludge, or other comparable wastes is known as vermistabilization, in contrast. Earthworms boost plant growth (39%) and grain yield (35%), particularly in the production of grains.

3 Vermicomposting

Vermicomposting is a straightforward biotechnological method of composting in which epigeic species of earthworms are employed to speed up the waste-to-product conversion process and provide a superior final product. During vermicomposting, earthworms ingest organic wastes and microorganisms by their mouth. The ingested organic matter is broken down into smaller particles in the earthworm's gizzard and pharynx, and then the broken-down material is mixed with microorganisms, such as bacteria and fungi, in the earthworm's gut for the release of nutrient (N, P, K) by microorganisms. The earthworm excretes the nutrient-rich waste, known as castings, through its anus. Microorganisms, however, continue to decompose the castings for more nutrient release and humus-rich vermicompost until maturity (Ansari et al., 2016). Vermicompost is an organic, nutrient-rich soil conditioner that can be used to enhance soil conditions in a variety of soil types. The usage of earthworms is crucial to this procedure since they serve as the composting agents converting organic material into a durable, nontoxic, well-structured material with a potential for great economic value. Additionally, it improves soil for plant growth (Abioye et al., 2020; Ansari et al., 2016). Vermicomposting is an established method for getting rid of rubbish waste that has a lot of positive environmental effects. This method also benefits the soil and reduces the need for artificial fertilizers (Ansari et al., 2016). Different types of earthworms play crucial roles in vermicomposting (Table 1) (Babaniyi et al., 2023).

Aransiola et al. (2022) reported the production of vermicompost from goat manure and chicken droppings to assist phytoextractors (*Sida acuta* and *Melissa officinalis* L.) to restore heavy metal polluted soils in Madaka district, Niger State, Nigeria.

Table 1 Types and roles of earthworms used in vermicomposting

S/N	Types of earthworms in vermicomposting	Role(s) of earthworms in vermicomposting
1	Canadian nightcrawler (*Lumbricus terrestris*)	✓ Can be used for vermicomposting but may not be as efficient as red wigglers ✓ Native to North America
2	Red wiggler (*Eisenia fetida*)	✓ Well adapted to breaking down organic matter ✓ Most commonly used species for vermicomposting ✓ Have high reproductive rate and can adapt to changes in the environment
3	European nightcrawler (*Eisenia hortensis*)	✓ Produce higher-quality vermicompost ✓ Similar to red wiggler but larger in size ✓ More tolerant of cooler temperatures
4	African nightcrawler (*Eudrilus eugeniae*)	✓ Native to tropical regions ✓ Well adapted to high temperatures and humidity ✓ Produce a large amount of castings (vermicompost)
5	Asian jumping worm (*Amynthas agrestis*)	✓ Native to Asia ✓ Can be invasive in some regions ✓ Can be used for vermicomposting but may require more maintenance

The investigators produced casts through vermicomposting and concluded that the physical and chemical properties (Plate 1) of the casts are rich in carbon, organic matter, nitrogen, and other constituents as found in any organic fertilizers (Aransiola et al., 2023). Therefore, vermicomposting is a good source of organic manure for agricultural purposes. In their conclusion, the polluted mining environment landscape was recovered for agricultural purposes, hence indicating the effectiveness of vermitechnology in the remediation of polluted soil.

4 Setting up a Vermicomposting Unit

There are numerous techniques to set up vermicomposting systems. This composter can be installed in a sizable box, a bucket, a cement bin, a bin or basket, or even a trench dug into the ground. It is crucial to remember that a vermicomposting unit should be at least 1 m deep but may be as wide as desired. The following are the steps needed for vermicomposting (Samal et al., 2019):

Step 1: Choose a Location

(i) Select a shaded area: Vermicomposting requires a shaded area to maintain optimal temperatures.

(ii) Ensure good ventilation: Adequate ventilation helps maintain oxygen levels and prevents anaerobic conditions.

Plate 1 **a** Earthworms (*Eisenia fetida*) for vermicomposting, **b** production stage of goat manure vermicast, **c** produced goat manure vermicast, **d** production stage of chicken droppings vermicast, and **e** produced chicken dropping vermicast (Aransiola et al., 2024)

(iii) Keep it accessible: Choose a location that is easily accessible for maintenance and harvesting.

Step 2: Select the Worms

(i) Red wiggler worms (*Eisenia fetida*): These worms are ideal for vermicomposting due to their high reproductive rate and ability to break down organic matter.
(ii) Purchase worms from a reputable supplier: Ensure the worms are healthy and suitable for vermicomposting.

Step 3: Prepare the Bedding

(i) Choose a carbon-rich bedding material: Coconut coir, shredded newspaper, or peat moss work well.
(ii) Add a nitrogen-rich component: Include a small amount of nitrogen-rich material, such as food scraps or manure.
(iii) Maintain a 2:1 carbon-to-nitrogen ratio: This ratio ensures optimal decomposition and worm health.

Step 4: Set Up the Vermicomposting System

(i) Choose a container: Select a container that is at least 6–8 inches deep and has adequate drainage.
(ii) Add a 4- to 6-inch layer of bedding: Place the prepared bedding material at the bottom of the container.
(iii) Add the worms: Gently add the worms to the bedding material.
(iv) Add food and maintain moisture: Provide a consistent food source and maintain optimal moisture levels.

Step 5: Maintain the Vermicomposting System

(i) Monitor temperature: Maintain a temperature range of 55–77 °F (13–25 °C).
(ii) Maintain moisture: Keep the bedding material consistently moist, like a damp sponge.
(iii) Add food regularly: Provide a consistent food source, such as fruit and vegetable scraps.
(iv) Harvest the vermicompost: After 2–3 months, harvest the vermicompost and repeat the process.

5 Earthworms in the Soil

Earthworms aid in the recycling of organic nutrients for plants' efficient growth. In addition to living in the soil, earthworms also help to change its physical and chemical composition, promoting soil fertility and plant growth (Aransiola et al., 2024). They play a major role in soil aeration, soil porosity, decomposition, aggregation, compaction, and stimulation (Jha et al., 2024). Earthworm casts, which are enriched with microorganisms, are found in soils where earthworms live. As an indicator, earthworms play a crucial role in these processes. In contrast to traditional farming, which uses chemical fertilizers, organic supplements such as vermicompost and the inoculation of earthworms can help the formation of humus and limit the loss of nutrients from the soil by their delayed release.

6 Economic Implications of Vermitechnology

In order to conduct sustainable agriculture, one must study new, cutting-edge techniques created by farm scientists and farmers alike. One must also learn from farmers' old knowledge and ways and put what was beneficial and timely into practice (Aransiola et al., 2022; Hussaini, 2013). Vermitechnology presents many benefits such as new business opportunities, job creation, improved waste management, increased crop yield, and enhanced food security (Hussaini, 2013; Jha et al., 2024). Hence, the economic implications of vermitechnology are as follows:

6.1 Reduced Usage of Chemical Pesticides and Cost Cutting

The development of high-yielding crop types that were more prone to pests and diseases required the widespread use of chemical pesticides. Crop pests and diseases developed "biological resistance" as a result of repeated use of chemical pesticides, necessitating progressively larger doses to remove them. Vermicompost has been shown in recent years to be effective at protecting plants against a variety of pests and diseases, either by repelling or suppressing them, by "inducing biological resistance" in plants to combat them, or by actually killing them through pesticide action (Jha et al., 2024). In agricultural settings where earthworms and vermicompost were used, pesticide application was greatly reduced (Aransiola et al., 2024). Studies have showed the use of vermicompost to help in the rate of disease by 75% which significantly cut down the cost of food production (Hussaini, 2013; Samal et al., 2019). Also, vermicompost may hold onto more soil moisture, which lowers the need for irrigation water by roughly 30 to 40%, thereby reducing the usage of water for irrigation and cutting costs.

6.2 Improvement of Growth and Higher Yield

According to studies, crops develop more successfully when only a tiny amount of vermicompost is used. The strongest growth responses were observed in all growth trials, according to Subler and Kirsch (1998), when the vermicompost made up between 10 and 20% of the total volume of the container medium. Valani (2009) discovered that compared to pot soils containing 400 g and 500 g of vermicompost, those with 200 g of vermicompost produced superior growth in wheat crops. Singh (2009) discovered that the growth and production of wheat crops grew steadily over the years in the agricultural plots where vermicompost was applied in the second, third, and fourth consecutive years at the same rate of vermicompost application, or a rate of 20 Q/ha. The yield was 38.8 Q/ha in the fourth consecutive year, which was extremely close to the yield (40.1 Q/ha) where vermicompost was applied at a rate of 25 Q/ha. As baby worms emerge from their cocoons over time, the use of vermicompost in farm soil eventually results in an increase in the number of earthworms in the farmland. This implies that the use of vermicompost can be gradually decreased over time as the worms improve the physical, chemical, and biological qualities of the soil.

As the soil's natural fertility is improved and restored, the yield per hectare can also rise even further. According to Webster (2005), a single application of vermicompost increased the yield of "cherries" over the next 3 years. When the vermicompost was covered in "mulch," the yield was substantially higher. Trees treated with 5- and 20-mm vermicompost + mulch produced cherries at the first harvest worth AU $63.92 and AU $0.42 per tree, respectively. With 20 mm of vermicompost, the plants produced cherries costing AU $36.46 per tree at the first harvest and AU $40.48 per tree after three harvests. The agronomic effects of compost in vineyards were also researched by Webster (2005), who found that vines treated with compost produced 23% more grapes due to an increase of 18% in bunch counts. The extra grape harvest was worth an extra AU $3400/ha.

7 Vermiculture as a Commercial Commodity

Vermiculture is a rising sector that not only manages trash and land extremely inexpensively, but also promotes "sustainable agriculture" by increasing crop yield in both quantity and quality at a far lower economic cost than pricey agrochemicals (Bogdanov et al., 1996). Earthworms not only turn "waste" into "wealth," but they also develop into a valuable resource known as worm biomass. Vermicompost production on a large scale has the potential to replace chemical fertilizers, and the utilization of protein-rich "earthworms" in agriculture and other related sectors presents a solid financial opportunity.

Every country with a rising population is producing enormous amounts of municipal solid waste (MSW), thus there will never be a shortage of raw materials for making vermicompost. Vermiculture has also improved the standard of living for the underprivileged in India and created chances for independent work for the jobless. To encourage rural people to collect garbage from villages and farmers, vermicompost it, and sell both worms and vermicompost to farmers, nongovernmental organizations (NGOs) are giving out cement tanks and 1000 worms in a number of Indian communities. People make between 5 and 6 lakh rupees (about AU $1,520,000) per year from selling worms and their vermicompost to farmers (Samal et al., 2019). According to estimates, one ton of earthworm biomass typically contains approximately one million worms. At the end of a year, one million worms that multiply by two every 2 months can number 64 million. Each adult worm, especially *Eisenia fetida*, can consume organic waste equivalent to around 1% of its body weight each day. This means that 64 million worms, each weighing 1% of a ton, can consume 1% of a ton of garbage each day.

Earthworm biomass is a desirable by-product of all vermiculture techniques and is a rich source of nutritional "worm meal." It is utilized as feed material in the poultry, dairy, and fisheries industries because it is high in protein (65%) and contains 70–80% of the high-quality essential amino acids "lysine" and "methionine." Additionally, as it has antipathogenic effects, the by-product of vermiculture techniques is employed in the production of medicines and in the creation of "antibiotics" from coelomic fluid.

8 Municipal Solid Waste Management by Vermitechnology

Vermicomposting as a method of recycling organic waste has been tested in several nations for a variety of objectives and at various operational sizes (Moledor et al., 2016). In India and the Philippines, it has been evaluated for processing human biosolids and organic industrial wastes, such as manure from cattle breeding facilities, coffee industry byproducts (Murthy & Naidu, 2012), and residues from palm oil mills (Jha et al., 2024; Murthy & Naidu, 2012; Singh et al., 2011a, 2011b).

Vermicomposting is used in other research to support the dairy, poultry, food, abattoir, and olive oil sectors (Munnoli et al., 2010a, 2010b). The method of vermicomposting has various advantages. Earthworm processing is a safe, sanitary, and scalable method for managing solid waste that decreases the volume of organic waste by around 50% (Adhikary, 2012; Singh et al., 2011a, 2011b). When there is a preference for a quicker decomposition rate (Sinha et al., 2002), greater reduction of heavy metals (Aransiola et al., 2021; Samal et al., 2019; Singh et al., 2011a, 2011b), pathogen stabilization, and/or lack of odors, there is evidence to suggest that it is preferable to the more popular and well-known practice of composting (Lazcano et al., 2008).

Second, adding vermicast to the soil boosts its physical, chemical, and biological qualities as well as its fertility (Singh et al., 2011a, 2011b), while also supplying plants with crucial nutrients and promoting plant development (Munnoli et al., 2010a, 2010b). The nitrogen, phosphorus, and potassium content of potting soil treated with a standard inorganic fertilizer and various composts is reported by Atiyeh et al. (2000) (Fig. 1).

In addition, vermicast's abilities as a pesticide are the subject of an increasing corpus of research. Many diseases, such as damping-off (*Pythium*, *Rhizoctonia*), root rot (*Phytophthora*), sugar beet cyst nematode (*Heterodera schachtii*), as well as pests including aphids, mealybugs, cucumber beetles, and tobacco hornworms, have been proven to be considerably suppressed by adding vermicast to growth media (Moledor, 2014). Vermicompost can increase the marketable fruit production by up to 58.6%, according to another study that examined the reduction in albinism, damage, deformity, and Botrytis rot signs in strawberries (Singh et al., 2008). Vermicast is effectively a two-in-one soil supplement since it has both fertilizer and insecticidal qualities.

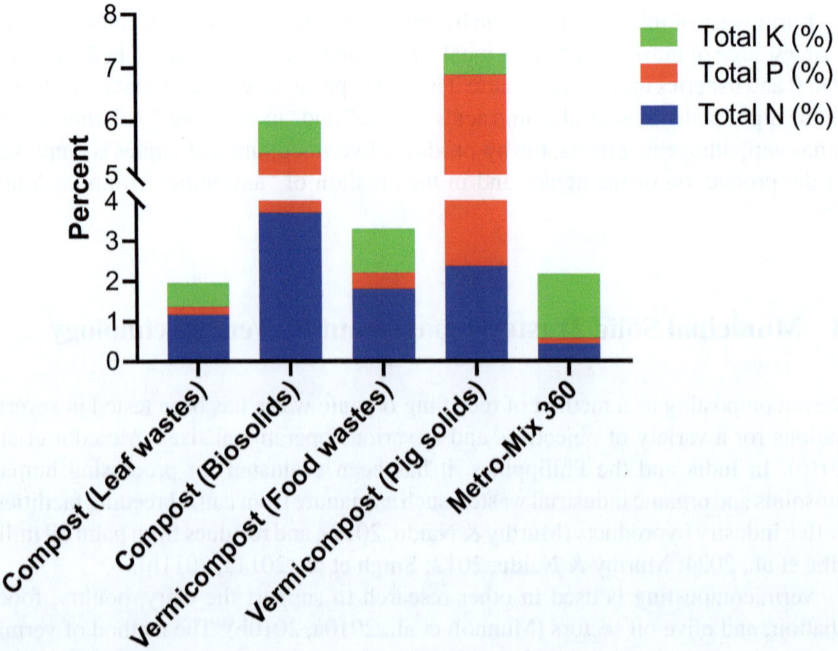

Fig. 1 Nitrogen, phosphorus, and potassium content of potting soil treated with a standard inorganic fertilizer and various composts

9 The Role of Vermitechnology in Agriculture

Composting is a long-standing agricultural practice that dates back to the existence of the first man, as food for the soil. Compost is an organic mass produced from organic waste like vegetable scraps, food waste, slaughterhouse waste, green waste, and human waste. It is an effective alternative to synthetic fertilizers and helps to improve soil structure and reduce erosion. The use of compost by farmers reduces their environmental impact and helps to preserve soil health and ensure sustainable agriculture for future generations, hence the use of Vermitechnology (Kamboj et al., 2022). Vermitechnology is the ability of earthworms to convert organic waste into high-quality compost, which serves as fertilizer for crops. This process helps reduce waste and improve soil fertility, promoting sustainable agriculture (Gopi, 2017). Vermitechnology is an environmentally friendly technology that utilizes earthworms to convert waste into valuable resources. It is a sustainable solution to managing waste, as it not only reduces the amount of waste in the environment but also produces a beneficial product in the form of compost. This technology has gained attention globally due to its potential to create wealth from waste while benefiting the environment and increasing microbial activity (Singh et al., 2022). According to Gopi

(2017), vermitechnology is a term that canopies several categories such as vermiremediation, vermifiltration, vermiculture, vermiprotection, vermicomposting, vermi-agro production and vermiproduction. There are two pillars into which all these categories are subclassified: vermicomposting and vermiculture.

Vermitechnology is incomplete without the invertebrate giants in the soil, the earthworms. Earthworms are distributed all over the world and play a very important role in soil diversity. Earthworms stimulate soil health by breaking down organic matter, improving soil structure and fertility, and increasing soil moisture retention through burrowing and excretion of castings (Ansari & Ismail, 2012). There are important factors that influence the use of earthworms in vermitechnology. The ecological strata, pH, organic matter, moisture, soil texture, and temperature all play a crucial role in the growth and survival of earthworms. These factors have to be observed to ensure a successful vermitechnology setup (Ansari & Ismail, 2012).

10 Vermitechnology in Agriculture

The excessive use of chemical fertilizers and pesticides has caused significant environmental harm, including water and land pollution. These chemicals can leach into waterways and soil, causing damage to wildlife and potentially affecting human health. Additionally, the overreliance on chemical fertilizers and pesticides can lead to soil degradation and reduce soil fertility over time, making it even more difficult for plants to grow. Therefore, finding sustainable alternatives, such as vermitechnology, which uses earthworms to improve soil health and fertility, is beneficial for the long-term health of the environment. Vermiculture and vermicomposting are emerging technologies that are gaining popularity for sustainable agriculture. Vermiculture is the breeding and cultivation of earthworms, while vermicomposting is the application of earthworms to break down organic matter into compost. Both vermiculture and vermicomposting are part of the broader field of vermitechnology, which encompasses the utilization of earthworms in various applications related to agriculture. Vermicomposting is a sustainable alternative to traditional composting methods. It can be done on a small scale and has several environmental benefits, such as reducing waste, improving soil health, and reducing greenhouse gas emissions (Bhagat et al., 2022; Kamboj et al., 2022). Vermiculture and vermicomposting technology promote sustainable agriculture and serve as a source of job creation. The wastewater produced during vermiculture, also known as vermiwash, can be filtered and used as a foliar spray (Sharma et al., 2009). This process provides additional benefits to the plants, such as improved growth and increased resistance to pests and diseases, and also creates new employment opportunities in the production and application of vermiwash. By promoting sustainable agriculture practices, vermitechnology can promote a more environmentally friendly and economically viable agricultural industry (Aransiola et al., 2022).

Perionyx excavatus, *Eisenia fetida*, *Eudrilus eugeniae*, and *Metaphire posthuma* are some of the most widely used species of earthworms for vermicomposting due to their tolerance to extreme atmospheric conditions. These species can tolerate high temperatures up to 42 °C and low soil temperatures below 5 °C, making them well suited for the vermicomposting of organic waste materials. The polyphenol concentration and carbon-to-nitrogen ratio are also important factors for determining the palatability of these detritivorous earthworm species. These factors play a crucial role in the growth and survival of earthworms and must be carefully monitored to ensure a successful vermicomposting setup (Boruah, 2019).

11 The Role of Vermitechnology in Phytoremediation of Polluted Soils

In the report of Aransiola et al. (2022), the physical and chemical properties of the vermicomposts from goat manure and chicken droppings were determined (Fig. 2). It was reported that vermicasts produced by vermitechnology are rich in carbon and nitrogen, and could be used as organic fertilizers.

12 Vermitechnology and Environmental Sustainability and Phytoremediation

Environmental problems today span across a wide range of human and animal activities. According to Samoraj et al. (2022), there is a meteoric increase in livestock rearing which has resulted in improper disposal and increased production of animal waste that can lead to water pollution, affecting the quality of drinking water and aquatic life. Excessive use of these animal wastes as manure and sewage sludge as fertilizer can lead to soil contamination and degradation. Hence, the application of vermitechnology is essential for a sustainable environment. Vermitechnology is key to improving the landscape quality and restoring the fertility of the soil (Ojuolape et al., 2015).

13 Latest Insights in Vermitechnology

Vermicomposting conditions such as moisture content, mixing ratio, and turning frequency play vital role in the production of high-quality vermicompost (Banda et al., 2023). In the same investigation, municipal sewage sludge amended with coffee husks and cow dung was found to be an excellent source for the production of high-quality vermicompost as an organic fertilizer. Similarly, the chemical composition of the substrates, the diversity of earthworm species and their stocking density, and the dilution parameters have a significant impact on the production of

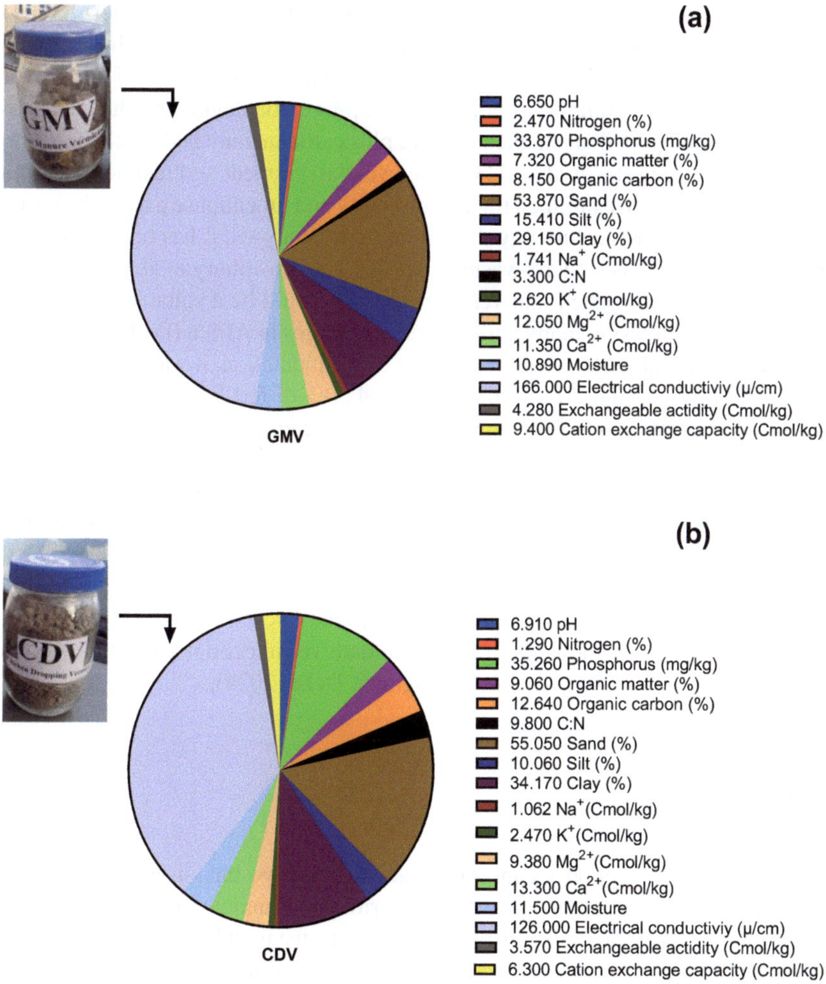

Fig. 2 Physical and chemical properties of the vermicomposts from goat manure and chicken droppings

quality vermicompost (Mupambwa et al. 2024). The addition of a bulking agent greatly improves the conditioning of wastes during vermicomposting (Samal et al., 2019). Fly ash and human excreta have also been utilized in the vermicompost technology (Kashyap et al., 2023). Vermicompost having C:N value below 15 is good for agronomic value (Sethi et al., 2023). Investigations have also reached the level of detecting microenvironmental conditions for sustainable operation of vermicomposting (Thirunavukkarasu et al., 2023). In improving agricultural soil fertility, vermicomposting was treated as a low-cost and environmentally friendly approach. Thus, vermicomposting is on an increasing trend in agro-food production and animal waste management (Bellitürk & Sundari, 2024). Vermicompost has also been successfully applied in organic farming (Siddiqui et al., 2022). In analyzing the effects of different substrates on physicochemical characteristics of vermicompost, it was found that compost of horse manure plus cattle manure produced a more efficient vermicompost than chicken manure plus cattle manure and chicken manure plus horse manure (Saba et al., 2023). Life cycle assessment and techno-economic analysis have also been examined for sustainability and greenhouse gas (GHG) emissions in vermicompost technology. Based on these analyses, it has been confirmed that vermitechnology is a "zero-waste technology" (Chowdhury et al., 2022). Dada and Balogun have suggested that vermitechnology could be a suitable agrotechnological tool for addressing many agricultural challenges in Africa (Dada & Balogun, 2023). In a recent laboratory- and field-scale experimentation, it was confirmed that a mixture of vermicompost, biofertilization, and 50% of fertilizer was found to be a recommended combination for achieving highly efficient soil fertility (Al-Maamori et al., 2023). Vermibag, vermifiltration, vermi-irrigation, vermiremediation, vermiculture (Hajam et al., 2023), and vermiwash (Shakya et al., 2023) are the latest innovations in the field of vermitechnology. Vermicomposting significantly ($p < 0.001$) reduced the concentrations of various potentially toxic elements, such as Cu^{2+}, Fe^{2+}, Pb^{2+}, Cd^{2+}, and Zn^{2+}; this ultimately resulted in improved soil chemical and biological properties (e.g., increased the growth of nitrogen-cycle microflora) (Huda et al., 2023). Nevertheless, methods that readily validate vermitechnology for large-scale implementation are greatly warranted (Singha & Deka, 2024).

14 Conclusion

Vermitechnology is a good technology that can improve the sustainability of the global environment. Different types of earthworms play a major role in breaking down organic wastes to produce useful by-products for sustainable agriculture. In fact, useful by-products such as vermiwash and vermicast from this technology contain high nutritional value and can be used as organic fertilizers to improve the plant growth for food safety and, in turn, repair the damaged landscape when used as enhancers in polluted soil remediation processes.

References

Abioye, O. P., Usman, A. U., Oyewole, O. A., & Aransiola, S. A. (2020). Application of Box-Behnken model to study biosorption of lead by *Saccharomyces Cerevisiae* and *Candida tropicalis* Isolated from electrical and electronic waste dumpsite. *GlobalNest, 22*(1), 95–101. https://journal.gnest.org/publication/gnest_02257

Adhikary, S. (2012). Vermicompost, the story of organic gold: A review.

Al-Maamori, H. A., Salman, A. D., Al-Budeiri, M., Al-Shami, Y. A. O., & Al-shaabani, E. M. (2023) Effect of vermicompost production on some soil properties and nutrients in plants. *IOP Conference Series: Earth and Environmental Science, 1214*(1), 012006.

Ansari, A. A., & Ismail, S. A. (2012). Role of earthworms in vermitechnology. *Journal of Agricultural Technology, 8*(2), 403–415. http://www.ijat-aatsea.com

Ansari, A. A., Jaikishun, S., Islam, M. A., Kuri, S. K., Fiedler, K., & Nandwani, D. (2016). Principles of vermitechnology in sustainable organic farming with special reference to Bangladesh. In *Organic farming for sustainable agriculture* (pp. 213–229). Springer.

Aransiola, S. A., Afolabi, F., Joseph, F., & Maddela, N. R. (2023). Soil enzymes: Distribution, interactions and influencing factors. In P. Jeschke, & E. B. Starikov (Eds.), *Agricultural biocatalysis: Enzymes in agriculture and industry* (1st ed., pp. 303–333). Jenny Stanford Publishing. ISBN 9789814968478. https://www.routledge.com/Agricultural-Biocatalysis-Enzymes-in-Agriculture-and-Industry/Jeschke-Starikov/p/book/9789814968478

Aransiola, S. A., Ijah, U. J. J., Abioye, O. P., & Bala, J. D. (2021). Microbial and heavy metal determination of contaminated soil using *Melissa officinalis* L. *International Journal of Environmental Planning and Management, 7*(3), 102–107. http://www.aiscience.org/journal/paperinfo/ijepm?paperid=5369

Aransiola, S. A, Ijah, U. J. J. Abioye, O. P., & Bala J. D. (2022). Vermicompost-assisted phytoremediation of toxic trace element-contaminated soil in Madaka, Nigeria using *Melissa officinalis* L and *Sida acuta. International Journal of Environmental Science and Technology.* https://doi.org/10.1007/s13762-022-04105-y

Aransiola, S. A., Ijah, U. J. J., Abioye, O. P., Bala, J. D., Rivadeneira-Mendoza, R. F., Luque, R., Rodríguez-Díaz, J. M., & Maddela, N. R. (2024). Micro and vermicompost assisted remediation of heavy metal contaminated soils using phytoextractors. *Case Studies in Chemical and Environmental Engineering., 9*(2024), Article 100755. https://doi.org/10.1016/j.cscee.2024.100755

Atiyeh, R. M., Subler, S., Edwards, C. A., Bachman, G., Metzger, J. D., & Shuster, W. (2000). Effects of vermicomposts and composts on plant growth in horticultural container media and soil. *Pedobiologia, 44*(5), 579–590.

Babaniyi, B. R., Ogundele, O. D., Bisi-omotosho, A., Babaniyi, E. E., & Aransiola, S. A. (2023). Remediation approaches in environmental sustainability. In N. R. Maddela, L. K. W. Eller, & R. Prasad (Eds.), *Microbiology for cleaner production and environmental sustainability.* CRC Press. ISBN: 9781032496061. https://www.routledge.com/Microbiology-for-Cleaner-Production-and-Environmental-Sustainability/Maddela-Eller-Prasad/p/book/9781032496061

Banda, B., Habtu, N. G., Gebreeyessus, G. D., & Meshesha, B. T. (2023). Vermicomposting as an effective approach to municipal sewage sludge management through optimization of the selected process variables. *Water Science & Technology, 88*(8), 1957–1973.

Bellitürk, K., & Sundari, R. S. (2024). The power of earthworm: Vermicompost drives to sustainable agriculture. In *Earthworm technology in organic waste management* (pp. 307–321). Elsevier.

Bhagat, A., Lal, M., Singh, J., & Kaur, A. (2022). Vermicomposting: A promising tool for sustainable agriculture. In A. Pal Vig, S. S. Suthar, & J. Singh (Eds.), *Earthworms and their ecological significance* (pp. 79–95). Nova Science Publishers, Inc. ISBN: 978-1-68507-567-5.

Bogdanov, M., Sun, J., Kaback, H. R., & Dowhan, W. (1996). A phospholipid acts as a chaperone in assembly of a membrane transport protein (∗). *Journal of Biological Chemistry, 271*(20), 11615–11618.

Boruah, T. (2019). Vermitechnology: An overview. *Researchgate.* https://www.researchgate.net/publication/333207735_Vermitechnology-_An_overview on 3/2/2023.

Chowdhury, S. D., Bandyopadhyay, R., & Bhunia, P. (2022). Techno-economic analysis and life-cycle assessment of vermi-technology for waste bioremediation. In *Biomass, biofuels, biochemicals* (pp. 315–349). Elsevier.

Dada, E. O., & Balogun, Y. O. (2023). Vermitechnology: An underutilised agro-tool in Africa. In *Vermicomposting for sustainable food systems in Africa* (pp. 127–143). Springer Nature Singapore.

Gopi, P. (2017). Vermitechnology, a scenario of sustainable agriculture-A mini review. *Vistas, 6*(1), 51–56. e-ISSN 2394-1138.

Hajam, Y. A., Kumar, R., and & Kumar, A. (2023). Environmental waste management strategies and vermi transformation for sustainable development. *Environmental Challenges*, 100747.

Huda, N., Rana, M. R., Rahman, M. M., Huq, M. A., Easmin, L., Rahman, S. T., Rahman, F., Rafi, M. H., Rauf, M., & Arif, M. (2023). Vermicompost and organic manure interactions: Effects on heavy metal concentrations, nitrification activity, Comammox Nitrospira inopinata, and Archaea/Bacteria.

Hussaini, A. (2013). Vermiculture bio-technology: An effective tool for economic and environmental sustainability. *African Journal of Environmental Science and Technology, 7*(2), 56–60.

Jha, S., Banerjee, S., Ghosh, S., Verma, A., & Bhattacharyya, P. (2024). *Eisenia fetida*-driven vermitechnology for the eco-friendly transformation of steel waste slag into organic amendment: An insight through microbial diversity and multi-model approach. *Environmental Research, 251*(2024), Article 118636. https://doi.org/10.1016/j.envres.2024.118636

Kamboj, A., Chouhan, A., Singh, S., Roy, J., Naorem, M., & Singh, A. (2022). Vermicomposting: Reusing wastes to create beneficial organic fertilizer. *Pharma Innovation Journal, 11*(11), 537–539.

Kashyap, S., Tharannum, S., Krishna Murthy, V., & Kale, R. D. (2023). Management of biomass residues using vermicomposting approach. *Bio-inspired land remediation* (pp. 261–286). Springer International Publishing.

Lazcano, C., Gómez-Brandón, M., & Domínguez, J. (2008). Comparison of the effectiveness of composting and vermicomposting for the biological stabilization of cattle manure. *Chemosphere, 72*(7), 1013–1019.

Moledor, S. (2014). *Exploring technical and economic aspects of vermicomposting as a microenterprise in rural communities of Lebanon* [Doctoral dissertation].

Moledor, S., Chalak, A., Fabian, M., & Talhouk, S. N. (2016). Socioeconomic dynamics of vermicomposting systems in Lebanon. *Journal of Agriculture, Food Systems, and Community Development, 6*(4), 145–168.

Munnoli, P. M., da Silva, J. A. T., & Bhosle, S. (2010a). Geotechnical properties of vermicomposts of press mud using Eisenia fetida, Eudrilus eugeniae and Megascolex megascolex. *Dynamic Soil and Dynamic Plant, 1*, 145–150.

Munnoli, P. M., Da Silva, J. A. T., & Saroj, B. (2010b). Dynamics of the soil-earthworm-plant relationship: A review. *Dynamic Soil, Dynamic Plant, 4*(1), 1–21.

Mupambwa, H. A., Muchara, B., Nyambo, P., & Nciizah, A. D. (2024). Utilization of vermicompost and vermileachate on plant growth and development: Aspects to consider. In *Earthworm technology in organic waste management* (pp. 323–337). Elsevier.

Murthy, P. S., & Naidu, M. M. (2012). Sustainable management of coffee industry by-products and value addition—A review. *Resources, Conservation and Recycling, 66*, 45–58.

Ojuolape, O. T., Aransiola, S. A., & Obideyi, O. A. (2015). Vermitechnology: A solution to environmental problems; a review. *Journal of Global Ecology and Environment, 3*(3), 127–135. http://www.ikpress.org/abstract/4516

Saba, Z., Magwaza, L. S., Sithole, N. J., Mditshwa, A., & Odindo, A. O. (2023). Physico-chemical analysis of vermicompost mixtures. *Agronomy, 13*(4), 1056.

Samal, K., Mohan, A. R., Chaudhary, N., & Moulick, S. (2019). Application of vermitechnology in waste management: A review on mechanism and performance. *Journal of Environmental Chemical Engineering, 7*(5), 103392. https://doi.org/10.1016/j.jece.2019.103392. Retrieved from https://www.sciencedirect.com/science/article/abs/pii/S2213343719305159

Samoraj, M., Mironiuk, M., Witek-Krowiak, A., Izydorczyk, G., Skrzypczak, D., Mikula, K., Baśladyńska, S., Moustakas, K., & Chojnacka, K. (2022). Biochar in environmental friendly fertilizers—Prospects of development products and technologies. *Chemosphere, 296*, Article 133975. https://doi.org/10.1016/j.chemosphere.2022.133975

Sethi, D., Kusumavathi, K., Ravindran, B., Panda, N., Padhan, K., Dash, S., Sahoo, T. R., Mangaraj, S., Dhal, A., Swain, S. K., & Sarkar, S. (2023). Bioconversion of organic wastes into wealth by vermitechnology: A review. In *Recent trends in solid waste management* (pp. 27–53).

Shakya, L., Babu, R., Singh, A., Dhiman, S., & Dhiman, S. K. (2023). Vermitechnology: A sustainable approach to manage organic waste in urban areas. *International Journal of Current Science Research and Review, 6*(8), 5373–5380. https://doi.org/10.47191/ijcsrr/V6-i8-05

Sharma, S., Kumar, A., Singh, A. P., & Vasudevan, P. (2009). Earthworm and vermitechnology—A review. *Dynamic Soil, Dynamic Plant, 3*(2), 1–12.

Siddiqui, M. A., Neeraj, A., & Hiranmai, R. Y. (2022). Vermitechnology: An eco-friendly approach for organic solid waste management and soil fertility improvement—A review. In *Strategies and tools for pollutant mitigation: Research trends in developing nations* (pp. 91–112).

Singh, A., Saman, Z., & Singh, K. (2022). Vermibiotechnology: A promising tool for waste management and organic farming. In A. Pal Vig, S. S. Suthar, & J. Singh (Eds.), *Earthworms and their ecological significance* (pp. 97–109). Nova Science Publishers, Inc. ISBN: 978-1-68507-567.

Singh, R. P., Embrandiri, A., Ibrahim, M. H., & Esa, N. (2011a). Management of biomass residues generated from palm oil mill: Vermicomposting a sustainable option. *Resources, Conservation and Recycling, 55*(4), 423–434.

Singh, R. P., Singh, P., Araujo, A. S., Ibrahim, M. H., & Sulaiman, O. (2011b). Management of urban solid waste: Vermicomposting a sustainable option. *Resources, Conservation and Recycling, 55*(7), 719–729.

Singh, R., Sharma, R. R., Kumar, S., Gupta, R. K., & Patil, R. T. (2008). Vermicompost substitution influences growth, physiological disorders, fruit yield and quality of strawberry (Fragaria x ananassa Duch.). *Bioresource Technology, 99*(17), 8507–8511.

Singh, K. (2009). Microbial and nutritional analysis of vermicompost, aerobic and anaerobic compost. *40 CP Honours Project for Master in Environmental Engineering,* Griffith University, Brisbane, Australia (Supervisors: Dr. Rajiv K. Sinha & Dr. Sunil Heart).

Singha, W. J., & Deka, H. (2024). Instrumental characterization of matured vermicompost produced from organic waste. In *Earthworm technology in organic waste management* (pp. 231–255). Elsevier.

Sinha, N. K., Hui, Y. H., Evranuz, E. Ö., Siddiq, M., & Ahmed, J. (2010). *Handbook of vegetables and vegetable processing.* John Wiley & Sons.

Sinha, R. K., Herat, S., Agarwal, S., Asadi, R., & Carretero, E. (2002). Vermiculture and waste management: Study of action of earthworms Elsinia foetida, Eudrilus euginae and Perionyx excavatus on biodegradation of some community wastes in India and Australia. *The Environmentalist, 22*(3), 261–268.

Sinha, R. K., Herat, S., Chauhan, K., & Valani, D. (2009). Earthworms vermicompost: A powerful crop nutrient over the conventional compost & protective soil conditioner against the destructive chemical fertilizers for food safety and security. *Journal of Agriculture & Environmental Sciences, 5,* 1–55.

Subler, S., & Kirsch, A. S. (1998). Spring dynamics of soil carbon, nitrogen, and microbial activity in earthworm middens in a no-till cornfield. *Biology and Fertility of Soils, 26*(3), 243–249.

Thirunavukkarasu, A., Sivashankar, R., Nithya, R., Sathya, A. B., Priyadharshini, V., Kumar, B. P., Muthuveni, M., & Krishnamoorthy, S. (2023). Sustainable organic waste management using vermicomposting: A critical review on the prevailing research gaps and opportunities. *Environmental Science: Processes & Impacts, 25*(3), 364–381.

Valani, D. (2009). Study of aerobic, anaerobic and vermicomposting systems for food and garden wastes and the agronomic impacts of composts on corn and wheat crops. *Report of 40 CP Honours Project for the Partial Fulfillment of Master of Environmental Engineering Degree,* Griffith University, Australia (Supervisors: Dr. Rajiv K. Sinha and Dr. Sunil Herat).

Webster, K. A. (2005). Vermicompost increases yield of cherries for three years after a single application. *EcoResearch, South Australia, 207*. https://www.ecoresearch.com.au

Weltzien, H. C. (1989). Some effects of composted organic materials on plant health. *Agriculture Ecosystems.*

Vermitechnology and the Influencing Factors

Innocent Ojeba Musa, Job Oloruntoba Samuel, Sikirulai Abolaji Akande, Udeme Joshua Josiah Ijah, Olabisi Peter Abioye, Ummulkhair Salama Ilyasu, Abd'Gafar Tunde Tiamiyu, and Asmau M. Maude

Abstract Vermitechnology is an eco-friendly and a sustainable approach that utilizes earthworms for waste management and soil improvement. It involves the use of earthworms to decompose organic matter and convert it into nutrient-rich vermicompost, which can be used as a natural fertilizer and a soil conditioner. Vermicomposting has numerous environmental and agricultural benefits, including reducing waste, enhancing soil fertility, improving plant growth, and reducing chemical fertilizer and pesticide use. Additionally, vermitechnology has been shown to remediate contaminated soils and wastewater, making it a promising solution to environmental cleanup efforts. Poorly managed vermicomposting can lead to issues such as unpleasant odours, pest infestations, and nutrient imbalances. The efficiency of vermitechnology depends on factors such as the availability of feedstock, the suitability of earthworm species, and various infrastructural constraints. Large-scale commercial operations may involve significant costs in terms of infrastructure and technology, making them less accessible to smallholders and communities. Additionally, the reliance on earthworms as primary decomposers raises concerns about potential disruptions to local ecosystems if non-native species are introduced. Careful consideration of these factors is essential to ensure that vermitechnology initiatives are sustainable and implemented responsibly. Overall, vermitechnology presents a promising solution for addressing environmental and agricultural problems, as it provides a sustainable and an eco-friendly approach to waste management and soil improvement. Its potential applications in reducing waste, enhancing soil fertility, and remediating contaminated soils and wastewater make it an important technology to consider for sustainable development.

I. O. Musa (✉)
Department of Microbiology, Skyline University Nigeria, Kano, Kano, Nigeria
e-mail: innocentmusa0011@gmail.com

J. O. Samuel · U. J. J. Ijah · O. P. Abioye · U. S. Ilyasu · A. M. Maude
Department of Microbiology, Federal University of Technology, Minna, Niger, Nigeria

S. A. Akande
Department of Mathematics, Federal University of Technology, Minna, Niger, Nigeria

A. T. Tiamiyu
Department of Mathematics, The Chinese University of Hong Kong, Hong Kong, China

S. A. Aransiola and N. R. Maddela (eds.), *Vermitechnology: Economic, Environmental and Agricultural Sustainability*, https://doi.org/10.1007/978-3-031-96212-7_2

Keywords Vermicompost · Environment · Remediating · Fertilizer

1 Introduction

Scientists have advanced the field of vermitechnology, which explores the potential of epigeic earthworm species, to gain deeper insights into these organisms. Vermitechnology has applications in waste stabilization (both industrial and domestic), organic farming, and wastewater treatment (Bhattacharya & Kim, 2016; Livinus et al., 2024). Our farmers are growing better at using sustainable practices and organic fertilizers to maximize crop yields without negatively impacting the environment. Long-term crop success depends on continuous work to keep organic matter in the soil (Aransiola et al., 2022). When biowaste, especially the organic fraction of municipal solid waste (MSW), goes through the composting process, it may be used by being added to the soil. Contributing to the reduction of environmental pollution via recycling the numerous types of biowaste is a viable option (Ijah et al., 2022; Samal et al., 2019a, b).

Charles Darwin made groundbreaking contributions to understanding the ecological significance of earthworms. In his 1881 publication, *The Formation of Vegetable Mould through the Action of Worms, with Observations on Their Habits*, Darwin highlighted the crucial role of earthworms in soil formation and the production of humus. While his research laid the foundation for understanding earthworm ecology, it is important to note that Darwin's work preceded the modern concept of vermitechnology, which involves the applied use of earthworms for composting and environmental management. Since then, vermitechnology has been integrated into the organic waste management infrastructures of both developing and developed countries. Its widespread usage in business contexts in both Asia and the Americas has been documented in several reports. Waste decomposition using vermitechnology, often known as worm farming, is an aerobic, a nonhaemophilic, a bio-oxidative process. For this technique to work, earthworms must be present. Created on the international holiday honouring factories and factories throughout the globe (Hussain et al., 2018). The purpose of this review was to investigate what variables impact the use of vermitechnology to address environmental issues.

2 Earthworms and Vermiculture

The term "vermiculture" is used to describe the practice of raising earthworms in a controlled environment for the benefit of bioremediation, increased soil fertility, and recycling of organic waste (Ojuolape et al., 2015). The finished product may be used as vermicompost, a soil amendment rich in nutrients (Vodounnou et al., 2016). There are more than 3000 species of earthworms in the globe, but only 384 of them can be found in India. The great majority of worm species do not go into the water at any point in their lives. However, other species, like *Pontodrilus burmudensis*, are confined to the estuary's unique aquatic environment. Earthworms may be found in a variety of environments, including hydrophilic conditions and very cold areas such as beneath snow, in addition to organic waste products like manure litter and compost (Byambas et al., 2019; Leena et al., 2023).

Earthworms are saprophages; however, because of their food, they are more accurately classified as detrivores or geophages. Detrivores typically inhabit the soil's uppermost layer but may also be found living in the first few centimetres of leaf litter, dead roots, and other plant detritus. *Octochaetona curensis, Perionyx excavates, Octochaetona serrata, Eisenia fetida, Polypheretima elongate, Lampito mauritii*, and *Eudrilus eugeniae* are all instances of detrivorous earthworms. Geophage earthworms, which live below and subsist by consuming dirt, consume enormous amounts of soil that is rich in microbial life. *Octochaetona thurstoni, E. eugeniae*, and *Metaphire posthuma* are only a few of the many geophages found in the world (Byambas et al., 2019; Zhong et al., 2023).

Earthworms are reared in climate-controlled environments away from direct sunlight and wet conditions to extend their lives (Domínguez, 2018; Leena et al., 2023). Culturing earthworms requires a container with a large enough volume to hold 100–500 worms for a period of 6–8 weeks (wooden boxes, cement tanks, plastic trays, or earthen pots with tiny holes in the bottom to allow excess water outflow work well). Around 3–4 cm of wet sawdust or shredded coconut coir at the very bottom of the bed should be put. Feed, such as cow or chicken dung or grass clippings, would be added around 5–6 inches higher (Domínguez, 2018). Worms break down a portion of the excrement from cattle, allowing room for the simple addition of other organic waste (Bart et al., 2018). An ideal moisture level may be maintained by watering regularly. To keep the contents within the container, a wet jute bag works wonders. This makes the surroundings dark, which is beneficial for worms since it discourages predators. Furthermore, it keeps humidity in, controls temperature, and permits enough ventilation (Musyoka et al., 2019). When the garbage has decomposed enough, a granular black substance will form at the bottom of the container, where the worms will have begun to congregate. The odourless compost on top is then scraped off and dried in the shade. Earthworms feast on organic garbage that is high in nitrogen (Xu et al., 2021; Zhong et al., 2023). Microbes may find it simple to get nourishment from animal faeces and other partly decomposed materials. Earthworms perform an excellent job of decomposing garbage and organizing the compost they produce. Municipal solid trash includes garbage generated from places like restaurants, butcheries, gardens, sugar refineries, dairies, cities, towns, breweries, distilleries, and hatcheries (Xu et al., 2021).

3 The Life Cycle of Earthworm

An egg is the first stage of an earthworm's life cycle. The earthworm goes through a period of development before it emerges from its egg. The cocoon acts as a shell for the egg. However, certain members of the family Lumbricidae, often known as cocoon worms, lay as many as 20 eggs in each cocoon, whereas most earthworm species lay only one egg (Bondhare & Desai, 2019). Although most cocoons take on a "lemon" shape, the precise shape may vary from species to species. Species and environmental conditions, among others, considerably affect how long it takes for an egg to hatch. Cocoons of certain species may "wait out" painful, dry conditions in the soil, whereas cocoons of other species may hatch more rapidly in warmer temperatures (Adeel et al., 2021; Wang et al., 2023). Baby earthworms seem like little, whitish replicas of their parents. Potworms (*Enchytraeidae*) are a family of tiny, segmented worms that are sometimes mistaken for these. As it matures and gains weight, the earthworm nymph will eventually assume the adult worm's colour pattern. Juvenile earthworms lack a saddle (or clitellum), which is the sole defining characteristic of their adult counterparts (Casquero et al., 2020; Leena et al., 2023). Saddles on the backs of adult earthworms indicate that they have reached sexual maturity. Earthworms are considered hermaphrodites because they have both male and female reproductive organs.

When two earthworms find each other, they engage in temporary pairing for reproduction rather than forming permanent bonds. During mating, earthworms align ventrally and exchange sperm, facilitated by their clitella (not jaws or a slime tunnel). The exchanged sperm is stored in special receptacles and used later for fertilization. A cocoon is eventually formed by the clitellum, where the sperm and eggs are deposited for fertilization, ensuring the continuation of their lifecycle. Due to their capacity to act as both male and female during reproduction, all earthworms are classified as simultaneous hermaphrodites. After this sperm exchange, the earthworms begin to split apart (Adeel et al., 2021). As the earthworm travels, the mucus-encased clitellum is carried along its body and finally discharged by its tail. Eggs and sperm are collected from male earthworms as they move. The mucous covering that nourishes the fertilized egg is what shields it within the cocoon (Casquero et al., 2020). Figure 1 displays an earthworm's life cycle stages and anatomy.

4 Vermitechnology of Potential Uses

4.1 Worm Composting

Organic waste is one kind of garbage that may be broken down by microbes. Earthworms are used in the vermicomposting process, which gradually breaks down organic waste into a humus-like, black, odourless material (Samal et al., 2018). In the process of vermicomposting, several different types of heat and greenhouse gas

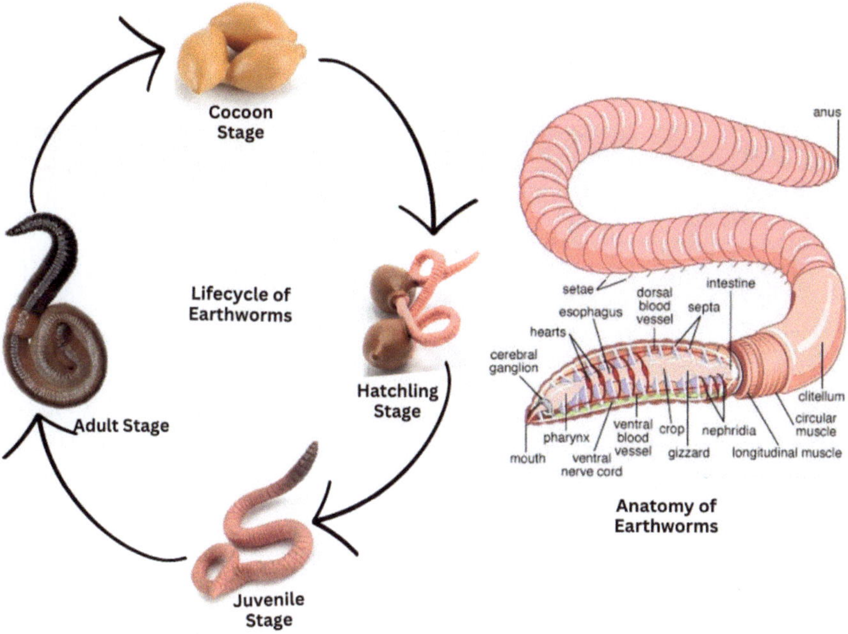

Fig. 1 Earthworm's life cycle stages and anatomy

(GHG) emissions are emitted into the air. Compared to raw materials, the particles in vermicompost are much smaller. In vermiculture, earthworms are provided with an abundance of plant and animal matter to ensure their health and development (Xie et al., 2016). Worm composting is a viable option for the management of both sewage sludge and industrial sludge. Earthworms, given the right circumstances (moisture, temperature, and pH), will excrete a homogeneous material known as cast or vermicast, which is the result of the organic stuff they have digested. Vermicast has a high moisture retention rate and a high concentration of beneficial microorganisms, nutrients, and ionic compounds (Aransiola et al., 2024; Singh et al., 2018). The practice of worm composting may be done on a small or large scale. Several physical, chemical, and biological indicators may be used to determine whether vermicompost is ready for use. Maturity may be assessed by a wide variety of instrumental means. Different approaches and parameters are employed to establish a "vermicompost readiness" state, as shown in Fig. 2.

4.1.1 Worms for Vermicomposting

The most critical aspect of vermitechnology is selecting the appropriate earthworm species for vermicomposting. This variety in earthworm species means that vermicompost produced by these worms may have a broad range of nutrient profiles. The

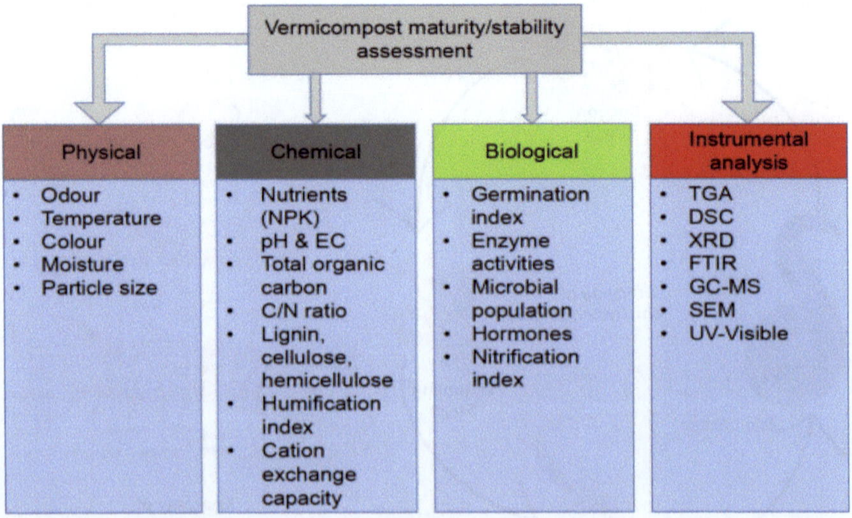

Fig. 2 Methods used to determine the maturity of vermicompost

ability of several detritivorous earthworm species to vermicompost has been the subject of several studies (Singh et al., 2020a, 2020b; Thakur et al., 2021; Vuković et al., 2021; Zhong et al., 2023). *Eisenia fetida, M. posthuman, E. eugeniae*, and *P. excavates* were shown to have the highest endurance to a wide variety of atmospheric conditions compared to other earthworm species. Thus, these worms are the most liked for use in vermicomposting. They can survive in temperatures as low as − 5 °C below the earth and as high as 42 °C above it. Detritivorous earthworm species place a high value on a high polyphenol content and low carbon to nitrogen ratio.

4.1.2 Advantages of Vermicomposting

Earthworm compost's primary selling point is the fact that it is made entirely of natural materials. Fast decomposition of waste into a nutrient-rich soil amendment that plants may utilize to flourish is possible thanks to earthworms (Aransiola et al., 2022; Bellitürk, 2018; Wang et al., 2023). This means that the final product of a vermicomposting operation is a plant aid that is naturally prepared and has numerous applications. The nutrients in earthworm compost are rapidly absorbed by plant roots, which is the most notable advantage. Worm mucus included in vermicompost makes it difficult to remove like chemical fertilizers (Leena et al., 2023; Yuvaraj et al., 2021).

Incorporating worms into the composting process results in a greater diversity of soil microorganisms and bacteria. A plant's resistance to disease and some pests may be boosted by these variables. An increase in the number of birds in a region may aid in pest management for plants if more beneficial microorganisms are present (Yuvaraj et al., 2021). It is the plant growth hormones in vermicompost that give plants a boost. It improves crop yields, speeds up plant development, and facilitates seed germination (Yuvaraj et al., 2021; Zhong et al., 2023). Because of its colloid status, vermicompost is capable of retaining nine times its weight in water. This may be especially important during drought. This slow evaporation occurs because the water is kept in the organic matter, making it available to the plants even as it evaporates (Coulibaly et al., 2018). Vermicomposting has the potential to provide those who are economically disadvantaged with an income if it is done correctly (Coulibaly et al., 2018; Leena et al., 2023).

4.2 Biodegradation and Vermitechnology for Environmental Cleanup

The adoption of new agricultural techniques has greatly reduced once-thriving ecosystems and animal populations. Using living organisms, the method of bioremediation restores polluted or otherwise damaged land. Using biological approaches to revive degraded land improves its fertility. Earthworms can tolerate and even bioaccumulate a wide range of chemical soil contaminants, including metals and organic pollutants (Aransiola et al., 2022). A variety of earthworm species, including *Eiseniella fetida, E. tetraedra, Lumbricus terrestris, Lumbricus rubellus*, and *Allolobophora chlorotica*, have been shown to remove a variety of contaminants from soil, including heavy metals (Cd, Pb, Cu, Hg, etc.), pesticides, and lipophilic organic micropollutants like polycyclic aromatic hydrocarbons (PAHs) (Chowdhury et al., 2022a, 2022b). Vermiremediation has the potential to be a low-cost and ecologically beneficial method for cleaning up polluted soils and locations (Fig. 3).

4.2.1 Heavy Metal

Toxic heavy metals are a major concern in sewage sludge, which endangers human health. Densities of heavy metals range from around 5.5 to 6.6 g/cm^3 (Aransiola et al., 2013; Bhat et al., 2017; Ijah et al., 2015). Sewage sludge's heavy metal content varies greatly by industry. Sewage from cities contains significant quantities of many elements, including aluminium, iron, zinc, copper, and chromium. Several species of earthworms are capable of detoxifying the soil of heavy metals, pesticides, and lipophilic organic micropollutants, including polycyclic aromatic hydrocarbons (PAHs). *Aporrectodea tuberculata, Eiseniella fetida, L. rubellus, Dendrobaena rubida, Eiseniellate, L. terrestris* are only a few examples (Singh et al., 2020a, 2020b). Bioaccumulation of copper (Cu) and zinc (Zn) in *E. fetida* was analysed after 10 weeks of testing by Sharma et al. (2009), concerning Malley et al. (2012).

Earthworms improve plant nutrient availability and soil pH, both of which may reduce heavy metal availability. As a result, earthworms play a critical role in lowering

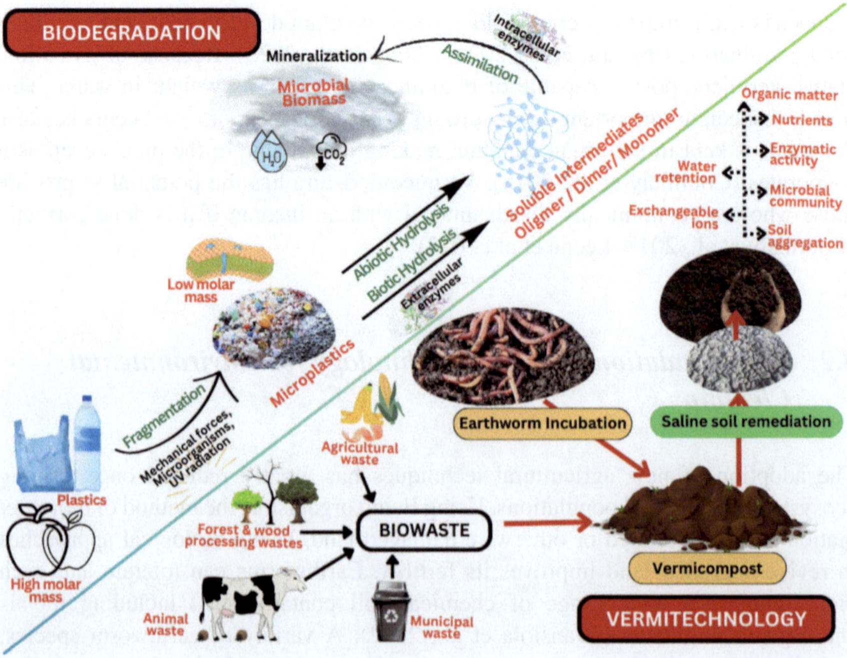

Fig. 3 Process of biodegradation and vermitechnology for environmental cleanup

soluble and transportable heavy metal concentrations in soil (Siddiqui et al., 2022). Given that certain earthworm species are known to be potential heavy metal accumulators, the vermicomposting technique's shown efficacy in lowering the toxicity of industrial and municipal wastes is particularly important. Similar to how introduced species have had to adjust to new conditions, native creatures have had to do the same (Yuvaraj et al., 2020). In environments where metals (Pb and Zn) pose a concern, *L. mauritii* excels because of its ability to metabolize electrophilic xenobiotics. Bioconversion efficiency of *Eisenia andrei* applied to a mixture of paper mill waste and primary sewage sludge. Recent research has shown that earthworms like *L. terrestris* and *E. fetida* may increase both microbial activity and oil breakdown (Roy et al., 2022a, 2022b). In addition, earthworms of the species *E. fetida* have been employed as a test organism for many different kinds of poisons. In contrast to *L. terrestris*, which can only tolerate concentrations of up to 0.5% crude oil, some sources suggest that *E. fetida* can tolerate concentrations of up to 1.5% (Roy et al., 2022a, 2022b). *E. fetida* improves the rate at which PAHs are eliminated. It is exciting to see a novel approach to wastewater treatment, like vermifiltration, which employs earthworms for the job. Due to the general mechanism of "ingestion" and biodegradation of organic wastes, heavy metals, and solids from wastewater, as well as "absorption" through body walls, earthworms function as a "biofilter," removing 90% of the biochemical oxygen demand (BOD) after 5 days, 80% of the COD, 90% of the total dissolved solids (TDS), and 90% of the total suspended solids (TSS) from wastewater. Earthworms are a useful "bioindicator" for detecting heavy metal pollution in soil since they have been shown to have many more of these elements than worms taken from other areas (Yuvaraj et al., 2020).

4.2.2 Bioaccumulation of Pesticides

Earthworms have been shown to accumulate large levels of pesticides and metals in their tissues, although they can withstand a wide range of these toxins. They serve as a soil detoxifier while also preventing the spread of illnesses that may be taken up from the ground. Earthworms are effective in cleaning polluted soil of a wide variety of contaminants, including hydrocarbons and the more stable polycyclic aromatic hydrocarbon (PAH) benzo(a)pyrene. After just one application, the pesticide aldrin was still present in the soil at over 34% concentration 5 years later. Ingestion and/or degradation of "organochlorine pesticides" and "polycyclic aromatic hydrocarbons" (PAHs) by earthworms has been established in several investigations (Chauhan & Punia, 2022). In approximately 12 weeks, with a loading rate of around 50 worms per kilogram of soil, 60–65% of the most abundant PAHs were eliminated. Vermiremediation dramatically improves soil and water quality. A significant increase in earthworm population is one of vermiremediation's most notable advantages to the soil (Zeb et al., 2020).

4.3 Vermitechnology as a Source of Nutrients for Fish and Poultry Feed Production

Earthworms are a good source of protein for a variety of animals, including chickens, pigs, rabbits, and the aquarium fish *Poecilia reticulata*. The use of vermitechnology—a safe organic manure that enhances both the pond's substrate and water quality—allows for the organic cultivation of aqua crops in a natural setting (Nalunga et al., 2021). According to common nutritional assessments, the protein content of *E. fetida* meal is rather high, ranging from 54.6 to 71.0% dry matter. Worm castings from the earthworm species *E. fetida* are a potential protein source for animals, with a dry matter protein level of 7.9%. Worm body fluid has high concentrations of protein (9.4%), vitamins, and minerals (particularly iron). With well-managed vermiculture, fish meal can be supplemented with worm protein, and the need for animal protein can be evaluated (Parolini et al., 2020). Worms have also been a source of food for Maoris in New Zealand and Papua New Guinea (Bhuvaneshwaran et al., 2019).

4.4 Vermitechnology for Improvement of Soil Fertility

Earthworms play a crucial role in boosting soil fertility. They not only help break up the soil and spread it around, but also bring oxygen to the ground. When an earthworm dies, it releases into the soil a cocktail of nutrients (N and P) and the millions of beneficial microorganisms (including nitrogen fixers) that dwell in its digestive system. It is true that both soil acidity and alkalinity may be neutralized by earthworms (Ahmed & Al-Mutairi, 2022). Through vermicomposting, nitrogen-fixing bacteria and actinomycetes may proliferate. The enzymes in earthworm vermicastings continue to break down organic materials in the soil, releasing nutrients that may then be absorbed by plant roots, even after earthworms have been removed (Siddiqui et al., 2022). The lignocellulolytic enzymes in vermicompost include glucose oxidase, phosphatase, and urease. Plant growth hormones and humic acids have been detected in vermicompost, which may explain why it improved germination, growth, and yields in test plots (Cheng et al., 2021). After microbial inoculation, the vermicomposting process significantly enhances the humic acid concentration and acid phosphatase activity of organic substrates (Iwai & Kruapukee, 2017). When pig dung vermicompost was introduced to the germination circumstances, the tomato and marigold plants had increased shoot and root weight, leaf area, and shoot to root ratios (Ansari et al., 2016).

4.5 Factors that Influence Vermitechnology

Multiple abiotic and biotic elements interact to determine the effectiveness of the vermicomposting process. Some are discussed below.

4.5.1 Abiotic Components

The moisture content, pH, aeration, feed quality, light, temperature, and C to N ratio are all highly essential abiotic elements that impact the vermicomposting process. Earthworms and other microorganisms in a vermicomposting system cannot do their jobs without a certain level of moisture, so that is another factor to consider. The blood capillaries at the skin's surface should be adequately moistened so that they may function properly while breathing (Ahmed et al., 2023). Earthworm activity may be used as a barometer for whether or not the soil is wet enough. Because earthworm activity increases with moisture, it is important to prevent the soil from drying out. The amount of moisture in the air is different for different species and different regions of the world. Earthworms have developed a variety of strategies to survive in arid environments, including underground migration, hibernation, and the formation of cocoons that may withstand the effects of dryness (Ahmed et al., 2023). For optimal results in vermicomposting, moisture levels should be between 60 and

80% (Roy et al., 2022a, 2022b). Physical and chemical changes in feed supplies may lead to subtle shifts in moisture content; even a 5% shift can have a major impact on clitellum growth in *E. fetida* worms (Wang et al., 2023).

An environment with a pH between 5.5 and 8.5 is ideal for microbial and earthworm life. Nonetheless, a pH of 7.0–7.4 is optimal (Chowdhury et al., 2022a, 2022b). During vermicomposting, there are dramatic changes in the pH value of feed substrates. Early on in the vermicomposting process, the pH of the feed substrate is low. The carbon dioxide and volatile fatty acids produced early on in the vermicomposting process are the cause of this. Gradually, when CO_2 is produced and volatile fatty acids are used, the pH rises (Lim et al., 2022).

The earthworm's environment has a significant impact on their ability to reproduce, behave, metabolize, develop, and breathe. In certain cases, temperatures exceeding a specific threshold may be fatal for earthworms. If given the choice between wet and warm or dry or dry and warm, earthworms would choose the former (Al-Maliki et al., 2021). It is between 12 and 28 °C that earthworms flourish. If temperatures continue to rise, earthworm activity will decrease dramatically. The optimum range for indoor temperatures is between 15 and 30 °C year-round (Ismail et al., 2022). Lower temperatures prevent earthworms from reproducing and slow their metabolic rate. Earthworms are cold-blooded creatures, and they cease feeding when the temperature drops below a certain point. Above 35 °C, earthworms' metabolic activity and reproductive ability begin to diminish, and they eventually die (Al-Maliki et al., 2021).

Feed quality is essential for the earthworm vermicomposting process. When it comes to diet, earthworms seem to have a wide range of capabilities. This is due to a variety of variables, including the food's degree of breakdown, its C to N ratio, its particle size, and its salt concentration (Vos et al., 2019). Worms will decompose more quickly if their feed excrement consists of particles no bigger than a dime. Worms can digest the trash because the tiny particle size of the feed waste allows for enough oxygenation throughout the mound of garbage (Patwa et al., 2020). A worm's daily caloric intake is, on average, between 100 and 300 mg per gram of body weight. Live microbes, decaying organic matter, and dead macrofauna all offer food for earthworms (Javed & Hashmi, 2021).

Earthworms thrive in oxygen-rich environments; hence, ventilation is crucial for the vermicomposting process. Oxygen consumption by earthworms serves a microbial function; oxygen concentrations are also correlated with substrate temperatures. Vermicomposting systems can be negatively impacted by a continuous supply of moisture due to poor aeration and the resulting reduced oxygen availability for the worms. When earthworms move around in their bedding, they create micro-vents that help oxygenate the soil. There have been reports of large-scale migrations of *E. fetida* from areas where oxygen levels are low, water is saturated, or carbon dioxide or hydrogen sulfide has built up (Kaur, 2020).

The carbon-to-nitrogen (C:N) ratio in their food sources also has an impact on earthworm development and reproduction (Biruntha et al., 2020). Worm growth and reproduction rates are increased by feed materials with a higher carbon-to-nitrogen (C:N) ratio. A high or low C:N ratio slows down waste degradation. Plants need a

C:N ratio between 25 and 20 to 1 to absorb mineral nitrogen (Samal et al., 2019, b). Neither earthworms nor any other invertebrate can make it in ammonia- and salt-rich organic wastes. In the same way that humans are sensitive to salts, worms prefer low-salt diets (less than 0.5% salt). However, many types of manures have high salt contents, so they must be leached before being used as bedding. This is done by simply running water through the material for a while, which reduces the salt content (Kaur, 2020).

Deposition of acid rain, biphenyls, polychlorinated biphenyls, heavy metals, and pesticides are all examples of agricultural chemicals that can harm soil quality. These chemicals have direct effects on earthworm abundance and distribution when present in soil, reducing earthworm productivity, activity, and ultimately death. As stated by Al-Maliki et al. (2021), large-scale contact with pesticides in action, even in deep soil, is lethal to earthworms regardless of the form or rates of pesticide usage, the earthworm's age, species, or the prevailing environmental conditions (Edwards & Arancon, 2022). Earthworms are particularly susceptible to the toxicity of fungicides and insecticides. However, pesticides are typically applied after canopy closure to keep the chemicals away from the earthworms. It is safe to eat or drink from treated plants or soil until the next rainstorm, but dumping contaminated plant matter onto the ground can be harmful to humans and animals. The increasing use of copper-based fungicides, which are known to be toxic to earthworms, poses a threat to organic farmers. Copper-based fungicides, widely used in organic farming, might be resistant to the nematodes that normally eat them (Ahmed & Al-Mutairi, 2022). The fact that copper oxychloride was directly toxic to earthworms in South African vineyards suggested that copper might accumulate in earthworms. If we compare a control plot where no herbicides were used to one where paraquat was used at the commercial dose, we find that paraquat significantly reduces cast formation. Before planting or the emergence of weeds and plants, herbicide treatments are applied to the soil's surface to discourage anecdotal and epiges feeding on surface litter. Determining the toxicity of pesticides to earthworms is challenging in laboratory studies due to the low success rates, such as the use of pesticides in the field, compared to the toxic levels used in laboratory studies. Under normal field conditions, standard application rates rarely result in fatal effects. However, a sublethal effect on reproduction and growth may occur depending on the earthworm species and the product used (Ahmed et al., 2022).

The soil type is a factor in how much of an effect earthworms have on population growth and distribution. Earthworm populations are influenced by soil texture because of the role that texture plays in regulating soil hydration, nutrient availability, and cation exchange capacity (cation exchange capacity). There are more earthworms in soils with a light to medium texture, such as loam than in those with a heavy to medium texture, such as hard clay, sandy, or alluvial soils (Nazeer & Al-Mutairi, 2022). Furthermore, the total number of *A. squirreli, Aporrectodea caliginosa, Amblyeleotris rosea,* and *Amblyeleotris trapezoids* was found to be correlated with the clay content of the soil over a week. Among these species, however, *A. caliginosa* displayed the most strikingly positive correlation with the clay (Al-Maliki et al., 2021).

Fertilizer Effect

Inorganic fertilizers have varying effects on earthworms depending on where you look. There is some debate over whether or not earthworms benefit from the use of inorganic fertilizers. Inorganic fertilizers alter the soil's pH and cause a change in plant diversity and yield when applied. One way in which chemical fertilizers affect earthworm populations is by lowering pH, but they can also increase earthworm populations by boosting plant growth. It was discovered that the number of earthworms in pasture increased by a factor of four after being treated with superphosphate and lime (Al-Maliki et al., 2021). Many kinds of earthworms prefer soils high in calcium and avoid those low in pH, another reason why lime is helpful. Grass plots have seen fewer earthworms after being fertilized with superphosphate fertilizers. Additionally, nitrogenous fertilizers can cause a large number of earthworms to congregate. The increased grass yield that followed the widespread application of nitro chalk to a variety of pastures was accompanied by a corresponding increase in earthworm populations. Furthermore, as the amount of nitrogen in arable land increases with chemical fertilizers, the number of earthworms increases as well. Due to the additional nutrition provided by organic manures, earthworm populations flourish, making crop residues more enticing to earthworms that prefer a high carbon-to-nitrogen ratio in their diet (Al-Maliki et al., 2021).

Constituents of a Biological Nature

Microorganisms, enzymes, earthworm stocking density, etc. are all examples of biotic factors that can affect the vermicomposting process. The earthworm population, that is, stocking density, is affected by a variety of factors during vermicomposting, including reproduction, feeding, respiration, and burrowing. The waste composition of those components that undergo vermicomposting is greatly influenced by the microorganisms present in the process. In the process of vermicomposting, organic matter is stabilized by earthworms and microorganisms through a mutualistic interaction (Ahmed et al., 2022; Musa et al., 2021). Earthworms consume fungi with organic substrates to fulfil their protein or nitrogen requirement, fungal population in earthworm casts was almost equal to or higher than that of initial substrates. Both complex substances and biologically active substances are synthesized by the microorganisms, making them available to the plants (Verma et al., 2022). By grazing microorganisms and the surrounding area vulnerable to microbial attacks after organic matter is comminuted, earthworms can influence the actions of microbial decomposers. Sometimes, earthworms will eat microorganisms because they are easy to digest. Different worm species, food substrates, and the presence of an ecosystem in which they reside can have a significant impact on the amounts and types of food that earthworms can assimilate (Ahmed & Al-Mutairi, 2022).

Biochemical conversion of proteinaceous and cellulosic materials occurs rapidly due to the worms' stimulation of enzymes in their gizzard and intestine (Sanchez-Hernandez et al., 2019). The most common enzymes required in vermicomposting are cellulases, amidohydrolase, proteases, urease, β-glucosidases, and phosphatases (Debnath, 2018). The intensity of microbial metabolism in soil can be understood by measuring enzyme activities, which are commonly used as indicators of microbial activity. In addition to catalysing many different metabolic processes, they degrade and detoxify many different pollutants (Cui et al., 2021).

The current status of earthworms and the environments in which they live are as follows: There are about 3320 different species of earthworms in the world. While there are approximately 590 species of earthworms, each with their ecological preferences, little is known about the functional role of many of these species or how they affect their environment (Ashwood et al., 2019). Earthworms come in a wide variety of sizes, shapes, colours, and shapes, and they live for varying amounts of time depending on species. In general, the physicochemical properties of soils affect the distribution of earthworms. Examples include carbon and nitrogen content, temperature, moisture, and the carbon-to-nitrogen ratio. Most species of earthworms thrive in soil with 12–34% moisture, 10–35–0 °C, a pH of 7, and a C:N ratio of 2–8. It is uncommon to find earthworms in soil that has a very rough texture and a high clay content or in soil that has a pH (Edwards & Arancon, 2022).

Measures to Ensure Food/Feed Quality

Maintaining high food and feed quality through vermitechnology requires the implementation of specific measures. First, the quality of feedstock is critical. Selecting organic waste materials with optimal characteristics, such as an ideal carbon-to-nitrogen (C:N) ratio, minimal salinity, and appropriate particle sizes, ensures efficient vermicomposting and supports a healthy earthworm population. Additionally, the choice of earthworm species is crucial. Species like red wigglers (*E. fetida*) and white worms (*Enchytraeus albidus*) are well suited for converting organic matter into nutrient-rich vermicompost.

Environmental conditions must also be carefully controlled. Temperature should be maintained within the optimal range of 10–30 °C to support earthworm survival and activity. Moisture levels are equally important, with an ideal content of around 60%, facilitating earthworm reproduction, efficient composting, and proper nutrient leaching. pH levels should remain between 5.5 and 8.5, preferably close to neutral, to optimize earthworm activity and microbial processes. Furthermore, adequate aeration must be provided to prevent anaerobic conditions, which are harmful to earthworms and can slow down vermicomposting.

Nutrient management is another essential aspect. Supplementing feedstock with essential macronutrients and micronutrients ensures the production of vermicompost rich in nitrogen, phosphorus, potassium, calcium, magnesium, and trace elements. Harvesting should occur when the vermicompost has matured, as indicated by reduced moisture content, the absence of visible earthworms, and a distinct earthy smell. Finally, the application of vermicompost should align with its composition and the specific nutrient requirements of target crops to achieve optimal results.

5 Conclusion

There are many different types of organic-rich solid and liquid waste, and vermicomposting technology has been proven to be an effective and environmentally friendly way to deal with these materials. The symbiotic and synergistic activity of earthworms and microorganisms degrades all organic pollutants and converts them to beneficial byproducts. The proper disposal of garbage is becoming a pressing issue. Waste management is one area where vermitechnology can be used as an alternate method. Vermicomposting is a process that can be used to create beneficial goods like earthworm biomass and vermicompost from a variety of organic solid wastes. The biomass produced by earthworms could be used in a wide variety of contexts. Making and managing large-scale worm cultures and collecting waste for vermicompost production are two promising cottage industries for developing countries. Since vermiculture and vermicomposting can be used for a variety of purposes besides just enhancing soil quality and crop productivity, they have a lot of untapped potential in developing countries that are still relying on traditional farming techniques. The widespread adoption of vermitechnology has the potential to boost employment opportunities and accelerate the rate at which the labour market expands.

References

Adeel, M., Shakoor, N., Shafiq, M., Pavlicek, A., Part, F., Zafiu, C., Raza, A., Ahmad, M. A., Jilani, G., White, J. C., Ehmoser, E. K., & Rui, Y. (2021). A critical review of the environmental impacts of manufactured nano-objects on earthworm species. *Environmental Pollution, 290*, 118041.

Ahmad, A., Aslam, Z., Bellitürk, K., Ullah, E., Raza, A., & Asif, M. (2022a). Vermicomposting by bio-recycling of animal and plant waste: A review on the miracle of nature. *Journal of Innovative Sciences, 8*(2), 175–187.

Ahmad, F. B., Cisewski, J. A., Xu, J., & Anderson, R. N. (2022b). Provisional mortality data—United States, 2022. *Morbidity and Mortality Weekly Report, 72*(18), 488.

Ahmed, N., & Al-Mutairi, K. A. (2022c). Earthworms effect on microbial population and soil fertility as well as their interaction with agriculture practices. *Sustainability, 14*(13), 7803.

Ahmed, S. K., Abdulqadir, S. O., Hussein, S. H., Omar, R. M., Ahmed, N. A., Essa, R. A.,... & Abdulla, A. Q. (2022d). The impact of monkeypox outbreak on mental health and counteracting strategies: A call to action. *International Journal of Surgery, 106*, 106943.

Ahmed, S. K., et al. (2023). Knowledge, attitude and worry in the Kurdistan Region of Iraq during the Mpox (Monkeypox) outbreak in 2022: An online cross-sectional study. *Vaccines, 11.3*, 610.

Al-Maliki, S., Al-Taey, D. K., & Al-Mammori, H. Z. (2021). Earthworms and eco-consequences: Considerations to soil biological indicators and plant function: A review. *Acta Ecologica Sinica, 41*(6), 512–523.

Ansari, A. A., Jaikishun, S., Islam, M. A., Kuri, S. K., Fiedler, K., & Nandwani, D. (2016). Principles of vermitechnology in sustainable organic farming with special reference to Bangladesh. In *Organic farming for sustainable agriculture* (pp. 213–229). Springer.

Aransiola, S. A., Ijah, U. J. J., & Abioye, O. P. (2013). Phytoremediation of lead polluted soil by *Glycine max L. Applied and Environmental Soil Science, 2013.* Article ID 631619. https://doi.org/10.1155/2013/631619. https://www.hindawi.com/journals/aess/2013/631619/abs/

Aransiola, S. A., Ijah, U. J. J., Abioye, O. P., Bala, J. D. (2022). Vermicompost-assisted phytoremediation of toxic trace element-contaminated soil in Madaka, Nigeria using *Melissa officinalis*

L. and *Sida acuta*. *International Journal of Environmental Science and Technology*. https://doi.org/10.1007/s13762-022-04105-y

Aransiola, S. A., Ijah, U. J. J., Abioye, O. P., Bala, J. D., Rivadeneira-Mendoza, R. F., Luque, R., Rodríguez-Díaz, J. M., & Maddela, N. R. (2024). Micro and vermicompost assisted remediation of heavy metal contaminated soils using phytoextractors. *Case Studies in Chemical and Environmental Engineering., 9*(2024), Article 100755. https://doi.org/10.1016/j.cscee.2024.100755

Ashwood, F., Vanguelova, E. I., Benham, S., & Butt, K. R. (2019). Developing a systematic sampling method for earthworms in and around deadwood. *Forest Ecosystems, 6*(1), 1–12.

Bart, S., Amossé, J., Lowe, C. N., Mougin, C., Péry, A. R., & Pelosi, C. (2018). Aporrectodea caliginosa, a relevant earthworm species for a posteriori pesticide risk assessment: Current knowledge and recommendations for culture and experimental design. *Environmental Science and Pollution Research, 25*(34), 33867–33881.

Bellitürk, K. (2018). Vermicomposting in Turkey: Challenges and opportunities in future. *Eurasian Journal of Forest Science, 6*(4), 32–41.

Bhat, S. A., Singh, J., Singh, K., & Vig, A. P. (2017). Genotoxicity monitoring of industrial wastes using plant bioassays and management through vermitechnology: A review. *Agriculture and Natural Resources, 51*(5), 325–337.

Bhattacharya, S. S., & Kim, K. H. (2016). Utilization of coal ash: Is vermitechnology a sustainable avenue? *Renewable and Sustainable Energy Reviews, 58*, 1376–1386.

Bhuvaneshwaran, T., Sanjay, G., Jayakumar, N., Ahilan, B., Felix, N., & Prabu, E. (2019). Potentiality of earthworm as a replacement for fish meal and its role in aquaculture. *Journal of Aquaculture in the Tropics, 34*(1/2), 115–127.

Biruntha, M., Karmegam, N., Archana, J., Selvi, B. K., Paul, J. A. J., Balamuralikrishnan, B., Chang, S. W., & Ravindran, B. (2020). Vermiconversion of biowastes with low-to-high C/N ratio into value added vermicompost. *Bioresource Technology, 297*, Article 122398.

Bondhare, S. O., & Desai, R. B. (2019). Comparison of life cycle of earthworms Eisenia fetida and Lumbricus rubellus under controlled condition, in Nanded district. *Waste Management, 5*, 6.

Byambas, P., Hornick, J. L., Marlier, D., & Francis, F. (2019). Vermiculture in animal farming: A review on the biological and nonbiological risks related to earthworms in animal feed. *Cogent Environmental Science, 5*(1), 1591328.

Casquero, S., Trigo, D., Guitarte, J. L. M., & Novo, M. (2020). When sunscreens reach the soil: Impacts of a UV filter on the life cycle of earthworms. *Applied Soil Ecology, 147*, Article 103354.

Chauhan, N. S., & Punia, A. (2022). Strategies for sustainable and ecofriendly pest management in Agroecosystem. In *Pesticides in the natural environment* (pp. 365–381). Elsevier.

Cheng, Q., Lu, C., Shen, H., Yang, Y., & Chen, H. (2021). The dual beneficial effects of vermiremediation: Reducing soil bioavailability of cadmium (Cd) and improving soil fertility by earthworm (Eisenia fetida) modified by seasonality. *Science of the Total Environment, 755*, Article 142631.

Chowdhury, S. D., Bandyopadhyay, R., & Bhunia, P. (2022a). Techno-economic analysis and life-cycle assessment of vermi-technology for waste bioremediation. In *Biomass, biofuels, biochemicals* (pp. 315–349). Elsevier.

Chowdhury, S. D., Bhunia, P., & Surampalli, R. Y. (2022b). Sustainability assessment of vermifiltration technology for treating domestic sewage: A review. *Journal of Water Process Engineering, 50*, Article 103266.

Coulibaly, S. S., Edoukou, F. E., Kouassi, K. I., Barsan, N., Nedeff, V., & Zoro, I. B. (2018). Vermicompost utilization: A way to food security in rural area. *Heliyon, 4*(12), Article e01104.

Cui, Y., Moorhead, D. L., Guo, X., Peng, S., Wang, Y., Zhang, X., & Fang, L. (2021). Stoichiometric models of microbial metabolic limitation in soil systems. *Global Ecology and Biogeography, 30*(11), 2297–2311.

Debnath, T. (2018). *Study on bioconversion of household waste and faecal sludge through various composting process* [Doctoral dissertation, Khulna University of Engineering & Technology (KUET), Khulna, Bangladesh].

Domínguez, J. (2018). *Earthworms and vermicomposting* (pp. 63–77). IntechOpen.

Earthworm Society of Britain. Life cycle of an earthworm. (2022). *Life cycle of an earthworm.* Retrieved December 17, 2022, from https://www.earthwormsoc.org.uk/lifecycle

Edwards, C. A., & Arancon, N. Q. (2022). The influence of environmental factors on earthworms. *Biology and ecology of earthworms* (pp. 191–232). Springer.

Hussain, N., Das, S., Goswami, L., Das, P., Sahariah, B., & Bhattacharya, S. S. (2018). Intensification of vermitechnology for kitchen vegetable waste and paddy straw employing earthworm consortium: Assessment of maturity time, microbial community structure, and economic benefit. *Journal of Cleaner Production, 182*, 414–426.

Ijah, U. J. J., Aransiola, S. A., & Abioye, O. P. (2015). Restoration of lead contaminated soil using *Arachis hypogaea. International Journal of Environmental Pollution and Control Research, 1*(1). http://epcr.scienceresearchlibrary.com/view_issue.php?id=NTE

Ijah, U. J. J., Musa, O. I., Amao, F., Maude A. M., & Victor-Ekwebelem, M. O. (2022). Wastewater quality and unhygienic practices in Minna Abattoir, North Central Nigeria. *Science World Journal, 17*(2). ISSN: 1597-6343.

Ismail, K. A., Lino, F. A., Machado, P. L. O., Teggar, M., Arıcı, M., Alves, T. A., & Teles, M. P. (2022). New potential applications of phase change materials: A review. *Journal of Energy Storage, 53*, Article 105202.

Iwai, C. B., & Kruapukee, A. (2017). Using vermitechnology in soil rehabilitation for rice production in salt-affected area of northeast Thailand. *International Journal of Environmental and Rural Development, 8*(2), 88–93.

Javed, F., & Hashmi, I. (2021). Vermiremediation–remediation of soil contaminated with oil using earthworm (*Eisenia fetida*). *Soil and Sediment Contamination: An International Journal, 30*(6), 639–662.

Kaur, T. (2020). Vermicomposting: An effective option for recycling organic wastes. *Organic Agriculture*, 1–10.

Leena S. R. B., Archana S., Srishti D., & Sunil K. D (2023). Vermitechnology: A sustainable approach to manage organic waste in urban areas. *International Journal of Current Science Research and Review, 6*. ISSN: 2581–8341.

Lim, E. Y., Lee, J. T. E., Zhang, L., Tian, H., Ong, K. C., Tio, Z. K., Zhang, J., & Tong, Y. W. (2022). Abrogating the inhibitory effects of volatile fatty acids and ammonia in overloaded food waste anaerobic digesters via the supplementation of nano-zero valent iron modified biochar. *Science of the Total Environment, 817*, Article 152968.

Livinus, M. U., Bala, S. Z., Abdulsalam, M., Innocent, M. O., Hassan, M., & Kini, P. (2024). Natural occurrences of soil dilapidation. In *Prospect for soil regeneration and its impact on environmental protection* (Chap. 9, pp. 205–223). Springer. https://doi.org/10.1007/978-3-031-53270-2_7 ISBN: 978-3-031-53270-2

Malley, J. D., Kruppa, J., Dasgupta, A., Malley, K. G., & Ziegler, A. (2012). Probability machines. *Methods of Information in Medicine, 51*(01), 74–81.

Musa, O. I., Ijah, U. J. J., Abioye, O. P., & Adebola, M. O. (2021). Assessment of the phytoremediation potentials of Melissia officinalis in removing spent engine oil from contaminated soils. *International Journal of Applied Biological Research., 12*(1), 136–149.

Musyoka, S. N., Liti, D. M., Ogello, E., & Waidbacher, H. (2019). Utilization of the earthworm, Eisenia fetida (Savigny, 1826) as an alternative protein source in fish feeds processing: A review. *Aquaculture Research, 50*(9), 2301–2315.

Nalunga, A., Komakech, A. J., Jjagwe, J., Magala, H., & Lederer, J. (2021). Growth characteristics and meat quality of broiler chickens fed earthworm meal from Eudrilus eugeniae as a protein source. *Livestock Science, 245*, Article 104394.

Nazeer, A., & Al-Mutairi, K. A. (2022). Earthworms effect on microbial population and soil fertility as well as their interaction with agriculture practices. *Sustainability, 14*(13), 7803.

Ojuolape, O. T., Aransiola, S. A., & Obideyi, O. A. (2015) Vermitechnology: A solution to environmental problems; A review. *Journal of Global Ecology and Environment, 3*(3), 127–135. http://www.ikpress.org/abstract/4516

Parolini, M., Ganzaroli, A., & Bacenetti, J. (2020). Earthworm as an alternative protein source in poultry and fish farming: Current applications and future perspectives. *Science of the Total Environment, 734*, Article 139460.

Patwa, A., Parde, D., Dohare, D., Vijay, R., & Kumar, R. (2020). Solid waste characterization and treatment technologies in rural areas: An Indian and international review. *Environmental Technology & Innovation, 20*, Article 101066.

Roy, G., Wasim, I., & Chattopadhyay, G. N. (2022a). Effect of moisture status on vermicomposting of organic waste amended fly ash. *The Journal of Solid Waste Technology and Management, 48*(3), 401–407.

Roy, S., Sarkar, D., Datta, R., Bhattacharya, S. S., & Bhattacharyya, P. (2022b). Assessing the arsenic-saturated biochar recycling potential of vermitechnology: Insights on nutrient recovery, metal benignity, and microbial activity. *Chemosphere, 286*, Article 131660.

Samal, K., Dash, R. R., & Bhunia, P. (2018). Design and development of a hybrid macrophyte assisted vermifilter for the treatment of dairy wastewater: A statistical and kinetic modelling approach. *Science of the Total Environment, 645*, 156–169.

Samal, K., Mohan, A. R., Chaudhary, N., & Moulick, S. (2019a). Application of vermitechnology in waste management: A review on mechanism and performance. *Journal of Environmental Chemical Engineering, 7*(5), Article 103392.

Samal, A. K., Srivastava, R. K., & Gautam, G. C. (2019b). Paleoproterozoic (~1.88–1.89 Ga) ultramafic–mafic sills, Cuddapah basin, India—revisited: Implications for interaction between mantle plume and metasomatized subcontinental lithospheric mantle. *Journal of Earth System Science, 128*(8), 232.

Sanchez-Hernandez, J. C., Ríos, J. M., Attademo, A. M., Malcevschi, A., & Cares, X. A. (2019). Assessing biochar impact on earthworms: Implications for soil quality promotion. *Journal of Hazardous Materials, 366*, 582–591.

Sharma, S., Kumar, A., Singh, A. P., & Vasudevan, P. (2009). Earthworms and vermitechnology–A review. *Dynamic Soil, Dynamic Plant, 3*(2), 1.

Siddiqui, M. A., Neeraj, A., & Hiranmai, R. Y. (2022). Vermitechnology: An eco-friendly approach for organic solid waste management and soil fertility improvement—A review. In *Strategies and tools for pollutant mitigation* (pp. 91–112).

Singh, R., Bhunia, P., & Dash, R. R. (2018). Understanding intricacies of clogging and its alleviation by introducing earthworms in soil biofilters. *Science of the Total Environment, 633*, 145–156.

Singh, S. I., Singh, S., & Vig, A. P. (2020). Earthworm-assisted bioremediation of agrochemicals. In *Agrochemicals detection, treatment and remediation* (pp. 307–327). Butterworth-Heinemann.

Singh, S., Singh, J., Kandoria, A., Quadar, J., Bhat, S. A., Chowdhary, A. B., & Vig, A. P. (2020b). Bioconversion of different organic waste into fortified vermicompost with the help of earthworm: A comprehensive review. *International Journal of Recycling Organic Waste in Agriculture, 9*(4), 423–439.

Thakur, A., Kumar, A., Kumar, C. V., Kiran, B. S., Kumar, S., & Athokpam, V. (2021). A review on vermicomposting: By-products and its importance. *Plant. Cell Biotechnology and Molecular Biology, 22*, 156–164.

Verma, K., Sharma, P., & Saini, S. P. (2022). Impact of integrated nutrient management in enhancing the growth and yield of crops. *Vigyan Varta, 3*(9), 83–93.

Vodounnou, D. S. J. V., Kpogue, D. N. S., Tossavi, C. E., Mennsah, G. A., & Fiogbe, E. D. (2016). Effect of animal waste and vegetable compost on production and growth of earthworm (Eisenia fetida) during vermiculture. *International Journal of Recycling of Organic Waste in Agriculture, 5*(1), 87–92.

Vos, H. M., Koopmans, G. F., Beezemer, L., de Goede, R. G., Hiemstra, T., & van Groenigen, J. W. (2019). Large variations in readily-available phosphorus in casts of eight earthworm species are linked to cast properties. *Soil Biology and Biochemistry, 138*, Article 107583.

Vuković, A., Velki, M., Ečimović, S., Vuković, R., Štolfa Čamagajevac, I., & Lončarić, Z. (2021). Vermicomposting—Facts, benefits and knowledge gaps. *Agronomy, 11*(10), 1952.

Wang, R., Yue, S., Huang, C., Shen, Z., Qiao, Y., Charles, S., Yu, J., Cao, Z., Li, Z., & Li, Z. (2023). Uptake, distribution, and elimination of selenite in earthworm *Eisenia fetida* at sublethal concentrations based on toxicokinetic model. *Science of the Total Environment, 858*, Article 159632.

Xie, D., Wu, W., Hao, X., Jiang, D., Li, X., & Bai, L. (2016). Vermicomposting of sludge from animal wastewater treatment plant mixed with cow dung or swine manure using Eisenia fetida. *Environmental Science and Pollution Research, 23*(8), 7767–7775.

Xu, T., Fei, F., Ding, Y., Liu, Y., Mao, G., Yang, L., Zhao, T., Liao, T., Feng, W., & Wu, X. (2021). Study on the comprehensive utilization of solid residues of Flammulina velutipes and vinegar and their application as feed in Eisenia fetida earthworm culture. *Environmental Science and Pollution Research, 28*(35), 49153–49165.

Yuvaraj, A., Karmegam, N., Ravindran, B., Chang, S. W., Awasthi, M. K., Kannan, S., & Thangaraj, R. (2020). Recycling of leather industrial sludge through vermitechnology for a cleaner environment—A review. *Industrial Crops and Products, 155*, Article 112791.

Yuvaraj, A., Thangaraj, R., Ravindran, B., Chang, S. W., & Karmegam, N. (2021). Centrality of cattle solid wastes in vermicomposting technology–A cleaner resource recovery and biowaste recycling option for agricultural and environmental sustainability. *Environmental Pollution, 268*, Article 115688.

Zeb, A., Li, S., Wu, J., Lian, J., Liu, W., & Sun, Y. (2020). Insights into the mechanisms underlying the remediation potential of earthworms in contaminated soil: A critical review of research progress and prospects. *Science of the Total Environment, 740*, Article 140145.

Zhong, Y., Wang, X. J., Yi, C., Qiong-E, D., Jiang-Yun, T., & Ming, H. M. (2023). Vermicomposting of *Pleurotus eryngii* spent mushroom substrates and the possible mechanisms of vermicompost suppressing nematode disease caused by Meloidogyne incognita. *Heliyon, 9*(4): https://doi.org/10.1016/j.heliyon.2023.e15111

Vermitechnology as a Sustainable Solution: Transforming Organic Waste into Economic and Environmental Wealth

Omolola Temilade Evinemi, Bukola Opeyemi Oluwarinde, and Ayomide Emmanuel Fadiji

Abstract Vermitechnology offers a practical and sustainable solution for managing organic waste while contributing to agricultural productivity and environmental conservation. This technique leverages the transformative capabilities of earthworms to convert organic waste into valuable resources. This method reduces landfill waste, eliminates harmful pathogens, and improves soil health. It also supports industries like food production, brewing, and textiles by turning their organic waste into valuable fertilizers and soil conditioners. This study explores the potential of vermitechnology to address environmental, economic, and social challenges. It highlights the role of this approach in recycling waste, enriching soil, and reducing the need for chemical fertilizers. We further demonstrate how vermitechnology can drive green entrepreneurship, support community development, and promote a circular economy. By focusing on waste-to-wealth initiatives, vermitechnology proves to be a powerful tool for sustainable development and resource optimization.

Keywords Earthworms · Environmental protection · Fertilizer · Organic wastes · Waste management

O. T. Evinemi (✉)
Bioresources Development Centre, National Biotechnology Development Agency (NABDA), Ogbomosho, Nigeria
e-mail: omololaojuolape@gmail.com

B. O. Oluwarinde
Department of Infectious Diseases and Public Health, Jockey Club College of Veterinary, City University of Hong Kong, Hong Kong, China
e-mail: boluwari@cityu.edu.hk

A. E. Fadiji
Hawkesbury Institute for the Environment, Western Sydney University, Penrith, Australia

Food Security and Safety Focus Area, North-West University, Mafikeng, South Africa

S. A. Aransiola and N. R. Maddela (eds.), *Vermitechnology: Economic, Environmental and Agricultural Sustainability*, https://doi.org/10.1007/978-3-031-96212-7_3

1 Introduction

Large quantities of organic waste are generated by human and animal activities, and the by-products of their decomposition significantly affect the quality of water, soil, and air. Due to the presence of diverse pathogenic bacteria, biosolid wastes are highly infectious (Law et al., 2021). When these wastes are improperly disposed of in the environment, they not only pose serious health and environmental risks but also contribute to the spread of antimicrobial resistance (AMR) by facilitating the proliferation of antibiotic-resistant bacteria and resistance genes in soil and water systems, which is a major public health concern (Oluwarinde et al., 2024). Such wastes also alter soil properties, including water-holding capacity, pH, conductivity, bulk density, and organic carbon levels (Tohumcu et al., 2023).

Effectively managing biosolid wastes is essential for maintaining a healthy environment and improving soil quality, which in turn boosts primary productivity. Composting is widely recognized as a safe and efficient method for recycling organic wastes, as the direct application of unstabilized waste can be both harmful and pathogenic (Butler et al., 2001). Composting involves the biological transformation of organic waste into stabilized products under aerobic conditions. This process results in the mineralization of organic matter into carbon dioxide and the formation of humic compounds (Senesi et al., 2007).

In Indian towns, solid waste is a major contributor to environmental problems, particularly during the rainy season when it obstructs water flow (Kumar & Henock, 2015). Several techniques, including composting, solid waste management, gasification, incineration, and waste-to-energy conversion, are used to address this issue. Organic waste can also be processed into vermicompost, which serves as a valuable fertilizer with significant economic potential (Dey Chowdhury et al., 2023). For instance, *Eudrilus* sp. has been shown to transform household waste such as paper, food, garden leaves, and vegetables into nutrient-rich vermicompost containing beneficial microbial communities, including fungi, bacteria, actinomycetes, pseudomonads, nitrogen fixers, and phosphate solubilizers (Sequeira & Chandrashekar, 2015).

Synthetic chemical misuse negatively impacts nontarget organisms, disrupting agroecosystem. A'ali et al. (2017) demonstrated that agricultural waste materials, such as pomegranate peels, sheep manure, sugar beet pulp, sawdust, and chopped corn can be used to create vermibeds. The resulting compost improved soil properties, including pH, electrical conductivity, and nutrient content (NPK), and served as an effective fertilizer.

During vermicomposting, earthworms interact closely with microorganisms in the decomposer community to accelerate the decomposition of organic waste (Khan et al., 2024). This bio-oxidative process stabilizes the waste, resulting in altered physical and biochemical properties. Unlike traditional composting, vermicomposting relies on the earthworms' digestive systems to break down organic material, producing nutrient-rich casts that enhance soil fertility and improve its physical characteristics.

Through physical and biological activity, earthworms actively decompose organic matter. Physically, they fragment the organic material, increasing its surface area and aeration, which facilitates microbial activity (Ojuolape et al., 2015; Rakkini et al., 2017). Biologically, microbes carry out biochemical changes, breaking down organic matter through enzymatic digestion, enriching it with nitrogen, and transporting essential organic and inorganic elements.

Earthworms significantly contribute to recycling organic waste and producing high-humic-content organic manure. This manure improves soil structure, aeration, and fertility, offering advantages over conventional fertilizers (Bhunia et al., 2021). Unlike regular fertilizers, humic acid fertilizer contains bioactive compounds that promote yield, growth, seed germination, and overall plant health. These fertilizers also help crops resist drought, cold stress, pathogenic bacteria, insect pests, and soil-borne diseases (Rakkini et al., 2017). Hence, this chapter gives an overview of the vermicomposting process using various earthworm species, highlighting its ecological and economic importance.

2 Historical Background of Vermitechnology

Vermitechnology is an important aspect of biotechnology involving the conversion of various types of organic waste into valuable resources using earthworms. It is one of the latest approaches that produces biofertilizers, such as vermicompost, for agricultural uses and high-quality protein from earthworm biomass to supplement the nutritional and energy needs of animals, all at a faster rate (Singh & Kothyari, 2019).

The vermicomposting process involves the degradation of organic waste by earthworms and the microorganisms in their environment (Huang et al., 2020). Earthworms consume a mixture of organic materials, such as food scraps, yard waste, and paper, breaking them down through a combination of mechanical and enzymatic processes. Their digestive systems introduce beneficial microorganisms into the mixture, which further degrade the organic matter.

The end product of the vermicomposting process is a nutrient-rich soil amendment containing high levels of organic matter, nitrogen, phosphorus, and potassium (Huang et al., 2020). This vermicompost also harbors beneficial microorganisms that can improve soil structure, water-holding capacity, and plant growth. Vermicompost is a nutrient-rich and readily available to plants, often referred to as the "gold of organic fertilizers."

3 Vermitechnology in Waste Management

3.1 Managing Solid Organic Waste

Solid waste generation is a significant environmental challenge globally, primarily generated by domestic and commercial activities (Temilade et al., 2015) (Fig. 1). Solid waste can be categorized into inorganic or organic materials, with the latter offering considerable potential for sustainable management through vermitechnology. Organic waste can be transformed into nutrient-rich biofertilizer via vermicomposting, yielding a product that is 4–5 times more potent than traditional compost and more beneficial for crop growth than chemical fertilizers (Aransiola et al., 2024; Rajiv et al., 2010).

There are huge quantities of waste generated by modern society, which end up in landfills every day, creating severe economic and environmental challenges. These include greenhouse gas emissions, the release of toxic gases, and leachate contamination of groundwater, all of which necessitate prolonged monitoring for environmental safety—often lasting decades (Rajiv et al., 2010). Vermitechnology offers an efficient and eco-friendly alternative to conventional waste management methods. The use of earthworms in waste degradation accelerates composting, significantly reduces odor, cuts composting time by more than half, and ensures that the final product is both sanitized and detoxified (Lee & Khor, 2022). Sharma (2003) discusses the benefits of vermicomposting as a waste management option. These benefits include the ability to reduce the volume and mass of organic waste, reduce greenhouse gas emissions, and produce valuable soil amendments. They also highlight the role of vermicomposting in improving soil quality and promoting plant growth. Vermicomposting solid organic

Fig. 1 Solid waste (Dam, 2021)

waste has been considered a feasible and sustainable waste management technique (Alshehrei & Ameen, 2021).

3.2 *Biodegradable Waste Suitable for Vermitechnology*

It has been shown that earthworms are capable of physically processing a wide variety of organic wastes from municipal, agricultural, and industrial sources (Datar et al., 1997; Edwards & Arancon, 2022). Vermicomposting is highly effective for managing biodegradable materials such as food waste, including fruit and vegetable scraps, coffee grounds, tea bags, and eggshells. Yard waste, such as grass clippings, leaves, and twigs, is all considered an excellent food source for earthworms. Additionally, municipal wastewater-derived "sewage sludge" (biosolids) can be digested by earthworms, converting a significant portion into nutrient-rich vermicompost (Panday et al., 2024).

Paper-based products, such as cardboard, newspaper, and paper towels, are another suitable feedstock, along with slaughterhouse waste, which can be efficiently processed by earthworms (Babaniyi et al., 2023; Datar et al., 1997; Edwards & Arancon, 2022;). Additionally, waste from agriculture and livestock rearing, including cattle dung, pig manure, and poultry litter, makes excellent feed material for vermicomposting.

Industrially generated organic waste provides additional opportunities for vermicomposting. Earthworms can process pulp and cardboard sludge, by-products from food processing industries such as distilleries, breweries, and potato or corn chip manufacturers, as well as organic waste from sugarcane and aromatic oil industries (Kale, 1998). These examples highlight the versatility and efficiency of vermitechnology in managing diverse biodegradable waste streams, turning them into valuable organic resources.

3.3 *Wastewater Treatment Through Vermifiltration*

Vermifiltration is an innovative and eco-friendly wastewater treatment method that employs earthworms to purify water. It is a type of constructed wetland system that utilizes the natural decomposition abilities of worms, such as earthworms, to break down organic matter in wastewater. In a vermifiltration system, wastewater is filtered through a bed of worms housed in containers, such as plastic tanks or trenches. The worms consume organic matter, breaking it down into simpler compounds while releasing treated water as the primary output (Singh et al., 2019). The by-product of the process is a mixture of worm castings and partially decomposed organic matter, which can be used as a nutrient-rich soil amendment for gardening and agriculture (Sinha et al., 2008).

Vermifiltration has several advantages over traditional wastewater treatment methods. These include minimal energy input, low maintenance, and the ability to remove a wide range of pollutants from wastewater. Additionally, it is a relatively cost-effective solution, making it well suited for use in developing countries and other regions with limited access to expensive treatment systems. This technology holds great potential for sustainable wastewater management, combining ecological benefits with economic practicality.

4 Vermiremediation: A Tool for Land Reclamation

Vermiremediation is the application of earthworms and other soil invertebrates to degrade and remove contaminants from contaminated soil (Shi et al., 2020). This approach offers an effective and sustainable method for cleaning up contaminated lands, restoring them to a state suitable for development or agricultural uses. It is a viable method of cleaning up sites contaminated with certain types of heavy metals and organic pollutants (Dabke, 2013).

Key applications of vermiremediation for soil cleanup include:

1. Bioremediation: Earthworms and other invertebrates can naturally degrade contaminants or be paired with genetically modified organisms to target specific contaminants. This process enhances the breakdown and removal of harmful compounds from the soil (Nisa et al., 2022).
2. Phytoremediation support: Vermicompost, rich in nutrients and beneficial microbes, can be used as a soil amendment to improve soil health and support the growth of phytoremediation plants. This synergy accelerates the uptake and removal of contaminants by plants while simultaneously enhancing soil properties (Aransiola et al., 2022).
3. Landfarming: Vermiremediation can also be used as a form of landfarming, in which contaminated soil is excavated and placed in a controlled area for treatment. Worms and other invertebrates are introduced to degrade and detoxify contaminants in the soil.

In recent years, vermicomposting has gained recognition as a technology for treating organically polluted soil. According to Rodriguez-Campos et al. (2014), earthworms can remove pesticides, herbicides, polycyclic aromatic hydrocarbons (PAHs), polychlorinated biphenyls (PCBs), and crude oil, making it a valuable tool for soil decontamination. Technology is relatively efficient compared with phytoremediation. Vermiremediation can achieve substantial accumulation and degradation of organic pollutants, unlike phytoaccumulation, offering a low-cost, eco-friendly alternative for managing soil contamination (Wu et al., 2018).

5 Vermitechnology in Agriculture

5.1 *Enhancing Modern Agriculture*

Agriculture has evolved differently across regions, shaped by diverse weather patterns and topographies. This evolution transformed human population enabling population growth far beyond what hunting and gathering could sustain. If the nutritional requirements and preferences of a growing population that is also becoming more affluent are to be satisfied, the globe will need to see a significant rise in the amount of food that is produced. This calls for the continued development of modern agriculture practices and innovations, including sustainable technologies like vermitechnology.

Modern agriculture relies on a dynamic method in which farmers apply technology and information to optimize productivity and use fewer natural resources like water, land, and energy. This helps the world satisfy its demands for food, fuel, and fiber. Aside from modern machines that make the strenuous work embedded in farming easier, the use of biofertilizers as a viable alternative to the ill effects of chemical fertilizers and pesticides for both the people and the environment is a breath of relief in modern agriculture, as it has also provided a great deal of comfort that has resulted in higher crop yields and increased agricultural productivity. There is no doubt that various forms of technology, including vermitechnology, are to a significant degree responsible for such an increase. The significant growth that was observed in the latter half of the twentieth century was primarily the result of intensification and corresponding increases in yields.

Vermicomposting is the process of converting organic matter into a more stable form by utilizing the joint efforts of microbes and earthworms (Siddiqui et al., 2022) (Fig. 2). Earthworms are the major drivers of the process, conditioning the substrate and changing biological activity. Bacteria are responsible for the biochemical breakdown of organic materials; nevertheless, earthworms are the key drivers of the process. Vermicomposting, which uses earthworms and the microorganisms that live in their digestive tracts, decomposes organic matter at a faster rate and with greater efficiency than aerobic composting, which uses traditional composting techniques (Lim et al., 2016). The structure of the soil is significantly altered as a result of the activity of earthworms, which also contributes to an increase in both fertility and production (Blouin et al., 2013). Earthworms are sometimes referred to as the "administrators of the soil" because of the manner in which they influence the soil's physical, chemical, and biological properties (Ansari & Ismail, 2012).

5.2 *Applications of Vermicompost in Plant Growth*

When vermicompost (cattle manure) was tested for its effectiveness on *Petroselinum crispum*, the results showed that it increased the yield, leaf size, and height of the plant (Peyvast et al., 2008). According to Kizilkaya et al. (2012), earthworms play a

Fig. 2 Application of vermicomposting in agriculture (Pierce, 2024)

crucial part in the management of organic waste by vermicomposting. The vermibed was made in several combinations using sewage sludge that had been treated with cow manure and hazelnut husk. After formulation, *Eisenia fetida* was added into the vermicompost, and its effects on *Triticum aestivum* were evaluated. It was discovered that all treatments increased the yield and growth of the tested plants compared to the control.

Every population, according to Ansari and Hanief (2015), has dumped a sizable amount of organic garbage into landfills (river or burn systems). Each day, a significant volume of market garbage and plant matter fouled the river. A significant amount of the trash was degraded while being turned into vermicompost (30% yield). The biggest technological transfer for turning biowaste into useful resources is due to earthworms. It also has bacteria that are good for plants, including *Actinomycetes*, *Aspergillus*, *Azotobacter*, and *Nitrobacter*. Vermicompost, chemical fertilizer, 50% manure + 50% fertilizer, and control were all studied by Karmakar et al. (2015) for their effects on rice fields. They discovered that the examined plants that received vermicompost grew well and received the most nutrients possible.

There is a demand for fertile land to cultivate due to the growing population. In both lab and field settings, *Mentha arvensis* was grown in salt-stressed circumstances. Plant growth was enhanced by the inclusion of the fungus *Glomus aggregatum* and *Exiguobacterium oxidotolerans* in vermicompost. Additionally, it was determined that vermicompost and multimicrobial inoculations were effective biofertilizers for the growth of *M. arvensis* (Bharti et al., 2016). Vermicompost, when cultivated in high salinity, exhibited complicated impacts on the antioxidant enzyme activities of plants, according to Xu et al. (2016). Compost from municipal solid waste and industrial sewage sludge was combined with arable soil. When industrial trash produced a larger yield than the treated municipal waste soil, both treatments increased organic matter and microorganisms but decreased the ability of the soil to store water (Aransiola et al., 2021; Zamani et al., 2016). Aransiola et al. (2022) reported the positive effects of vermicast in the phytoremediation of heavy-metal-polluted soil.

According to Masullo (2017), waste materials were digested under anaerobic conditions and subsequently turned into vermicompost utilizing earthworms. Vermicompost was put in the field, reducing the frequency of watering and promoting plant growth. The vermicompost made from biochar, wood chips, and sewage sludge increased the rate of earthworm reproduction (cocoon and juveniles) and decreased the levels of Cd and Zn (Malińska et al., 2017). According to Maji et al. (2017), vermicompost that is humic-acid-rich caused an increase in the height, dry weight, and fresh weight of plants. When compared to chemical fertilizer, the highest quantity and density of microorganisms (fungus and bacteria) were also noted. Vermicompost, a potent biofertilizer in sustainable agriculture that reduces the usage of conventional agrochemicals, is produced using a variety of earthworm species. The worms participate in the recycling of organic waste and in waste management (Bhat et al., 2018).

6 Environmental and Economic Benefits of Vermitechnology

6.1 Environmental Impact

Earthworms are crucial and do an excellent job of breaking up compacted soil in preparation for plant development (Adams, 2018). Results from several research support Darwin's theory that earthworms improve soil fertility, which is essential for plant development (Alban & Berry, 1994; Decaëns, 1999; Lee & Foster, 1991). The positive effects of earthworms on plant development and agricultural production have been the subject of much study. It has been proven that earthworms can decompose a wide variety of organic waste, including that generated from homes, cities, and businesses involved in the paper, wood, and food sectors; also, earthworms improve soil quality (Decaëns, 1999; Lalitha et al., 2000).

Auxins and cytokinins, two hormones beneficial to plant development, are thought to be secreted by earthworms (Krishnamoorthy & Vajranabhaiah, 1986). There is evidence to suggest that earthworms help plants expand by producing metabolites that stimulate expansion (Wang et al., 2021; Wong et al., 2020). It is believed that these metabolites are responsible for this outcome. However, there are more benefits of earthworms to plant growth than only the vast quantities of macronutrients and micronutrients in vermicast and the secretions they produce.

Furthermore, earthworms are used to clean up heavy metals, like cadmium (Cd), chromium (Cr), and lead (Pb); they do this by eating them or absorbing the metals through their skin (Homa et al., 2010; Yuvaraj et al., 2018). Morgan and Morgan (1988) found that some types of earthworms can live in places with a lot of metal pollution. For example, *Lumbricus rubellus* lived in lead-contaminated soil, while *Dendrodrilus rubidus* could live in copper-polluted soil, and *Eisenia fetida* could live in cadmium-polluted soil (Langdon et al., 2001). Earthworms make more metallothionein (MT) protein when they are stressed by heavy metals; these MT proteins can get rid of a number of metal ions in the environment (Lionetto et al., 2012). Earthworms and microorganisms can stabilize certain chemical elements in substrate materials during the waste degradation process (Pathma & Sakthivel, 2012).

In thermal power plants (TPPs), vermitechnology has been used as a method for reducing the number of pollutants released into the atmosphere. Pollution of the environment has been a significant barrier to achieving a healthy ecosystem for the purpose of ensuring the survival of all living things (Destoumieux-Garzón et al., 2018), which is one of the fundamental requirements of human rights. Despite the fact that national initiatives have been developed to reduce the negative effects of industrialization, manufacturing, and other operations on the environment, pollutants resulting from human activity are still, for the most part, a significant problem (Häder et al., 2020). The combustion of coal to produce energy, for example, results in the generation of vast quantities of fly ash (FA). Fly ash has been recognized as a significant pollutant across the globe for the past two decades (Satapathy et al., 2020). It creates a barrier to photosynthesis, particularly in aquatic plants, which in turn disrupts the food chain. Furthermore, it is harmful to the environment because it lowers pH, increases turbidity in water bodies, and contributes to increased sedimentation. As a result, it is necessary to dispose of FA in an appropriate and risk-free manner to get around unavoidable issues and turn these materials into a product with additional value. Conventional methods of fly ash disposal have resulted in the contamination of groundwater and landfills (Haynes, 2009). Vermitechnology has gained attention as an active and faster method for converting the heavy metal toxic content of FA into organic manure by the action of certain species of earthworm. Species that belong to the phylum Annelida and the subclass Oligochaeta have the potential to decompose the fly ash's heavy metal content in a quick and effective manner (Sohal & Vig, 2020). During the process of converting fly ash into vermicompost, earthworms are responsible for consuming the heavy metals that are present in FA. The addition of FA and organic material, in turn, increases the operation of microbes, and the addition of FA contents enhances the plant nutrients for agricultural purposes. As a direct consequence of this, FA that has been subjected to biological modification is now capable of being utilized in agricultural settings (Usmani & Kumar, 2017).

During the process of making leather, the leather industry uses a large amount of fresh water and different chemicals. They also throw away solid waste like hides, buffing dust, and wastewater sludge. Also, leather industry waste has a lot of dangerous compounds in it, like cadmium, chromium, lead, nickel, cobalt, aluminum sulfate, and magnesium oxide. When wastewater sludge is thrown away in an unsafe way, it harms the soil and groundwater. This situation calls for the fastest and most environmentally friendly way to solve these kinds of conflicts. Other technologies used in the field to get rid of trash have problems like taking a long time and being expensive. On the other hand, vermitechnology can be used to lower the level of toxicity in waste from the leather industry. Xiang et al. (2015) have done a good job of keeping track of most of the earthworm researchers working today. They did this by looking at the overall citation data (SCI category—Science Citation Index). According to the data collected, developed countries did a lot more research on earthworms from 2000 to 2015. This showed the importance of stabilizing leather industry organic solid wastes with earthworms and the role earthworms play in getting nutrients out of these wastes and using them.

Furthermore, improving vermicomposting procedures would be of tremendous help in both the transition to a circular economy and the acquisition of high-quality natural fertilizers for sustainable farming operations (Awasthi et al., 2022). Vermicomposting is already being investigated as a viable technology that can provide high-quality fertilizers for the cultivation of vegetable crops like lettuce. This is because new technologies are brought about as a result of scientific research and creative endeavors. It is anticipated that urban farming will continue to develop in the coming years with the assistance of new technologies (Schröder et al., 2021).

The importance of organic matter and the notion of sustainable agriculture is gradually being brought to the attention of farmers. Maintaining the health of the organic matter in the soil is particularly essential for ensuring continued productivity over the long term. After going through the composting process, biowaste, particularly the organic component of municipal solid waste, can be utilized to improve the soil. The recycling of biowaste from a variety of sources can contribute to a reduction in the amount of pollution in the environment. When organic worm waste is added to the soil, it naturally improves the health of the soil, makes plants grow better, provides plants with a nutrient-rich and strong environment, makes plants less likely to be damaged by insects, and makes plants better able to handle harsh weather. Organic worm waste is a valuable addition to all soils because it improves the health of the soil and makes plants more resistant to damage.

6.2 Revenue Generation Opportunities

Vermicomposting is a growing global movement and booming business, even though farmers have been using it for centuries. Vermicomposting of municipal solid waste, including sewage sludge on a commercial scale, is a recent development to divert it from landfills and is now a global movement. In 1970, Holland began experimenting with the management of municipal/industrial organic wastes. This was followed by England and Canada, and it was later adopted in the United States, Italy, the Philippines, Thailand, China, Korea, Japan, Brazil, France, Australia, Israel, and

Fig. 3 Ways to generate revenue from vermicomposting

Russia. Vermicomposting can be a source of revenue for individuals or businesses that produce and sell vermicompost as a soil amendment (Acabal, 2022). Vermicompost is in high demand due to its nutrient-rich nature and the beneficial microorganisms it contains, which can improve soil health and increase crop yields.

There are several ways in which revenue can be generated through vermicomposting (Fig. 3). These include:

1. Selling vermicompost directly to consumers: Vermicompost can be sold to home gardeners, farmers, and landscapers as a soil amendment. This can be done through online marketplaces, at farmers' markets, or through retail outlets.
2. Partnering with local waste management companies: Vermicomposting facilities can partner with local waste management companies to process and compost organic materials. This can generate revenue from the sale of finished vermicompost and from the processing fees paid by the waste management company.
3. Selling vermicompost to commercial growers: Commercial growers of vegetables, flowers, and other crops can use vermicompost as a soil amendment to improve soil health and increase crop yields. Vermicomposting facilities can sell vermicompost to these growers on a wholesale basis.

7 Regional Perspective: Vermicomposting in Africa

Recently, vermitechnology has gained popularity in multiple continents, including Africa. Ghana has implemented vermicomposting as a sustainable waste management technique (Sharma & Garg, 2023) and as a source for organic waste for urban and peri-urban farmers (Mainoo et al., 2008). An initiative worth mentioning is the application of vermicomposting as a technique for handling organic waste in urban settings, like Accra (Acquah et al., 2021). As an alternative to traditional waste disposal methods, organizations such as Waste Enterprisers advocate vermicomposting. Furthermore, in several Kenyan initiatives, the viability of vermicomposting for agricultural and waste management purposes has been examined. In Nairobi, vermicomposting initiatives employing household and market organic detritus have been implemented. The objective of these initiatives is to decrease the amount of waste sent to landfills while concurrently producing high-quality organic fertilizer. Vermitechnology has been implemented by farmers and agricultural organizations in Africa as well, with the intention of enhancing soil fertility and crop yields (Dada & Balogun, 2023). In regions afflicted with degradation and nutrient depletion, the efficacy of vermicompost in replenishing soil nutrients and promoting soil health has been demonstrated. In Zimbabwe, vermitechnology has been implemented by a number of community-based initiatives to combat food insecurity and waste management. Local farmers and their families are assisted by these programs through the recycling of organic refuse into nutrient-rich vermicompost. Vermicomposting has been adopted as a method by small-scale producers in Tanzania to increase soil fertility and harvest yields. The central initiatives of these programs have been the education and instruction of farmers regarding the production methods of vermicompost, which is promoted as an organic fertilizer (Shovon et al., 2024). These case studies exemplify the diverse approaches that vermitechnology has employed to address agricultural, social, and environmental challenges in Africa. Their efforts illustrate the potential of earthworm-based systems to facilitate efficient resource management and promote sustainable development in the region.

8 Industrial and Commercial Aspects of Vermitechnology

The prospects of the vermiculture sector have been the subject of extensive discussion and reporting. It is currently unknown how much of the market potential is based on a true comprehension of business realities and how much is hype. Indeed, many backyard-style enterprises have been drawn to the market due to the prospect of rapid profits. It is time to begin evaluating the commercial realities. The recycled waste product markets that may be sold to the horticultural sector and for large-scale applications have received a lot of interest; however, the commercial potential for worms has received far less attention. In certain circumstances, the need for vermicast is what drives the compost worm markets. It should be noted that vermitechnology competes with other waste management technologies like conventional chemical conversion and composting systems. Vermiculture-based contract waste management is one sector that might grow in the future (Tripathi et al., 2005).

The average worm farmers operate on a modest scale and with limited resources. It would be challenging for them to create significant worm markets on their own. An active industry association is necessary to connect with waste management groups, government environmental agencies, and organizations that represent the interests of the pig, poultry, dairy, and other industries. Before the full potential of compost worms can be realized, vermiculture technology and processes need to be improved, as well as the data used to estimate prospective supply and demand. Although employing worms as a source of high-protein feed for cattle has been discussed, there are not many examples of this practice being commercially successful (Tripathi et al., 2005).

The market size for vermicast and combinations including vermicast determines the patronage for compost worms from this source. The ability of worms to be exported has been discussed. There have been a lot of claims made in this regard, including by a few unsuccessful buyback programs, but there does not seem to be any concrete proof of successful export projects employing vermicast or worms. The greatest promise appears to be in selling breeding stock and knowledge to nations looking to launch a budding vermiculture business. Worms as a source of high-quality protein may also be exported.

Worm composting systems are available for purchase by homeowners, and they have the potential to be employed in industrial waste or municipal conversion systems. Governments have made a commitment to lowering the quantity of trash dumped in landfills. Thus, there are grounds for an increase in demand for compost worms from environmental, economic, and regulatory perspectives. Such systems demand large upfront financial commitments. Their capacity to make a profit is partially based on the fees that a waste producer will pay for this type of waste management, as well as the price that can be acquired for the vermicast and related goods that are produced as a result of the process (Tripathi et al., 2005).

Vermitech Australia's and Vermiculture Resources International's (VRI) facilities created for Redland City Council are examples of what this technology can be used for. Another example of how nonindustrial waste is converted using vermiculture is the windrow-based facility at the Grace McKellar Institute in Geelong. One company has created medium-sized mechanical equipment that can utilize vermiculture. It should be noted that vermitechnology competes with other waste management technologies like conventional composting and chemical conversion systems. Vermiculture-based contract waste management is one sector that might grow in the future. The prospect for exporting worms has been discussed. There have been a lot of claims made in this regard, including by a few unsuccessful buyback programs, but there does not seem to be any concrete proof of successful export projects employing worms. The greatest promise appears to be in exporting breeding stock and knowledge to nations looking to launch a budding vermiculture business. Worms may also be exported as a source of high-quality protein to poor nations (Tripathi et al., 2005).

9 Strengths of Vermitechnology

There are several strengths of vermitechnology as depicted in Fig. 4.

1. Versatility: Vermitechnology can process a wide range of organic materials, including food scraps, yard waste, paper products, etc. This makes it a versatile waste management option for households, businesses, and industries.

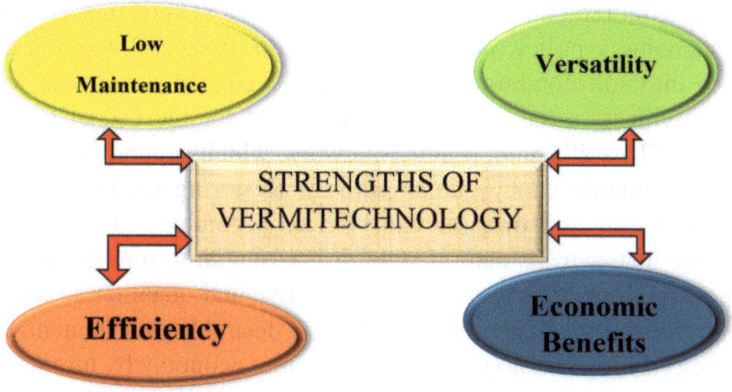

Fig. 4 Strengths of vermitechnology

2. Efficiency: Worms can break down organic materials much faster than traditional composting methods, and the process produces high-quality compost that is rich in nutrients and beneficial microorganisms.
3. Low maintenance: Vermicomposting systems are relatively simple to set up and maintain and do not require large amounts of space or specialized equipment.
4. Sustainability: Vermicomposting is a viable waste management option that diverts organic materials from landfills and reduces greenhouse gas emissions. It is also an effective way to improve soil health and support sustainable agriculture and horticulture practices.
5. Economic benefits: The strength of the vermicomposting program is that recycling is a business opportunity (Moledor et al., 2016). It is economically valuable as its demand has increased rapidly as an alternative to chemical fertilizers. Hence commercialization of vermicomposting is on the verge of tackling the demand and supply gap. Vermicomposting can help to reduce waste management costs and generate revenue through the sale of vermicompost as a soil amendment. In addition to improving soil health, it can boost crop yields and benefit communities economically.

10 Conclusion

Earthworms are used in the biotechnological process involving vermicomposting, which uses them as natural bioreactors to break down organic materials and preserve soil fertility. Worms were used to recycle organic waste and promote plant development. The significance of vermicompost is increased by the fact that it also has additional benefits; extra worms may be utilized as protein-rich animal feed and in medicinal applications. Vermitechnology is applicable across different sectors, including agriculture, waste management, and environmental remediation.

This technology has long been recognized for its ability to recycle and manage organic wastes, produce high-quality fertilizer, and provide protein for food and animal feed. The bioremediation of organic wastes is a method that, from the standpoint of sustainability policy, prevents the creation of several waste categories by turning them into soil conditioners, foods, and feeds. More so than for agriculture, human civilization has a greater requirement for the use of bioremediated organic wastes. The soil's natural nutrient cycles, which support life on our planet, are increasingly threatened by human activity. Restoring the balance of nutrient cycles and combating soil diseases can both be accomplished by returning soil-derived organic matter to its natural state. It is essential to monitor the process of vermicomposting's development at both qualitative and quantitative levels since it contributes significantly to the provision of many desirable environmental and agronomic features. Vermitechnology supports a circular economy by turning waste into valuable resources (waste to wealth), creating revenue opportunities, and fostering green entrepreneurship. Moving forward, investments in research, awareness, and education will be critical to advancing vermitechnology and unlocking its full potential for environmental, economic, and social benefits. Embracing this solution not only contributes to a healthier planet but also paves the way for a more resilient and sustainable future.

References

A'ali, R., Jafarpour, M., Kazemi, E., & Pessarakli, M. (2017). Effects of raw materials on vermicompost qualities. *Journal of Plant Nutrition, 40*(11), 1635–1643.

Acabal, C. D. (2022). Profitability and sustainability of vermi composting business toward social entrepreneurship in Negros oriental, Philippines. *Sustainable Development, 10*(1), 17–26.

Acquah, M. N., Essandoh, H. M. K., Oduro-Kwarteng, S., Appiah-Effah, E., & Owusu, P. A. (2021). Degradation and accumulation rates of fresh human excreta during vermicomposting by *Eisenia fetida* and *Eudrilus eugeniae. Journal of Environmental Management, 293,* Article 112817.

Adams, C. E. (2018). Urban soils. In *Urban wildlife management* (pp. 117–134). CRC Press.

Alban, D. H., & Berry, E. C. (1994). Effects of earthworm invasion on morphology, carbon, and nitrogen of a forest soil. *Applied Soil Ecology, 1*(3), 243–249.

Alshehrei, F., & Ameen, F. (2021). Vermicomposting: A management tool to mitigate solid waste. *Saudi Journal of Biological Sciences, 28*(6), 3284–3293.

Ansari, A., & Hanief, A. (2015). Microbial degradation of organic waste through vermicomposting. *International Journal of Sustainable Agricultural Research, 2*(2), 45–54.

Ansari, A., & Ismail, S. (2012). Role of earthworms in vermitechnology. *Journal of Agricultural Technology, 8*(2), 403–415.

Aransiola, S. A., Ijah, U. J. J., Abioye, O. P., Victor-Ekwebelem, M. O. (2021). ANAMMOX in wastewater treatment. In N. R. Maddela, L. C. García Cruzatty, & S. Chakraborty (Eds.), *Advances in the domain of environmental biotechnology. Environmental and microbial biotechnology*. Springer. https://doi.org/10.1007/978-981-15-8999-7_15

Aransiola, S. A., Ijah, U. J. J., Abioye, O. P., & Bala, J. D. (2022). Vermicompost-assisted phytoremediation of toxic trace element-contaminated soil in Madaka, Nigeria using *Melissa officinalis* L and *Sida acuta. International Journal of Environmental Science and Technology*. https://doi.org/10.1007/s13762-022-04105-y

Aransiola, S. A., Ijah, U. J. J., Abioye, O. P., Bala, J. D., Rivadeneira-Mendoza, R. F., Luque, R., Rodríguez-Díaz, J. M., & Maddela, N. R. (2024). Micro and vermicompost assisted remediation of heavy metal contaminated soils using phytoextractors. *Case Studies in Chemical and Environmental Engineering., 9*(2024), Article 100755. https://doi.org/10.1016/j.cscee.2024.100755

Awasthi, M. K., Singh, E., Binod, P., Sindhu, R., Sarsaiya, S., Kumar, A., Chen, H., Duan, Y., Pandey, A., & Kumar, S. (2022). Biotechnological strategies for bio-transforming biosolid into resources toward circular bio-economy: A review. *Renewable and Sustainable Energy Reviews, 156*, Article 111987.

Babaniyi, B. R., Ogundele, O. D., Bisi-omotosho, A., Babaniyi, E. E., & Aransiola, S. A. (2023). Remediation approaches in environmental sustainability. In N. R. Maddela, L. K. W. Eller, & R. Prasad (Eds.), *Microbiology for cleaner production and environmental sustainability*. CRC Press. ISBN: 9781032496061. https://www.routledge.com/Microbiology-for-Cleaner-Production-and-Environmental-Sustainability/Maddela-Eller-Prasad/p/book/9781032496061

Bharti, N., Barnawal, D., Shukla, S., Tewari, S. K., Katiyar, R., & Kalra, A. (2016). Integrated application of *Exiguobacterium oxidotolerans, Glomus fasciculatum*, and vermicompost improves growth, yield and quality of *Mentha arvensis* in salt-stressed soils. *Industrial Crops Products, 83*, 717–728.

Bhat, S. A., Singh, J., & Vig, A. P. (2018). Earthworms as organic waste managers and biofertilizer producers. *Waste Biomass Valorization, 9*(7), 1073–1086.

Bhunia, S., Bhowmik, A., Mallick, R., & Mukherjee, J. (2021). Agronomic efficiency of animal-derived organic fertilizers and their effects on biology and fertility of soil: A review. *Agronomy, 11*(5), 823.

Blouin, M., Hodson, M. E., Delgado, E. A., Baker, G., Brussaard, L., Butt, K. R., Dai, J., Dendooven, L., Pérès, G., & Tondoh, J. (2013). A review of earthworm impact on soil function and ecosystem services. *European Journal of Soil Science, 64*(2), 161–182.

Butler, T., Sikora, L., Steinhilber, P., & Douglass, L. (2001). Compost age and sample storage effects on maturity indicators of biosolids compost. *Journal of Environmental Quality, 30*(6), 2141–2148.

Dabke, S. V. (2013). Vermi-remediation of heavy metal-contaminated soil. *Journal of Health and Pollution, 3*(4), 4–10.

Dada, E. O., & Balogun, Y. O. (2023). Vermitechnology: An underutilised agro-tool in Africa. In *Vermicomposting for sustainable food systems in Africa* (pp. 127–143). Springer Nature Singapore.

Dam, A. V. (2021). *Turning mixed waste into energy: Separate organics from solid and fibrous waste*. Royaldutchkusters.com. https://www.royaldutchkusters.com/blog/turning-mixed-waste-into-energy-separate-organics-from-solid-and-fibrous-waste. Accessed on January 21, 2025.

Datar, M., Rao, M., & Reddy, S. (1997). Vermicomposting- a technological option for solid waste management. *Journal of Solid Waste Technology and Management, 24*(2), 89–93.

Decaëns, T. (1999). Effect of exclusion of the anecic earthworm Martiodrilus carimaguensis Jimenez and Moreno on soil properties and plant growth in grasslands of the eastern plains of Colombia. *Pedobiologia, 43*(6), 835–841.

Destoumieux-Garzón, D., Mavingui, P., Boetsch, G., Boissier, J., Darriet, F., Duboz, P., Fritsch, C., Giraudoux, P., Le Roux, F., & Morand, S. (2018). The one health concept: 10 years old and a long road ahead. *Frontiers in Veterinary Science, 14.*

Dey Chowdhury, S., Suhaib, K. H., Bhunia, P., & Surampalli, R. Y. (2023). A critical review on the vermicomposting of organic wastes as a strategy in circular bioeconomy: mechanism, performance, and future perspectives. *Environmental Technology*, 1–38.

Edwards, C. A., & Arancon, N. Q. (2022). The use of earthworms in organic waste management and vermiculture. In *Biology and ecology of earthworms* (pp. 467–527). Springer.

Häder, D.-P., Banaszak, A. T., Villafañe, V. E., Narvarte, M. A., González, R. A., & Helbling, E. W. (2020). Anthropogenic pollution of aquatic ecosystems: Emerging problems with global implications. *Science of the Total Environment, 713*, Article 136586.

Haynes, R. (2009). Reclamation and revegetation of fly ash disposal sites–Challenges and research needs. *Journal of Environmental Management, 90*(1), 43–53.

Homa, J., Klimek, M., Kruk, J., Cocquerelle, C., Vandenbulcke, F., & Plytycz, B. (2010). Metal-specific effects on metallothionein gene induction and riboflavin content in coelomocytes of Allolobophora chlorotica. *Ecotoxicology and Environmental Safety, 73*(8), 1937–1943.
Huang, K., Xia, H., Li, F., & Bhat, S. A. (2020). Recycling of municipal sludge by vermicomposting. In *Earthworm assisted remediation of effluents and wastes* (pp. 55–67).
Kale, R. (1998). Earthworms: Nature's gift for utilization of organic wastes.
Karmakar, S., Adhikary, M., Gangopadhyay, A., & Brahmachari, K. (2015). Impact of vermicomposting in agricultural waste management vis-à-vis soil health care. *Journal of Environmental Science and Natural Resources, 8*(1), 99–104.
Khan, I., Kurovsky, A., & Babenko, A. (2024). The role of microbial enzymes in vermicomposting of organic wastes. *Science and Innovation, 3*(D4), 74–85.
Kizilkaya, R., Hepsen Turkay, F. S., Turkmen, C., & Durmus, M. (2012). Vermicompost effects on wheat yield and nutrient contents in soil and plant. *Archives of Agronomy Soil Science, 58*(sup1), S175–S179.
Krishnamoorthy, R., & Vajranabhaiah, S. (1986). Biological activity of earthworm casts: An assessment of plant growth promotor levels in the casts. *Proceedings: Animal Sciences, 95*(3), 341–351.
Kumar, K., & Henock, T. (2015). Conversion of solid waste into bio fertilizer by vermicomposting a case study of Padmanadapuram. *International Journal of Innovative Research in Science, Engineering Technology, 4*, 3801–3808.
Lalitha, R., Fathima, K., & Ismail, S. (2000). Impact of biopesticides and microbial fertilizers on productivity and growth of Abelmoschus esculentus. *Vasundhara the Earth, 1*(2), 4–9.
Langdon, C. J., Piearce, T. G., Meharg, A. A., & Semple, K. T. (2001). Survival and behaviour of the earthworms Lumbricus rubellus and Dendrodrilus rubidus from arsenate-contaminated and non-contaminated sites. *Soil Biology and Biochemistry, 33*(9), 1239–1244.
Law, A., Solano, O., Brown, C. J., Hunter, S. S., Fagnan, M., Top, E. M., & Stalder, T. (2021). Biosolids as a source of antibiotic resistance plasmids for commensal and pathogenic bacteria. *Frontiers in Microbiology, 12*, Article 606409.
Lee, B. H., & Khor, S. M. (2022). Biodegradation versus composting. *Handbook of biodegradable materials* (pp. 1–34). Springer International Publishing.
Lee, K., & Foster, R. (1991). Soil fauna and soil structure. *Soil Research, 29*(6), 745–775.
Lim, S. L., Lee, L. H., & Wu, T. Y. (2016). Sustainability of using composting and vermicomposting technologies for organic solid waste biotransformation: Recent overview, greenhouse gases emissions and economic analysis. *Journal of Cleaner Production, 111*, 262–278.
Lionetto, M. G., Calisi, A., & Schettino, T. (2012). Earthworm biomarkers as tools for soil pollution assessment. *Soil Health and Land Use Management, 16*, 305–331.
Mainoo, N. K., Barrington, S., & Whalen, J. K. (2008). Vermicompost as a fertilizer for urban and peri-urban farms: Perceptions of farmers in Accra, Ghana. *Ghana Journal of Agricultural Science, 41*(2).
Maji, D., Misra, P., Singh, S., & Kalra, A. (2017). Humic acid rich vermicompost promotes plant growth by improving microbial community structure of soil as well as root nodulation and mycorrhizal colonization in the roots of *Pisum sativum*. *Applied Soil Ecology, 110*, 97–108.
Malińska, K., Golańska, M., Caceres, R., Rorat, A., Weisser, P., & Ślęzak, E. (2017). Biochar amendment for integrated composting and vermicomposting of sewage sludge–the effect of biochar on the activity of Eisenia fetida and the obtained vermicompost. *Bioresource Technology, 225*, 206–214
Masullo, A. (2017). Organic wastes management in a circular economy approach: Rebuilding the link between urban and rural areas. *Ecological Engineering, 101*, 84–90.
Morgan, J., & Morgan, A. (1988). Calcium-lead interactions involving earthworms. Part 2: The effect of accumulated lead on endogenous calcium in Lumbricus rubellus. *Environmental Pollution, 55*(1), 41–54.

Moledor, S., Chalak, A., Fabian, M., & Talhouk, S. N. (2016). Socioeconomic dynamics of vermicomposting systems in Lebanon. *Journal of Agriculture, Food Systems, and Community Development, 6*(4), 145–168.

Nisa, R. U., Maqbool, S., & Nisa, A. U. (2022). Bioremediation on the crossroads of technology for environmental clean-up: an overview. In *Microbial and biotechnological interventions in bioremediation and phytoremediation* (pp. 3–25).

Ojuolape, O. T., Aransiola, S. A., & Obideyi, O. A. (2015). Vermitechnology: A solution to environmental problems; a review. *Journal of Global Ecology and Environment, 3*(3), 127–135. http://www.ikpress.org/abstract/4516

Oluwarinde, B. O., Ajose, D. J., Abolarinwa, T. O., Montso, P. K., Njom, H. A., & Ateba, C. N. (2024). Molecular characterization and safety properties of multi drug-resistant *Escherichia coli* O157: H7 bacteriophages. *BMC Microbiology, 24*(1), 528.

Panday, D., Bhusal, N., Das, S., & Ghalehgolabbehbahani, A. (2024). Rooted in nature: The rise, challenges, and potential of organic farming and fertilizers in agroecosystems. *Sustainability, 16*(4), 1530.

Pathma, J., & Sakthivel, N. (2012). Microbial diversity of vermicompost bacteria that exhibit useful agricultural traits and waste management potential. *Springerplus, 1*(1), 1–19.

Peyvast, G., Olfati, J.-A., Madeni, S., Forghani, A., & Samizadeh, H. (2008). Vermicompost as a soil supplement to improve growth and yield of parsley. *International Journal of Vegetable Science, 14*(1), 82–92.

Pierce, R. (2024). Vermicompost worm deaths: Common causes & what to do. Accessed on January 25, 2025.

Rajiv, K. S., Sunita, A., Krunal, C., Vinod, C., & Brijal Kiranbhai, S. (2010). Vermiculture technology: reviving the dreams of Sir Charles Darwin for scientific use of earthworms in sustainable development programs. *Technology and Investment*.

Rakkini, V. M., Vincent, S., Kumar, A. S., & Baskar, K. (2017). An overview: Organic waste management by earthworm. *Journal of Civil Engineering and Environmental Sciences, 3*(1), 013–017.

Rodriguez-Campos, J., Dendooven, L., Alvarez-Bernal, D., & Contreras-Ramos, S. M. (2014). Potential of earthworms to accelerate removal of organic contaminants from soil: A review. *Applied Soil Ecology, 79*, 10–25.

Satapathy, S., Nayak, Y., & Satapathy, K. B. (2020). Fly ash amendment for sustainable agriculture through vermicomposting. *PalArch's Journal of Archaeology of Egypt/egyptology, 17*(9), 2274–2287.

Schröder, C., Häfner, F., Larsen, O. C., & Krause, A. (2021). Urban organic waste for urban farming: Growing lettuce using vermicompost and thermophilic compost. *Agronomy, 11*(6), 1175.

Senesi, N., Plaza, C., Brunetti, G., & Polo, A. (2007). A comparative survey of recent results on humic-like fractions in organic amendments and effects on native soil humic substances. *Soil Biology and Biochemistry, 39*(6), 1244–1262.

Sequeira, V., & Chandrashekar, J. (2015). Vermicomposting of biodegradable municipal solid waste using indigenous Eudrilus sp. earthworms. *International Journal of Current Microbiology and Applied Sciences, 4*(4), 356–365.

Sharma, S. (2003). Municipal solid waste management through vermicomposting employing exotic and local species of earthworms. *Bioresource Technology, 90*(2), 169–173.

Sharma, K., & Garg, V. K. (2023). Vermicomposting technology for organic waste management. In *Current developments in biotechnology and bioengineering* (pp. 29–56). Elsevier.

Shi, Z., Liu, J., Tang, Z., Zhao, Y., & Wang, C. (2020). Vermiremediation of organically contaminated soils: Concepts, current status, and future perspectives. *Applied Soil Ecology, 147*, Article 103377.

Shovon, S. M., Akash, F. A., Rahman, W., Rahman, M. A., Chakraborty, P., Hossain, H. Z., & Monir, M. U. (2024). Strategies of managing solid waste and energy recovery for a developing country–A review. *Heliyon, 10*(2), Article e24736.

Siddiqui, M. A., Neeraj, A., & Hiranmai, R. (2022). Vermitechnology: An eco-friendly approach for organic solid waste management and soil fertility improvement—A review. In *Strategies and tools for pollutant mitigation* (pp. 91–112).

Singh, P., & Kothyari, H. S. (2019). Vermi-technology: (vermiculture, methods, beneficial effects). *International Journal of Research in Engineering, Science and Management, 2*(11).

Singh, R., Samal, K., Dash, R. R., & Bhunia, P. (2019). Vermifiltration as a sustainable natural treatment technology for the treatment and reuse of wastewater: A review. *Journal of Environmental Management, 247*, 140–151.

Sinha, R. K., Bharambe, G., & Ryan, D. (2008). Converting wasteland into wonderland by earthworms—A low-cost nature's technology for soil remediation: A case study of vermiremediation of PAHs contaminated soil. *The Environmentalist, 28*, 466–475.

Sohal, B., & Vig, A. P. (2020). Earthworm-assisted amelioration of thermal ash. In *Earthworm assisted remediation of effluents and wastes* (pp. 281–295). Springer.

Temilade, O., Abiodun, A., & Adeola, O. (2015). Vermitechnology: A solution to environmental problems; a review. *Journal of Global Ecology and Environment*, 127–135.

Tohumcu, F., Aydin, A., & Simsek, U. (2023). The effects of organic wastes applied to alkaline soils on some physical and chemical properties of the soil. *Eurasian Soil Science, 56*(3), 387–403.

Tripathi, Y., Hazarika, P., Kaushik, P., & Arvind, K. (2005). Verms and vermitechnology (pp. 10–21). A.P.H. Publishing Corporation.

Usmani Z, Kumar V (2017) The implications of fly ash remediation through vermicomposting: A review. *Nature Environment & Pollution Technology, 16*(2).

Wang, G., Wang, L., Ma, F., Yang, D., & You, Y. (2021). Earthworm and arbuscular mycorrhiza interactions: Strategies to motivate antioxidant responses and improve soil functionality. *Environmental Pollution, 272*, Article 115980.

Wong, W. S., Zhong, H. T., Cross, A. T., & Yong, J. W. H. (2020). Plant biostimulants in vermicomposts: Characteristics and plausible mechanisms. In *The chemical biology of plant biostimulants* (pp. 155–180).

Wu, Y., Ding, Q., Zhu, Q., Zeng, J., Ji, R., Dumont, M. G., & Lin, X. (2018). Contributions of ryegrass, lignin and rhamnolipid to polycyclic aromatic hydrocarbon dissipation in an arable soil. *Soil Biology and Biochemistry, 118*, 27–34.

Xiang, H., Zhang, J., & Zhu, Q. (2015). Worldwide earthworm research: A scientometric analysis, 2000–2015. *Scientometrics, 105*(2), 1195–1207.

Xu, L., Yan, D., Ren, X., Wei, Y., Zhou, J., Zhao, H., & Liang, M. (2016). Vermicompost improves the physiological and biochemical responses of blessed thistle (*Silybum marianum* Gaertn.) and peppermint (*Mentha haplocalyx* Briq) to salinity stress. *Industrial CropsProducts, 94*, 574–585.

Yuvaraj, A., Karmegam, N., & Thangaraj, R. (2018). Vermistabilization of paper mill sludge by an epigeic earthworm Perionyx excavatus: Mitigation strategies for sustainable environmental management. *Ecological Engineering, 120*, 187–197.

Zamani, J., Afyuni, M., Sepehrnia, N., & Schulin, R. (2016). Opposite effects of two organic wastes on the physical quality of an agricultural soil. *Archives of Agronomy and Soil Science, 62*(3), 413–427.

Vermicomposting Process—Optimisation

Anu Bala Chowdhary, Rahil Dutta, Raman Tikoria, Jahangeer Quadar, Surbhi Sharma, Jaswinder Singh, and Adarsh Pal Vig

Abstract In the modern era, solid waste management has become a severe ecological challenge. In order to manage the different types of wastes, there are so many traditional treatment techniques that are used including incineration, landfill, anaerobic digestion and composting. In comparison to these procedures, vermicomposting is an effective waste stabilisation method by the combined activity of earthworms and microbes. It is an innovative ecotechnology that converts waste into organically enriched products known as vermicompost. Vermicompost is a humus-like, finely granulated and stabilised material that can be used as a soil conditioner to reintegrate the organic matter into the agricultural soils. It functions as an organic fertiliser and biological control agent, eradicating many plant ailments caused by soil-borne plant pathogens and pests. Moreover, other soil indicators like nitrogen and C:N ratio can be enhanced by using vermicomposting. The success of vermicomposting process is dependent on numerous parameters such as quality of raw material, pH, temperature, moisture, aeration and so on, as well as the type of earthworm species used. Thus, this chapter focuses mainly on vermicomposting and the role of various factors in the vermicomposting process.

Keywords Abiotic · Biotic · Composting · Earthworms · Vermicomposting

---

A. B. Chowdhary (✉) · R. Dutta · J. Quadar · S. Sharma · A. P. Vig (✉)
Department of Botanical and Environmental Sciences, Guru Nanak Dev University, Amritsar, Punjab, India
e-mail: anuchowdhary240@gmail.com

A. P. Vig
e-mail: dr.adarshpalvig@gmail.com

R. Tikoria
Department of Zoology, Guru Nanak Dev University, Amritsar, Punjab, India

J. Singh
Post Graduate Department of Zoology, Khalsa College, Amritsar, Punjab, India

A. P. Vig
Punjab Pollution Control Board (PPCB), Vatavaran Bhawan, Patiala, Punjab, India

R. Dutta
Department of Environmental Sciences, GDC Reasi, Reasi, India

S. A. Aransiola and N. R. Maddela (eds.), *Vermitechnology: Economic, Environmental and Agricultural Sustainability*, https://doi.org/10.1007/978-3-031-96212-7_4

1 Introduction

Numerous human activities, such as increasing urbanisation, industrialisation and economic expansion, are causing massive amounts of solid waste to be produced across the world. The disposal of this solid waste has now turned into an environmental and technical issue for everyone (Yadav & Garg, 2011). To maintain the environment clean and healthy, sustainable solid waste management procedures are required (Singh et al., 2011). According to several research, 23 developing nations produce an average of 0.77 kg/person/day of solid waste (Troschinetz & Mihelcic, 2009). It is now expected that the world's solid waste output will rise to 3 billion tonnes by 2025 (Charles et al., 2009). Organic waste accounts for 44 to 46% (by mass) of total municipal solid waste production globally (Leh-Togi Zobeashia et al., 2018; Ricci-Jürgensen et al., 2020). On a worldwide scale, the major waste types of organic solid waste are food and green waste, accounting for 44% of total waste. Dry recyclables such as plastic, paper, cardboard, metal and glass account for another 38% of waste (Kaza et al., 2018). Apart from municipal solid waste, other waste streams also exist around the world, such as industrial and agricultural waste. These wastes are produced in far greater amounts than municipal waste. On a worldwide platform, industrial and agricultural waste generation rates are 18 times and four times higher, respectively, than municipal solid waste (Sharma & Jain, 2020). Agricultural waste, particularly crop residue burning, and industrial waste disposal in freshwater bodies are common practices across the world. Agricultural residue burning emits enormous amounts of particulate pollutants and gases, especially during and after harvest, which typically increases pollution at the local and regional levels (Chowdhary et al., 2022b; Wang et al., 2013). Direct discharge of industrial waste, especially sludge, pollutes water and harms aquatic life. Another major organic waste is animal manure, which is mostly utilised as organic fertiliser on agricultural land (Aransiola et al., 2022; Maharjan et al., 2022). Although various solutions for proper solid waste management have been suggested and executed, like source reduction, kerbside recycling, material recovery, waste to energy, landfill disposal, incineration and composting (Chang & Davila, 2008). Some of these treatment and disposal techniques may have major environmental consequences. Several studies have found that waste disposed of in landfills or open dumps contaminates groundwater owing to leachate of organic and inorganic substances found in waste (Cadena et al., 2009; Mor et al., 2006; Tsai et al., 2007). Landfill dumping also contributes to the greenhouse effect. Incineration treatment is also confined due to its poor fattening value and the cost of fuel additives (Lee et al., 2009; Tsai et al., 2007). Sewage sludge, which is directly dumped on agricultural land due to the presence of high content of nitrogen and phosphorus, may, however, be hazardous to plants and soil and may also have an adverse impact on the metabolism of soil microorganisms (Ayuso et al., 1996; Chowdhary et al., 2022b). Under these conditions, vermicomposting may be a feasible and environmentally sound process for converting solid waste into organic-rich manure. Vermicomposting is a waste management process that involves the environmentally benign breakdown of organic fractions of solid waste to a level at which it can be readily stored, handled and used on agricultural land without causing any harm (Aira et al., 2002; Khwairakpam & Bhargava, 2009; Singh et al., 2011). Table 1 shows the vermicomposting capability of earthworms for decomposing various types of wastes amended with acceptable co-substrates.

2 Vermicomposting Process

Vermicomposting has been described as a low-cost, feasible and quick method of using organic wastes and agricultural leftovers. It is a non-thermophilic biodegradation process that uses earthworms and microorganisms to break down organic waste (Suthar, 2009a). Earthworms are said to operate as mechanical blenders, breaking down organic matter and changing its physical, chemical and biological properties while gradually decreasing its C:N ratio. It increases the surface area of organic materials, making them more accessible to microbes and hence more suitable for microbial activities and subsequent breakdown (Aransiola et al., 2022; Domínguez et al., 2017; Yadav & Garg, 2011). The process of vermicomposting of different types of wastes is presented in Fig. 1. Vermicomposting is faster than traditional composting because the material goes through the guts of earthworms, which speeds up the process. Earthworm castings are rich in plant growth regulators, microbiological activity and pest-repellent attributes (Gandhi, 1997). According to various studies, vermicast also contains plant growth-stimulating components such as ethylene, auxin and gibberellin along with enzymes like cellulase, nitrogenase and phosphatase (Aslam & Ahmad, 2020; Aslam et al., 2020). When biodegradable material passes through the intestines of earthworms, nutrients are produced which are also considered to stimulate plant growth while minimising pathogen invasion. Organic fertilisation and soil conditioning characteristics of vermicast are highly recommended. Concerned researchers have also revealed their findings that vermicomposting has plant-beneficial impacts (Aslam & Ahmad, 2020; Aslam et al., 2020; Bellitürk et al., 2020; Wang et al., 2007). Vermicomposting provides a number of advantages over conventional waste management technologies, including the ability to compost both indoors and outdoors, allowing for year-round composting. This procedure enables the production of organic nutrients for crops in shorter time, which are more nutritionally, physiologically and biochemically efficient than other composts (Yadav et al., 2010). It is also known as a low-cost technological procedure for processing or treating organic waste. According to a comparative study between traditional composting and vermicomposting, it has been found that vermicomposting produces enriched compost with high amounts of N, P and K content, as well as reduction in heavy metal content (Cardosa Vigueros and Ramírez Camperos, 2002). Solid waste can be transformed into beneficial compost through the vermicomposting process, offering an effective substitute for chemical fertilisers and also reducing pollution (Monroy et al., 2009).

Table 1 Vermicomposting of different types of wastes with amendments or substrates employing various earthworm species

S. No.	Type of waste	Amendment or substrate	Earthworm species	References
1	Agro-industrial sludge	Cow dung, biogas plant slurry and wheat straw	*Eisenia fetida*	Suthar (2010)
2	Tannery sludge	Cattle dung	*Eisenia fetida*	Vig et al. (2011)
3	Rice husk	Market-refused banana, honeydew or papaya	*Eudrilus eugeniae*	Lim et al. (2012)
4	Garden waste, kitchen waste and cow dung	Used separately	*Eisenia fetida*	Wani and Rao (2013)
5	Vegetable waste	Paddy straw, cow dung and sawdust-based spent mushroom compost	*Lumbricus rubellus*	Abu Bakar et al. (2014)
6	Sugarcane bagasse and sugarcane press mud	Farm manure	*Lumbricus rubellus*	Shah et al. (2015)
7	Sewage sludge	Mown grass, sawdust and organic fraction of municipal wastes	*Eisenia fetida* and *Eisenia andrei*	Sosnecka et al. (2016)
8	Food and vegetable processing waste	Buffalo dung	*Eisenia fetida*	Sharma and Garg (2017)
9	Distillery sludge	Tea leaf residues	*Eisenia fetida*	Mahaly et al. (2018)
10	Bakery industry sludge	Cow dung	*Eisenia fetida*	Yadav and Garg (2019)
11	Onion waste	Cow dung	*Eisenia fetida*	Pellejero et al. (2020)
12	Fly ash	Cattle dung	*Eisenia fetida*	Sohal et al. (2021b)
13	Biomedical waste ash	Cattle dung	*Eisenia fetida*	Sohal et al. (2021a)
14	Allopathic pharmaceutical industry sludge	Cattle dung	*Eisenia fetida*	Singh et al. (2022)
15	Coconut husk	Cattle dung	*Eisenia fetida*	Quadar et al. (2022)
16	Milk processing industry sludge	Cattle dung, paddy straw biochar	*Eisenia fetida*	Dutta et al. (2023)

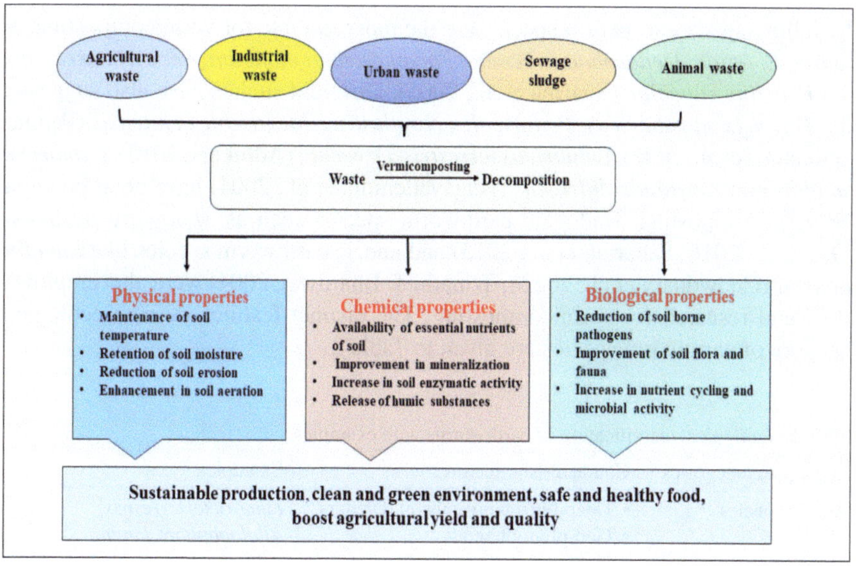

Fig. 1 Vermicomposting of different types of wastes

3 Earthworm Species Used for Vermicomposting

Earthworms have been in the Earth's biosphere since the pre-Cambrian era, some 600 million years ago. These are invertebrates belonging to the phylum Annelida, order Oligochaeta and class Chaetopoda (Piearce et al., 1990). They are hermaphrodites in nature, and the existence of clitellum shows that they are sexually mature. Cross-fertilisation and copulation are the most prevalent reproductive procedures. They lay their eggs in a cocoon, which is a small round-shaped capsule that takes 3–6 weeks to incubate (Aira et al., 2002; Brown et al., 2000; Domfnguez, 2004). Bouché (1977) classified earthworms into three groups based on their feeding behaviour and ecological nature: epigeic, anecic and endogeic. There are currently around 3000 species of earthworms in the environment; however, not all of them are suitable for vermicomposting (Wu et al., 2014). Earthworms utilised in the vermicomposting process are classified as 'epigeic' species, which means 'surface dwellers' (Lee, 1985; Wu et al., 2014). The species from epigeic ecological groups, such as *Eisenia fetida* and *Eisenia andrei*, are the most commonly employed for vermicomposting (Domínguez & Ray, 2018). Epigeic species feed on the surface and have a high bioconversion rate. They can withstand a variety of poor environmental conditions, such as decreased food supply, predatory pressure and so on. When compared with the anecic and endogeic ecological categories of earthworms, epigeic species have a higher reproduction and feeding rate, allowing earthworm species to expand their population for vermiculture while also producing vermicompost in a shorter time interval (Domínguez & Ray, 2018; Julka, 2008).According to one study, the following six earthworm species are the most suitable for vermicomposting: *E. andrei*, *E. fetida*, *Dendrobaena veneta*, *Polypheretima elongata*, *Perionyx excavatus* and *Eudrilus eugeniae* (Sharma et al., 2005). Different studies have also suggested that *Perionyx sansibaricus*, *Pontoscolex corethrurus*, *Megascolex chilensis* (Padmavathiamma et al., 2008), *Lumbricus terrestris*, *D. veneta* (Adhikary, 2012), *Lumbricus rubellus* and *Amynthas diffrigens* (Nagavallemma et al., 2004) have great potential for vermicomposting. Endogeic earthworm species such as *Metaphire posthuma* (Das et al., 2016; Sahariah et al., 2015) and anecic earthworm species like *Lampito mauritii* (Chowdhary et al., 2022a; Tripathi & Bhardwaj, 2004) were also employed in several research for vermicomposting. The distinct features of each ecological category of earthworm species are given in Table 2.

Table 2 Ecological classification of earthworms with examples

Ecological categories	Characteristic features	Examples
Anecic species	• Deep-burrowing subsoil dwellers • Geo-phytophagous • Dig permanent vertical burrows • Dark brown in appearance • Slow-moving animals with a low reproduction rate • Enhance the pedological processes in soil • Very active agents in decomposing organic matter and cycling nutrients	*Lumbricus terrestris*, *Allolobophora longa*, *Allolobophora chlorotica*, etc.
Endogeic species	• Topsoil dwellers • Geophagous • Make non-permanent horizontal burrows • Small, whitish, slow-moving animals of different sizes • Promote aeration, water drainage and sunlight accessibility to crop roots	*Aporrectodea caliginosa*, *Aporrectodea rosea*, *Aporrectodea icterica*, etc.
Epigeic species	• Surface dwellers • Phytophagous • Do not create permanent burrows • Fast-moving, small, reddish animals • Eat organic matter and are significant in composting	*Eisenia fetida*, *Lumbricus castaneus*, *Lumbricus rubellus*, etc.

Source Singh et al. (2020), Yatoo et al. (2022)

4 Factors Affecting Vermicomposting Process

A variety of biotic and abiotic factors impact the efficiency of the vermicomposting process. Figure 2 summarises the various factors that influence the process of vermicomposting.

4.1 Abiotic Factors

Abiotic factors such as moisture content, pH, temperature, aeration, feed quality, light and C:N ratio are among the most crucial ones that influence the vermicomposting process (Table 3).

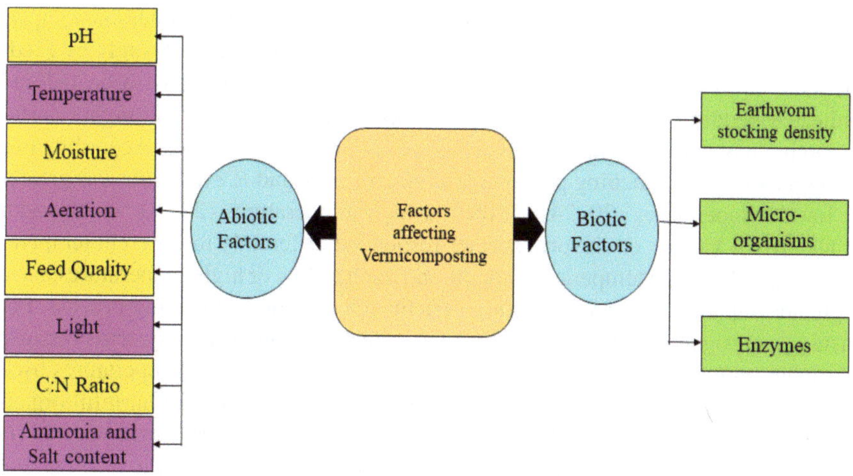

Fig. 2 Various factors influencing vermicomposting process

Table 3 Optimum abiotic factors for earthworms

Abiotic factor	Optimum value	References
Temperature	25–37 °C	Kauser and Khwairakpam (2022)
pH	6.5 and 7.5	Abdulla-Al-Mamun et al. (2023)
C:N ratio	Less than 20	Xie et al. (2022)
Moisture content	50 and 80%	Barik et al. (2010)
Light	Photophobic (do not like light)	Raza et al. (2022)
Ammonia and salt content	Less than 0.5%	Eijsackers (2011)

4.1.1 Temperature

The ideal temperature range for earthworms is 25–37 °C, which is favourable for their activity, growth, metabolism, respiration, reproduction and cocoon formation. It is also favourable for the microorganisms that are linked with earthworms. Researchers have identified various tolerable temperature ranges between 0 and 40 °C for earthworms (Kauser & Khwairakpam, 2022). Chemical and microbiological activity in the substrate increases at higher temperatures (over 30 °C), which lowers the oxygen content and negatively affects earthworms (Bo et al., 2021). Temperature has a big impact on worm behaviour. The temperature must be kept above 10 °C in winter and below 35 °C in summer for the system to stay operational (Stoknes et al., 2016). Various earthworm species responded to temperature in different ways. For instance, *D. veneta* demonstrates optimal development at lower temperatures and shows less tolerance to high temperatures than *E. fetida*, which grows best at 25 °C with a temperature tolerance of 0–35 °C. While the optimal development temperature for *E. eugeniae* and *P. excavatus* is likewise about 25 °C, their typical tolerable temperature range is between 9 and 35 °C (Edwards & Arancon, 2022). A study by Frederickson and Howell (2003) comparing the earthworm populations in heated and unheated bedding at 13.7 and 6.3 °C revealed that heated beds had higher earthworm biomass and more hatchlings and cocoons than unheated beds. According to a different study, the hatching percentage of *E. eugeniae* and *P. excavatus* was greater at lower temperatures (20–24 °C) than at high temperatures (27–30 °C) (Giraddi et al., 2010). Various earthworm species responded differently to different temperature ranges. Extreme temperature conditions, such as low or high temperatures, have a greater impact on vermicomposting systems than on the composting process. For instance, nitrogen is lost as NH_3 volatilisation in vermicomposting systems because of higher temperatures (Meng et al., 2020). Nevertheless, lower temperatures used during the vermicomposting process do not successfully eliminate harmful organisms (Amuah et al., 2022). According to Edwards (1998), earthworm species (*P. excavatus*) grew faster at temperatures up to 30 °C and took less time to reach sexual maturity. Also, they came to the conclusion that *P. excavatus* reproduced at their best rate at 25 °C, while *E. fetida* showed a wide range of temperature tolerance between 15 and 20 °C. The weight of *L. terrestris* increased while the temperature was between 15 and 17.5 °C indicating that 12–28 °C is the ideal temperature range for these earthworms during the vermicomposting process (Valckx et al., 2011). The earthworms' ability to reproduce and their metabolic activity both decrease when the temperature in the vermicomposting system drops and completely stops when the temperature is very low. Their metabolic activity and reproduction start reducing with increase in their mortality rate at higher temperatures (over 35 °C) (Arora & Kazmi, 2015). Temperature has a significant impact on an earthworm's activity, metabolism, growth, respiration and reproduction (Ansari & Ismail, 2012). Although *E. fetida* cocoons may endure prolonged periods of sub-freezing and stay alive, they are unable to reproduce and cannot ingest enough food at temperatures below 0 °C (Appelhof & Olszewski, 2017).

4.1.2 pH

One of the key elements influencing the vermicomposting process is the pH level (Appelhof & Olszewski, 2017). Worms can function properly in neutral pH; however, in the pH range of 5 to 9, epigeic worms can persist, reproduce and effectively move throughout the vermibed (Gaupp-Berghausen et al., 2015). Worm bedding pH tends to decrease with time. If the food source is acidic, then the pH of the beds can go well below 7, and if the bedding or food source is alkaline, the pH of the bed will drop to neutral or slightly alkaline (Abdulla-Al-Mamun et al., 2023). Calcium carbonate can be added to raise the pH, while peat moss can be used to lower the pH. Although the pH range for compost should be between 6.5 and 7.5, the active microorganisms in vermicomposting can continue to function even at a lower pH of approximately 4 (Li et al., 2022). The pH of the vermicomposting process is mostly changed by variations in the physicochemical properties of waste. During the breakdown phase, microbial activity alters the physicochemical properties of waste along with the degradation of nitrogen and phosphorus into nitrites/nitrates and orthophosphates (Suthar, 2009b). While negatively and positively charged groups result in either a neutral or acidic pH, some intermediates created by the breakdown of organic materials, such as ammonium and humic acids, affect the change of pH (Pereira et al., 2022). Because different intermediate species are produced by different substrate types, the vermicomposting system's pH is also affected by these differences. As a result, different waste types exhibit different pH-shifting behaviours, and the process's overall pH shifts from an alkaline to an acidic state (Sun et al., 2019). Numerous studies have suggested that the pH of the vermicomposting process is acidic (Garg et al., 2006; Khwairakpam & Bhargava, 2009; Tikoria et al., 2022c; Yadav & Garg, 2011). However, the pH value is initially in the alkaline range (8.3–7.2) and only slightly shifts to the acidic or neutral range (6.3–7.1) at the end of the process as a result of the intermediate products produced during vermicomposting (Tikoria et al., 2022a). If the pH of a feed material is neutral, the worms take lesser time to convert it into vermicompost compared with those which have acidic or basic pH (Khwairakpam & Bhargava, 2009; Yadav & Garg, 2011). The presence of CO_2 and the build-up of organic acids within the feed material were the main causes of waste pH which ultimately affects the activity of worms (Yu & Huang, 2009).

4.1.3 Moisture

One of the most crucial elements required for the operation of earthworms and microorganisms in a vermicomposting system is adequate moisture content. Because earthworms breathe through their skin; the system has to be adequately wet. The amount of moisture in the vermicomposting system has been linked to the growth rate of earthworms (Ali et al., 2015). Nevertheless, up to 90% of water content has also been deemed effective for vermicomposting process. However, an ideal moisture range of around 50–80% has been recommended for efficient vermicomposting (Barik et al., 2010). Earthworms' sexual development is delayed in low-moisture environments. In accordance with Reinecke and Venter (1987), *E. fetida* should have moisture content of 70% or higher. Development and reproduction of *E. fetida* were observed to occur at a moisture content rate of 82% by Tomlin and Miller (1980) and 70–80% by Neuhauser et al. (1979) in cow dung, respectively. Certain earthworm species, such as *L. terrestris*, do well in dry environments, while other species, including *Allolobophora chlorotica, Allolobophora caliginosa* and *Aporrectodea rosea*, did not grow enough in dry conditions (Lowe & Butt, 2005). According to Reinecke and Venter (1987), even a 5% change in moisture content has a substantial impact on how the clitellum develops. Throughout the process, water serves as a medium for a variety of chemical reactions as well as for the transport of nutrients. The moisture content of organic wastes and the speed of earthworm movement are closely correlated (Yadav & Garg, 2011). According to comparative research on vermicomposting process and earthworm growth at various temperature and moisture ranges, 65–75% moisture is the best range for all vermicomposting temperature conditions, compared with lower or higher moisture levels (Gajalakshmi & Abbasi, 2004). Epigeic species, *E. fetida* and *E. andrei*, can withstand moisture levels between 50 and 90%, although they develop more quickly between 80 and 90% (Singh et al., 2011). Bacteria are essential to vermicomposting as well. The speed of vermicomposting drops during its action when moisture level was lower than 40% and completely stops when it was below 10% (Yang et al., 2017).

4.1.4 Aeration

As worms are aerobic creatures, vermicomposting needs oxygen to function. The amount of oxygen consumed depends on microbial and earthworm activity, while oxygen levels are influenced by substrate temperature. In a vermicomposting system, too much moisture might result in inadequate aeration and limit the worms' access to oxygen. High concentrations of greasy and oily wastes in the feed substrate can further reduce oxygen flow (Sim & Wu, 2010). This thus serves as justification for not introducing fatty and oily wastes to feedstock without first precomputing them. Mechanical aeration methods or manual turning are used to improve aeration in challenging vermicomposting conditions (Lee et al., 2018). Since they need oxygen to breathe, earthworms cannot thrive in anaerobic environments. They work well with porous, well-aerated compost material. Moreover, earthworms benefit themselves by aerating their bedding as they move around in it. *E. fetida* moves in large numbers from substrates with low oxygen levels, water saturation or an accumulation of carbon dioxide or hydrogen sulphide (Kiyasudeen et al., 2016).

4.1.5 Feed Quality

Feeding is crucial for the development and reproduction of earthworms throughout vermicomposting as well as the pace at which cocoons are produced. The feeding rate is affected by a number of variables, including substrate organic content, particle size and moisture (Ali et al., 2015). According to Do et al. (2021), feeding the substrate in a methodical manner is crucial to preventing anaerobic conditions. The worms' ability to accelerate vermicomposting is enhanced by the feed's small particle size. Its small particle size makes it more accessible to worms and enables optimum aeration through the waste pile. A worm consumes between 100 and 300 mg of food per gram of body weight per day (Burgos-Díaz et al., 2022). Organic matter, live microbes and decaying macrofauna provide food for earthworms. While deep-soil-living earthworms consume dirt as such, surface-living earthworms preferentially consume food items. Salts are extremely irritating to worms. The feed should not include more than 0.5% salt (Addy et al., 2021). There should be no hazardous or non-biodegradable items in the feed (such as inert materials, plastic, glass, metal objects, detergents, etc.) that might harm the earthworms or their metabolic by-products (Varjani et al., 2021). In a vermicomposting system, worms devour both organic materials and anaerobic microbes. Increased organic content decreases worm activity, which increases the anaerobic activity of microorganisms and leads to anaerobic conditions with bad odours (Moghadam et al., 2022). If toxic metals are present in the organic meal, earthworms will die. With 60% feed as fly ash, worms show resisted growth; when the concentration reaches 80%, they show a progressive decline in the worms' ability to break down the material (Cai et al., 2022). When several varieties of dungs were utilised as feed, it was observed that cow, sheep, horse and goat excrement had more acceptable feed for the manufacture of value-added fertiliser than buffalo, camel and donkey dung (Garg et al., 2006). One of the most important requirements for the vermicomposting process is suitable earthworm feed. Almost anything that is organic in nature may be eaten by earthworms. The quantity of food which an earthworm can eat each day depends on a variety of variables, including the feed's salt content, C:N ratio, particle size and degree of decomposition (Haynes & Zhou, 2016).

4.1.6 Light

Earthworms are naturally photophobic (Lin et al., 2018). Thus, it is best to keep them away from light. Long exposures to sunlight are fatal to earthworms, whereas short exposures result in partial to total paralysis. They detect light and migrate away from it using skin cells concentrated at the tip of their bodies. The photophobic nature of earthworms can be utilised for the separation of topmost layer of vermibed from the earthworms (Raza et al., 2022).

4.1.7 C:N Ratio

In earthworm cell emergence, development and metabolism, the C:N ratio is crucial which is frequently employed as an index for compost maturation. Carbon and nitrogen should be provided as substrates in the right proportions for optimal nutrition

(Mothe & Polisetty, 2021). The C:N ratio of the feed material impacts the development and reproduction of earthworms. The development and reproduction of worms are accelerated by a higher C:N ratio in the feedstock. It was observed that waste degradation is delayed by either a high or low C:N ratio (Shak et al., 2014). A high C:N ratio and a low nitrogen content in the organic feed material will cause a decline in microbial activity in the feed substrate (Kumar et al., 2010). When the substrate's initial C:N ratio is 25, the increased compost maturity is indicated by a C:N ratio of less than 20 (Xie et al., 2022). Rapid mineralisation and organic matter breakdown cause carbon to be lost as carbon dioxide in microbial respiration, while worms increase nitrogen by producing mucus, which leads to a reduction in the ratio of carbon to nitrogen (C:N) (Condron et al., 2010). Yet, the initial nitrogen levels in the substrate have a major role in determining the vermicompost's ultimate N content and the overall level of breakdown. As nitrogen is lost as volatile ammonia at high pH levels, the pH drop also has a significant impact on nitrogen retention (Wong et al., 2001).

4.1.8 Ammonia and Salt Content

High-ammonia organic wastes are inhospitable to earthworm survival. Moreover, worms are extremely salt sensitive and prefer salt concentrations of less than 0.5% (Eijsackers, 2011). However, because many types of manures have high salt concentrations, they must first be leached to lower the salt content before being used as bedding. This is done by simply running water through the material for a while (Niedzialkoski et al., 2021).

4.2 *Biotic Factors*

Vermicomposting is influenced by a variety of biotic parameters, such as the earthworm density, microorganisms, enzymes, etc.

4.2.1 Earthworm Stocking Density

The quality and amount of the original substrate, temperature, moisture content, and soil texture and structure are all variables that affect earthworm density and its vermicomposting efficiency (Ali et al., 2015). When populations are sparse, earthworms are more likely to copulate often; however, this tendency declines when populations get closer to the substrate's carrying capacity (Edwards et al., 2022). According to some reports, 1.60 kg of worms/m^2 is the ideal stocking density for vermicomposting (Buzie-Fru, 2010). Earthworms are well known for having the most significant impact on the vermicomposting system's microbial populations and nutrient dynamics (Domínguez & Gómez-Brandón, 2013). The number of earthworms in a vermicomposting system (stocking density) influences a number of physiological functions, including respiration rate, reproduction, eating rate and digging activity. According to Dominguez and Edwards (1997), the ideal stocking density for sexual development is eight earthworms (*E. andrei*) per 43.61 g dry material of pig dung. The impact of population density on physiological functions may vary depending on the type of earthworm, their population densities, growth rate, cocoon output per earthworm and mortality (Uvarov, 2021). Despite similar and perfect physical circumstances, earthworms develop more slowly and with less biomass at increasing population densities (Opute & Maboeta, 2022). Fresh organic matter quickly turns into earthworm casts in vermicomposting systems with high earthworm population numbers (Domínguez et al., 2010). In order to achieve maximal population growth and reproduction in the shortest amount of time, it is crucial to maintain optimal earthworm density while setting up a vermicomposting system (Rodríguez-Canché et al., 2010).

4.2.2 Microorganisms

Microorganisms naturally occupy the biodegradable organic waste materials, and in the right environmental circumstances, they aid in the breakdown of organic wastes. The make-up of the microorganism populations in a vermicomposting system is influenced by the make-up of the waste materials being composted. Organic matter is stabilised during vermicomposting by the interaction of earthworms and microbes (Ojuolape et al., 2015; Villar et al., 2016). The fungal density in earthworm castings was almost equivalent to or higher than that of the original substrates, despite the fact that earthworms consume fungus along with organic substrates to meet their protein or nitrogen requirements (Gomez-Brandon et al., 2012). The microorganisms produce physiologically active chemicals in addition to mineralising complicated compounds into plant-available forms (Dhiman et al., 2021; Kaur et al., 2022). Further, certain microorganisms that earthworms engulf during vermicomposting with organic substrates are not completely destroyed during stomach degradation and pass out in the excreted material as such (Pramanik, 2010). In reality, spore development is aided by the favourable conditions of earthworm guts. This is likely to be the cause of the rising microbial population in vermicompost (Ali et al., 2015).

4.2.3 Enzymes

Organic wastes are quite complicated chemically, and enzymatic activity is required for full stabilisation. The worms release enzymes in their gizzard and gut that quickly biochemically convert the proteinaceous and cellulose components of the organic wastes (Manikanta et al., 2023). Cellulases, which depolymerise cellulose, b-glucosidases, which hydrolyse glucosides, amidohydrolase, proteases and urease, which are involved in N mineralisation, and phosphatases, which remove phosphate groups from organic matter are a few of the key enzymes utilised in the vermicomposting process (Dutta et al., 2023; Tikoria et al, 2022b). Indicators of microbial activity, such as enzyme activities, are frequently utilised. They may also be used to evaluate how much microbial metabolism is occurring in soil. In actuality, enzymes serve as the drivers of key metabolic processes such as the breakdown and detoxification of pollutants (Wang et al., 2021).

5 Conclusion

In the vermicomposting process, a wide range of organic wastes such as livestock, municipal, agricultural, industrial and wastewater residuals can be managed using engineered earthworm systems. The resulting vermicompost has good physical and chemical attributes that compare favourably to conventionally prepared composts. It can be used in agriculture as a source of plant nutrients and has also been shown to be an effective soil conditioner. Moreover, vermicompost has 'high porosity', 'aeration', 'drainage' and 'water-holding capacity', which also enhance soil texture. The waste substrate composition, moisture content and aeration are all important factors in the vermicomposting process. As a result, changes in agro-climatic and edaphic variables might have an impact on the efficacy of the vermicomposting process and the survival of earthworms.

References

Abdulla-Al-Mamun, M., Hossain, N., Hossain, M. I., & Sultana, R. (2023). Leather industries solid waste conversion to organic fertilizer through vermicomposting: utilization for plant growth.

Abu Bakar, A., Syed Mohd Gawi, S. N. A., Mahmood, N. Z., & Abdullah, N. (2014). Vermicomposting of vegetable waste amended with different sources of agro-industrial by-product using Lumbricus rubellus. *Polish Journal of Environmental Studies*, *23*(5), 1491–149.

Addy, M., Huo, S., Liu, J., Li, K., Cheng, P., Schiappacasse, C., Chen, D., Cheng, Y., Liu, Y., Ma, Y., & Wang, L. (2021). Bioconversion technologies: Insect and worm farming. In *Current developments in biotechnology and bioengineering* (pp. 235–256). Elsevier.

Adhikary, S. (2012). Vermicompost, the story of organic gold: A review. *Agricultural Science, 3*, 905–917.

Aira, M., Monroy, F., Domínguez, J., & Mato, S. (2002). How earthworm density affects microbial biomass and activity in pig manure. *European Journal of Soil Biology*, *38*(1), 7–10.

Ali, U., Sajid, N., Khalid, A., Riaz, L., Rabbani, M. M., Syed, J. H., & Malik, R. N. (2015). A review on vermicomposting of organic wastes. *Environmental Progress & Sustainable Energy, 34*(4), 1050–1062.

Amuah, E. E. Y., Fei-Baffoe, B., Sackey, L. N. A., Douti, N. B., & Kazapoe, R. W. (2022). A review of the principles of composting: Understanding the processes, methods, merits, and demerits. *Organic Agriculture, 12*(4), 547–562.

Ansari, A. A., & Ismail, S. A. (2012). Earthworms and vermiculture biotechnology. *Management of Organic Waste, 87*, 87–96.

Appelhof, M., & Olszewski, J. (2017). *Worms eat my garbage: How to set up and maintain a worm composting system: Compost food waste, produce fertilizer for houseplants and garden, and educate your kids and family*. Storey Publishing.

Aransiola, S. A., Ijah, U. J. J., Abioye, O. P., & Bala, J. D. (2022). Vermicompost-assisted phytoremediation of toxic trace element-contaminated soil in Madaka, Nigeria using Melissa officinalis L and Sida acuta. *International Journal of Environmental Science and Technology*. https://doi.org/10.1007/s13762-022-04105-y

Arora, S., & Kazmi, A. A. (2015). The effect of seasonal temperature on pathogen removal efficacy of vermifilter for wastewater treatment. *Water Research, 74*, 88–99.

Aslam, Z., & Ahmad, A. (2020). Effects of vermicompost, vermi-tea and chemical fertilizer on morpho-physiological characteristics of maize (*Zea mays* L.) in Suleymanpasa District, Tekirdag of Turkey. *Journal of Innovative Sciences, 6*(1), 41–46.

Aslam, Z., Ahmad, A., Bellitürk, K., Iqbal, N., Idrees, M., Rehman, W.U., Akbar, G., Tariq, M., Raza, M., Riasat, S., & ur Rehman, S. (2020). 26. Effects of vermicompost, vermi-tea and chemical fertilizer on morpho-physiological characteristics of tomato (*Solanum lycopersicum*) in Suleymanpasa District, Tekirdag of Turkey. *Pure and Applied Biology (PAB), 9*(3), 1920–1931.

Ayuso, M., Pascual, J. A., García, C., & Hernández, T. (1996). Evaluation of urban wastes for agricultural use. *Soil Science and Plant Nutrition, 42*(1), 105–111.

Barik, T., Gulati, J. M. L., Garnayak, L. M., & Bastia, D. K. (2010). Production of vermicompost from agricultural wastes-A review. *Agricultural Reviews, 31*(3), 172–183.

Bellitürk, K., Aslam, Z., Ahmad, A., & Rehman, S. U. (2020). Alteration of physical and chemical properties of livestock manures by *Eisenia fetida* (Savigny, 1926) and developing valuable organic fertilizer. *Journal of Innovative Sciences, 6*(1), 47–53.

Bo, Z., Yiyong, C., Zhang, C., Jianlong, L., Hao, T., Jiayu, L., Jun, D., & Jinchi, T. (2021). Earthworm biomass and population structure are negatively associated with changes in organic residue nitrogen concentration during vermicomposting. *Pedosphere, 31*(3), 433–439.

Bouché, M. B. (1977). Strategies lombriciennes. *Ecological Bulletins*, 122–132.

Brown, G. G., Barois, I., & Lavelle, P. (2000). Regulation of soil organic matter dynamics and microbial activity in the drilosphere and the role of interactions with other edaphic functional domains. *European Journal of Soil Biology, 36*(3–4), 177–198.

Burgos-Díaz, C., Opazo-Navarrete, M., Palacios, J. L., Verdugo, L., Anguita-Barrales, F., & Bustamante, M. (2022). Food-grade bioactive ingredient obtained from the *Durvillaea incurvata* brown seaweed: Antibacterial activity and antioxidant activity. *Algal Research, 68*, Article 102880.

Buzie-Fru, C. A. (2010). *Development of a continuous single chamber vermicomposting toilet with urine diversion for on-site application*. Technische Universität Hamburg.

Cadena, E., Colón, J., Sánchez, A., Font, X., & Artola, A. (2009). A methodology to determine gaseous emissions in a composting plant. *Waste Management, 29*(11), 2799–2807.

Cai, L., Gong, X., Ding, H., Li, S., Hao, D., Yu, K., Ma, Q., Sun, X., & Muneer, M. A. (2022). Vermicomposting with food processing waste mixtures of soybean meal and sugarcane bagasse. *Environmental Technology & Innovation, 28*, Article 102699.

Cardosa Vigueros, L., & Ramírez Camperos, E. (2002). Vermicomposting of sewage sludge: A new technology for Mexico. *Water Science and Technology, 46*(10), 153–158.

Chang, N. B., & Davila, E. (2008). Municipal solid waste characterizations and management strategies for the Lower Rio Grande Valley, Texas. *Waste Management, 28*(5), 776–794.

Charles, W., Walker, L., & Cord-Ruwisch, R. (2009). Effect of pre-aeration and inoculum on the start-up of batch thermophilic anaerobic digestion of municipal solid waste. *Bioresource Technology, 100*(8), 2329–2335.

Chowdhary, A.B., Singh, J., Quadar, J., Singh, S., Dutta, R., Angmo, D., & Vig, A. P. (2022a). Earthworm's show tolerance and avoidance response to pesticide clothianidin: Effect on antioxidant enzymes. *International Journal of Environmental Science and Technology*, 1–10.

Chowdhary, A. B., Singh, J., Quadar, J., Singh, S., Singh, A., Dutta, R., Angmo, D., & Vig, A. P. (2022b). Metsulfuron-methyl induced physiological, behavioural and biochemical changes in exotic (*Eisenia fetida*) and indigenous (*Metaphire posthuma*) earthworm species: Toxicity and molecular docking studies. *Pesticide Biochemistry and Physiology, 188*, Article 105276.

Condron, L., Stark, C., O'Callaghan, M., Clinton, P., & Huang, Z. (2010). The role of microbial communities in the formation and decomposition of soil organic matter. In *Soil microbiology and sustainable crop production* (pp. 81–118).

Das, S., Deka, P., Goswami, L., Sahariah, B., Hussain, N., & Bhattacharya, S. S. (2016). Vermiremediation of toxic jute mill waste employing *Metaphire posthuma. Environmental Science and Pollution Research, 23*, 15418–15431.

Dhiman, S., Ibrahim, M., Devi, K., Sharma, N., Kapoor, N., Kaur, R., Sharma, N., Tikoria, R., Ohri, P., Mir, B. A., & Bhardwaj, R. (2021). Biochar assisted remediation of toxic metals and metalloids. In *Handbook of assisted and amendment: Enhanced sustainable remediation technology* (pp. 131–162).

Do, Q., Ramudhin, A., Colicchia, C., Creazza, A., & Li, D. (2021). A systematic review of research on food loss and waste prevention and management for the circular economy. *International Journal of Production Economics, 239*, Article 108209.

Domfnguez, J. (2004). *20 state-of-the-art and new perspectives on vermicomposting research* (pp. 401–424). CRC Press.

Dominguez, J., & Edwards, C. A. (1997). Effects of stocking rate and moisture content on the growth and maturation of *Eisenia andrei* (Oligochaeta) in pig manure. *Soil Biology and Biochemistry, 29*(3–4), 743–746.

Domínguez, J., & Gómez-Brandón, M. (2013). The influence of earthworms on nutrient dynamics during the process of vermicomposting. *Waste Management & Research, 31*(8), 859–868.

Domínguez, J., & Ray, S. (2018). *Earthworms: The ecological engineers of soil*. Intech Open.

Domínguez, J., Aira, M., & Gómez-Brandón, M. (2010). Vermicomposting: Earthworms enhance the work of microbes. In *Microbes at work: From wastes to resources* (pp. 93–114).

Domínguez, J., Sanchez-Hernandez, J. C., & Lores, M. (2017). Vermicomposting of winemaking by-products. In *Handbook of grape processing by-products* (pp. 55–78). Academic Press.

Dutta, R., Angmo, D., Singh, J., Chowdhary, A. B., Quadar, J., Singh, S., & Vig, A. P. (2023). Synergistic effect of biochar amendment in milk processing industry sludge and cattle dung during the vermiremediation. *Bioresource Technology*, 128612.

Edwards, C. A., & Arancon, N. Q. (2022). The influence of environmental factors on earthworms. In *Biology and ecology of earthworms* (pp. 191–232). Springer US.

Edwards, C. A. (1998). The use of earthworms in the breakdown and management of organic wastes.

Edwards, C. A., Arancon, N. Q., Bohlen, P. J., & Hendrix, P. (2022). *Biology and ecology of earthworms*. Springer.

Eijsackers, H. (2011). Earthworms as colonizers of natural and cultivated soil environments. *Applied Soil Ecology, 50*, 1–13.

Frederickson, J., & Howell, G. (2003). Large-scale vermicomposting: Emission of nitrous oxide and effects of temperature on earthworm populations: The 7th international symposium on earthworm ecology, Cardiff, Wales, 2002. *Pedobiologia, 47*(5–6), 724–730.

Gajalakshmi, S., & Abbasi, S. A. (2004). Vermiconversion of paper waste by earthworm born and grown in the waste-fed reactors compared to the pioneers raised to adulthood on cowdung feed. *Bioresource Technology, 94*(1), 53–56.

Gandhi, M. (1997). Composting of household wastes with and without earthworms. *Environment and Ecology, 15*, 432–434.

Garg, P., Gupta, A., & Satya, S. (2006). Vermicomposting of different types of waste using *Eisenia foetida*: A comparative study. *Bioresource Technology, 97*(3), 391–395.

Gaupp-Berghausen, M., Hofer, M., Rewald, B., & Zaller, J. G. (2015). Glyphosate-based herbicides reduce the activity and reproduction of earthworms and lead to increased soil nutrient concentrations. *Scientific Reports, 5*(1), 1–9.

Giraddi, R. S., Gundannavar, K. P., Tippannavar, P. S., & Sunitha, N. D. (2010). Reproductive potential of vermicomposting earthworms, *Eudrilus eugeniae* (Kinberg) and *Perionyx excavatus* (Perrier) as influenced by seasonal factors. *Karnataka Journal of Agricultural Sciences, 21*(1).

Gomez-Brandon, M., Lores, M., & Domínguez, J. (2012). Species-specific effects of epigeic earthworms on microbial community structure during first stages of decomposition of organic matter. *PLoS ONE, 7*(2), Article e31895.

Haynes, R. J., & Zhou, Y. F. (2016). Comparison of the chemical, physical and microbial properties of composts produced by conventional composting or vermicomposting using the same feedstocks. *Environmental Science and Pollution Research, 23*, 10763–10772.

Julka, J. M. (2008). Know your earthworms. In *Foundation for life sciences and business management, Solan* (51pp).

Kaur, R., Sharma, N., Tikoria, R., Ali, M., Kour, S., Kumar, D., & Ohri, P. (2022). Insights into biosynthesis and signaling of cytokinins during plant growth, development and stress tolerance. *Auxins, cytokinins and gibberellins signaling in plants* (pp. 153–187). Springer International Publishing.

Kauser, H., & Khwairakpam, M. (2022). Organic waste management by two-stage composting process to decrease the time required for vermicomposting. *Environmental Technology & Innovation, 25*, Article 102193.

Kaza, S., Yao, L., Bhada-Tata, P., & Van Woerden, F. (2018). *What a waste 2.0: A global snapshot of solid waste management to 2050*. World Bank Publications.

Khwairakpam, M., & Bhargava, R. (2009). Vermitechnology for sewage sludge recycling. *Journal of Hazardous Materials, 161*(2–3), 948–954.

Kiyasudeen, S. K., Ibrahim, M. H., Quaik, S., Ahmed Ismail, S., Ibrahim, M. H., Quaik, S., & Ismail, S. A. (2016). Optimal conditions and environmental factors involved in breeding earthworms for vermicomposting. In *Prospects of organic waste management and the significance of earthworms* (pp. 147–165).

Kumar, M., Ou, Y. L., & Lin, J. G. (2010). Co-composting of green waste and food waste at low C/N ratio. *Waste Management, 30*(4), 602–609.

Lee, D. H., Behera, S. K., Kim, J. W., & Park, H. S. (2009). Methane production potential of leachate generated from Korean food waste recycling facilities: A lab-scale study. *Waste Management, 29*(2), 876–882.

Lee, K. E. (1985). *Earthworms: Their ecology and relationships with soils and land use*. Academic Press Inc.

Lee, L. H., Wu, T. Y., Shak, K. P. Y., Lim, S. L., Ng, K. Y., Nguyen, M. N., & Teoh, W. H. (2018). Sustainable approach to biotransform industrial sludge into organic fertilizer via vermicomposting: A mini-review. *Journal of Chemical Technology & Biotechnology, 93*(4), 925–935.

Leh-Togi Zobeashia, S. S., Aransiola, S. A., Ijah, U. J. J., & Abioye, O. P. (2018). Anaerobic digestion and agricultural application of organic wastes. *Advances in Environmental Research, 7*(2), 73–85. http://www.techno-press.org/content/?page=article&journal=aer&volume=7&num=2&ordernum=1

Li, H., Zhang, T., Shaheen, S. M., Abdelrahman, H., Ali, E. F., Bolan, N. S., Li, G., & Rinklebe, J. (2022). Microbial inoculants and struvite improved organic matter humification and stabilized phosphorus during swine manure composting: Multivariate and multiscale investigations. *Bioresource Technology, 351*, Article 126976.

Lim, S. L., Wu, T. Y., Sim, E. Y. S., Lim, P. N., & Clarke, C. (2012). Biotransformation of rice husk into organic fertilizer through vermicomposting. *Ecological Engineering, 41*, 60–64.

Lin, J., Liu, Z., Xing, H., Luo, S., Yuan, Q., & Cao, H. (2018). Earthworm photophobic movement under different light conditions and quantitative analysis of mechanical separating vermicompost parameters. *Transactions of the Chinese Society of Agricultural Engineering, 34*(2), 235–241.

Lowe, C. N., & Butt, K. R. (2005). Culture techniques for soil dwelling earthworms: A review. *Pedobiologia, 49*(5), 401–413.

Mahaly, M., Senthilkumar, A. K., Arumugam, S., Kaliyaperumal, C., & Karupannan, N. (2018). Vermicomposting of distillery sludge waste with tea leaf residues. *Sustainable Environment Research, 28*(5), 223–227.

Maharjan, K. K., Noppradit, P., & Techato, K. (2022). Suitability of vermicomposting for different varieties of organic waste: A systematic literature review (2012–2021). *Organic Agriculture*, 1–22.

Manikanta, L., Sudheer, S. V. S., Dhasmana, A., & Sudheer, B. (2023). The science of vermiculture: Use of earthworms in organic waste management.

Meng, L., Li, W., Zhang, X., Zhao, Y., Chen, L., & Zhang, S. (2020). Influence of spent mushroom substrate and molasses amendment on nitrogen loss and humification in sewage sludge composting. *Heliyon, 6*(9), Article e04988.

Moghadam, H. N., Banaei, A., & Bozorgian, A. (2022). Biological adsorption for removal of hydrogen sulfide from aqueous solution by live *Eisenia foetida* worms. *Advanced Journal of Chemistry, Section B: Natural Products and Medical Chemistry, 4*, 144–157.

Monroy, F., Aira, M., & Domínguez, J. (2009). Reduction of total coliform numbers during vermicomposting is caused by short-term direct effects of earthworms on microorganisms and depends on the dose of application of pig slurry. *Science of the Total Environment, 407*(20), 5411–5416.

Mor, S., Ravindra, K., Dahiya, R. P., & Chandra, A. (2006). Leachate characterization and assessment of groundwater pollution near municipal solid waste landfill site. *Environmental Monitoring and Assessment, 118*, 435–456.

Mothe, S., & Polisetty, V. R. (2021). Review on anaerobic digestion of rice straw for biogas production. *Environmental Science and Pollution Research, 28*, 24455–24469.

Nagavallemma, K. P., Wani, S. P., Lacroix, S., Padmaja, V. V., Vineela, C., Rao, M. B., & Sahrawat, K. L. (2004). Vermicomposting: Recycling wastes into valuable organic fertilizer. *Global Theme on Agroecosystems Report No. 8*.

Neuhauser, E. F., DL, K., & Hartenstein, R. (1979). Life history of the earthworm Eudrilus eugeniae.

Niedzialkoski, R. K., Marostica, R., Damaceno, F. M., de Mendonça Costa, L. A., & de Mendonça Costa, M. S. S. (2021). Combination of biological processes for agro-industrial poultry waste management: Effects on vermicomposting and anaerobic digestion. *Journal of Environmental Management, 297*, Article 113127.

Ojuolape, O. T., Aransiola, S. A., & Obideyi, O. A. (2015). Vermitechnology: A solution to environmental problems; A review. *Journal of Global Ecology and Environment, 3*(3), 127–135. http://www.ikpress.org/abstract/4516

Opute, P., & Maboeta, M. (2022). A review of the impact of extreme environmental factors on earthworm activities and the feedback on the climate.". *Applied Ecology and Environmental Research, 20*, 3277–3297.

Padmavathiamma, P. K., Li, L. Y., & Kumari, U. R. (2008). An experimental study of vermi-biowaste composting for agricultural soil improvement. *Bioresource Technology, 99*(6), 1672–1681.

Pellejero, G., Rodriguez, K., Ashchkar, G., Vela, E., García-Delgado, C., & Jiménez-Ballesta, R. (2020). Onion waste recycling by vermicomposting: Nutrients recovery and agronomical assessment. *International Journal of Environmental Science and Technology, 17*, 3289–3296.

Pereira, M. M. A., Moraes, L. C., Mogollón, M. C. T., Borja, C. J. F., Duarte, M., Buttrós, V. H. T., Luz, J. M. Q., Pasqual, M., & Dória, J. (2022). Cultivating biodiversity to harvest sustainability: Vermicomposting and inoculation of microorganisms for soil preservation and resilience. *Agronomy, 13*(1), 103.

Piearce, T. G., Oates, K., & Carruthers, W. J. (1990). A fossil earthworm embryo (Oligochaeta) from beneath a Late Bronze Age midden at Potterne, Wiltshire, UK. *Journal of Zoology, 220*(4), 537–542.

Pramanik, P. (2010). Changes in microbial properties and nutrient dynamics in bagasse and coir during vermicomposting: Quantification of fungal biomass through ergosterol estimation in vermicompost. *Waste Management, 30*(5), 787–791.

Quadar, J., Chowdhary, A. B., Dutta, R., Angmo, D., Rashid, F., Singh, S., Singh, J., & Vig, A. P. (2022). Characterization of vermicompost of coconut husk mixed with cattle dung: Physico-chemical properties, SEM, and FT-IR analysis. *Environmental Science and Pollution Research, 29*(58), 87790–87801.

Raza, S. T., Wu, J., Rene, E. R., Ali, Z., & Chen, Z. (2022). Reuse of agricultural wastes, manure, and biochar as an organic amendment: A review on its implications for vermicomposting technology. *Journal of Cleaner Production*, 132200.

Reinecke, A. J., & Venter, J. M. (1987). Moisture preferences, growth and reproduction of the compost worm *Eisenia fetida* (Oligochaeta). *Biology and Fertility of Soils, 3*, 135–141.

Ricci-Jürgensen, M., Gilbert, J., & Ramola, A. (2020). Global assessment of municipal organic waste production and recycling. ISWA.

Rodríguez-Canché, L. G., Vigueros, L. C., Maldonado-Montiel, T., & Martínez-Sanmiguel, M. (2010). Pathogen reduction in septic tank sludge through vermicomposting using Eisenia fetida. *Bioresource Technology, 101*(10), 3548–3553.

Sahariah, B., Goswami, L., Kim, K. H., Bhattacharyya, P., & Bhattacharya, S. S. (2015). Metal remediation and biodegradation potential of earthworm species on municipal solid waste: A parallel analysis between *Metaphire posthuma* and *Eisenia fetida. Bioresource Technology, 180*, 230–236.

Shah, R. U., Abid, M., Qayyum, M. F., & Ullah, R. (2015). Dynamics of chemical changes through production of various composts/vermicompost such as farm manure and sugar industry wastes. *International Journal of Recycling of Organic Waste in Agriculture, 4*, 39–51.

Shak, K. P. Y., Wu, T. Y., Lim, S. L., & Lee, C. A. (2014). Sustainable reuse of rice residues as feedstocks in vermicomposting for organic fertilizer production. *Environmental Science and Pollution Research, 21*, 1349–1359.

Sharma, K., & Garg, V. K. (2017). Management of food and vegetable processing waste spiked with buffalo waste using earthworms (*Eisenia fetida*). *Environmental Science and Pollution Research, 24*, 7829–7836.

Sharma, K. D., & Jain, S. (2020). Municipal solid waste generation, composition, and management: The global scenario. *Social Responsibility Journal, 16*(6), 917–948.

Sharma, S., Pradhan, K., Satya, S., & Vasudevan, P. (2005). Potentiality of earthworms for waste management and in other uses–A review. *The Journal of American Science, 1*(1), 4–16.

Sim, E. Y. S., & Wu, T. Y. (2010). The potential reuse of biodegradable municipal solid wastes (MSW) as feedstocks in vermicomposting. *Journal of the Science of Food and Agriculture, 90*(13), 2153–2162.

Singh, R. P., Embrandiri, A., Ibrahim, M. H., & Esa, N. (2011). Management of biomass residues generated from palm oil mill: Vermicomposting a sustainable option. *Resources, Conservation and Recycling, 55*(4), 423–434.

Singh, S., Singh, J., Kandoria, A., Quadar, J., Bhat, S. A., Chowdhary, A. B., & Vig, A. P. (2020). Bioconversion of different organic waste into fortified vermicompost with the help of earthworm: A comprehensive review. *International Journal of Recycling Organic Waste in Agriculture, 9*(4), 423–439.

Singh, S. I., Singh, W. R., Bhat, S. A., Sohal, B., Khanna, N., Vig, A. P., Ameen, F., & Jones, S. (2022). Vermiremediation of allopathic pharmaceutical industry sludge amended with cattle dung employing *Eisenia fetida. Environmental Research, 214*, Article 113766.

Sohal, B., Bhat, S. A., & Vig, A. P. (2021a). Vermiremediation and comparative exploration of physicochemical, growth parameters, nutrients and heavy metals content of biomedical waste ash via ecosystem engineers Eisenia fetida. *Ecotoxicology and Environmental Safety, 227*, Article 112891.

Sohal, B., Singh, S., Singh, S. I. K., Bhat, S. A., Kaur, J., Singh, J., & Vig, A. P. (2021b). Comparing the nutrient changes, heavy metals, and genotoxicity assessment before and after vermicomposting of thermal fly ash using Eisenia fetida. *Environmental Science and Pollution Research, 28*, 48154–48170.

Sosnecka, A., Kacprzak, M., & Rorat, A. (2016). Vermicomposting as an alternative way of biodegradable waste management for small municipalities. *Journal of Ecological Engineering, 17*(3).

Stoknes, K., Scholwin, F., Krzesiński, W., Wojciechowska, E., & Jasińska, A. (2016). Efficiency of a novel "Food to waste to food" system including anaerobic digestion of food waste and cultivation of vegetables on digestate in a bubble-insulated greenhouse. *Waste Management, 56*, 466–476.

Sun, L. C., Lin, Y. C., Liu, W. F., Qiu, X. J., Cao, K. Y., Liu, G. M., & Cao, M. J. (2019). Effect of pH shifting on conformation and gelation properties of myosin from skeletal muscle of blue round scads (*Decapterus maruadsi*). *Food Hydrocolloids, 93*, 137–145.

Suthar, S. (2009a). Vermicomposting of vegetable-market solid waste using Eisenia fetida: Impact of bulking material on earthworm growth and decomposition rate. *Ecological Engineering, 35*(5), 914–920.

Suthar, S. (2009b). Vermistabilization of municipal sewage sludge amended with sugarcane trash using epigeic *Eisenia fetida* (Oligochaeta). *Journal of Hazardous Materials, 163*(1), 199–206.

Suthar, S. (2010). Recycling of agro-industrial sludge through vermitechnology. *Ecological Engineering, 36*(8), 1028–1036.

Tikoria, R., Kaur, A., & Ohri, P. (2022a). Potential of vermicompost extract in enhancing the biomass and bioactive components along with mitigation of *Meloidogyne incognita*-induced stress in tomato. *Environmental Science and Pollution Research, 29*(37), 56023–56036.

Tikoria, R., Kaur, A., & Ohri, P. (2022b). Modulation of various phytoconstituents in tomato seedling growth and *Meloidogyne incognita*–Induced stress alleviation by vermicompost application. *Frontiers in Environmental Science, 10*, Article 891195.

Tikoria, R., Sharma, N., Kour, S., Kumar, D., & Ohri, P. (2022c). Vermicomposting: An effective alternative in integrated pest management. In *Earthworm engineering and applications* (pp. 103–118). Nova Science Publishers.

Tomlin, A. D., & Miller, J. J. (1980). Development and fecundity of the manure worm, *Eisenia foetida* (Annelida: Lumbricidae), under laboratory conditions.

Tripathi, G., & Bhardwaj, P. (2004). Comparative studies on biomass production, life cycles and composting efficiency of *Eisenia fetida* (Savigny) and *Lampito mauritii* (Kinberg). *Bioresource Technology, 92*(3), 275–283.

Troschinetz, A. M., & Mihelcic, J. R. (2009). Sustainable recycling of municipal solid waste in developing countries. *Waste Management, 29*(2), 915–923.

Tsai, S. H., Liu, C. P., & Yang, S. S. (2007). Microbial conversion of food wastes for biofertilizer production with thermophilic lipolytic microbes. *Renewable Energy, 32*(6), 904–915.

Uvarov, A. V. (2021). The overwinter survival of three earthworm species in mono-and multispecific assemblages. *Biology Bulletin, 48*, 821–828.

Valckx, J., Pina, A. C., Govers, G., Hermy, M., & Muys, B. (2011). Food and habitat preferences of the earthworm Lumbricus terrestris L. for cover crops. *Pedobiologia, 54*, S139–S144.

Varjani, S., Shah, A. V., Vyas, S., & Srivastava, V. K. (2021). Processes and prospects on valorizing solid waste for the production of valuable products employing bio-routes: A systematic review. *Chemosphere, 282*, Article 130954.

Vig, A. P., Singh, J., Wani, S. H., & Dhaliwal, S. S. (2011). Vermicomposting of tannery sludge mixed with cattle dung into valuable manure using earthworm *Eisenia fetida* (Savigny). *Bioresource Technology, 102*(17), 7941–7945.

Villar, I., Alves, D., Pérez-Díaz, D., & Mato, S. (2016). Changes in microbial dynamics during vermicomposting of fresh and composted sewage sludge. *Waste Management, 48*, 409–417.

Wang, C., Sun, Z., Liu, Y., Zheng, D., Liu, X., & Li, S. (2007). Earthworm polysaccharide and its antibacterial function on plant-pathogen microbes in vitro. *European Journal of Soil Biology, 43*, S135–S142.

Wang, F., Zhang, W., Miao, L., Ji, T., Wang, Y., Zhang, H., Ding, Y., & Zhu, W. (2021). The effects of vermicompost and shell powder addition on Cd bioavailability, enzyme activity and bacterial community in Cd-contaminated soil: A field study. *Ecotoxicology and Environmental Safety, 215*, Article 112163.

Wang, Y., Zhang, Q. Q., He, K., Zhang, Q., & Chai, L. (2013). Sulfate-nitrate-ammonium aerosols over China: Response to 2000–2015 emission changes of sulfur dioxide, nitrogen oxides, and ammonia. *Atmospheric Chemistry and Physics, 13*(5), 2635–2652.

Wani, K. A., & Rao, R. J. (2013). Bioconversion of garden waste, kitchen waste and cow dung into value-added products using earthworm *Eisenia fetida*. *Saudi Journal of Biological Sciences, 20*(2), 149–154.

Wong, J. W. C., Mak, K. F., Chan, N. W., Lam, A., Fang, M., Zhou, L. X., Wu, Q. T., & Liao, X. D. (2001). Co-composting of soybean residues and leaves in Hong Kong. *Bioresource Technology, 76*(2), 99–106.

Wu, T. Y., Lim, S. L., Lim, P. N., & Shak, K. P. Y. (2014). Biotransformation of biodegradable solid wastes into organic fertilizers using composting or/and vermicomposting. *Chemical Engineering Transactions, 39*, 1579–1584.

Xie, Y., Zhou, L., Dai, J., Chen, J., Yang, X., Wang, X., Wang, Z., & Feng, L. (2022). Effects of the C/N ratio on the microbial community and lignocellulose degradation, during branch waste composting. *Bioprocess and Biosystems Engineering, 45*(7), 1163–1174.

Yadav, A., & Garg, V. K. (2011). Recycling of organic wastes by employing *Eisenia fetida*. *Bioresource Technology, 102*(3), 2874–2880.

Yadav, A., & Garg, V. K. (2019). Biotransformation of bakery industry sludge into valuable product using vermicomposting. *Bioresource Technology, 274*, 512–517.

Yadav, K. D., Tare, V., & Ahammed, M. M. (2010). Vermicomposting of source-separated human faeces for nutrient recycling. *Waste Management, 30*(1), 50–56.

Yang, F., Li, G., Zang, B., & Zhang, Z. (2017). The maturity and CH_4, N_2O, NH_3 emissions from vermicomposting with agricultural waste. *Compost Science & Utilization, 25*(4), 262–271.

Yatoo, A. M., Ali, M. N., Zaheen, Z., Baba, Z. A., Ali, S., Rasool, S., Sheikh, T. A., Sillanpää, M., Gupta, P. K., Hamid, B., & Hamid, B. (2022). Assessment of pesticide toxicity on earthworms using multiple biomarkers: A review. *Environmental Chemistry Letters, 20*(4), 2573–2596.

Yu, H., & Huang, G. H. (2009). Effects of sodium acetate as a pH control amendment on the composting of food waste. *Bioresource Technology, 100*(6), 2005–2011.

Molecules of Therapeutic and Prophylactic Value in the Earthworm

Adekunle Babajide Rowaiye, Doofan Bur, Tarimoboeme Agbalalah, Joseph Akwoba Ogugua, Gordon C. Ibeanu, and Jaime Humberto Flores Garcia

Abstract Background: Earthworms, beyond being farmers' friend, has been utilized in the maintenance of human health. Extracts from these animals offer alternatives to the treatment of many conditions where chemotherapy has not offered optimum protection without the usual side effects. It therefore becomes necessary to reorganize the roles this all-important specie could play in the provision and maintenance of public health. Objective: This study aims at showcasing how different molecules present in earthworms could be of value in keeping a healthy population in therapeutics and prophylaxis. Methods: A search of the relevant original research and review publications on the therapeutic effects of earthworms, mostly published between 2012 and 2022 in Google Scholar and PubMed using relevant words, was conducted. A total of 95 articles published in English were selected. Results: Records available showed that earthworms processed in the forms of crude extracts or bioactive compounds provided ethno-medical solutions to various ailments of humans, plants, and animals, including: antimicrobial, anti-inflammatory, antioxidant, anticoagulative, neuro-regenerative, antihyperlipidemic, antidiabetic, antihypertensive,

A. B. Rowaiye (✉)
Department of Agricultural Biotechnology, National Biotechnology Development Agency, Abuja, Nigeria
e-mail: adekunlerowaiye@gmail.com

A. B. Rowaiye · G. C. Ibeanu
Department of Pharmaceutical Science, North Carolina Central University, Durham, NC, USA

D. Bur · T. Agbalalah
Department of Medical Biotechnology, National Biotechnology Development Agency, Abuja, Nigeria

T. Agbalalah
Department of Human Anatomy, Baze University, Abuja, Nigeria

J. A. Ogugua
Department of Veterinary Public Health and Preventive Medicine, University of Nigeria, Nsukka, Nigeria

J. H. F. Garcia
Departamento Salud Publica, Facultad de Ciencias de La Salud, Universidad Técnica de Manabí, Portoviejo, Ecuador

S. A. Aransiola and N. R. Maddela (eds.), *Vermitechnology: Economic, Environmental and Agricultural Sustainability*, https://doi.org/10.1007/978-3-031-96212-7_5

anticancer, antiaging, and many other medical and agricultural benefits. Conclusion: Information available in literature shows that, at optimum form, temperature, dosage, and usage, earthworms can provide cost cost-effective alternative treatment to many ailments with the advantage of having negligible side effects. This calls for rigorous research and more clinical trial efforts in this direction to validate their therapeutic potentials, thereby improving public health, agricultural, and economic benefits obtainable from earthworms, especially in developing economies.

Keywords Earthworm · Prophylactic · Therapeutic · Chemotherapy · Health

1 Introduction

Animals have been discovered to be excellent sources of bioactive substances, which have a wide range of biological effects on human health. These bioactive chemicals may be required for an animal to survive or may be produced in significant amounts for the benefit of other creatures. Natural substances of animal origin have been extracted, identified, and used as dietary or therapeutic supplements in recent years to prevent, lessen, or treat a variety of diseases and their accompanying symptoms (Grdisa, 2013).

Earthworms have been used medically for ages, as recorded in the history of ancient Southeast Asian medicine (Chinese, Japanese, and Vietnamese). Proteins, peptides, enzymes, and biologically active compounds are found in earthworms. As a result, a variety of ailments have been treated using extracts made from earthworm tissue (Fiołka et al., 2019). Like other invertebrates, earthworms have different types of leukocytes that produce and secrete a wide range of immunoprotective molecules (Grdisa, 2013).

In ancient Chinese manuscripts, there are records of earthworms being used to treat hypertension, arthritis, scabies, boils, and erysipelas to facilitate the birth process, as well as neutralize toxins. They have also been used as antipyretic, analgesic, and anti-inflammatory agents. Earthworms are soil-dwelling animals, so they have oxygen exchange and antioxidant properties (Annakulovna et al., 2022). Some of the bioactive molecules found in earthworms are antitumor protein, fibrinolytic enzymes, lumbrokinase (LK), cholinesterase, glycosidases, lysenin, collagenase, calmodulin-binding protein, metallothionein, eiseniapore, superoxide dismutase, catalases, glycoprotein extract, gut mobility regulation peptide, antibacterial peptide, carbamidine, lumbrofobrim, terrestrolumbrolysin, lumbritin, purin, vitamin B, tyrosine, lauric acid, succinic acid, unsaturated fatty acid, etc. (Sun, 2015).

Studies have shown that the long-ignored earthworm's innate immune system contains the key to apparent medical characteristics, in addition to ecological benefits, notably soil preservation. Using earthworms substantially avoids limitations and inhibitions, which are present in many other alternative medical practices due to ethical and financial objections to animal experimentation (Cooper et al., 2012a, 2012b). Therefore, there is a need to revisit the potential benefits of the utilization of earthworms and their constituents optimally to solve important health and economic problems of society.

This chapter enriches the book by showcasing the broader applicability of earthworms beyond soil and agriculture. It contributes to the narrative that earthworms are not just vital for ecological balance but also hold promise in economic and health-related contexts, reinforcing the book's central theme of sustainability across economic, environmental, and agricultural domains.

2 Materials and Methods

A search of the relevant original research and review publications on the therapeutic effects of earthworms, mostly published between 2012 and 2022, was done in Google Scholar and PubMed. The search was conducted using keyword combinations such as "earthworm AND anti-inflammatory," "earthworm AND antimicrobial," "earthworm AND antioxidative," earthworm AND anti-coagulative," "earthworm AND neuroregenerative," "earthworm AND anti-hyperlipidemia," earthworm AND anti-diabetic," "earthworm AND antihypertensive," "earthworm AND anticancer," and "earthworm AND anti-aging." All related articles were selected and reviewed. A total of 95 articles published in English were selected.

The selected articles were carefully analyzed and categorized based on the therapeutic effects of earthworms. Data extraction was conducted to collate detailed information on study objectives, methodologies, key findings, and therapeutic outcomes reported in each article. Particular attention was given to experimental designs, model systems used, and the bioactive molecules isolated from earthworms. The results were then organized into thematic sections corresponding to the therapeutic categories, highlighting key discoveries and the potential mechanisms underlying the therapeutic effects of earthworm-derived molecules.

3 Ethnomedicinal Value of the Earthworm

There are roughly 4000 different species of earthworm (Karthick et al., 2020). Earthworms are significant invertebrates that have been used for thousands of years as food and a source of traditional medicine in the treatment of a variety of diseases (Ding et al., 2019). For over 2300 years, Southeast Asian nations have used and studied them the most (Annakulovna et al., 2022). Traditional oriental populations have used earthworms as anticonvulsants, analgesics, and sedatives to treat neurological problems (Moon & Kim, 2018). Specifically, the use of earthworms in China as traditional medicines both for prophylaxis and therapy of diseases is an agelong practice (Deng et al., 2018). The use of earthworms as food and medicine in Taiwan, as well as Guangdong and Fujian provinces, is well documented in the book *The Eu Yan Sang Heritage: An Anthology of Chinese Herbs and Medicines* (Cooper et al., 2012a,

2012b). For example, as a component of traditional Chinese medicine (TCM), the red earthworm *Lumbricus rubellus* has been used to treat a variety of health challenges such as inflammation, fever, hematological disorders, hepatic disorders, joint discomfort, and high blood pressure (Dewi et al., 2017).

Dried earthworm powder has also been used in TCM to treat atherosclerosis linked to tinnitus and vertigo (Annakulovna et al., 2022). In Burma and Laos, South-east Asia, earthworms have historically been used to cure several diseases. It is produced in liquid or powder form and used to treat diseases with symptoms similar to pyorrhea as well as postpartum women who are unable to milk their infants (Cooper et al., 2012a, 2012b).

Earthworms have long been utilized in Ayurvedic medicine. It is well known that many tribes and residents of isolated Indian villages used earthworms to treat a variety of illnesses and wounds (Balamurugan et al., 2009). Crushed earthworm is used by traditional healers in Tamil Nadu, India, to cure fever, stomach pain, neck pain, neurological problems, and digestive issues (Balamurugan et al., 2009). Also, the Madura tribe in India's Bangkalan Regency has a folk remedy for typhoid fever that involves treating youngsters in particular with an earthworm decoction (made from boiling water with earthworms) (Radina & Adi, 2020).

Earthworms are regarded as a highly successful remedy in a number of African and Middle Eastern cultures. The earthworms are used in a variety of ways and are used to cure a variety of illnesses, including alopecia, jaundice, and bladder stones (Cooper et al., 2012a, 2012b). Iranians believe that earthworms are a powerful and effective medicine. In this culture, earthworms are cooked and eaten with bread in order to reduce the amount of bladder stones, and dried earthworms are also eaten to treat people with jaundice. Applying earthworm ashes to the scalp while using rose oil is also said to help treat alopecia (Cooper et al., 2012a, 2012b). In a similar fashion, zootherapy with invertebrate-based remedies is a common practice in Africa. In the Plateau Department of South-eastern Benin, the African earthworm *Eudrilus eugeniae* K. is popular for its medicinal uses (Loko et al., 2019).

The mythical lore and tribal customs in Native America involved the use of earthworms as traditional medicines. The Cherokee Indians of North America employed earthworm poultices to get rid of thorns (Cooper et al., 2012a, 2012b). Earthworms have also been used by the Nanticoke Tribe of Delaware to relieve rheumatic discomfort. Though the molecular mechanisms underlying these treatments are not completely known, there is some indication that earthworm lipids include particular fatty acids that have a therapeutic effect (Cooper et al., 2012a, 2012b).

4 Bioactive Molecules of Earthworms

Nutritional composition of some earthworms: Proximate analysis reveals that *Eisenia fetida* contains crude protein (64.61%), crude ash (10.16%), crude fat (12.29%), crude fiber (0.27%), and N extract (12.67%). *Euglandina rosea* contains crude protein (63.71%), crude fat (12.29%), crude ash (10.66%), crude fiber (0.21%), and N extract (12.67%). *Allolobophora caliginosa* contains crude protein (57.96%), crude fat (6.53%), crude ash (21.09%), crude fiber (0.36%), and N extract (14.06%) (Afreen & Shaikh, 2020; Sun & Jiang, 2017).

The earthworm consists of metabolites (such as lumbritin, carbamidine, lumbrofebrine (tyrosine derivative), and terrestrolumbrolysin) (Sun, 2015; Sun & Jiang, 2017); special organic acids (such as lauric acid, succinic acid, and unsaturated fatty acid) (Li et al., 2011b; Sun, 2015); and other components such as pyrimidines, purines, guanidine, choline, vitamin B, and tyrosine (Afreen & Shaikh, 2020; Li et al., 2011b; Sun, 2015; Sun & Jiang, 2017). The blood and body fluids of the earthworm, *Lumbricus terrestris* consists of lipids which include neutral fat (lauric acid, oleate, myristic acid, and decanoic acid); glucolipid (decanoic acid and some short-chain fatty acids); and phosphatide (behenic, decanoic, linoleate, and oleate acids) (Afreen & Shaikh, 2020; Sun & Jiang, 2017). The tissues of *Pheretima* species of earthworms contain microelements such as zinc, copper, iron, chromium, molybdenum, calcium, and manganese (Afreen & Shaikh, 2020; Sun & Jiang, 2017).

Earthworms also consist of many bioactive peptides. Among these active ingredients are the Lumbricus metalchelatins (LMT), which have a detoxifying effect by the chelation of soft metal and metalloid ions like zinc, cadmium, and arsenic by short, cysteine-rich, non-ribosomal peptides (Sun, 2015). According to their structure and antibacterial properties, the antibacterial Verm peptides family (AVPF) was named after the six antimicrobial peptides that were isolated and purified from the earthworm's coelomic fluid and the liquid homogenate of its tissue. It contained 5–50 amino acid residues with similar or the same sequences as Ala–Met–Val–Ser–Gly (ACSAG) (Sun & Jiang, 2017).

Several other antimicrobial peptides have been isolated from earthworms, and they carry out antimicrobial, gut mobility, and other functions (Li et al., 2011b). The OEP3121, a new antibacterial peptide having a molecular weight of 510.8 Da with the ACSAG sequence, was isolated from the earthworm, *E. fetida* (Liu et al., 2004). Perinerin is found in abundance in clam worms, *Marphysa sanguinea* (Park et al., 2020). Lumbricin I, a 62 amino acid antibacterial peptide with a molecular mass of 7231 Da and a proline content of 15% in molar ratio was extracted from the earthworm *L. rubellus* (Cho et al., 1998). Lumbricin and its orthologue (477 and 575 nucleotide mRNA sequences, respectively) antimicrobial peptides have been identified in *Eisenia andrei* (Bodó et al., 2019). The earthworm *Pheretima guillelmi* has also been shown to secrete a novel lumbricin-like antimicrobial peptide called lumbricin-PG (Li et al., 2011a). According to analysis, its amino acid composition is FSRYARMRDSRPWSDRKNNYSGPQFTYPPEKAPPEKLIKWNN EGSPIFEMPAEGGHIEP (Li et al., 2011a). Also, $A_{3–4}$ and $A_{3\text{-}4\text{-}2}$ are antibacterial peptides found in the earthworms (Sun & Jiang, 2017).

Earthworms contain enzymes such as collagenase, catalase, superoxide dismutase, cholinesterase, glycosidase (Li et al., 2011b), esterase, porphyrin synthetase, β-D-glucosyl enzyme, peroxidase, S-amino-γ-ketoglutaric dehydrogenase, alkaline phosphatase, fibrinolytic enzymes (Li et al., 2011b), and lumbrokinase. (Sun, 2015), and metal-binding protein such as calmodulin-binding protein and metallothionein

(Li et al., 2011b). Other active protein contents include lysenin, which is a pore-forming protein found in *E. fetida's* coelomic fluid (Bruhn et al., 2006), eisenia-pore, the hemolytic protein (Yamaji-Hasegawa et al., 2003), antitumor proteins, and glycoprotein (Li et al., 2011b; Sun, 2015).

5 Medicinal Effect of Earthworms

Earthworms possess a wide range of medicinal effects because they contain many bioactive molecules ranging from metabolites to peptides and proteins. The bioactivities of many of these molecules have been established (Fig. 1, Table 1).

5.1 *Antimicrobial Effect*

In a recent study, the antibacterial activity of *L. rubellus* earthworms was evaluated by measuring the diameter of the inhibitory zone to the growth of the bacterium *Porphyromonas gingivalis*, which is the primary cause of periodontitis (Dharmawati et al., 2019). The biggest diameter of the inhibitory zone on the growth of the *P. gingivalis* bacteria was found in a 50% concentration of *L. rubellus* earthworm extract. A

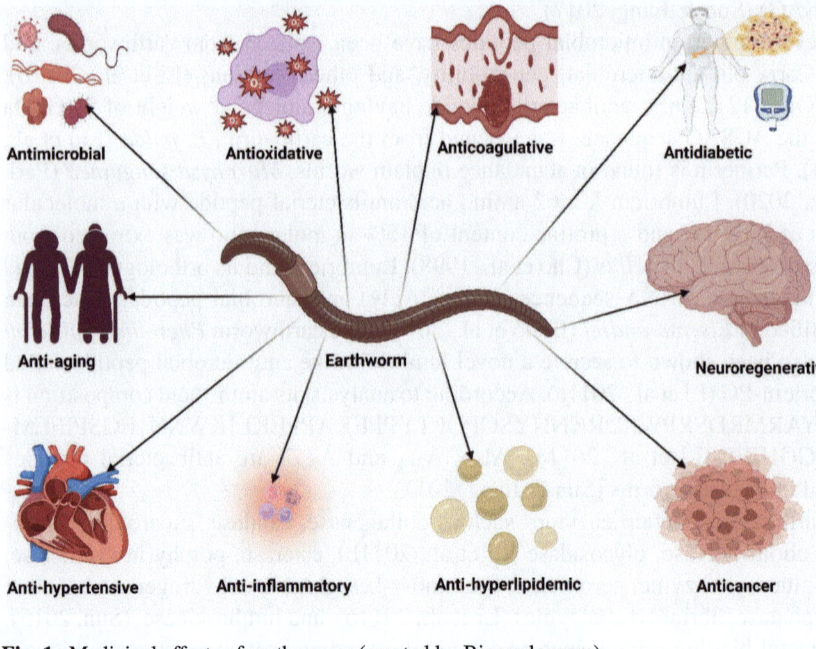

Fig. 1 Medicinal effects of earthworms (created by Biorender.com)

Table 1 The bioactive molecules found in different species of earthworms and their bioactivities

Bioactive molecule	Source	Bioactivity
Metabolites		
Lumbritin	*Pheretima* spp.	Unspecified (Sun, 2015)
Carbamidine	*Pheretima* spp.	Unspecified (Sun, 2015)
Lumbofebrine	*Pheretima* spp.	Unspecified (Sun, 2015)
Terrestrolumbrolysin	*Pheretima* spp.	Unspecified (Sun, 2015)
Peptides		
Lumbricus metalchelatins	*Dendrobaena veneta*	Detoxification (Sun, 2015)
Gut mobility regulation peptide	*Pheretima* spp.	Modulate gut motility (Sun, 2015)
AVPF	*Eisenia fetida*	Antimicrobial (Sun & Jiang, 2017)
A3-4	*Eisenia fetida*	Antimicrobial (Sun & Jiang, 2017)
A3-4-2	*Eisenia fetida*	Antimicrobial (Sun & Jiang, 2017)
Lumbricin PG	*Pheretima guillelmi*	Antimicrobial (Li et al., 2011a, 2011b)
OEP3121	*Eisenia fetida*	Antimicrobial (Liu et al., 2004)
Perinerin	*Marphysa sanguinea*	Antimicrobial (Park et al., 2020)
NCW	*Marphysa sanguinea*	Anti-inflammatory (Park et al., 2020)
Lumbricin	*Lumbricus rubellus*	Antimicrobial (Cho et al., 1998)
Lumbricin orthologue	*Eisenia andrei*	Antimicrobial (Bodó et al., 2019)
VQ-5	*Eisenia fetida*	Anti-inflammatory (Li et al., 2017)
AQ-5	*Eisenia fetida*	Anti-inflammatory (Li et al., 2017)
Lumbricusin	*Lumbricus terrestris*	Anti-inflammatory (Seo et al., 2017)
LumA5	*Synthesized*	Anti-inflammatory (Seo et al., 2017)
Col4a1	*Pheretima aspergillum*	Wound healing (Du et al., 2021)
Lumbricin I	*Lumbricus rubellus*	Antimicrobial (Lestari et al., 2019)
U3EE	*Eisenia fetida*	Inhibit DPP IV (Ogasawara et al., 2020)
Antimicrobial peptide I (PP-I)	*Pheretima tschiliensis*	Antimicrobial (Hussain et al., 2022)
Fetidin	*Eisenia fetida*	Hemolysis (Hussain et al., 2022)
CCF1	*Eisenia fetida*	Anticancer, antibacterial (Ghosh, 2020)
GGNG	*Eisenia fetida*	Modulate gut motility (Oumi et al., 1995)
Protein		
Lumbrokinase	*Lumbricus rubellus*	Fibrinolytic enzyme (Sun, 2015)
Superoxide dismutase	*Pheretima* spp.	Antioxidation (Sun, 2015)

(continued)

Table 1 (continued)

Bioactive molecule	Source	Bioactivity
Cholinesterase	*Pheretima* spp.	Hydrolysis of acetylcholine (Sun, 2015)
Catalases	*Pheretima* spp.	Antioxidation (Sun, 2015)
Glycosidases	*Pheretima* spp.	Hydrolysis of glycosidic linkages (Sun, 2015)
Metallothionein	*Pheretima* spp.	Detoxification (Sun, 2015)
Calmodulin-binding protein	*Pheretima* spp.	Intracellular Ca^{2+}-binding (Sun, 2015)
Antitumor protein	*Pheretima* spp.	Anticancer (Sun, 2015)
Lysenin	*Eisenia fetida*	Hemolysis (Bruhn et al., 2006)
Eiseniapore	*Eisenia fetida*	Hemolysis (Yamaji-Hasegawa et al., 2003)
Eisenia fetida proteases	*Eisenia fetida*	Antiviral (Wang et al., 2018)
G-90	*Eisenia fetida*	Antioxidative (Grdisa et al., 2001)
Alcalase hydrolysate	*Pheretima vulgaris*	Enzymatic hydrolysis (Feng et al., 2022)
Antihypertensive protein	*Pheretima* spp.	Reduce angiotensin II level (Li et al., 2005)
Coelomic cytolytic factor	*Eisenia fetida*	Anticancer (Hussain et al., 2022)

6.25% concentration of the *L. rubellus* earthworm extract did not inhibit the growth of *P. gingivalis* (Dharmawati et al., 2019). Also, the antibacterial activity of several solvent extracts of dried *Lampito mauritii* earthworm powder was evaluated using preliminary disc diffusion screening. The 95% ethanolic extract of the earthworm showed antibacterial activity and antifungal activity against *Candida albicans.* In comparison to the ethanolic extract of the earthworm powder, the petroleum ether and aqueous extracts had the least antifungal efficacy (Bhorgin & Uma, 2014). In contrast to *Streptococcus pyogenes,* the petroleum ether extract of earthworms showed the greatest inhibitory activity against *Staphylococcus aureus.* In contrast to *C. albicans*, the petroleum ether extract of earthworm was found to have the highest antifungal activity against *Aspergillus niger*. Ethanol and petroleum ether extracts had the least antibacterial action against *Escherichia coli.* Earthworm extract in phosphate buffer lacked inhibitory activity against bacterial and fungal cultures (Mathur et al., 2010).

Earthworms' paste and coelomic fluid were tested for their antibacterial and antifungal abilities against a number of pathogens, including *Vibrio parahaemolyticus, Bacillus subtilis, S. aureus, E. coli, C. albicans,* and *A. niger.* According to the findings of this study, the *Eudrilus eugeniae* paste and coelomic fluid exhibited antibacterial and antifungal activity against a number of different bacterial and fungal isolates suggesting that the constituents of the coelomic fluid and paste constituents have medicinal uses (Sethulakshmi et al., 2018). Also, ten different earthworms, including *Amynthas corticis, Amynthas gracilis, Pheretima posthuma, E. fetida, Aporrectodea rosea*, *Allolobophora chlorotica*, *Aporrectodea trapezoides*, *Polypheretima elongata*, *Aporrectodea caliginosa*, and *Pheretima hawayana*, were tested. Numerous bioactive substances, enzymes, and antioxidants found in the coelomic fluid of the earthworm function as anticoagulants and play a significant role in the suppression of bacterial growth. Therefore, it may be possible to avoid the evolution of multidrug-resistant bacteria by developing new drugs from earthworms (Mustafa et al., 2022).

It has been discovered that the coelomic fluid of the earthworm *Dendrobaena veneta* also contains an antifungal active fraction (AAF), which showed activity against clinical isolates of *C. albicans*, *Candida krusei* ATCC 6258, and *C. albicans* ATCC 10231 (Fiołka et al., 2019). AAF also significantly decreased the metabolic activity of *C. albicans* cells, caused a loss of integrity in the cell wall, and consequently caused apoptosis and necrosis. AAF contains some natural compounds and proteins and could be utilized to treat skin and mucous membrane candidiasis, as it demonstrated no endotoxicity or cytotoxicity toward normal skin fibroblasts (Fiołka et al., 2019).

Crop diseases are caused by phytopathogenic fungi, which also cause significant economic losses. The coelomic fluid of earthworms has been shown to inhibit the development of phytopathogenic fungi (Ečimović et al., 2021). In a research study conducted by Ečimović et al., the growth of six phytopathogenic fungus species (*Macrophomina phaseolina*, *Fusarium culmorum*, *Berkeleyomyces basicola*, *Rhizoctonia solani*, *Globisporangium irregulare*, and *Sclerotinia sclerotiorum*) was inhibited by the CF extract of the three earthworm species (*A. chlorotica, D. veneta, and E. andrei*) under investigation (Ečimović et al., 2021). The strongest inhibitory effect was demonstrated by the *E. andrei* CF extract, which reduced the growth of *R. solani* fungus. The study reported the antifungal activity of coelomic fluid collected from earthworm species belonging to several ecological categories, and also suggested a potential application in the protection of crops from phytopathogenic fungi (Ečimović et al., 2021). In another study, it was also discovered that the extracts of earthworms *D. veneta* and *E. fetida* coelomic fluid have activity against a phytopathogenic fungus *Fusarium oxysporum* in vitro (Plavšin et al., 2017).

The antibacterial activity of the different solvent extracts of the earthworm *E. eugeniae* has been used in the treatment of important human diseases, including Hepatitis B. A major therapeutic component of these extracts is the *E. fetida* proteases (Ef Ps), which exhibited fibronectin proteolytic activity and showed confirmed efficacy in the treatment of HBV infection in HepG2.2.15 cells (Wang et al., 2018). The study revealed that Ef Ps significantly altered HBV infection in animal models by degrading HBeAg in addition to proteolyzing fibronectin and which could be due to the presence of HBeAgases, which is one of the eight isozymes present in Ef Ps. This finding offers important insight into the therapeutic function and probable therapeutic mechanism of Ef Ps as an efficient antiviral drug for the treatment of chronic hepatitis B (Wang et al., 2018). The antibacterial activities of earthworms (*E. eugeniae, E. fetida,* and *P. posthuma*) against pathogenic microbes have also been established, as they have shown the potential to treat many fish ailments. It has been shown that *Bacillus megaterium* and *Aeromonas hydrophila* are sensitive to earthworm coelomic fluid (Kumar et al., 2022).

5.2 Anti-Inflammatory

Inflammation is a physiological response that is thought to be a component of a complex biological mechanism to remove irritants, pathogens, or harmful stimuli from the body. It is characterized by the activation of inflammatory signaling pathways and the release of a variety of proinflammatory mediators (Rowaiye et al., 2022). The in vitro anti-inflammatory activity of the aqueous extract of the Pre-Clitellar area of *E. eugeniae* (PAE) was investigated. The standard used (Aspirin; 100 g/ml) provided $61.2 \pm 2.76\%$ protection, but the test sample demonstrated strong membrane-stabilizing activities (50.3 ± 0.53, 62.23 ± 0.43, and $99.43 \pm 0.11\%$ at 10, 20, and 30 mg/ml concentrations of PAE, respectively). The results suggest that the PAE has anti-inflammatory properties, and the purification of its key ingredients could increase its ability to stabilize membranes (Falak et al., 2021).

A unique peptide, NCWPFQGVPLGFQAPP (NCW), obtained by high-performance liquid chromatography from the clam worm (*M. sanguinea*) extract, decreased catalase (CAT), superoxide dismutase (SOD), glutathione peroxidase (GSH-Px), and malondialdehyde (MDA) activities in LPS-stimulated RAW264. 7 cells (Park et al., 2020). Also, the NCW peptide inhibited the production of pro-inflammatory cytokines such as Interleukin-1 (IL-1) and tumor necrosis factor (TNF), nitric oxide (NO), inducible nitric oxide synthase (iNOS), and cyclooxygenase-2 (COX-2) in LPS-stimulated RAW264.7 macrophages, suggesting it could be a potential drug for diseases associated with inflammation (Park et al., 2020).

Studies have been conducted to determine the anti-inflammatory and anti-pyretic activity of the extract obtained from the earthworm *L. mauritii* (Balamurugan et al., 2009). In rats, inflammation induced by histamine and turpentine as well as pyrexia induced by Brewer's yeast were reduced and restored to normal conditions in a dose-dependent manner by the administration of indomethacin (10 mg/kg), paracetamol (150 mg/kg), and/or various doses of earthworm extract (EE) (50, 100, and 200 mg/kg) (Balamurugan et al., 2009).

The most significant inhibition of paw edema and granuloma, as well as the significant reduction in hyperpyrexia, when rats were administered conventional drugs and varying dosages of EE, suggest the presence of anti-inflammatory and antipyretic characteristics of EE similar to the glycoprotein complex (G-90) (Balamurugan et al., 2009).

In another study, the effect of the extract of the earthworm *L. rubellus* (EEW) on the reduction of the number of osteoclasts in chronic periodontitis in Wistar rats was investigated. It was demonstrated that EEW possesses anti-inflammatory properties that can inhibit NF-kB and reduce the secretion of pro-inflammatory cytokines (Dharmawati, 2020). Though the osteoclasts did not decrease when EEW was administered for up to 7 days, significant reductions were observed at Days 14 and 21. On Days 14 and 21, the quantity of osteoclasts decreased both orally and topically, with the same results (Dharmawati, 2020). The oral administration of EEW also significantly reduced neutrophil count (Dharmawati et al., 2022).

The methanolic earthworm extract (EE) has been demonstrated to have a protective effect against acrylamide (ACR)-induced reproductive dysfunction. The rats treated with ACR showed reduced sperm quantity, viability, and overall motility and altered the architecture of the testicular tissue (Ahmed et al., 2022). However, ACR had no impact on FSH or LH levels, reduced the glutathione (GSH) level in testicular tissues, increased testosterone, nitric oxide (NO), and malondialdehyde (MDA) levels and showed elevated expression of p53 and Ki-67 in the hyperplastic Leydig cells and the degenerating spermatogenic cells, respectively (Ahmed et al., 2022). However, the methanolic EE repaired the testicular histological structures, reversed the biochemical changes induced by ACR, and restored the sperm parameters. The EE reduced ACR-induced reproductive damage by reestablishing the antioxidant balance in the testicles and reducing the expression of p53 and Ki-67 in the testicular tissues (Ahmed et al., 2022).

Empirical data suggest that the Ohira II earthworm extract (OEE) had a beneficial effect on the process of skin wound healing in Kunming mice. OEE quickened wound healing and reduced inflammation as determined by the histopathologic, macroscopic, hematologic, and immunohistochemistry properties (Deng et al., 2018). Accelerated secretion of hydroxyproline and TGF-β was believed to be the potential mechanism responsible for that, thus enhancing collagen synthesis and promoting the proliferation of blood capillaries and fibroblasts. Accelerating the production of interleukin-6, white blood cells, and platelets hastens the clearance of necrotic tissue and foreign objects. As a result, it raises immunity, lowers the chance of infection, and speeds up wound healing (Deng et al., 2018).

The antipyretic and analgesic effects of earthworm extracts have also been investigated. In a study of the antipyretic impact on rabbits having pyrogen-induced fevers generated by *E. coli* pyrogen and chromatographic separations, the antipyretic components in the Japanese earthworms (*Lumbricus spencer*, *Perichaeta communishima, Goto,* and *Hatai*) were discovered (Hori et al., 1974). All-cis-5,8,11,14-eicosatetraenoic acid and all-cis-5,8,11,14,17-eicosapentaenoic acid are believed to be the key antipyretic components with demonstrated effectiveness (Hori et al., 1974). The extract of *Ohira II* (*E. fetida*) earthworms raised pain threshold and had peripheral but not central analgesic effects in mice. In a similar manner as morphine and aspirin, the extract of *Ohira II* reduced the levels of the enzymes that synthesize nitric oxide, norepinephrine, and 5-hydroxytryptamine in the blood, suggesting that it possesses peripheral analgesic qualities and may act as a potential analgesic drug (Luo et al., 2018).

From the coelomic fluid of *E. fetida*, two novel analgesic and anti-inflammatory peptides, VQ-5 and AQ-5, were isolated and characterized, with their primary structures identified as VSSVQ and AMADQ. In animal models of chronic inflammation and neuropathic pain, both peptides, but particularly AQ-5, demonstrated analgesic efficacy. Additionally, AQ-5 reduced the generation of cyclooxygenase-2 and tumor necrosis factor alpha. AQ-5 suppressed the mitogen-activated protein kinase signaling pathway, which has anti-inflammatory and analgesic effects (Li et al., 2017).

A novel role for earthworm peptide Lumbricusin as a regulator of neuroinflammation has been discovered[45]. A study demonstrated that the antimicrobial peptide Lumbricusin, which was obtained from the earthworm *L. terrestris*, improved motor dysfunction and dopaminergic neurodegeneration while enhancing neuronal growth (Seo et al., 2017). One of the 9-mer Lumbricusin analogues, LumA5 (QLICWRRFR-NH2), created based on the amino acid sequence of Lumbricusin, was also investigated for its inhibitory activity on lipopolysaccharide (LPS)-induced microglial activation and its ability to reduce neuroinflammation (Seo et al., 2017). LumA5 significantly decreased the expression of enzymes (COX-2, iNOS), cytokines (IL-6, IL-1β, TNF-α), and signal transduction factors (AKT, MAPKs, NF-B) associated with inflammation brought on by LPS in vitro and in vivo. Additionally, LumA5 increased cell survival and reduced the cytotoxicity of conditioned media produced by LPS-activated BV-2 microglia to neuronal SH-SY5Y cells. These findings suggest that LumA5 may have therapeutic promise in the management of a range of neuroinflammatory diseases (Seo et al., 2017).

In a recent study, Col4a1, a newly discovered collagen-like peptide, was cloned and expressed to thoroughly examine the role of wound healing and the underlying mechanism. Both in vitro and in vivo, it had considerable impacts on wound healing, including improved viability, proliferation, fibroblast migration, granulation, and collagen deposition. Additionally, the col4a1 worked by interacting with integrin $\alpha_2\beta_1$ and enhancing the RAS/MAPK signaling pathway. The study showed how the unique collagen-like peptide col4a1, which was extracted from severed earthworms, promoted faster wound healing and opened up new possibilities for wound treatment (Du et al., 2021).

5.3 Antioxidative

The excessive production of free radicals in cells, which outweighs the normal operation of exogenous and endogenous antioxidants, leads to oxidative stress. Free radicals accumulate in the body and destroy vital biological components like lipids, proteins, and DNA molecules because of their highly reactive nature (Rowaiye et al., 2020). In a recent study, the antioxidative properties of the earthworm, *L. rubellus,* were investigated. It was discovered that the ethanolic extract of *L. rubellus* powder had an IC 50% of 12.33 mg/mL and a total phenolic content of 1016.31 mg/100 g gallic acid equivalent (GAE). The *L. rubellus* powder ethanolic extract contains phenolic acid and has an in vitro antioxidant effect, suggesting it could be used as a natural source of antioxidants to treat disorders linked to oxidative stress (Dewi et al., 2017).

Glycolipoprotein extract (G-90) obtained from the earthworm *E. fetida* has been shown to exert some in vitro antioxidative activity in cultured human fibroblasts and epithelial cells. G-90 protected the cells from the toxicity induced by H_2O_2 and stimulated their proliferation 4 h after H_2O_2 treatment. G-90 was proven to perform better than a well-known antioxidant, ascorbic acid, which neither protected nor promoted cell development after injury unless it was administered at the same time as H_2O_2 (Grdisa et al., 2001).

The extract of *L. rubellus,* which contains Lumbricin I (peptide), glycoprotein G-90, and polyphenols, also demonstrated its antibacterial and antioxidant properties in male Wistar Rats infected with *Salmonella typhimurium.* The extract of *L. rubellus* lowered the levels of ALT, AST, and the number of bacterial colonies of *S. typhimurium* (Lestari et al., 2019). Also, in male Wistar rats exposed to *Salmonella typhi* infection, the extract of *L. rubellus* lowered the levels of malondialdehyde (MDA) and 8-hydroxy-deoxyguanosine (8-OHdG), suggesting that it could ameliorate the oxidative damage induced by typhoid fever (Samatra et al., 2017).

The effects of ethanolic EE of *E. fetida* and the earthworm meal (EM) on the oxidative parameters of native roosters from Western Azerbaijan was studied. It was discovered that MDA levels considerably decreased in the EE and EM treated groups in the blood, liver, and testis. Alkaline phosphatase and AST (not ALT) levels also decreased significantly. The results suggest that EE and EM may help improve male reproductive performance by protecting the blood, liver, and testicles from oxidative stress and having antioxidant activity (Shokouh et al., 2018). Also, the effects of earthworm extract on the levels of antioxidant enzymes, total antioxidant capacity, and oxidative damage in the midgut of *Bombyx mori*, a silkworm, were studied. It was demonstrated that dietary EE could increase *B. mori*'s antioxidant capability through lowering the amount of Lactate Dehydrogenase (LDH) and raising the expression of genes for antioxidant enzymes (Xu et al., 2020). The EE also increased Superoxide Dismutase (SOD) and the spleen index (Liu et al., 2019).

An antioxidant assay was conducted to determine the total phenolic content of the coelomic fluid of *E. fetida* using two solvents. According to the analysis, phenolic component extraction with 85% ethanol produced a high phenol content of 208.6 mg GAE/L. On the other hand, the extraction with 85% methanol produced 189.1 mg GAE/LL, suggesting that the coelomic fluid of *E. fetida* contains a large amount of phenols in its ethanol extract (Pinky et al., 2020). Furthermore, a novel peptide, NCW (NCWPFQGVPLGFQAPP) with a molecular weight of 1757.86 kDa was obtained from the extract of clam worm (*M. sanguinea*) using high-performance liquid chromatography and the structure elucidated by tandem mass spectrometry. NCW demonstrated notable antioxidant properties, causing a 50% inhibition of the DPPH radical at a concentration of 20 M without cytotoxicity (Park et al., 2020).

5.4 Anticoagulative Activity

Cardiovascular disorders (CVDs) are responsible for more than 31% of deaths globally (Joyia et al., 2018), and they include deep vein thrombosis, stroke, myocardial infarction, and pulmonary embolism (Feng et al., 2022; Kumar & Sabu, 2019). The development of blood clots inside the vessel is one cause of CVD. The dissolution of intraluminal blood clots involves numerous protein molecules known as thrombolytics (Joyia et al., 2018). Conventional thrombolytics are not only effective but also have a primary side effect is non-specific fibrinolysis, which causes extensive internal bleeding (hemorrhage) and significant blood loss, which finally results in death (Joyia et al., 2018). Certain enzymes found in earthworms can be used as thrombolytic agents and can destroy already existing thrombus (Kumar & Sabu, 2019; Malik & Afsheen, 2021).

The in vitro thrombolytic and fibrinolytic activity of the crude EE of *Eutyphoeus gammiei* has been reported in Tripura, Northeast India (Debnath et al., 2018). *E. gammiei* crude homogenate produced results that were comparable to those of streptokinase in both the thrombolytic assay with whole blood and the fibrin plate assay. Fibrin zymography also produced clear bands, indicating dose and time-dependent activity. The results demonstrated strong fibrinolytic and thrombolytic activity on human blood and suggests that *E. gammiei* might be a useful alternative source for the production of thrombolysis (Debnath et al., 2018). Biochemical and pharmaceutical investigations reveal that earthworm protease has anti-thrombosis and anti-fibrosis properties. With a wide range of substrate specificity, earthworm protease works against thrombosis by preventing platelet aggregation and exhibiting fibrinolytic activity. It also works against fibrosis by reducing fibronectin, collagen, and laminin. The protease regulator (U3EE) from earthworms performs both activator and inhibitor roles on a variety of target proteins (Wang et al., 2019).

To investigate the Chinese ethnomedicinal practice of administering the dry earthworm powder for the treatment of cardiovascular disorders, a study was performed to purify and identify the thrombolytic peptides from the enzymatic hydrolysate of *Pheretima vulgaris* (Feng et al., 2022). Eight different commercial proteases were used to hydrolyze the total active proteins from *P. vulgaris,* and the alcalase hydrolysate exhibited the most potent thrombolytic activity. Using bioactivity-directed fractionation of the active hydrolysate, four unique thrombolytic peptides were recovered and identified through nano-LC-ESI-Orbitrap mass spectrometry, and they are HEPLPEP, EYPLPEP, LGEPSVP, and LLAPP. In both the plasmin assay and the fibrinogen-thrombin time assay, HEPLPEP and EYPLPEP, which both contain the same PLPEP residues, demonstrated greater thrombolytic activity. The study presented new opportunities for thrombolytic drugs and confirmed that *P. vulgaris* was a possible source of active peptides with thrombolytic properties (Feng et al., 2022).

Also known as earthworm fibrinolytic enzymes (EFE), lumbrokinases (LK) are protease enzymes that have been isolated and characterized from the gut of *L. rubellus* (Verma & Pulicherla, 2017) and other different species of earthworms (Karthick et al., 2020) and used to treat blood clots (Cooper et al., 2012a, 2012b). The LK is a combination of six serine protease isomers, each of which has a unique molecular weight between 14 and 1 kDa and a range of fibrinolytic activity (Verma & Pulicherla, 2017). LK, which is easy to absorb and stable against pH and temperature fluctuations, has demonstrated anti-ischemic activity by enhancing the activity of adenylate cyclase, which caused an increase in c-AMP levels. The Glycoprotein IIB/IIIA (GPIIB/IIIA) and P-selectin, necessary for brain trauma, are inhibited by these changes in c-AMP expression (Verma & Pulicherla, 2017). The functional characterization of a purified LK obtained from an Earthworm *E. fetida* revealed its ability to dissolve fibrin or convert plasminogen to plasmin by activating endogenous tissue plasminogen activator (t-PA) to break down fibrin clots without causing any negative side effects. All these suggest that LK can be used as a perfect therapeutic molecule for oral administration in CVD patients (Joyia et al., 2018).

In a study, it was discovered that the plasma fibrinogen level and euglobulin lysis time could both be significantly reduced by the thrombolytic impact of an earthworm crude extract. The crude extract, which contains many fibrinolytic enzymes, caused an improvement in hemorheology when administered to rabbits. The experiment demonstrates that the enzymatic preparation may significantly lower the index of erythrocyte rigidity, lower the viscosity of whole blood and plasma, and obviously lessen platelet aggregation. These findings demonstrated its capability to increase blood flow and disperse stasis (Zhang & Wang, 1992). In another study using column chromatography, the thrombolytic enzyme was extracted and purified from the supernatant of the earthworm *Aporrectodea longa* and the blood clot lysis method was used to confirm thrombolytic activity. Results indicated that the extracted elute could act as a suitable therapeutic agent and showed potential fibrinolytic activity (Malik & Afsheen, 2021).

5.5 Neuroregenerative Activity

In several preclinical neuronal injury models, EE and its components have been demonstrated to preserve nerve cells and restore nerve function. This is an outcome of research on the involvement of numerous biomolecules produced by earthworms as antioxidant and anti-inflammatory agents (Moon & Kim, 2018).

In a recent study, the preventive and therapeutic benefits of earthworm extracts and their components were explored in several neuropathic models. Earthworm extracts were used as a preventative measure and treatment for several neurodegenerative conditions, including Parkinson's disease, mild cognitive impairment, cerebral infarction, and peripheral nerve injury (Moon & Kim, 2018).

Furthermore, the extract of *Lumbricus* promoted the regeneration of injured peripheral nerve in Sprague–Dawley (SD) rats. The study revealed a higher nerve function index value, higher conduction velocity of the injured sciatic nerve, and a higher number of regenerated myelinated nerve fibers in treatment groups than in the control group (Wei et al., 2009). The clinical potential of *Lumbricus* extract on the treatment of peripheral nerve damage in humans was suggested, given that it appeared to have improved sciatic nerve regeneration and function recovery after injury (Wei et al., 2009). Additionally, the earthworm aqueous extract promoted axonal sprouting and PC12 cell differentiation in peripheral nerve damage. It greatly encouraged PC12 cell production of GAP-43 and synapsin I, known to prompt NGF-induced neurite outgrowth (Chen et al., 2010).

It has been established that the bioactive ingredient IGF-1 (insulin-like growth factor 1) found in the earthworm, *Pheretima aspergillum,* promotes Schwann cell proliferation, survival, and migration into the distal end of the wounded nerve area to enhance axonal regrowth (Kristianto & Mardiati, 2017). To better understand how the earthworm (*P. aspergillum*) ethanol extract affects the augmentation of nerve fiber density in rats with stage II diabetic ulcers, a study was conducted. The findings of the study indicate that, as compared to the control groups, the groups that got the EE had a much higher density of nerve fibers. Applying earthworm extract topically had the best impact on increasing the density of the nerve fibers. In a diabetic rat model, earthworm extract promoted peripheral nerve regeneration (Kristianto & Mardiati, 2017).

With an intact nervous system, the earthworm has been shown to regenerate amputated parts due to the presence of biochemical substances. The recovery of damaged peripheral nerves can be greatly aided by the earthworm (Moon & Kim, 2018). The administration of the earthworm extracts has been shown to induce ERK1/2 and p38 activation (Chang et al., 2011). Because Schwann cells build the myelin sheath and protect axons in the peripheral nervous system, the earthworm extract increased Schwann cell migration. MMP2/9 expression was mediated by MAPK pathways that comprised ERK1/2 and p38. These findings suggest that the earthworm extract may have a significant impact on the migration of Schwann cells and neuro-regeneration via the MAPKs signaling pathway (Chang et al., 2011; Moon & Kim, 2018).

Using specific sciatic nerve lesion paradigms in diabetic rats generated by streptozotocin injection, a study evaluated the therapeutic effects of lumbrokinase on peripheral-nerve regeneration. After nerve transection, it was discovered that lumbrokinase therapy might enhance the rats' circulatory blood flow and encourage the regeneration of axons in a silicone rubber conduit. Treatment with lumbrokinase was believed to potentially enhance neuromuscular functioning and nerve conductivity (Lee et al., 2015). According to immunohistochemical labelling, lumbrokinase significantly increased the expression of the calcitonin gene-related peptide (CGRP) in the lamina I-II areas of the dorsal horn next to the lesion and significantly increased the number of macrophages drawn to the distal nerve stumps. Additionally, in dissected diabetic sciatic nerve segments, the lumbrokinase could promote the release of nerve growth factor (NGF), interleukin-1 (IL-1), transforming growth factor (TGF), and platelet-derived growth factor (PDGF). Finally, it was discovered that giving lumbrokinase to diabetic rats after their nerves were repaired during surgery had significant effects on boosting peripheral nerve regeneration and functional recovery (Lee et al., 2015).

5.6 Antihyperlipidemic Activity

Hyperlipidemia is a disorder of lipid metabolism characterized by increased blood total cholesterol, triglycerides, LDL (Low Density Lipoprotein), or diminished levels of HDL (High Density Lipoprotein) or a combination of both aberrations (Nalurika, 2015; Rochma, 2016).

The modulatory effect of lyophilized earthworm powder on the level of serum lipids in experimental hyperlipidemic mice has been studied. The results showed that different dosages of earthworm lyophilized powder significantly improved the level of blood HDL-C and decreased the level of serum TCTG and LDL-C within 8 weeks (Jin-xia et al., 2008). Another study that used a guinea pig model shows that EE can ameliorate high-fat diet (HFD)-induced fatty liver and modulate lipid profile. This study demonstrated that the induction of serum TC, TG, and LDL-C in response to high-fat diet (HFD) was decreased by the administration of earthworm extract. In guinea pigs fed the HFD, EE also lessened liver damage. This result implies that EE can reduce dyslipidemia and liver damage brought about by nonalcoholic fatty liver disease (NAFLD) (Deng et al., 2021). In Kunming mice fed for 10 weeks with high-level lipid feeds, it was demonstrated that one of the factors that contributed to the reduction of serum cholesterol in hyperlipidemic mice treated with earthworm lyophilized powder was the substantial upregulation of Lecithin cholesterol acyltransferase and Lipoprotein Lipase mRNA expression (Jinxia et al., 2015).

Malondialdehyde (MDA) levels rise as a result of lipid peroxidation caused by hyperlipidemia, which also damages the duodenum (Nalurika, 2015). The oil obtained from the Earthworm, *L. rubellus,* contains long-chain omega-3 unsaturated fatty acids, which can be used as an antioxidant in the treatment of hyperlipidemia. A study demonstrated that in hyperlipidemia model rats, earthworm oil can significantly lower MDA and reverse the duodenal histopathology, as shown by the return of normal goblet cells and epithelial cell growth (Nalurika, 2015). In a similar experiment, the omega-3 fatty acids found in oil earthworm (*L. rubellus*) lowered blood lipid levels and prevented oxidative damage brought on by free radicals. At the most effective dose of 400 mg/kg BW, the earthworm oil lowered blood triglyceride and LDL levels, and increased SOD activity. The earthworm oil also improved the histological appearance of the abdominal aorta by reducing foam cells and repairing the connective tissue and smooth muscle cells (Rochma, 2016).

5.7 *Antidiabetic Effects*

Alpha amylase and alpha glycosidase are directly involved in the development of diabetes. They are the main enzymes for controlling glucose in the human system, and inhibiting them ameliorates diabetes. The inhibitory activities of different solvent EE on these two enzymes have been evaluated (Mir et al., 2018). Of all the extracts, the water extract had the highest inhibitory activity, followed by the ethanol extract. Between water extract (highest inhibition) and ethyl acetate extract (lowest inhibition), DMSO and acetone extracts also demonstrated the ability to inhibit. The study revealed that generally, polar solvents better inhibit the glucose-controlling enzymes than the lesser polar solvents (Mir et al., 2018).

In diabetic rats, EE has been demonstrated to improve organ function and restore the physiological and histological changes associated with the disease. EE significantly reduced the levels of glucose and increased the levels of insulin and glucose-6-phosphate dehydrogenase (Abdelaziz et al., 2022). The histological analysis showed a clear improvement in the liver, heart, kidney, and testis architecture as well as the regeneration of injured pancreatic beta cells. The study showed that EE was effective in reversing the histological and biochemical alterations in the organs of diabetic rats. The study showed that the therapeutic efficacy of EE on diabetes complications is due to its hypoglycemic activity, antioxidant effects, and the ability to regenerate damaged tissues (Abdelaziz et al., 2022).

The G-90, the glycolipoprotein obtained from the earthworm *E. fetida,* possesses a variety of biological activities (Goodarzi et al., 2016). Given the biological properties of G-90, a study was conducted to ascertain how *E. fetida* homogenate extract affected the rate of wound healing in rats with alloxan-induced diabetes. The findings showed that D-panthenol treatment in rats had an exact replica of the impact of utilizing G-90 to speed up wound healing. In comparison to D-panthenol treatment, G-90 treatment reduced the risk of infection at the wound site and showed improved extracellular matrix production with increased neovascularization, fibroblast proliferation, and collagen synthesis, and early epithelial layer development. As a result, the G-90 may be viewed as a novel wound healing agent that introduces promising treatment modalities in both human and veterinary medicine (Goodarzi et al., 2016).

The effect of Earthworm-containing composite powder (CEP) on improving lipid metabolism and increasing fibrinolytic activity in Zucker diabetic fatty (ZDF) rats to protect against diabetic complications has been established. Following a feeding regimen of ZDF rats for 10 weeks, CEP significantly reduced euglobulin clot lysis time (ECLT), decreased HbA1c, hepatic fat buildup, and urine albumin excretion, while also enhancing the glomerular mesangial matrix score. This shows the potential of CEP to ameliorate diabetes and diabetic nephropathy (Kawakami et al., 2016). The CEP also significantly reduced ECLT in Sprague Dawley rats at 4 and 24 h after consumption (Kawakami et al., 2016). Furthermore, in an in vitro study, it was discovered that CEP possessed a high level of urokinase-type plasminogen activator-like activity (Kawakami et al., 2016).

Human dipeptidyl-peptidase IV (DPP IV) is inhibited in vitro by the aqueous extract of the earthworm, *E. fetida.* A fraction of this extract, under 3KDa, U3EE, has been demonstrated to possess inhibitory activity. U3EE seems to inhibit DPP IV due to the synergistic effects of the inhibitory amino acids (methionine, histidine, leucine, and isoleucine), suggesting that it may be effective as a medication and dietary supplement for the prevention of diabetes (Yoshii et al., 2020). U3EE has also demonstrated inhibitory activity against porcine pancreatic α-amylase (PPA) with an IC50 of 73.7 4.0 mg/mL (Ogasawara et al., 2020).

5.8 *Antihypertensive Activity*

In a recent study, hypertensive patients who had failed to respond to previous drugs were treated with an earthworm extract tincture. The earthworm treatment typically resulted in a reduction in high blood pressure within 4 to 10 days. When "Earth dragon B1," an EE, was used to treat hypertension, the results indicated a 90.9% success rate in preventing hypertension without causing any overt negative effects. With very positive treatment outcomes, an extract of an earthworm mixture was also employed to treat hypertension cases. When an earthworm K factor was injected intramuscularly to lower high blood pressure, 30 cases showed an improvement of 86.6%, which is more effective than the majority of pharmacological therapies for the condition (Afreen & Shaikh, 2020).

Spontaneously hypertensive rats (SHR) were used to study the effects of the antihypertensive protein obtained from *Pheretima* on blood pressure, angiotensin II, and angiotensin II AT1 receptor. The antihypertensive protein from *Pheretima,* either administered as a single intravenous injection or in several doses to significantly lowered blood pressure in SHR. Angiotensin II AT1 receptor expression in the kidney of the SHR model significantly increased as compared to the normal control (Wistar rats), but it decreased with the addition of the antihypertensive protein from *Pheretima.* The mechanism of action was believed to be connected to the decrease in angiotensin II levels and the downregulation of angiotensin II AT1 receptor expression in the kidney (Li et al., 2005).

The extract of *E. fetida* (EFE) has hypotensive and angiotensin-converting enzyme inhibitory effects in SHR. In vitro, EFE demonstrated significant ACE inhibitory activity ($IC_{50} = 2.5$ mg/mL) (Mao & Li, 2015). In another study, the extract of *Lamnodrilus gotai* (LGE) reduced spontaneous motor activity and blood pressure. The LGE was also found to have sedative and antihypertensive effects, notably in SHR (Wie et al., 1992).

5.9 *Anticancer*

Natural available extracts have been shown to have protective effects against the genotoxicity induced in normal cells by the traditional anticancer drugs (Kour et al., 2017). These extracts are sought for in the treatment of malignancies because of their antioxidant effects. One natural extract known for its anticancer activity is the earthworm extract (EE). The EEs contain proteases that possess anti-proliferative properties in vitro. The fluid obtained from the coelom (CF) of various species of earthworms was found to have a cytotoxic effect against the Squamous Cell Carcinoma cell line SCC-9, especially the CF of *E. eugeniae* (Augustine et al., 2019). The EE also showed suppressive effects against oral cancer squamous cell line carcinoma as well as breast, liver, gastrointestinal, and brain cancers (Augustine et al., 2017).

Although the EE are known to prevent tissue damage by excess reactive oxygen species (ROS), it is believed that the antioxidant enzymes may promote invasive cancer (Hawk et al., 2016). This is because antioxidant activity brings about loss of cell membrane integrity and the release of mediators associated with inflammation and immunosuppression into the extracellular environment, thereby promoting carcinogenesis (Terzić et al., 2010). The EE was found to reduce cancer cell lines multiplication in inverse relation with dosage; elevate spleen and thymus activities, and raise RBC and WBC counts at high dosage (90 mg/kg). The EE regulatory effects were found to be by increasing Bax, which promotes apoptosis, while decreasing Bcl-2, which inhibits cancer cells while not exerting immune suppression (Deng et al., 2019).

Testing the effect of coelomic fluid (CF) of *D. veneta* in colonic cancer (CC), Czerwonka et al. (2020) found that active protein-carbohydrate fraction (AF) obtained from thermally treated (70 °C) CF of the earthworm *D. veneta* inhibited damage to normal cell membrane (thereby preventing the possible side effect of promoting carcinogenesis) but retained the cancer cell apoptotic activities. The CF cell membrane damage activity was believed to be based on lysenin (protein) (Czerwonka et al., 2020). Lysozymes, also present in the worm, possess antitumor activities by the activation of the immune system of the host, achieved by inducing regulatory and helper T lymphocytes, and in addition, increase tumor cell immunogenicity. The lysozymes oligomerize and form complexes with different compounds, including sugars and peptides, which increase their in vitro anticancer activities by entering the cancer cells, resulting in the formation of plasmic granules and damaging the cell membrane (Czerwonka et al., 2020).

AF activity is also related to mitochondrial-associated events that play its role in the activation of the intrinsic pathway of apoptosis. Here, permeabilization of the outer membrane of mitochondria by factors such as the Bcl-2 family of proteins leads to the release of cytochrome c, which binds the caspase adaptor molecules Apaf-1 and procaspase-9, to form the apoptosome. This results in the assembling of apoptosome complexes, thereby triggering the activation of procaspase-3 to active caspase-3, leading to cell death (Tait & Green, 2013). Therefore, the ability of AF to inhibit the proteasome is believed to be the pathway responsible or co-responsible for the apoptosis observed in the CC cells (Czerwonka et al., 2020).

5.10 Antiaging

Earthworm extracts have been reported to exert antiaging effects. However, to date, only a few studies have shown that earthworms possess properties useful in fighting against skin aging. In a study, the earthworm extracts were shown to have significant anti-tyrosinase, anti-elastase, and matrix metalloproteinase-1 inhibitory activity as compared with the control, N-isobutyl-N-(4-methoxyphenylsulfonyl)-glycylhydroxamic acid (Azmi et al., 2014). These bioactivities were investigated in the earthworms *L. rubellus*, *E. fetida,* and *E. eugeniae*. The results support the traditional cosmetic use of earthworm extracts as an anti-wrinkle agent, as these enzymes prevent the loss of skin elasticity (Azmi et al., 2014). The earthworm mucus is highly effective at moisturizing, protecting, and exfoliating the skin and could potentially be commercialized for its variety of effects on skin regeneration, anti-aging, and antioxidant protection (Jeong et al., 2013). Furthermore, phytohormones such as cytokinins and trans-zeatin, which are components of earthworm casting (Puga-Freitas et al., 2012), are reported to reduce several aging markers in human fibroblasts (Fathy et al., 2022; Kadlecová et al., 2018).

5.11 Others

The EE was recorded to significantly reduce arginase, alanine aminotransferase, aspartate aminotransferase, alkaline phosphatase, low-density lipoprotein, total cholesterol, triglycerides, creatinine, uric acid, malondialdehyde, and nitric oxide (Abdelaziz et al., 2022). Earthworm extract was also reported to significantly increase albumin, high-density lipoprotein, total protein, testosterone, follicle-stimulating hormone, and luteinizing hormone; reduce glutathione S-transferases, glutathione, and catalase. These bioactivities were all found to significantly increase in response to EE exposure (Abdelaziz et al., 2022).

In addition to demonstrating many bioactivities, earthworms were found to be rich in proteins and other nutrients. With advances in biotechnology, the nutritional composition of some useful functional components of earthworms has been deciphered. Consequently, eating earthworms could be a novel nutritional supplement (Ding et al., 2019).

A study discovered that the CF of the earthworm *Metaphire peguana* instantly immobilized human, hamster, goat, and rat sperm in a dose-dependent manner. At 20 s, 300 g of CF immobilized 100% of human and other mammalian sperm. A 47-kD protein, Immotilin, obtained from the earthworm is said to be responsible for this spermicidal activity without affecting the other cells. Immotilin thus exhibits promise as a noninvasive anti-fertilizing agent (Mukherjee et al., 2003).

6 Conclusion

The medicinal importance of earthworms and the growing research interest in their use as sources of drugs are discussed in this review. The bioactivities of numerous earthworm species have been confirmed through evidence-based studies. Potentially, earthworms are cheap and safe sources of nutraceuticals as they contain substances that are vital for the human body and its systemic functions and also treat a variety of human disorders. However, more clinical trials are required to validate their therapeutic potentials and also to translate them into health care and cosmetic products for humans.

References

Abdelaziz, M. H., Abdelfattah, M. A., Bahaaeldine, M. A., Rashed, A. R., Mohamed, A. S., Ali, M. F., Elbatran, M. M., & Saad, D. Y. (2022). Earthworm extract enhanced organ functions in diabetic rats by ameliorating physiological and structural changes. *Biointerface Research in Applied Chemistry, 13*(5), 1–19. https://doi.org/10.33263/BRIAC135.445

Afreen, S., & Shaikh, A. (2020). Therapeutic uses of earthworm–A review. *International Journal of Advanced Ayurveda, Yoga, Unani, Siddha and Homeopathy, 9*(1), 571–580. https://doi.org/10.23953/cloud.ijaayush.469

Ahmed, M. M., Hammad, A. A., Orabi, S. H., Elbaz, H. T., Elweza, A. E., Tahoun, E. A., Elseehy, M. M., El-Shehawi, A. M., & Mousa, A. A. (2022). Reproductive injury in male rats from acrylamide toxicity and potential protection by earthworm Methanolic extract. *Animals, 12*(13), 1723. https://doi.org/10.3390/ani12131723

Annakulovna, N. Z., Abdulfayiz o'g'li, T. A., & Urmonovna, U. Z. (2022). Medicinal properties of earthworms and their importance in medicine. *Eurasian Medical Research Periodical, 8*, 52–54.

Augustine, D., Rao, R. S., Anbu, J., & Murthy, K. C. (2017). In vitro antiproliferative effect of earthworm coelomic fluid of Eudrilus eugeniae, Eisenia foetida, and Perionyx excavatus on squamous cell carcinoma-9 cell line: A pilot study. *Pharmacognosy Research, 9*(Suppl 1), S61–S66. https://doi.org/10.4103/pr.pr_52_17

Augustine, D., Rao, R. S., Anbu, J., & Murthy, K. C. (2019). In vitro cytotoxic and apoptotic induction effect of earthworm coelomic fluid of Eudrilus eugeniae, Eisenia foetida, and Perionyx excavatus on human oral squamous cell carcinoma-9 cell line. *Toxicology Reports, 6*, 347–357. https://doi.org/10.1016/j.toxrep.2019.04.005

Azmi, N., Hashim, P., Hashim, D. M., Halimoon, N., & Majid, N. M. (2014) Anti–elastase, anti–tyrosinase and matrix metalloproteinase–1 inhibitory activity of earthworm extracts as potential new anti–aging agent. *Asian Pacific Journal of Tropical Biomedicine, 4*, S348–S352. https://doi.org/10.12980/APJTB.4.2014C1166

Balamurugan, M., Parthasarathi, K., Cooper, E. L., & Ranganathan, L. S. (2009). Anti-inflammatory and anti-pyretic activities of earthworm extract—Lampito mauritii (Kinberg). *Journal of Ethnopharmacology., 121*(2), 330–332. https://doi.org/10.1016/j.jep.2008.10.021

Bhorgin, A. J., & Uma, K. (2014). Antimicrobial activity of earthworm powder (Lampito mauritii). *International Journal of Current Microbiology and Appl Science, 3*(1), 437–443.

Bodó, K., Boros, Á., Rumpler, É., Molnár, L., Böröcz, K., Németh, P., & Engelmann, P. (2019). Identification of novel lumbricin homologues in Eisenia andrei earthworms. *Developmental & Comparative Immunology, 90*, 41–46. https://doi.org/10.1016/j.dci.2018.09.001

Bruhn, H., Winkelmann, J., Andersen, C., Andrä, J., & Leippe, M. (2006). Dissection of the mechanisms of cytolytic and antibacterial activity of lysenin, a defence protein of the annelid Eisenia fetida. *Developmental & Comparative Immunology, 30*(7), 597–606. https://doi.org/10.1016/j.dci.2005.09.002

Chang, Y. M., Kuo, W. H., Lai, T. Y., Shih, Y. T., Tsai, F. J., Tsai, C. H., Shu, W. T., Chen, Y. Y., Chen, Y. S., Kuo, W. W., & Huang, C. Y. (2011). RSC96 Schwann cell proliferation and survival induced by dilong through PI3K/Akt signaling mediated by IGF-I. *Evidence-Based Complementary and Alternative Medicine, 2011*(216148). https://doi.org/10.1093/ecam/nep216

Chen, C. T., Lin, J. G., Lu, T. W., Tsai, F. J., Huang, C. Y., Yao, C. H., & Chen, Y. S. (2010). Earthworm extracts facilitate PC12 cell differentiation and promote axonal sprouting in peripheral nerve injury. *The American Journal of Chinese Medicine, 38*(03), 547–560. https://doi.org/10.1142/S0192415X10008044

Cho, J. H., Park, C. B., Yoon, Y. G., & Kim, S. C. (1998). Lumbricin I, a novel proline-rich antimicrobial peptide from the earthworm: purification, cDNA cloning and molecular characterization. *Biochimica et Biophysica Acta (BBA)-Molecular Basis of Disease, 1408*(1), 67–76. https://doi.org/10.1016/S0925-4439(98)00058-1

Cooper, E. L., Balamurugan, M., Huang, C. Y., Tsao, C. R., Heredia, J., Tommaseo-Ponzetta, M., & Paoletti, M. G. (2012). Earthworms dilong: Ancient, inexpensive, noncontroversial models may help clarify approaches to integrated medicine emphasizing neuroimmune systems. *Evidence-Based Complementary and Alternative Medicine, 2012*(164152). https://doi.org/10.1155/2012/164152

Cooper, E. L., Hirabayashi, K., & Balamurugan, M. (2012b). Dilong: Food for thought and medicine. *Journal of Traditional and Complementary Medicine., 2*(4), 242–248. https://doi.org/10.1016/S2225-4110(16)30110-9

Czerwonka, A., Fiołka, M. J., Jędrzejewska, K., Jankowska, E., Zając, A., & Rzeski, W. (2020). Pro-apoptotic action of protein-carbohydrate fraction isolated from coelomic fluid of the earthworm Dendrobaena veneta against human colon adenocarcinoma cells. *Biomedicine & Pharmacotherapy, 126*, Article 110035. https://doi.org/10.1016/S2225-4110(16)30110-9

Debnath, M., Saha, S., & Sil, S. K. (2018). First report on fibrinolytic and thrombolytic activity of Eutyphoeus Gammiei an earthworm species collected from Tripura, Northeast India. *Asian Journal of Pharmaceutical and Clinical Research, 11*(11), 20236–20240. https://doi.org/10.22159/ajpcr.2018.v11i11.27739

Deng, Z., Gao, S., An, Y., Huang, Y., Liu, H., Zhu, W., Lu, W., He, M., Xie, W., Yu, D., & Li, Y. (2021). Effects of earthworm extract on the lipid profile and fatty liver induced by a high-fat diet in guinea pigs. *Annals of Translational Medicine, 9*(4), 292. https://doi.org/10.21037/atm-20-5362

Deng, Z., Gao, S., Xiao, X., Yin, N., Ma, S., Li, W., & Li, Y. (2019). The effect of earthworm extract on mice S180 tumor growth and apoptosis. *Biomedicine & Pharmacotherapy., 115*, Article 108979. https://doi.org/10.1016/j.biopha.2019.108979

Deng, Z. H., Yin, J. J., Luo, W., Kotian, R. N., Gao, S. S., Yi, Z. Q., Xiao, W. F., Li, W. P., & Li, Y. S. (2018). The effect of earthworm extract on promoting skin wound healing. *Bioscience Reports, 38*(2), BSR20171366. https://doi.org/10.1042/BSR20171366

Dewi, N. W., Mahendra, A. N., Putra, G. W., Jawi, I. M., Sukrama, D. M., & Kartini, N. L. (2017). Ethanolic extract of the powder of red earthworm (Lumbricus rubellus) obtained from several organic farmlands in Bali, Indonesia: Analysis of total phenolic content and antioxidant capacity. *Bali Medical Journal, 6*(3), S80–S83. https://doi.org/10.15562/bmj.v313.730

Dharmawati, I. (2020). Effectiveness of Lumbricus rubellus earthworm extract against the number of osteoclasts in Wistar periodontitis rat. *European Journal of Molecular & Clinical Medicine, 7*(11), 1584–1592.

Dharmawati, I. A., Sukrama, I. D., Thahir, H., & Manuaba, I. B. (2022). Utilization of rhizosphere earthworm extracts to support the immune system. *KnE Life Sciences*. https://doi.org/10.18502/kls.v7i3.11114

Dharmawati, I. G., Mahadewa, T. G., & Widyadharma, I. P. (2019). Antibacterial activity of Lumbricus Rubellus earthworm extract Against Porphyromonas Gingivalis as the bacterial cause of periodontitis. *Open Access Macedonian Journal of Medical Sciences, 7*(6), 1032. https://doi.org/10.3889/oamjms.2019.222

Ding, S., Lin, X., & He, S. (2019). Earthworms: A source of protein. *Journal of Food Science and Engineering, 9*(2019), 159–170. https://doi.org/10.17265/2159-5828/2019.05.001

Du, C., Li, Y., Xia, X., Du, E., Lin, Y., Lian, J., Ren, C., Li, S., Wei, W., & Qin, Y. (2021). Identification of a novel collagen-like peptide by high-throughput screening for effective wound-healing therapy. *International Journal of Biological Macromolecules, 173*, 541–553. https://doi.org/10.1016/j.ijbiomac.2021.01.104

Ečimović, S., Vrandečić, K., Kujavec, M., Žulj, M., Ćosić, J., & Velki, M. (2021). Antifungal activity of earthworm coelomic fluid obtained from Eisenia andrei, Dendrobaena veneta and Allolobophora chlorotica on six species of phytopathogenic fungi. *Environments, 8*(10), 102. https://doi.org/10.3390/environments8100102

Falak, Y. Y., Iqbal, U. N., Mahajan, N. G., & Chopda, M. Z. (2021). In vitro anti-inflammatory activity of pre-clitellar extract of Eudrilus eugeniae. *BIOINFOLET-A Quarterly Journal of Life Sciences, 17*(1b), 190–191. https://doi.org/10.3390/environments8100102

Fathy, M., Saad Eldin, S. M., Naseem, M., Dandekar, T., & Othman, E. M. (2022). Cytokinins: wide-spread signaling hormones from plants to humans with high medical potential. *Nutrients, 14*(7), 1495. https://doi.org/10.3390/nu14071495

Feng, T., Zhang, J., Wang, Y., Wei, D., Sun, J., Yu, H., Tao, X., Mao, X., Hu, Q., & Ji, S. (2022). Purification and identification of thrombolytic peptides from enzymatic hydrolysate of Pheretima vulgaris. *Journal of Food Biochemistry, 19*, Article e14414. https://doi.org/10.1111/jfbc.14414

Fiołka, M. J., Czaplewska, P., Macur, K., Buchwald, T., Kutkowska, J., Paduch, R., Kaczyński, Z., Wydrych, J., & Urbanik-Sypniewska, T. (2019). Anti-Candida albicans effect of the protein-carbohydrate fraction obtained from the coelomic fluid of earthworm Dendrobaena veneta. *PLoS ONE, 14*(3), Article e0212869. https://doi.org/10.1371/journal.pone.0212869

Ghosh, S. (2020). In-silico study of earthworm CCF1 peptides in earthworm. *International Journal of Peptide Research and Therapeutics, 26*, 2213–2224. https://doi.org/10.1007/s10989-020-10014-w

Goodarzi, G., Qujeq, D., Elmi, M. M., Feizi, F., & Fathai, S. (2016). The effect of the glycolipoprotein extract (G-90) from earthworm Eisenia foetida on the wound healing process in alloxan-induced diabetic rats. *Cell Biochemistry and Function, 34*(4), 242–249. https://doi.org/10.1002/cbf.3186

Grdisa, M. (2013). Therapeutic properties of earthworms. *Bioremediation, Biodiversity and Bioavailability, 7*, 1–5.

Grdisa, M., Popovic, M., & Hrzenjak, T. (2001). Glycolipoprotein extract (G-90) from earthworm Eisenia foetida exerts some antioxidative activity. *Comparative Biochemistry and Physiology Part a: Molecular & Integrative Physiology., 128*(4), 821–825. https://doi.org/10.1016/S1095-6433(00)00323-8

Hawk, M. A., McCallister, C., & Schafer, Z. T. (2016). Antioxidant activity during tumor progression: A necessity for the survival of cancer cells? *Cancers, 8*(10), 92. https://doi.org/10.3390/cancers8100092

Hori, M., Kondo, K., Yoshida, T., Konishi, E., & Minami, S. (1974). Studies of antipyretic components in the Japanese earthworm. *Biochemical Pharmacology, 23*(11), 1583–1590. https://doi.org/10.1016/0006-2952(74)90370-0

Hussain, M., Liaqat, I., Hanif, U., Sultan, A., Ara, C., Aftab, N., & Butt, A. (2022). Medicinal perspective of antibacterial bioactive agents in earthworms (Clitellata, Annelida): A comprehensive review. *Journal of Oleo Science, 71*(4), 563–573. https://doi.org/10.5650/jos.ess21379

Jeong, Y. G., Park, S. Y., & Lee, D. W. (2013). Identification of regeneration and anti-wrinkle effect on skin using Jangheung mud and earthworm mucus. *Korean Journal of Aesthetics and Cosmetology, 11*(4), 685–691.

Jinxia, W. U., Jian, W. U., Siqiang, S. U., Hongyue, W. U., & Heying, Z. H. (2015). Regulation of earthworm lyophilized powder on reverse cholesterol transport genes expression in hyperlipidemic mice. *Journal of Hebei University (Natural Science Edition), 35*(2), 159. https://doi.org/10.3969/j.issn.1000-1565.2015.02.009

Jin-xia, W. U., Xing-hang, Z. H., Li-jun, L. I., Yang, X. I., & Peng, Y. I. (2008). Modulating effects of earthworm lyophilized powder on serum lipids in experimental hyperlipidemic mice. *Journal of Hebei University (Natural Science Edition), 28*(6), 652.

Joyia, F. A., Zia, M. A., Mustafa, G., Faheem, A., Raana, H. T., & Khan, M. S. (2018). Isolation, purification and functional characterization of fibrinolytic Protease from an earthworm Eisenia foetida. *Pure and Applied Biology (PAB), 7*(2), 906–909. https://doi.org/10.19045/bspab.2018.700110

Kadlecová, A., Jirsa, T., Novák, O., Kammenga, J., Strnad, M., & Voller, J. (2018). Natural plant hormones cytokinins increase stress resistance and longevity of Caenorhabditis elegans. *Biogerontology, 19*, 109–120. https://doi.org/10.1007/s10522-017-9742-4

Karthick, P. J., Hemavathi, A., Bhuvanasree, M., Angumani, A., & Prabhu, N. (2020). Isolation and characterisation of lumbrokinase from different species of earthworm. *Research Journal of Agriculture and Forestry Sciences, 8*(3), 57–60.

Kawakami, T., Fujikawa, A., Ishiyama, Y., Hosojima, M., Saito, A., Kubota, M., Fujimura, S., & Kadowaki, M. (2016). Protective effect of composite earthworm powder against diabetic complications via increased fibrinolytic function and improvement of lipid metabolism in ZDF rats. *Bioscience, Biotechnology, and Biochemistry., 80*(10), 1980–1989. https://doi.org/10.1080/09168451.2016.1166932

Kour, J., Ali, M. N., Ganaie, H. A., & Tabassum, N. (2017). Amelioration of the cyclophosphamide induced genotoxic damage in mice by the ethanolic extract of Equisetum arvense. *Toxicology Reports, 4*, 226–233. https://doi.org/10.1016/j.toxrep.2017.05.001

Kristianto, H., & Mardiati, N. P. (2017). The effects of earthworm (pheretima aspergillum) ethanol extract toward the improvement of nerve fibers density in diabetic ulcers care degree II of rats wistar. *MNJ (Malang Neurology Journal), 3*(2), 61–72. https://doi.org/10.21776/ub.mnj.2017.003.02.3

Kumar, R., Yadav, R., & Kumar Gupta, R. (2022). A sustainable way for fish health management by replacement of chemical and drugs by earthworm. In *Environmental degradation in Asia: Land degradation, environmental contamination, and human activities* (pp. 329–352). Springer International Publishing. https://doi.org/10.1007/978-3-031-12112-8_16

Kumar, S. S., & Sabu, A. (2019). Fibrinolytic enzymes for thrombolytic therapy. *Therapeutic Enzymes: Function and Clinical Implications, 1148*, 345–381. https://doi.org/10.1007/978-981-13-7709-9_15

Lee, H. C., Hsu, Y. M., Tsai, C. C., Ke, C. J., Yao, C. H., & Chen, Y. S. (2015). Improved peripheral nerve regeneration in streptozotocin-induced diabetic rats by oral lumbrokinase. *The American Journal of Chinese Medicine, 43*(02), 215–230. https://doi.org/10.1142/S0192415X15500147

Lestari, A. A., Sukrama, I. D., & Nurmansyah, D. (2019). The earthworm (Lumbricus Rubellus) extract decreased amino transaminase enzyme level and number of bacterial colony in male wistar rats infected with Salmonella Typhimurium. *Biomedical and Pharmacology Journal, 12*(1), 325–332. https://doi.org/10.13005/bpj/1643

Li, C., Chen, M., Li, X., Yang, M., Wang, Y., & Yang, X. (2017). Purification and function of two analgesic and anti-inflammatory peptides from coelomic fluid of the earthworm, Eisenia foetida. *Peptides, 89*, 71–81. https://doi.org/10.1016/j.peptides.2017.01.016

Li, C., Mao, S., & Kang, B. (2005). Effects of antihypertensive protein from pheretima on Angiotensin II and Angiotensin II AT_1 receptor in spontaneously hypertensive rats. *China Pharmacy, 2005.*

Li, W., Li, S., Zhong, J., Zhu, Z., Liu, J., & Wang, W. (2011a). A novel antimicrobial peptide from skin secretions of the earthworm, Pheretima guillelmi (Michaelsen). *Peptides, 32*(6), 1146–1150. https://doi.org/10.1016/j.peptides.2011.04.015

Li, W., Wang, C., & Sun, Z. (2011b). Vermipharmaceuticals and active proteins isolated from earthworms. *Pedobiologia, 54*, S49-56. https://doi.org/10.1016/j.pedobi.2011.09.014

Liu, H., Wu, Y., Zhu, J., Chen, Y., Xu, Q., Xu, Z., Yang, X., Qiu, S., Qi, G., Li, J., & Sun, Y. (2019). Physiological effects of earthworm extracts in mice. *Chinese Journal of Animal Nutrition, 31*(1), 437–443.

Liu, Y. Q., Sun, Z. J., Wang, C., Li, S. J., & Liu, Y. Z. (2004). Purification of a novel antibacterial short peptide in earthworm Eisenia foetida. *Acta Biochimica Et Biophysica Sinica., 36*(4), 297–302. https://doi.org/10.1093/abbs/36.4.297

Loko, L. E., Medegan Fagla, S., Orobiyi, A., Glinma, B., Toffa, J., Koukoui, O., Djogbenou, L., & Gbaguidi, F. (2019). Traditional knowledge of invertebrates used for medicine and magical–religious purposes by traditional healers and indigenous populations in the Plateau Department,

Republic of Benin. *Journal of Ethnobiology and Ethnomedicine, 15*, 1–21. https://doi.org/10.1186/s13002-019-0344-x

Luo, W., Deng, Z. H., Li, R., Cheng, G., Kotian, R. N., Li, Y. S., & Li, W. P. (2018). Study of analgesic effect of earthworm extract. *Bioscience Reports, 38*(1), BSR20171554. https://doi.org/10.1042/BSR20171554

Malik, M. F., & Afsheen, S. (2021). Isolation and purification of thrombolytic enzyme extracted from earthworm Punjab, Pakistan. *Abasyn Journal of Life Science, 4*(1), 127–135.

Mao, S., & Li, C. (2015). Hypotensive and angiotensin-converting enzyme inhibitory activities of Eisenia fetida extract in spontaneously hypertensive rats. *Evidence-Based Complementary and Alternative Medicine, 2015*(349721). https://doi.org/10.1155/2015/349721

Mathur, A., Verma, S. K., Bhat, R., Singh, S. K., Prakash, A., Prasad, G. B., & Dua, V. K. (2010). Antimicrobial activity of earthworm extracts. *Journal of Chemical and Pharmaceutical Research, 2*(4), 364–370.

Mir, M. A., Upadhay, S., & Mir, B. A. (2018). Inhibition of alpha amylase and alpha glycosidase enzymes by various earth worm extracts. *Biomedical and Pharmacology Journal, 11*(3), 1261–1268. https://doi.org/10.13005/bpj/1487

Moon, B. C., & Kim, J. S. (2018). The potential of earthworm and its components as a therapeutic agent for neuronal damage. *Journal of Biomedical Translational Research, 19*(3), 58–64. https://doi.org/10.12729/jbtr.2018.19.3.058

Mukherjee, M., Datta, M., Biswas, S., Pal, A. K., Malakar, D., Bhattacharyya, A. K., Bhattacharya, S., & Kobayashi, H. (2003). Immotilin, a novel sperm immobilizing protein. *Fertility and Sterility, 79*(Suppl 3), 1673–1675. https://doi.org/10.1016/S0015-0282(03)00371-6

Mustafa, R. G., Saiqa, A., Domínguez, J., Jamil, M., Manzoor, S., Wazir, S., Shaheen, B., Parveen, A., Khan, R., Ali, S., & Ali, N. M. (2022) Therapeutic values of earthworm species extract from Azad Kashmir as anticoagulant, antibacterial, and antioxidant agents. *Canadian Journal of Infectious Diseases and Medical Microbiology, 2022* (6949117). https://doi.org/10.1155/2022/6949117

Nalurika, A. (2015). *Pengaruh Pemberian Minyak Cacing Tanah (Lumbricus Rubellus) Terhadap Kadar Malondialdehida (Mda) Dan Gambaran Histopatologi Duodenum Pada Tikus (Rattus Novergicus) Model Hiperlipidemia Dengan Indu* [Doctoral dissertation, Universitas Brawijaya].

Ogasawara, M., Yoshii, K., Wada, J., Yamamoto, Y., & Inouye, K. (2020). Identification of guanine, guanosine, and inosine for α-amylase inhibitors in the extracts of the earthworm Eisenia fetida and characterization of their inhibitory activities against porcine pancreatic α-amylase. *Enyzme and Microbial Technology, 142*, Article 109693. https://doi.org/10.1016/j.enzmictec.2020.109693

Oumi, T., Ukena, K., Matsushima, O., Ikeda, T., Fujita, T., Minakata, H., & Nomoto, K. (1995). The GGNG peptides: Novel myoactive peptides isolated from the gut and the whole body of the earthworms. *Biochemical and Biophysical Research Communications, 216*(3), 1072–1078. https://doi.org/10.1006/bbrc.1995.2730

Park, Y. R., Park, C. I., & Soh, Y. (2020). *Antioxidant and Anti-Inflammatory Effects of NCW Peptide from Clam Worm (Marphysa Sanguinea)., 30*(9), 1387–1394. https://doi.org/10.4014/jmb.2003.03050

Pinky, D., Yadav, Y., & Shukla, V. (2020). Evaluation of antioxidant activity of Eisenia fetida. *International Journal of Biological Innovations, 2*(2), 109–116. https://doi.org/10.46505/IJBI.2020.2205

Plavšin, I., Velki, M., Ečimović, S., Vrandečić, K., & Ćosić, J. (2017). Inhibitory effect of earthworm coelomic fluid on growth of the plant parasitic fungus Fusarium oxysporum. *European Journal of Soil Biology, 78*, 1–6. https://doi.org/10.1016/j.ejsobi.2016.11.004

Puga-Freitas, R., Barot, S., Taconnat, L., Renou, J. P., & Blouin, M. (2012). Signal molecules mediate the impact of the earthworm Aporrectodea caliginosa on growth, development and defence of the plant Arabidopsis thaliana. *PLoS ONE, 7*(12), Article e49504. https://doi.org/10.1371/journal.pone.0049504

Radina, D. F., & Adi, A. C. (2020). Factors causing Bangkalan communities to consume earthworm boiled water (drugs/traditional remedies) for typhoid handling. *Indian Journal of Public Health Research and Development, 11*(4), 576–580.

Rochma, O. (2016). *Pengaruh Terapi Minyak Cacing Tanah (Lumbricus Rubellus) Terhadap Aktivitas Enzim Superoksida Dismutase Dan Gambaran Histopatologi Aorta Abdominalis Pada Tikus Putih (Rattus Novergicus) Model Hiper* [Doctoral dissertation, Universitas Brawijaya].

Rowaiye, A., Wilfred, O. I., Onuh, O. A., Bur, D., Oni, S., Nwonu, E. J., Ibeanu, G., Oli, A. N., & Wood, T. T. (2022). Modulatory effects of mushrooms on the inflammatory signaling pathways and pro-inflammatory mediators. *Clinical Complementary Medicine and Pharmacology, 2*(4), Article 100037. https://doi.org/10.1016/j.ccmp.2022.100037

Rowaiye, A. B., Onuh, O. A., Oli, A. N., Okpalefe, O. A., Oni, S., & Nwankwo, E. J. (2020) The pandemic COVID-19: A tale of viremia, cellular oxidation and immune dysfunction. *Pan African Medical Journal, 36*(1). https://doi.org/10.11604/pamj.2020.36.188.23476

Samatra, D. P., GB, M. T., Sukrama, I. D., Dewi, N. W., Praja, R. K., & Nurmansyah, D. (2017). Extract of earthworms (Lumbricus rubellus) reduced malondialdehyde and 8-hydroxy-deoxyguanosine level in male wistar rats infected by salmonella typhi. *Biomedical and Pharmacology Journal, 10*(4), 1765–1771. https://doi.org/10.13005/bpj/1290

Seo, M., Lee, J. H., Baek, M., Kim, M. A., Ahn, M. Y., Kim, S. H., Yun, E. Y., & Hwang, J. S. (2017). A novel role for earthworm peptide Lumbricusin as a regulator of neuroinflammation. *Biochemical and Biophysical Research Communications., 490*(3), 1004–1010. https://doi.org/10.1016/j.bbrc.2017.06.154

Sethulakshmi, K. C., Ranilakshmi, K. C., & Thomas, A. P. (2018). Antibacterial and antifungal potentialities of earthworm Eudrilus eugeniae paste and coelomic fluid. *Asian Journal of Biology, 5*, 2456–7124.

Shokouh, M. R., Torshizi, M. A., & Alizadeh, A. R. (2018). Evaluation of the effects of earthworm (Eisenia fetida) meal and ethanol extract on antioxidant status, liver enzymes and some blood parameters in Western Azarbaijan native roosters. *Iranian Journal of Animal Science Research, 10*(1), Pe47–Pe59.

Sun, Z. (2015). Earthworm as a biopharmaceutical: From traditional to precise. *European Journal of BioMedical Research, 1*(2), 28–35.

Sun, Z., & Jiang, H. (2017). Nutritive evaluation of earthworms as human food. *Future Food, 37*. https://doi.org/10.5772/intechopen.70271

Tait, S. W., & Green, D. R. (2013). Mitochondrial regulation of cell death. *Cold Spring Harbor Perspectives in Biology, 5*(9), Article a008706. https://doi.org/10.1101/cshperspect.a008706

Terzić, J., Grivennikov, S., Karin, E., & Karin, M. (2010). Inflammation and colon cancer. *Gastroenterology, 138*(6), 2101–2114. e.5. https://doi.org/10.1053/j.gastro.2010.01.058

Verma, M. K., & Pulicherla, K. K. (2017). Broad substrate affinity and catalytic diversity of fibrinolytic enzyme from Pheretima posthumous—purification and molecular characterization study. *International Journal of Biological Macromolecules, 95*, 1011–1021. https://doi.org/10.1016/j.ijbiomac.2016.10.090

Wang, X., Fan, S., & He, R. (2018). Earthworm proteases multiply function in protein fibrils degradation involving several diseases including Hepatitis B. In 中国生物化学与分子生物学会第十二届全国会员代表大会暨 2018 年全国学术会议摘要集 2018.

Wang, X. M., Fan, S. C., Chen, Y., Ma, X. F., & He, R. Q. (2019). Earthworm protease in anti-thrombosis and anti-fibrosis. *Biochimica et Biophysica Acta (BBA)-General Subjects, 1863*(2), 379–383. https://doi.org/10.1016/j.bbagen.2018.11.006

Wei, S., Yin, X., Kou, Y., & Jiang, B. (2009). Lumbricus extract promotes the regeneration of injured peripheral nerve in rats. *Journal of Ethnopharmacology, 123*(1), 51–54. https://doi.org/10.1016/j.jep.2009.02.030

Wie, M. B., Jung, J. S., Song, D. K., & Kim, Y. H. (1992). Lamnodrilus gotai extract lowers blood pressure and spontaneous motor activity in spontaneously hypertensive rats. *Planta Medica, 58*(S 1), 644–645. https://doi.org/10.1055/s-2006-961651

Xu, Q., Qi, G., Jiang, Y., Ni, K., Ya-Ying, C., Liu, H., Zhu, J. W., & Wu, Y. (2020). Effects of earthworm extract on the oxidative damage, total antioxidant capacity and antioxidant enzyme contents in the midgut of the silkworm, Bombyx mori. *Acta Entomologica Sinica, 63*(3), 278–284. https://doi.org/10.16380/j.kcxb.2020.03.004

Yamaji-Hasegawa, A., Makino, A., Baba, T., Senoh, Y., Kimura-Suda, H., Sato, S. B., Terada, N., Ohno, S., Kiyokawa, E., Umeda, M., & Kobayashi, T. (2003). Oligomerization and pore formation of a sphingomyelin-specific toxin, lysenin. *Journal of Biological Chemistry, 278*(25), 22762–22770. https://doi.org/10.1074/jbc.M213209200

Yoshii, K., Ogasawara, M., Wada, J., Yamamoto, Y., & Inouye, K. (2020). Exploration of dipeptidyl-peptidase IV (DPP IV) inhibitors in a low-molecular mass extract of the earthworm Eisenia fetida and identification of the inhibitors as amino acids like methionine, leucine, histidine, and isoleucine. *Enyzme and Microbial Technology, 137*, Article 109534. https://doi.org/10.1016/j.enzmictec.2020.109534

Zhang, Z. X., & Wang, F. F. (1992). Effects of crude extract of earthworm on promoting blood circulation to removing stasis. Zhongguo Zhong xi yi jie he za zhi Zhongguo Zhongxiyi Jiehe Zazhi= *Chinese Journal of Integrated Traditional and Western Medicine, 12*(12), 741–743.

Vermitechnology—Application in Environmental Sustainability

Assisted Phytoremediation of Trace Toxic Elements (TTE): The Role of Vermicompost

S. A. Aransiola, S. S. Leh-Togi Zobeashia, A. E. Oyewumi, O. P. Abioye, U. J. J. Ijah, and Naga Raju Maddela

Abstract Several anthropogenic activities, such as the use of pesticides, hazardous chemicals, industrial effluents, and metals, cause environmental pollution. The health of plants, animals, and people is seriously threatened by pollution, which affects both aquatic and terrestrial environments. The conventional methods for cleaning up contaminants from soil and water are seen to be ineffective, costly, and detrimental to the environment. The effect of trace toxic elements (TTE) can be mitigated by a process referred to as phytoremediation. It involves the use of green plants to remove pollutants such as trace toxic elements from contaminated sites, and the method is also used to eliminate toxins from the environment. Phytoremediation, which is environmentally friendly could be difficult to achieve many a times because of the level of pollutions in the surrounding environment. Therefore, plants could be assisted by products like vermicast. Vermicasts are produced from vermicomposting process which involves the use of earthworms in degrading organic substances and turning them into useful products. The accumulation and tolerance capacities of plants used in phytoremediation can be greatly improved with advancements in vermicomposting. This paper reviews phytoremediation techniques for TTE with consideration of vermicompost in an assisted capacity.

Keywords Phytoremediation · Trace toxic elements · Vermicompost · Contamination · Pollution · Soil

S. A. Aransiola (✉)
Department of Microbiology, University of Abuja, Abuja, Nigeria
e-mail: blessedabiodun@gmail.com

Bioresources Development Centre, National Biotechnology Development Agency, Ogbomoso, Nigeria

S. S. Leh-Togi Zobeashia
National Biotechnology Development Agency, Abuja, Nigeria

A. E. Oyewumi · O. P. Abioye · U. J. J. Ijah
Department of Microbiology, Federal University of Technology, Minna, Nigeria

N. R. Maddela
Department of Biological Sciences, Faculty of Health Sciences, Universidad Técnica de Manabí, Portoviejo, Ecuador

S. A. Aransiola and N. R. Maddela (eds.), *Vermitechnology: Economic, Environmental and Agricultural Sustainability*, https://doi.org/10.1007/978-3-031-96212-7_6

1 Introduction

The main environmental pollutants are trace toxic elements (TTE), which because of their protracted environmental persistence seriously endanger both human and animal health (Subhashini and Swamy). The remediation of trace toxic element-contaminated soils is an expensive and technically challenging process (Montinaro et al., 2012; Xu & Lu, 2012). Biological, physical, and chemical techniques are the basis of traditional remediation approaches, which can be combined to reduce contamination to a level that is safe and acceptable (Aransiola & Maddela, 2024; Jadia & Fulekar, 2009). Although effective, these strategies are costly, time-consuming, and environmentally harmful (Ahmadpour et al., 2012). They also typically produce significant amounts of waste and are detrimental to the natural soil ecosystem (Cunningham et al., 1995). Due to its long-term relevance, cost-effectiveness, and benefits attached to it, phytoremediation is a promising method and should be taken into consideration for the remediation of contaminated sites (Moosavi & Seghatoleslami, 2013).

The crust of the planet naturally contains trace toxic elements (Ismail et al., 2013). They have an atomic density of more than 5 g/cm^3 (Alloway & Ayres, 1997) and an atomic number of more than 20, which are their main properties. Cd, Cr, Cu, Hg, Pb, and Zn are the heavy metal pollutants that are most frequently found (Aransiola et al., 2021). Metals whose fraction in the composition of rocks is less than or equal to 0.1% are considered trace elements from a geochemical perspective (Zovko & Romic, 2011). Steps are required to reduce the harmful impacts of pollution due to the increasing burden of contaminants. Regarding the selection of plants for an efficient and cost-effective phytoremediation approach in contaminated environments, knowledge of the potential of various plants to absorb, accumulate, and translocate metals under diverse conditions is vital. Because both soil contaminants and soil minerals possess small electrical charges that cause them to bind with one another, repairing polluted soil may also be extremely difficult (Araansiola et al., 2024; Daza et al., 2024). It is generally known that trace toxic elements preferably physically removed or immobilized because they cannot be destroyed by chemicals. (Kroopnick, 1994). In the past, treating trace toxic element-contaminated soils on the spot or excavating them before disposing of them at a lowland location were the two main methods of remediation (Parker, 1994). However, this technique of disposal only moves the negative effects of pollution elsewhere. An alternative to excavation and disposal in lowlands for the removal of contaminated soil is soil washing. This procedure is both expensive and results in significant metal-laden waste that may require further processing or burial. Furthermore, because they eliminate all biological activity, these physicochemical techniques for soil remediation make the land unusable as a medium for plant growth. On the other hand, other techniques, such as vitrification, leaching, electrokinetic soil vapor extraction, thermal natural process, chemical process, etc., need a lot of labor and require a lot of upkeep (Danh et al., 2009; Haque et al., 2008).

Fig. 1 Enhanced phytoremediation of trace toxic element (TTE) with vermicast

Phytoremediation involves the use of plants for the removal of trace toxic elements from the soil, and it can be achieved either by in situ or ex situ methods of phytoremediation (Fig. 1). In situ method of phytoremediation is more often used because it reduces the penetration of contaminants into airborne waste and into aquatic and terrestrial environments, which automatically reduces the risk to the neighboring environment (Aransiola et al., 2024; Raskin & Ensley, 2000).

Compared to other remediation methods, phytoremediation seems to be economically effective after treatment (Cristaldi et al., 2017), as it is a simple, nonlaborious technique requiring no installation of special equipment. Where other regularly used approaches are ineffective and costly, the process can be used to a great extent (Leguizamo et al., 2017). A recent analysis of the effectiveness of hyperaccumulator plants sparked an additional investigation into the molecular basis of phytoremediation (Abioye et al., 2017). Phytoremediation assisted by vermicompost is novel research where plants have enough strength to withstand the effects of contaminants (Aransiola et al., 2022a). This chapter, therefore, reviews the role of vermicompost in the phytoremediation of TTE in the environments.

2 Environmental Contamination by Trace Toxic Elements

Trace toxic elements have significant effects on both the aquatic and terrestrial ecosystem. Trace toxic elements pollute natural water bodies, such as oceans and seas, as well as sediments and soils, after being released from both anthropogenic and natural sources. When trace toxic elements are discharged into the environment during volcanic eruptions and through industrial pollutants, they eventually come down to earth and contaminate the soil and water. Because they linger in the environment, trace toxic elements can build up in biota or seep into groundwater (Aransiola et al., 2013).

Potentially harmful trace toxic element contamination of biota and groundwater has significant effects on human health. Investigating the amounts and distribution of these elements helps determine the extent of trace toxic element pollution in aquatic environments, sediments, and soils (Islam et al., 2018).

3 New Innovations in Phytoremediation of Contaminants (TTE)

Increased accumulation of toxic trace elements resulting from industrial and natural processes have become a serious environmental issue worldwide, which is due to their negative impacts on food chain and toxic effects on ecosystems and human health. The use of phytoremediation to mitigate toxic trace elements is a sustainable, environmentally friendly, and cost-effective process used to treat and curtail environmental contaminants. To enhance and assist the process of phytoremediation, new efficient innovations such as genetic engineering and soil microbial symbiotic interactions, among others, are been employed.

3.1 Genetic Engineering

Genetic engineering, also known as recombinant DNA technology, is a promising innovation used to enhance plant's ability to remediate toxic trace elements in polluted environments. Although, the technology is an advanced method of traditional plant breeding (e.g., crossing) that involves the selection of wild plants that are able to survive and grow in toxic trace element-contaminated soil (Roccotiello et al., 2015). However, genetic engineering genetically modifies plants by selecting toxic trace element-accumulating and tolerant gene from wild hyperaccumulator plant species or organisms and incorporating the gene into the genome of a targeted plant species with increased biomass, which confers the plant species with desirable traits for enhanced phytoremediation (Dushenkov et al., 2002). The modified plants (transgenic plants) are conferred with the ability to accumulate, uptake, tolerate, and transport toxic trace elements. Poplar plant species has been genetically modified for the recovery of TTE. Lyyra et al. (2007) reported improved phytoremediation of mercury from a contaminated environment by genetically coupling the bacterial merA (mercuric ion reductase that reduces Hg^{2+} to less toxic Hg^0, which can be volatilized easily by plants) and merB (organomercury lyase that converts organic Hg to Hg^{2+}) genes isolated from *Escherichia coli* and incorporating them into eastern cottonwood trees (*Populus deltoides* Bartr. ex Marsh.). The modified merA/merB plants showed the ability to detoxify organic mercury and were further resistant to phenylmercuric acetate than the undomesticated type.

To enhance the phytoremediation of TTE accumulation, genetic engineering involves the over expression and introduction of genes that can not only uptake TTE but also translocate and sequestrate TTE. Recently, most genetically engineered phytoremediation methods make use of genes that encode transporters of toxic trace elements. These genes, encoding TTE transporters such as metal tolerance proteins (MTPs), zinc–iron permeases (ZIPs), heavy metal ATPases (HMAs), and multidrug and toxin extrusion proteins (MATEs), are transferred and overexpressed in target plants to improve the accumulation of TTE. As metal chelators, they act as metal-binding ligands to increase the bioavailability of TTE, promote the uptake of TTE, improve root-to-shoot translocation, and also facilitate intracellular sequestration of TTE in organelles. For instance, transgenic poplar tree genetically engineered with *Saccharomyces cerevisiae* yeast cadmium factor 1 (yeast ScYCF1 gene) encodes vacuolar transporter involved in sequestration of toxic trace elements into vacuole. The use of transgenic poplar plants in TTE-contaminated soil from a mining site in South Korea indicated a decreased cadmium (Cd) toxicity and higher accumulation of Cd in contrast to wild plants. When plants were tested, the dry weight showed higher accumulation of zinc, lead, and cadmium in the transgenic root compared to the wild type, which therefore demonstrates the potential use for phytostabilization and phytoextraction (Fasani et al., 2018; Shim et al., 2013).

Clustered regularly interspaced short palindromic repeats (CRISPR)-Cas9 system is a genetically engineered tool used to introduce an extensive variety of genes in host organisms. It is a gene-editing tool applied in phytoremediation to improve the transfer and expression of a desired set of genes in the genome of plants for effective phytoremediation of TTE-contaminated sites. For instance, a study by Tang et al. (2017) reported a knockout for the TTE transporter gene OsNramp5 using the CRISPR/Cas9 system, generated low Cd-accumulating indica rice without compromising yield.

3.2 Interactions Between Plants and Microorganisms

The interaction of plants and associated microbes, especially those bacteria that inhabit the roots, has recently been studied for phytostabilization and extraction of toxic elements. Research has shown that plant–microbial interaction has the potential to not only enhance the efficiency of plant tolerance, uptake, and translocation of toxic elements but also promote the growth and fitness of plants and protect the plants from pathogenic organisms (Ma et al., 2011; Sessitsch et al., 2013).

The symbiotic relationship between microbes and plant roots can be applied to improve phytoremediation processes such as phytoextraction (for instance, plant growth-promoting rhizo- and endobacteria) and phytostabilization (e.g., arbuscular mycorrhizal fungi). The presence of arbuscular mycorrhizal fungi (AMF) in the rhizosphere increases the adsorptive surface area of the plant roots, which is solely due to their hyphal network and their ability to produce phytohormones that can assist phytoremediation (Göhre & Paszkowski, 2006; Vamerali et al., 2010). Arbuscular mycorrhizae form a synergy with other higher plants and advance the uptake and availability of nutrients by enhancing the texture of the soil via steady aggregation of the particles of the soil by binding the trace elements into the plant roots, preventing the translocation of the trace elements to the shoot tissues (Hassan et al., 2011).

Ker and Charest (2010) reported the remediation of nickel by AMF-colonized sunflower plants. Whereas, plant growth-promoting rhizobacteria and endophytes (PGPR and PGPE) can increase the remediation ability of plants by advancing the plant growth in the presence of TTE through the production of certain compounds such as siderophores, organic acids, antibiotics, phytohormones, indole-3-acetic acids, and enzymes. For example, PGPR can synthesize 1-aminocyclopropane-1-carboxylate (ACC) deaminase which can break down ethylene precursor ACC by lowering ethylene production, thereby promoting plant growth (Ma et al., 2011).

In toxic trace element-contaminated soil, toxic elements attach to the particles of the soil, which restricts their uptake by plants. However, the interaction between plant growth-promoting rhizo- and endobacteria, like *Pseudomonas* spp., *Bacillus, Microbacterium*, *Arthrobacter*, among others, aids in the solubilization of water-insoluble toxic elements (copper, zinc, and nickel) by means of soil acidification via the secretion of proton or organic anions such as lactate, succinate, gluconate acetate, glycolate, citrate, and ketogluconate. Becerra-Castro et al. (2011) reported the solubilization competence of nickel and the characteristic features of rhizobacteria isolated from hyperaccumulating and non-hyperaccumulating subspecies of *Alyssum serpyllifolium.*

The availability of trace elements can be improved by using PGPR that are able to produce biosurfactants, which help release TTE from particles of the soil. For instance, PDPR synthesize low molecular weight siderophores, which are chelators that accelerate the uptake of iron in an iron-reduced condition, thereby making it available for the microbes and plants (Gamalero & Glick, 2012; Schalk et al., 2011; Van Ginneken et al., 2007). They are also capable of chelating other toxic elements like lead, nickel, copper, zinc, magnesium, manganese, chromium, and cadmium to varying extents (Ikhumetse et al., 2019). Kumar et al. (2008) reported an increased production of biomass and phytoremediation (extraction) of toxic elements (chromium, nickel, and zinc) resulting from improved production of siderophore when *Brassica juncea* plants were inoculated with plant growth-promoting bacteria of the phosphate-solubilizing *Enterobacter* sp. and its mutant.

The interaction can also result in the production of bacterial auxin by PGPR to stimulate lateral root initiation, root hair development, and root cell proliferation. Therefore, the bacterial auxin enables plants to adapt to TTE-contaminated soil through physiological changes in the cell metabolism of the plants, thereby allowing the plants adapt to TTE stress and higher TTE concentrations (Glick, 2010). Most microbes used to improve phytoremediation are TTE-tolerant species linked with hyperaccumulator plants that can tolerant and grow in the presence of TTE. Plant growth-promoting rhizobacteria and endophytes isolated from TTE-tolerant plants have been used as bioinocula for remediation of toxic trace element-contaminated soil. Endophytic *Rahnella* sp. JN6 isolated from *Polygonum pubescens* has the potential to advance growth and enhance the uptake of Pb, Cd, and Zn by *Brassica napus* (He et al., 2013; Sheng et al., 2008; Visioli et al., 2015).

3.3 Alter Growth Conditions

Although this process is applied in wastewater treatment, growth bed materials such as constructed wetlands can also absorb and retain different TTE (Marchand et al., 2010). This approach depends on associated microbiota and rooted hydrophytes to adsorb and get rid of TTE via rhizofiltration, phytostabilization, and phytoextraction. Rezania et al. (2016) demonstrated the potential of different free-floating macrophyte aquatic plant species, such as *Eichhornia crassipes* and *Lemna* spp., among others to remove TTE. However, the major setback of this remediation approach is the massive invasive growth of these aquatic macrophyte species (Newete & Byrne, 2016), which has propelled the ample use of rooted hydrophyte plant species such as *Typha* spp., *Phragmites* spp., among others and also species that can tolerate flooding (*Chrysopogon zizanioides*) for creation of flow wetland which is important for TTE removal. As reported by Bavandpour et al. (2018), crushed seashell grits and plant biomass used in a wetland column almost completely removed dissolved toxic elements.

Another successful advance to alter the growth condition in phytoremediation is the use of wastewater, but it has a major drawback of potential environmental impacts and groundwater contamination. This can be remedied by the use of plant covers which prevent TTE from spreading and leaching to groundwater and also remediate TTE by detoxification and uptake (Aronsson et al., 2010; Burges et al., 2018). Phytoremediation of TTE-polluted sites can also be assisted by organic alterations which include addition of waste compost to enhance the growth of plant and alter the solubility and bioavailability of TTE (Aransiola et al., 2019; Burges et al., 2018). Karczewska et al. (2017) demonstrated the solubility and uptake of arsenic by ryegrass from a contaminated soil augmented with sewage sludge, cow manure, and litter from the forest, while Alvarenga et al. (2009) reported the immobilization of Cu, Pb, and Zn when municipal waste and sewage sludge were applied on trace toxic element-polluted soils, which helped phytostabilization.

4 Increase TTE Bioavailability

Another potential innovative approach of phytoremediation is to increase the bioavailability of TTE (Fig. 2). Toxic trace elements in soil are rarely available for absorption by plants, as only a small amount of TTE is present as soluble component in the soil for bioaccumulation by plants. For example, cadmium and zinc are readily available for absorption by plants (Blaylock & Huang, 2000; Lasat, 1999) Bioavailability of TTE in soil differs; some trace elements are readily available (copper, cadmium, nickel, zinc, among others) while others are only moderately or poorly accessible. Low bioavailability of lead can prevent the uptake of trace elements from the soil and thereby decreasing phytoremediation. Increase in bioavailability is determined by inherent solubility, physiochemical properties of the soil such as pH of the soil, chelating agents, and microbial activities (Wang et al., 2006). Plants can increase bioavailability of TTE by lowering the pH of the soil; this is obtainable through acidification of the rhizosphere by root exudates, which results in the formation of free ions by release of toxic elements from insoluble complexes. Plants also produce mobilizing compounds in the rhizosphere (carboxylates) that can affect the soil and trigger toxic element chelation which can enhance solubility, bioavailability, and mobility; therefore, these chelating agents assist to increase the uptake and translocation of TTE (Padmavathiamma & Li, 2012). These chelating agents can be applied to soil to increase bioavailability of TTE due to the formation of water-soluble toxic element–chelate complexes, which are mobile and easily absorbed by plants (Wuana & Okieimen, 2011). Sarwar et al. (2017) demonstrated how different synthetic chelating agents (ethylenediaminetetraacetic acid and diethylenetriamine-pentaacetic acid) and organic chelating agents (malic acid, citric acid, oxalic acid, and acetic acid) can be used to modify soil and increase TTE bioavailability, thereby facilitating effective phytoremediation.

5 The Role of Vermicompost in the Phytoremediation of Trace Toxic Elements (TTE)

Vermicompost, also known as natural fertilizer, is a developing green innovation produced from the biodegradation of organic wastes by the activities of earthworms that results in stabilization and biooxidation of the wastes. Vermicompost is a vital source of nutrients, immobilized microflora, enzymes, antibiotics, and growth hormones such as gibberellin that helps to control plant and microbial growth (Aransiola et al., 2022a, 2022b; Babaniyi et al., 2023). Vermicompost is used as a conditioner or fertilizer for soil due to high nutrient content and beneficial microbes (Suthar et al., 2005). As shown by Jadia and Fulekar (2008), the application of vermicompost on contaminated soil to remove TTE (zinc, cadmium, copper, nickel, and lead) by sunflower plant improved soil physical properties and fertility, in addition to

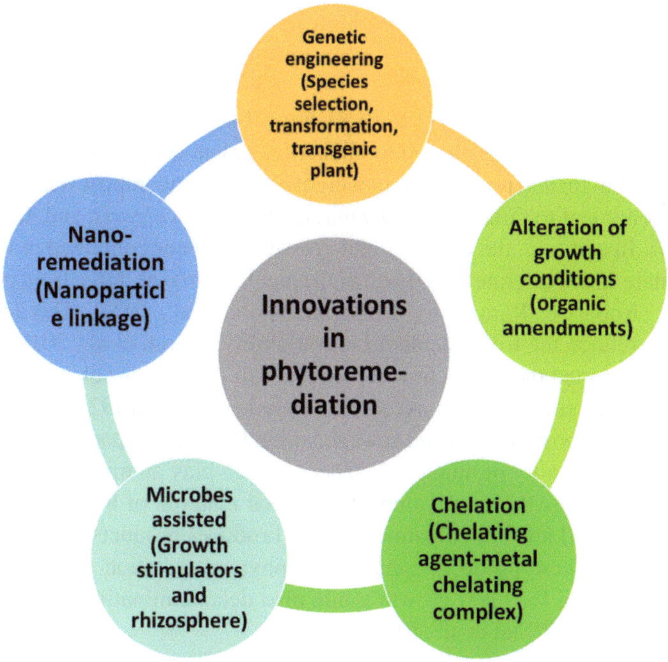

Fig. 2 Schematic diagram showing current innovations employed in phytoremediation

facilitating efficient phytoremediation. The use of vermicompost in phytoremediation also helps reduce the amount of organic waste in the environment. Furthermore, it improves the quality of the growing plants and increases the production of plant biomass, thereby enhancing plant growth which thus suggests better tolerance and uptake of toxic elements (Suthar et al., 2005).

5.1 Increase Plant Growth

Vermicompost is an organic material used to increase the productivity and growth rate of plant species used in phytoremediation which is a process of remediation using plants and soil microorganisms to minimize the negative effects of toxic elements in the soil (Ali et al., 2013). The vital role of vermicompost, when applied in a toxic element-polluted soil, is to enhance the soil fertility through physical (e.g., water retention, porosity), biological (enzymes, biomass), and chemical (organic content, pH) properties that can boost the growth rate and remediation potential of the plant species used for phytoremediation (Prabha, 2009; Suthar et al., 2005; Tejada et al., 2010). This is achieved by increasing the speed of organic material utilization by the plants and also by improving the amount of humification (Garg & Gupta, 2011; Jadia & Fulekar, 2008). Thus, vermicompost increases the quality of the soil by managing the soil nutrients; they are high in nutrients and retain nutrient in the soil without negative effects on the environment (Padmavathiamma et al., 2008; Pattnaik & Reddy, 2010).

When vermicompost is used for soil amendment in phytoremediation, they promote the overall growth and development of the plant. Roy et al. (2010) reported a surge in plant height and root and shoot weight when vermicompost was applied as an organic amendment to three crops (*Zea mays, Phaseolus vulgaris*, and *Abelmoschus esculentus*). In addition, the enzymes and growth hormones contained in vermicompost stimulate and accelerate the growth of the plant and makes it free from pathogenic organisms (Abbasi & Ramasamy, 1999; Aransiola et al., 2023). The activities of the enzyme reduce the toxic element bioavailability (Garau et al., 2019). Muscolo et al. (1999) reported the production of plant growth-promoting substances and plant growth hormones such as auxin. According to Senesi et al. (1992) and Garcia et al. (1995), vermicompost gotten from sewage, animal dung, and paper industry sludge consists of high quantity of humic substances, which play an important role in plant productivity and growth. Vermicompost is a good material of homogeneous nature due to its decreased level of contaminants and produces products such as vermicast and vermiwash. These products complement phytoremediation with nutrients free from contamination for plant growth while also decontaminating the environment from toxic elements (Bhat et al., 2016).

5.2 *Nutrient Cycling*

Vermicompost is a nutrient-rich organic amendment, high in both macro- and micronutrients, used to regenerate, prevent, and remediate TTE-polluted soils (Bhat et al., 2013; Zhang et al., 2022). When applied in phytoremediation of a TTE-polluted soil, it enhances nutrient cycling, that is, the mineralization of organic nutrients and toxic elements into forms utilizable by both microbes and plants. The nutrient provided can increase soil fertility, plant growth, and productivity, thus improving the uptake and tolerance of TTE by plants (Jadia & Fulekar, 2008; Tejada et al., 2010). Vermicompost increases the availability of potassium, nitrogen, phosphorus, calcium, iron, copper, sodium, manganese, zinc, magnesium, among others (Abioye et al., 2018; Manivannan et al., 2009), through direct and indirect effects on the soil microbiome. The release of nutrients such as nitrogen and organic carbon resulting from application of vermicompost to enhance phytoremediation facilitates stabilization and bioaccumulation by plants. For example, the earthworms in vermicompost stabilize organic matter via incorporation and protection in their casts (Bossuyt et al., 2005). These casts are abundant in plant nutrients, when disposed in the soil bind with microbial products and mucilage of earthworms to form very stable aggregates. Studies have shown, once the organic matter in the casts is stabilized, it can retain its stabilization for years (Mariani et al., 2007; Shipitalo & Protz, 1989). The earthworms also enhance mineralization by fragmentation of the organic matter and then mixing it with microbes and mineral particles which leads to the creation of large surface area for contact for both microbes and organic matter. Their contributive effect in the concentration of high amount of nutrient that are easily integrated by plants and used by microorganisms in the soil to enhance their activity (Bhadauria & Ramakrishnan, 1989; Parmelee et al., 1998).

5.3 *Immobilization of Toxic Elements*

Vermicompost also functions as a TTE immobilizer when used as a bio-conditioner in contaminated soils, which is attributed to high cation-exchange ability, functional groups, and large surface area resulting from biodegradation and mineralization by earthworms (Wang et al., 2018a, 2018b, 2018c). Vermicompost changes the TTE speciation, decreases the bioavailability and solubility of TTE, and also modifies the redox position of the soil (Burges et al., 2018). When applied in a TTE-polluted soil, it increases organic matter content and the availability of macro- and micronutrients such as nitrogen, potassium, phosphorus, calcium, magnesium, manganese, iron, zinc, copper, and sodium in the soil (Manivannan et al., 2009), and also improves soil properties which can help in the colonization of the plant and can also enhance water-holding capacity of the soil.

In addition, vermicompost is rich in microbial community composition which is an important indicator of the impact of vermicompost on TTE. Studies by Kelly et al. (2014) and Wang et al. (2015) showed that the use of vermicompost as bio-conditioner can significantly increase enzyme activity, influence soil microbial community structure and diversity, and result in sorption of toxic elements. For instance, microbial diversity in the rhizosphere (mycorrhiza and bacteria) can contribute in immobilization of TTE by the adsorption of toxic elements in their cell walls, thereby advancing the production of chelators and precipitation processes, which lessen their toxicity and limit their bioavailability (Dalvi & Bhalerao, 2013). In toxic element-polluted soil, vermicompost application increases the plant root surface area and depth to enable immobilization of TTE and also functions as a filtration barrier against the toxic element ion translocation from the roots to the shoots. Several researchers have shown the potential of vermicompost to immobilize toxic elements in polluted soil (Wang et al., 2018d; Zhang et al., 2019a, 2019b).

Another role of vermicompost in phytoremediation is the improvement of enzymatic activity in both the soil and the gut of earthworms by reducing the TTE bioavailability and blocking the mobility of the toxic elements by altering the toxic elements, that is, converting them to less toxic forms, thus immobilizing them in the soil and preventing the contamination of food chain, groundwater, among others (Eapen & Dsouza, 2005; Garau et al., 2019). For example, vermicompost contains extracellular detoxifying redox enzymes (dehydrogenase, peroxidase, carboxylesterase, nitroreductase, laccase, etc.) excreted in the plant rhizosphere. These enzymes can convert Cr(VI), which is toxic and bioavailable in soil, to Cr(III), thereby decreasing the toxic effects and mobility (Aransiola et al., 2022a, 2022b; Garau et al., 2019; Jabeen et al., 2009; Wu et al., 2010).

5.4 Bioaccumulation of TTE

Vermicompost is produced from biodegradative interaction of microbes and earthworms with organic matter. The earthworms themselves are bioaccumulators and can be used to remediate toxic elements (Sinkakarimi et al., 2020). Studies have shown that the concentration of TTE can be reduced via accumulation by earthworms (Swati & Hait, 2017; Richardson et al., 2017, 2020). Azhar-u-ddin et al. (2020) also reported an improved selenium uptake and mobility in bean plant *P. vulgaris* L. by 4% in the presence of the earthworm *Eisenia fetida.* Earthworms are able to accumulate toxic elements due to their developed species-specific detoxification systems such as synthesis of toxic element-binding proteins (metallothioneins), cytochrome P450 enzymes, and antioxidants (Hussain et al., 2021; Swati & Hait, 2017; Yuvaraj et al., 2021), which are in the earthworms' gut. The earthworm gut is high in digestive enzymes and microbial flora, which produces fine granular products that are packed with microflora and nutrients. These products are involved in detoxification reaction and metal-specific distribution by controlling the toxicity and element fractionation through redox reactions (Srut et al., 2019). When TTE such as Zn, Cu, Pb and Mn transit in the gut of earthworm, the possibility of change in specie in the soil resulting from decomposition of organic matter and change in soil properties such as soil texture and soil pH can affect toxic element accumulation as well as the population and composition of the earthworm population (Duarte et al., 2012; Huang et al., 2021). For example, Wang et al. (2019a) reported that an earthworm gut enriched with microbiome reduced As(V) and released As(III). Vijver et al. (2006) reported increased bioaccumulation and elimination of toxic elements by *Aporrectodea caliginosa* earthworms exposed to Cu, Cd, Ca, Pb, and Zn. Huang et al. (2009) also showed an increased concentration of Cu accumulation in *Eisenia fetida* earthworm from less than 50 μmol kg^{-1} fresh weight to around 125 μmol kg^{-1} fresh weight when exposed to increasing Cu concentrations (4, 20, 50, and 100 μM). While Sinkakarimi et al. (2020) reported an increased toxic element accumulation in three different earthworm species (*E. fetida*, *Aporrectodea rosea*, and *Aporrectodea trapezoides*) when exposed to increasing Cd and Pb concentrations in soil.

Apart from bioaccumulation, earthworms in vermicompost also affect soil content and TTE bioavailability. According to Lemtiri et al. (2016), *E. fetida* in Cd-, Cu-, Pb-, and Zn-contaminated soil decreased the bioavailability of Zn and Cd but increased the uptake of Cd by *Z. mays.* From the research, the concentration of Pb increased with exposure to high content of Pb (Fig. 3). *Vicia faba* and *Z. mays* plants were able to reduce the accumulation of Pb and Cd in earthworms and also enhance the reproduction activity of earthworms in the contaminated soils. Studies carried out by Wang et al. (2019b) showed that the activities of *E. fetida* earthworm in addition with biochar and *Brassica chinensis* L. plants increased the content of toxic elements in shoots of Bok choy by elevating the bioavailability of Cd (9.5%), Pb (20.6%), and Zn (22.8%).

Other studies have also reported increased bioavailability of Zn content in soils through reduction of the element fraction in the Fe–Mn oxides. The pH of the soil

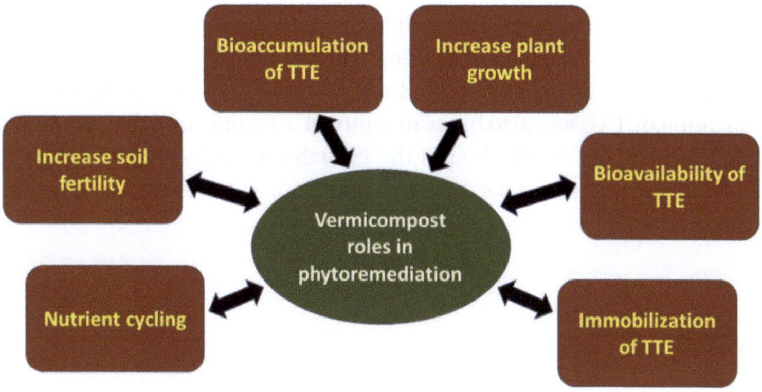

Fig. 3 Roles of vermicompost in phytoremediation

reduced, while the dissolved organic carbon (DOC) and the microbial biomass carbon content increased (Dehghanian et al., 2018). Sahariah et al. (2015) also showed activities of earthworm species on toxic element remediation and biodegradation of Cu, Pb, Mn, and Zn during vermicomposting.

Although changes in the speciation of toxic elements affect bioaccumulation by earthworms, they also facilitate sorption of the TTE onto organic ligands in the vermicompost, which has been demonstrated by several studies, reduced fraction of Cu, Ni, Cd, As Cr, Zn, and Pb irrespective of the substrate and earthworm species (He et al., 2016; Lv et al., 2016).

6 Conclusion

The contamination of TTE is a major problem in agriculture and food production due to their negative effects on the environment and human health, posing a severe danger to future generations. The need for reliable and environmentally friendly techniques to mitigate the toxic effects of TTE in the environment has given rise to the development of several techniques. However, phytoremediation using natural hyperaccumulators still suffers from few drawbacks, such as the longer period required to remediate highly and moderately TTE-polluted soils, which may be attributed to low biomass production and growth rate of the hyperaccumulators plants. Therefore, to achieve faster and effective phytoremediation, there is a need to enhance the technology. Innovations used to enhance phytoremediation include altering the soil conditions, plant and microbial interactions, or the use of agronomic products such as application of vermicompost. Vermicompost, produced from the joint biodegradation of organic matter by earthworms and microorganisms, have been utilized for detoxification and remediation of TTE-contaminated soils. This eco-friendly, sustainable method is cost-effective and generally accepted as the best remediation approach. When applied in phytoremediation, vermicompost plays several roles in enhancing the process, such as improving soil fertility, promoting nutrient cycling, and boosting microbial and enzymatic biodegradative activities, which thus lead to immobilization of TTE and also bioavailability of TTE in a form that can be used and accumulated by plants. It also boosts the growth rate of hyperaccumulator plants, thus facilitating the uptake, tolerance, translocation, and detoxification potentials in phytoremediation.

References

Abbasi, S. A., & Ramasamy, E. V. (1999). *Biotechnological methods of pollution control* (p. 168). Universities Press India Ltd.

Abioye, O. P., Abdulkareem, B. Y., Aransiola, S. A., & Bala, J. D. (2018). Assessment of Manganese biosorption efficacy of Bacillus subtilis and Pseudomonas aeruginosa isolated from waste dump site. *Nigerian Journal of Technological Research, 13*(2).

Abioye, O. P., Ijah, U. J. J., & Aransiola, S. A. (2017). Phytoremediation of soil contaminants by the biodiesel plant Jatropha curcas. In *Phytoremediation potential of bioenergy plants* (pp. 97–137). Springer: Berlin/Heidelberg.

Ahmadpour, P., Ahmadpour, F., Mahmud, T. M. M., Abdu, A., Soleimani, M., & Hosseini Tayefeh, F. (2012). Phytoremediation of heavy metals: A green technology. *African Journal of Biotechnology, 11*, 14036–14043.

Ali, H., Khan, E., & Sajad, M. A. (2013). Phytoremediation of heavy metals—Concepts and applications. *Chemosphere, 91*, 869–881.

Alloway, B. J., & Ayres, D. C. (1997). *Chemical principles of environmental pollution* (p. 168). Blackie Academic.

Alvarenga, P., Gonçalves, A. P., Fernandes, R. M., De Varennes, A., Vallini, G., Duarte, E., & Cunha-Queda, A. C. (2009). Organic residues as immobilizing agents in aided phytostabilization: (I) Effects on soil chemical characteristics. *Chemosphere, 74*, 1292–1300.

Aransiola, S. A., Afolabi, F., Joseph, F., & Maddela N. R. (2023). Soil enzymes: Distribution, interactions and influencing factors. In P. Jeschke, & E. B. Starikov (Eds.), *Agricultural biocatalysis: Enzymes in agriculture and industry* (1st ed., pp. 303–333). ISBN 9789814968478, Jenny Stanford Publishing.

Aransiola, S. A, Ijah, U. J. J. Abioye, O. P., & Bala J. D. (2022a). Vermicompost-assisted phytoremediation of toxic trace element-contaminated soil in Madaka, Nigeria using Melissa officinalis L and Sida acuta. *International Journal of Environmental Science and Technology*. https://doi.org/10.1007/s13762-022-04105-y

Aransiola, S. A., Ijah, U. J. J., & Abioye, O. P. (2013). Phytoremediation of lead polluted soil by *Glycine max L. Applied and Environmental Soil Science, 2013*, Article ID 631619. https://doi.org/10.1155/2013/631619

Aransiola, S. A., Ijah, U. J. J., Abioye, O. P., & Bala, J. D. (2019). Microbial-aided phytoremediation of heavy metals contaminated soil: A review. *European Journal of Biological Research, 9*(2), 104–125.

Aransiola, S. A., Ijah, U. J. J., Abioye, O. P., & Bala, J. D. (2021). Microbial and heavy metal determination of contaminated soil using Melissa officinalis L. *International Journal of Environmental Planning and Management, 7*(3), 102–107.

Aransiola, S. A., Ijah, U. J. J., Abioye, O. P., Bala, J. D., Rivadeneira-Mendoza, R. F., Luque, R., Rodríguez-Díaz, J. M., & Maddela, N. R. (2024). Micro and vermicompost assisted remediation of heavy metal contaminated soils using phytoextractors. *Case Studies in Chemical and Environmental Engineering., 9*(2024), Article 100755. https://doi.org/10.1016/j.cscee.2024.100755

Aransiola, S. A., Joseph, F., Oyedele, O. J., & Maddela, N. R. (2022b). Editorial—Ecological interplays in microbial enzymology. In N. R. Maddela, S. A. Aransiola, & R. Prasad (Eds.), *Ecological interplays in microbial enzymology* (1st ed.) Springer Nature Singapore Pte Ltd. ISSN: 2662-1681.

Aransiola, S. A., & Maddela, N. R. (Eds.). (2024). *Phytoremediation in food safety: Risks and prospects* (1st ed.). CRC Press, Taylor and Francis Group.

Aronsson, P., Dahlin, T., & Dimitriou, I. (2010). Treatment of landfill leachate by irrigation of willow coppice—Plant response and treatment efficiency. *Environmental Pollution, 158*, 795–804.

Azhar-u-ddin, Huang, J. C., Gan, X. Y., He, S. B., & Zhou, W. L. (2020). Interactive effects of earthworm *Eisenia fetida* and bean plant *Phaseolus vulgaris* L on the fate of soil selenium. *Environmental Pollution, 260*, 114048.

Babaniyi, G. G., Olagoke, O. E., & Aransiola, S. A. (2023). Extracellular enzymatic activity of bacteria in aquatic ecosystems. In N. R. Maddela, L. K. W. Eller, & R. Prasad (Eds.), *Microbiology for cleaner production and environmental sustainability*. CRC Press. ISBN: 9781032496061. https://www.routledge.com/Microbiology-for-Cleaner-Production-and-Environmental-Sustainability/Maddela-Eller-Prasad/p/book/9781032496061

Bavandpour, F., Zou, Y., He, Y., Saeed, T., Sun, Y., & Sun, G. (2018). Removal of dissolved metals in wetland columns filled with shell grits and plant biomass. *Chemical Engineering Journal, 331*, 234–241.

Becerra-Castro, C., Prieto-Fernández, Á., Álvarez-López, V., Cabello-Conejo, M. I., Acea, M. J., & Kidd, P. S. (2011). Nickel solubilizing capacity and characterization of rhizobacteria isolated from hyperaccumulating and non-hyperaccumulating subspecies of Alyssum serpyllifolium. *International Journal of Phytoremediation, 13*, 229–244.

Bhadauria, T., & Ramakrishnan, P. S. (1989). Earthworm population dynamics and contribution to nutrient cycling during cropping and fallow phases of shifting agriculture (jhum) in northeast India. *Journal of Applied Ecology, 26*(2), 505–520.

Bhat, S., Singh, J., & Vig, A. P. (2016). Management of sugar industrial wastes through vermitechnology. *International Letters of Natural Sciences, 55*, 35–43.

Bhat, S. A., Singh, J., & Vig, A. P. (2013). Vermiremediation of dyeing sludge from textile mill with the help of exotic earthworm *Eisenia fetida* Savigny. *Environmental Science and Pollution Research, 20*, 5975–5982.

Blaylock, M., & Huang, J. (2000). Phytoextraction of metals. In I. Raskin, & B. D. Ensley (Eds.), *Phytoremediation of toxic metals: Using plants to clean-up the environment* (p. 303). JohnWiley & Sons, Inc.

Bossuyt, H., Six, J., & Hendrix, P. F. (2005). Protection of soil carbon by microaggregates within earthworm casts. *Soil Biology & Biochemistry, 37*(2), 251–258.

Burges, A., Alkorta, I., Epelde, L., & Garbisu, C. (2018). From phytoremediation of soil contaminants to phytomanagement of ecosystem services in metal contaminated sites. *International Journal of Phytoremediation, 20*, 384–397.

Cristaldi, A., Conti, G. O., Jho, E. H., Zuccarello, P., Grasso, A., Copat, C., & Ferrante, M. (2017). Phytoremediation of contaminated soils by heavy metals and PAHs. A brief review. *Environmental Technology and Innovation, 8*, 309–326.

Cunningham, S. D., Berti, W. R., & Huang, J. W. (1995). Phytoremediation of contaminated soils. *Trends in Biotechnology, 13*, 393–397. https://doi.org/10.1016/S0167-7799(00)88987-8

Dalvi, A. A., & Bhalerao, S. A. (2013). Response of plants towards heavy metal toxicity: An overview of avoidance, tolerance and uptake mechanism. *Annals Plant Science, 2*, 362–368.

Danh, L. T., Truong, P., Mammucari, R., Tran, T., & Foster, N. (2009). Vetiver grass, Vetiveria zizanioides: A choice plant for phytoremediation of heavy metals and organic wastes. *International Journal of Phytoremediation, 11*, 664–691.

Daza, B. X. D., Mendoza, A. J. D., Zambrano, J. J. Z., Aransiola, S. A., & Maddela, N. R. (2024). Effects of soil contaminants on soil microbiome. In S. A. Aransiola, H. I. Atta, & N. R. Maddela,

(Eds.), *Soil microbiome in green technology sustainability*. Springer. https://doi.org/10.1007/978-3-031-71844-1_7

Dehghanian, H., Halajnia, A., Lakzian, A., & Astaraei, A. R. (2018). The effect of earthworm and arbuscular mycorrhizal fungi on availability and chemical distribution of Zn, Fe and Mn in a calcareous soil. *Applied Soil Ecology, 130*, 98–103.

Duarte, A. P., Melo, V. F., Brown, G. G., & Pauletti, V. (2012). Changes in the forms of lead and manganese in soils by passage through the gut of the tropical endogeic earthworm (*Pontoscolex corethrurus*). *European Journal of Soil Biology, 53*, 32–39.

Dushenkov, S., Skarzhinskaya, M., Glimelius, K., Gleba, D., & Raskin, I. (2002). Bioengineering of a phytoremediation plant by means of somatic hybridization. *International Journal of Phytoremediation, 4*, 117–126.

Eapen, S., & Dsouza, S. F. (2005). Prospects of genetic engineering of plants for phytoremediation of toxic metals. *Biotechnology Advances, 23*, 97–114.

Fasani, E., Manara, A., Martini, F., Furini, A., & DalCorso, G. (2018). The potential of genetic engineering of plants for the remediation of soils contaminated with heavy metals. *Plant, Cell and Environment, 41*, 1201–1232. https://doi.org/10.1111/pce.12963

Gamalero, E., & Glick, B. R. (2012). Plant growth-promoting bacteria and metals phytoremediation. In N. A. Anjum, M. E. Pereira, I. Ahmad, A. C. Duarte, S. Umar, & N. A. Khan (Eds.) *Phytotechnologies: Remediation of environmental contaminants* (pp. 361–376). CRC Press.

Garau, G., Porceddu, A., Sanna, M., Silvetti, M., & Castaldi, P. (2019). Municipal solid wastes as a resource for environmental recovery: Impact of water treatment residuals and compost on the microbial and biochemical features of As and trace metal-polluted soils. *Ecotoxicology and Environmental Safety, 174*, 445–454.

Garcia, C., Ceccanti, B., Masciandaro, G., & Hernandez, T. (1995). Phosphatase and β-glucosidase activities in humic substances from animal wastes. *Bioresource Technology, 53*, 79–87.

Garg, V. K., & Gupta, R. (2011). Optimization of cow dung spiked pre-consumer processing vegetable waste for vermicomposting using Eisenia fetida. *Ecotoxicology and Environmental Safety, 74*, 19–24.

Glick, B. R. (2010). Using soil bacteria to facilitate phytoremediation bacteria. *Biotechnology Advances, 28*, 367–374.

Haque, N., Peralta-Videa, J. R., Jones, G. L., Gill, T. E., & Gardea-Torresdey, J. L. (2008). Screening the phytoremediation potential of desert broom (Baccharis sarothroides Gray) growing on mine tailings in Arizona, USA. *Environmental Pollution, 153*, 362–368.

Hassan, S. E. D., Boon, E., St-Arnaud, M., & Hijri, M. (2011). Molecular biodiversity of arbuscular mycorrhizal fungi in trace metal-polluted soils. *Molecular Ecology, 20*, 3469–3483.

He, H., Ye, Z., Yang, D., Yan, J., Xiao, L., Zhong, T., Yuan, M., Cai, X., Fang, Z., & Jing, Y. (2013). Characterization of endophytic Rahnella sp. JN6 from Polygonum pubescens and its potential in promoting growth and Cd, Pb Zn Uptake by Brassica Napus. *Chemosphere, 90*, 1960–1965.

He, X., Zhang, Y., Shen, M., Zeng, G., Zhou, M., & Li, M. (2016). Effect of vermicomposting on concentration and speciation of heavy metals in sewage sludge with additive materials. *Bioresource Technology, 218*, 867–873.

Huang, C. D., Ge, Y., Yue, S. Z., Qiao, Y. H., & Liu, L. S. (2021). Impact of soil metals on earthworm communities from the perspectives of earthworm ecotypes and metal bioaccumulation. *Journal of Hazardous Materials, 406*, Article 124738.

Huang, R., Wen, B., Pei, Z., Shan, X. Q., Zhang, S., & Williams, P. N. (2009). Accumulation, subcellular distribution and toxicity of copper in earthworm (*Eisenia fetida*) in the presence of ciprofloxacin. *Environmental Science and Technology, 43*(10), 3688–3693.

Hussain, N., Chatterjee, S. K., Maiti, T. K., Goswami, L., Das, S., Deb, U., & Bhattcharya, S. S. (2021). Metal induced non-metallothionein protein in earthworm: A new pathway for cadmium detoxification in chloragogenous tissue. *Journal of Hazardous Materials, 401*, Article 123357.

Ikhumetse, A. A., Abioye, O. P., & Aransiola, S. A. (2019). Biosorption potential of bacteria on lead and chromium in groundwater obtained from mining community. *Acta Scientific Microbiology, 2*(6), 123–137.

Islam, M. S., Proshad, R., & Ahmed, S. (2018). Ecological risk of heavy metals in sediment of an urban river in Bangladesh. *Human and Ecological Risk Assessment: An International Journal, 24*(3), 699–720.

Ismail, S., Khan, F., & Zafar Iqbal, M. (2013). Phytoremediation: Assessing tolerance of tree species against heavy metal (PB and CD) toxicity. *Pakistan Journal of Botany, 45*, 2181–2186.

Jabeen, R., Ahmad, A., & Iqbal, M. (2009). Phytoremediation of heavy metals: Physiological and molecular mechanisms. *Botanical Review, 75*, 339–364.

Jadia, C. D., & Fulekar, M. H. (2008). Phytoremediation: The application of Vermicompost to remove zinc, cadmium, copper, nickel and lead by sunflower plant. *Environmental Engineering and Management Journal, 7*(5), 547–558.

Jadia, C. D., & Fulekar, M. H. (2009). Phytoremediation of heavy metals: Recent techniques. *African Journal of Biotechnology, 8*, 921–928.

Karczewska, A., Gałka, B., Dradrach, A., Lewińska, K., Mołczan, M., Cuske, M., Gersztyn, L., & Litak, K. (2017). Solubility of arsenic and its uptake by ryegrass from polluted soils amended with organic matter. *Journal of Geochemical Exploration, 182*, 193–200.

Kelly, C. N., Peltz, C. D., Stanton, M., Rutherford, D. W., & Rostad, C. E. (2014). Biochar application to hardrock mine tailings: Soil quality, microbial activity, and toxic element sorption. *Applied Geochemistry, 43*, 35–48.

Ker, K., & Charest, C. (2010). Nickel remediation by AM-colonized sunflower. *Mycorrhiza, 20*, 399–406.

Kroopnick, P. M. (1994). Vapor abatement costs analysis methodology for calculating life cycle costs for hydrocarbon vapour extracted during soil venting. In D. L. Wise & D. J. Trantolo (Eds.), *Remediation of hazardous waste* (pp. 779–790). Marcel Dekker.

Kumar, K. V., Singh, N., Behl, H. M., & Srivastava, S. (2008). Influence of plant growth promoting bacteria and its mutant on heavy metal toxicity in Brassica juncea grown in fly ash amended soil. *Chemosphere, 72*, 678–683.

Lasat, M. (1999). Phytoextraction of metals from contaminated soil: A review of plant/soil/metal interaction and assessment of pertinent agronomic issues. *Journal of Hazardous Substance and Research, 2*, 5. https://doi.org/10.4148/1090-7025.1015

Leguizamo, M. A. O., Gómez, W. D. F., & Sarmiento, M. C. G. (2017). Native herbaceous plant species with potential use in phytoremediation of heavy metals, spotlight on wetlands—A review. *Chemosphere, 168*, 1230–1247.

Lemtiri, A., Linenard, A., Alabi, T., Brostaux, Y., Cluzeau, D., Francis, F., & Colinet, G. (2016). Earthworms *Eisenia fetida* affect the uptake of heavy metals by plants Vicia faba and Zea mays in metal-contaminated soils. *Applied Soil Ecology, 104*, 67–78.

Lv, B., Xing, M., & Yang, J. (2016). Speciation and transformation of heavy metals during vermicomposting of animal manure. *Bioresource Technology, 209*, 397–401.

Lyyra, S., Meagher, R. B., Kim, T., Heaton, A., Montello, P., Balish, R. S., & Merkle, S. A. (2007). Coupling two mercury resistance genes in Eastern cottonwood enhances the processing of organomercury. *Plant Biotechnology Journal, 5*, 254–262.

Ma, Y., Rajkumar, M., Luo, Y., & Freitas, H. (2011). Inoculation of endophytic bacteria on host and non-host plants—Effects on plant growth and Ni uptake. *Journal of Hazardous Materials, 195*, 230–237.

Manivannan, S., Balamurugan, M., Parthasarathi, K., Gunasekaran, G., & Ranganathan, L. S. (2009). Effect of vermicompost on soil fertility and crop productivity—Beans (*Phaseolus vulgaris*). *Journal of Environmental Biology, 30*, 275–281.

Marchand, L., Mench, M., Jacob, D. L., & Otte, M. L. (2010). Metal and metalloid removal in constructed wetlands, with emphasis on the importance of plants and standardized measurements: A review. *Environmental Pollution, 158*, 3447–3461.

Mariani, L., Jimenez, J. J., Asakawa, N., Thomas, R. J., & Decaens, T. (2007). What happens to earthworm casts in the soil? A field study of carbon and nitrogen dynamics in Neotropical savannahs. *Soil Biology & Biochemistry, 39*(3), 757–767.

Montinaro, S., Concas, A., Pisu, M., & Cao, G. (2012). Remediation of heavy metals contaminated soils by ball milling. *Chemical Engineering Transactions, 28*, 187–192.

Moosavi, S. G., & Seghatoleslami, M. J. (2013). Phytoremediation: A review. *Advance in Agriculture and Biology, 1*, 5–11.

Muscolo, A., Bovolo, F., Gionfriddo, F., & Nardi, S. (1999). Earthworm humic matter produces auxins-like effect on Daucus carota cell growth and nitrate metabolism. *Soil Biology & Biochemistry, 31*, 1303–1311.

Newete, S. W., & Byrne, M. J. (2016). The capacity of aquatic macrophytes for phytoremediation and their disposal with specific reference to water hyacinth. *Environmental Science and Pollution Research, 23*, 10630–10643.

Padmavathiamma, P. K., & Li, L. Y. (2012). Rhizosphere influence and seasonal impact on phytostabilisation of metals—A field study. *Water, Air, and Soil Pollution, 223*, 107–124. https://doi.org/10.1007/s11270-011-0843-4

Padmavathiamma, P. K., Li, L. Y., & Kumari, U. R. (2008). An experimental study of vermi-biowaste composting for agricultural soil improvement. *Bioresource Technology, 99*, 1672–1681.

Parker, R. (1994). Environmental restoration technologies. In *EMIAA yearbook* (pp. 169–171).

Parmelee, R. W., Bohlen, P. J., & Blair, J. M. (1998). Earthworms and nutrient cycling processes: Integrating across the ecological hierarchy. In C. Edwards (Ed.), *Earthworm ecology* (pp. 179–211). St. Lucie Press.

Pattnaik, S., & Reddy, M. V. (2010). Nutrient status of vermicompost of urban green waste processed by three earthworm species: Eisenia fetida, Eudrilus eugeniae, and Perionyx excavates. *Applied and Environmental Soil Science.*

Prabha, M. L. (2009). Waste management by vermitechnology. *Indian Journal of Environmental Protection, 29*, 795–800.

Raskin, I., & Ensley, B. D. (2000). *Phytoremediation of toxic metals*. John Wiley & Sons.

Rezania, S., Taib, S. M., Md Din, M. F., Dahalan, F. A., & Kamyab, H. (2016). Comprehensive review on phytotechnology: Heavy metals removal by diverse aquatic plants species from wastewater. *Journal of Hazardous Materials, 318*, 587–599.

Richardson, J. B., Görres, J. H., & Sizmur, T. (2020). Synthesis of earthworm trace metal uptake and bioaccumulation data: Role of soil concentration, earthworm ecophysiology, and experimental design. *Environmental Pollution, 262*, 114126.

Richardson, J. B., Görres, J. H., & Friedland, A. J. (2017). Exotic earthworms decrease Cd, Hg, and Pb pools in upland forest soils of vermont and New Hampshire USA. *Bulletin of Environmental Contamination and Toxicology, 99*(4), 428–432.

Roccotiello, E., Marescotti, P., Di Piazza, S., Cecchi, G., Mariotti, M. G., & Zotti, M. (2015). Biodiversity in metal-contaminated sites-problem and perspective—A case study. In J. A. Blanco (Ed.), *Biodiversity in ecosystems—Linking structure and function* (pp. 563–582). IntechOpen.

Roy, S., Arunachalam, K., Dutta, B. K., & Arunachalam, A. (2010). Effect of organic amendments of soil on growth and productivity of three common crops *viz. Zea mays, Phaseolus vulgaris and Abelmoschus esculentus. Applied Soil Ecology, 45*, 78–84.

Sahariah, B., Goswami, L., Kim, K. H., Bhattacharyya, P., & Bhattacharya, S. S. (2015). Metal remediation and biodegradation potential of earthworm species on municipal solid waste: A parallel analysis between *Metaphire posthuma* and *Eisenia fetida. Bioresource Technol., 180*, 230–236.

Sarwar, N., Imran, M., Shaheen, M. R., Ishaque, W., Kamran, M. A., Matloob, A., et al. (2017). Phytoremediation strategies for soils contaminated with heavy metals: Modifications and future perspectives. *Chemosphere, 171*, 710–721. https://doi.org/10.1016/j.chemosphere.2016.12.116

Schalk, I. J., Hannauer, M., & Braud, A. (2011). New roles for bacterial siderophores in metal transport and tolerance. *Environmental Microbiology, 13*, 2844–2854.

Senesi, S., Saiz, J. C., & Miano, T. M. (1992). Spectroscopic characterization of metal humic acid like complexes of earthworms composed organic wastes. *Science of the Total Environment, 117*, 111–120.

Sessitsch, A., Kuffner, M., Kidd, P., Vangronsveld, J., Wenzel, W. W., Fallman, K., & Puschenreiter, M. (2013). The role of plant-associated bacteria in the mobilization and phytoextraction of trace elements in contaminated soils. *Soil Biology & Biochemistry, 60*, 182–194.

Sheng, X. F., Xia, J. J., Jiang, C. Y., He, L. Y., & Qian, M. (2008). Characterization of heavy metal-resistant endophytic bacteria from rape (Brassica napus) roots and their potential in promoting the growth and lead accumulation of rape. *Environmental Pollution, 156*, 1164–1170.

Shim, D., Kim, S., Choi, Y. I., Song, W. Y., Park, J., Youk, E. S., Jeong, S. C., Martinoia, E., Noh, E. W., & Lee, Y. (2013). Transgenic poplar trees expressing yeast cadmium factor 1 exhibit the characteristics necessary for the phytoremediation of mine tailing soil. *Chemosphere, 90*, 1478–1486.

Shipitalo, M. J., & Protz, R. (1989). Chemistry and micromorphology of aggregation in earthworm casts. *Geoderma, 45*(3–4), 357–374.

Sinkakarimi, M. H., Solgi, E., & Colagar, A. H. (2020). Interspecific differences in toxicological response and subcellular partitioning of cadmium and lead in three earthworm species. *Chemosphere, 238*, Article 124595.

Srut, M., Menke, S., Höckner, M., & Sommer, S. (2019). Earthworms and cadmium—Heavy metal resistant gut bacteria as indicators for heavy metal pollution in soils? *Ecotoxicology and Environmental Safety, 171*, 843–853.

Suthar, S. S., Watts, T., Sandhu, M., Rana, S., Kanwal, A., Gupta, D., & Meena, M. S. (2005). Vermicomposting of kitchen waste by using Eisenia foetida (SAVIGNY). *Asian Journal of Microbiology, Biotechnology and Environmental Science, 7*, 541–544.

Swati, A., & Hait, S. (2017). Fate and bioavailability of heavy metals during vermicomposting of various organic wastes—A review. *Process Saf. Environ., 109*, 30–45.

Tang, L., Mao, B., Li, Y., Lv, Q., Zhang, L. P., Chen, C., He, H., Wang, W., Zeng, X., Shao, Y., et al. (2017). Knockout of OsNramp5 using the CRISPR/Cas9 system produces low Cd-accumulating indica rice without compromising yield. *Science and Reports, 7*, 14438.

Tejada, M., Gomez, I., Hernandez, T., & Garcla, C. (2010). Utilization of vermicomposts in soil restoration: Effects on soil biological properties. *Soil Biology and Biochemistry, 74*(2), 525–532.

Vamerali, T., Bandiera, M., & Mosca, G. (2010). Field crops for phytoremediation of metal-contaminated land. A review. *Environmental Chemistry Letters, 8*, 1–17.

Van Ginneken, L., Meers, E., Guisson, R., Ruttens, A., Elst, K., Tack, F. M. G., Vangronsveld, J., Diels, L., & Dejonghe, W. (2007). Phytoremediation for heavy metal-contaminated soils combined with bioenergy production. *Journal of Environmental Engineering and Landscape Management, 15*, 227–236.

Vijver, M. G., van Gestel, C. A. M., van Straalen, N. M., Lanno, R. P., & Peijnenburg, W. J. G. M. (2006). Biological significance of metals partitioned to subcellular fractions within earthworms (*Aporrectodea caliginosa). Environmental Science and Technology, 25*(3), 807–814.

Visioli, G., Vamerali, T., Mattarozzi, M., Dramis, L., & Sanangelantoni, A. M. (2015). Combined endophytic inoculants enhance nickel phytoextraction from serpentine soil in the hyperaccumulator Noccaea caerulescens. *Frontiers in Plant Science, 6*, 638.

Wang, A. S., Angle, J. S., Chaney, R. L., Delorme, T. A., & Reeves, R. D. (2006). Soil pH effects on uptake of Cd and Zn by Thlaspi caerulescens. *Plant and Soil, 281*, 325–337. https://doi.org/10.1007/s11104-005-4642-9

Wang, H. T., Zhu, D., Li, G., Zheng, F., Ding, J., O'Connor, P. J., Zhu, Y. G., & Xue, X. M. (2019a). Effects of arsenic on gut microbiota and its biotransformation genes in earthworm *Metaphire sieboldi. Enviromental Science and Technology, 53*(7), 3841–3849.

Wang, J., Shi, L., Zhang, X. Z., Zhao, X., Zhong, K. C., Wang, S. X., Zou, J. W., Shen, Z. G., & Chen, Y. H. (2019b). Earthworm activities weaken the immobilizing effect of biochar as amendment for metal polluted soils. *Science of the Total Environment, 696*, Article 133729.

Wang, K., Qiao, Y., Zhang, H., Yue, S., Li, H., Ji, X., & Crowley, D. (2018a). Influence of cadmium-contaminated soil on earthworm communities in a subtropical area of China. *Applied Soil Ecology, 127*, 64–73.

Wang, K., Qiao, Y., Zhang, H., Yue, S., Li, H., Ji, X., & Liu, L. (2018b). Bioaccumulation of heavy metals in earthworms from field contaminated soil in a subtropical area of China. *Ecotoxicology and Environmental Safety, 148*, 876–883.

Wang, K., Qiao, Y., Zhang, H., Yue, S., Li, H., Ji, X., & Liu, L. (2018c). Influence of metal-contamination on distribution in subcellular fractions of the earthworm (*Metaphire californica*) from Hunan Province, China. *Journal of Environmental Science, 73*, 127–137.

Wang, X. B., Song, D., Liang, G. Q., Zhang, Q., Ai, C., & Zhou, W. (2015). Maize biochar addition rate influences soil enzyme activity and microbial community composition in a fluvo-aquic soil. *Applied Soil Ecology, 96*, 265–272.

Wang, Y., Xu, Y. A., Li, D., Tang, B. C., Man, S. L., Jia, Y. F., & Xu, H. (2018d). Vermicompost and biochar as bio-conditioners to immobilize heavy metal and improve soil fertility on cadmium contaminated soil under acid rain stress. *Science of the Total Environment, 621*, 1057–1065.

Wu, G., Kang, H., Zhang, X., Shao, H., Chu, L., & Ruan, C. A. (2010). Critical review on the bio-removal of hazardous heavy metals from contaminated soils: Issues, progress, eco-environmental concerns and opportunities. *Journal of Hazardous Materials, 174*, 1–8.

Wuana, R. A., & Okieimen, F. E. (2011). Heavy metals in contaminated soils: A review of sources, chemistry, risks and best available strategies for remediation. *Communications in Soil Science and Plant Analysis, 42*, 111–122. https://doi.org/10.5402/2011/402647

Xu, M., & Lu, N. (2012). Research on removing heavy metals from mine tailings. *Disaster Advances, 5*, 116–120.

Yuvaraj, A., Govarthanan, M., Karmegam, N., Biruntha, M., Kumar, D. S., Arthanari, M., Govindarajan, R. K., Tripathi, S., Ghosh, S., Kumar, P., Kanan, S., & Thangaraj, R. (2021). Metallothionein dependent-detoxification of heavy metals in the agricultural field soil of industrial area: Earthworm as field experimental model system. *Chemosphere, 267*, Article 129240.

Zhang, L. W., Shang, Z. B., Guo, K. X., Chang, Z. X., Liu, H. L., & Li, D. L. (2019a). Speciation analysis and speciation transformation of heavy metal ions in passivation process with thiol-functionalized nano-silica. *Chemical Engineering Journal, 369*, 979–987.

Zhang, Y., Tian, Y., Hu, D., Fan, J., Shen, M., & Zeng, G. (2019b). Is vermicompost the possible in situ sorbent? Immobilization of Pb, Cd and Cr in sediment with sludge derived vermicompost, a column study. *Journal of Hazardous Materials, 367*, 83–90.

Zhang, Z., Liu, B., He, Z., Pan, P., Wu, L., Lin, B., Li, Q., Zhang, X., & Wang, Z. (2022). The synergistic effect of biochar-combined activated phosphate rock treatments in typical vegetables in tropical sandy soil: Results from nutrition supply and the immobilization of toxic metals. *International Journal of Environmental Research and Public Health, 19*(11), 6431.

Zovko, M., & Romic, M. (2011). Soil contamination by trace metals: Geochemical behavior as an element of risk assessment. In I. A. Dar (Ed.), *Earth and environmental sciences* (pp. 437–456), InTech.

Vermiconversion and Vermifiltration of Wastewater Treatment

J. V. Addy, C. O. Aguoru, R. D. Akogwu, B. T. Buukume, and B. A. Ella

Abstract Wastewater and sludge are produced globally in large quantity. The lack of proper wastewater management has a direct impact on water quality and on the diversity of biological aquatic ecosystems. Vermifiltration is an earthworm-assisted process for the treatment of liquid and solid wastes with the use of a vermifilter. A vermifilter is constructed using different types of soil incorporated with earthworms that are capable of treating wastewater. The worm species, biomass, hydraulic retention time, hydraulic loading rate, type of wastewater, and seasonal variation greatly influence the performance of vermifiltration. In the alimentary canal of earthworms, enzymes and microorganisms synergize for considerable improvement in the decomposition of liquid/solid waste without sludge formation. The earthworm bulk densities per cubic meter (m^{-3}) of bedding during vermifiltration assist in reduction efficiencies of 41%–89% (total nitrogen [TN]), 46%–86% (NH_4^+–N), 34%–74% (NO_3^-–N), 3%–17% (total phosphorous [TP]), 18%–38% (ortho-P), 35%–66% (total solids [TS]), 90%–95% (total suspended solids [TSS]), 88–92% (biochemical oxygen demand [BOD]), 80–90% (chemical oxygen demand [COD]), and 90–92% (total dissolved solids [TDS]) and reduction of pathogens that is evident with no odor formation after vermifiltration. Heavy metals are also reduced when a specific metal-binding protein, metallothionein, inside the chloragogenous tissue links heavy metals to form protein–metal complexes that can accumulate in the tissues of earthworms. The advantages of vermifiltration technology over conventional systems of sewage treatment are a low-energy system, a source of animal feeds, no formation of sludge, and an odorless system. It is recommended that vermifiltration that is environmentally friendly be used in the treatment of wastewater for being cost-effective without the production of sludge and odor as compared to conventional methods.

Keywords Vermifilter · Vermicast · Wastewater · Enzymes · Microbes · Sludge

---

J. V. Addy (✉) · C. O. Aguoru · R. D. Akogwu · B. T. Buukume
Department of Botany, Federal University of Agriculture, Makurdi, North Central, Nigeria
e-mail: addy.jose@uam.edu.ng

B. A. Ella
Department of Biology, Benue State Polytechnic, Ugbokolo, North Central, Nigeria

S. A. Aransiola and N. R. Maddela (eds.), *Vermitechnology: Economic, Environmental and Agricultural Sustainability*, https://doi.org/10.1007/978-3-031-96212-7_7

1 Introduction

A large amount of solid waste, wastewater, and sludge has been produced worldwide as a result of urbanization (Aransiola et al., 2021; Suthar, 2009). The accumulation of solid waste has become a persistent crisis around the world (UNEP, 2015, 2024; Oliveira et al., 2024). Studies have shown that municipal solid waste (MSW) and agricultural waste contain 42% and 80% organic matter, respectively, making both appropriate biodegradable matters (Raphela et al., 2024). It is estimated that global solid waste could reach 7 megatons per day in 2025 if no improvement is made to the present solid waste disposal methods (Hoornweg & Pope, 2016; UNEP, 2024). Such an enormous increase in MSW raises the urgency for an efficient, eco-friendly process to treat MSW into an environmentally friendly product (Oliveira et al., 2024). Many developing countries cannot deal with this remarkable increase. Different MSW management and handling methods were produced to overcome the difficulty of municipal solid waste and agricultural waste buildup. Nevertheless, all of those treatment methods have some limitations. One of the modern environmentally friendly trends to resolve the waste accumulation setback is vermicomposting. It involves the synergy of both earthworms and microorganisms for waste biodegradation (Singh & Kalamdhad, 2016).

In developing countries, an estimated 90 percent of the total untreated wastewater is currently discharged directly into rivers, lakes, or oceans (Kirschner et al., 2024; Cohen et al., 2010). Lack of proper wastewater management has a direct impact on the diversity of biological aquatic ecosystems, altering the fundamental integrity of our life support systems, on which a wide variety of sectors from urban progress to food production and industry depend (Lorena et al., 2025). It is important that wastewater management is valued as part of integrated, ecosystem-based management that functions across different sectors of the environment (Kirschner et al., 2024). Water is very essential to all aspects of life, the defining feature of our planet. Water is one of the most important substances for the sustainability of the ecosystem. It is indispensable on earth for the maintenance of life forms. Access to safe water, to meet various human needs, is a human right. It is estimated that over one billion people lack access to safe drinking water with an estimated 2.6 billion persons lack access to basic sanitation (WATER, 2017). Unsafe water supply and unhygienic sanitation conditions are responsible for over 90% of diarrhea diseases worldwide (WHO, 2019). Only one percent of freshwater is accessible for extraction and use; the remaining ninety-seven and a half percent of all water is found in the oceans. Healthy and functioning aquatic ecosystems provide us with a dazzling array of great benefits, such as food, medicines, and tourism. The degradation of surface and groundwater water quality has increased due to discharges of inadequately treated wastewater (Aguoru et al., 2015). Water availability is critically affected by water pollution, and it needs to be managed properly so as to mitigate the negative impacts of increasing water scarcity (WATER, 2017). Global populations are rapidly increasing, and with the rise in population growth so does wastewater production and the increase in the number of people vulnerable to the impacts of severe wastewater pollution. Sewage need to be properly treated before discharging into the environment to reduce the organic loads, or else, the more dissolved oxygen (DO) will be consumed by aerobic bacteria, thereby reducing the DO values (Sinha et al., 2008). This would adversely affect the survival of aquatic organisms.

Vermifiltration (VF), which means the introduction of earthworms to the filtration systems, was first advocated by José Toha in 1992, which is a novel technology that is newly conceived with various advantages over the conventional wastewater treatment systems (Li et al., 2008). It is a technology with no energy requirement or very low energy and zero-waste production technology. It is cheap and easy to construct, operate, and maintain, and most importantly, it is environmentally friendly. Earthworms and microorganisms are jointly involved in the process of vermiconversion with considerable improvement in decomposition of waste without sludge formation (Xing et al., 2010).

Earthworms serve as mechanical blenders by modifying the physical and chemical status of organic matter. This process gradually reduces the carbon–to-nitrogen ratio and increases the surface area that favors microbial activities for further decomposition (Domínguez et al., 2003). Earthworms provide the required conditions for the proliferation of microorganisms and biodegradation of wastes in their gut and are therefore considered as natural bioreactors. The soil and gravel particles that constitute part of the bed in the vermifilter (VF) help in the filtration of the wastewater through the adsorption of organic impurities. Environmental, social, economic, and legal factors are the challenges facing wastewater treatment and disposal. The technology of vermifiltration can effectively control liquid and solid waste. This practice allows composting of biodegradable materials and, at the same point, utilizes their products to improve crop production, which eliminates the use of chemical fertilizer. The prominence of chemical fertilizers usage has led to the long-term discrepancy in soil pH and fertility, which has caused severe damage to the ecosystem. To cope with these insightful problems, the vermiculture technology has become a viable alternative as it synchronizes with nature and not against it. The utmost benefit of a vermifiltration system is that there is no development of "sewage sludge," and stinking odor is also removed (Hughes, 2013). The sludge is required to be treated prior to being discharged into the environment. Earthworms feed willingly upon the sludge components, speedily convert them into vermicompost without odor with a decline in pathogens to safe levels (Zhao et al., 2010).

2 Concept of Wastewater

Wastewater can be defined as "a mixture of one or more of sewage (excreta, urine, and fecal sludge) and gray water (bathing and kitchen wastewater), effluent from commercial and industrial firms, and other urban effluent runoffs (agricultural, horticultural, and aquaculture effluent, either dissolved or as suspended matter) (Aransiola et al., 2021; Jayakody, 2014). Contamination of wastewater can be with pathogens, organic matter, organic compounds, synthetic chemicals, nutrients, and heavy metals. The water contaminants are either in particulate matter or solution and are transported along in the water from different sources and affect the quality of water. These components can possess (bio-)cumulative, synergistic, and persistent characteristics affecting ecosystem health and functionality, production of food, and human well-being (Pimentel, 2009).

Wastewater generally may contain unsafe dissolved or suspended matter. Unchecked discharge of wastewater undermines biodiversity, natural resilience, and the capability of the planet to provide essential ecosystem services, impacting equally rural and urban populations and affecting various sectors ranging from health to industry, agriculture, fisheries, and tourism. The less privileged are the most severely affected (Auta et al., 2022).

2.1 *Impact of Wastewater on Ecosystem Function and Human Health*

All waterways in the environment are connected. The unregulated discharge of untreated wastewater therefore has implications for the health of aquatic ecosystems, which then threatens the well-being of humans that depends on the resilience of biodiversity and ecosystem services. Eutrophication is one of the most widespread global problems. It is a process by which water bodies are increasingly rich in plant nutrients, primarily nitrogen and phosphorus, originating from agricultural and urban areas, all the way through the earth or straight into rivers and oceans (Glibert et al., 2008). More than two-thirds of this nitrogen makes its way into waterways, exceeding all natural inputs to the nitrogen cycle. Almost half of phosphorus mined annually for fertilizers returns to the ocean—about eight times the natural input (Steffen et al., 2009). Together, the excess nitrogen and phosphorus cause potentially toxic algal blooms and biodiversity changes, which in turn lead to overwhelming hypoxic actions and promote dead zones, ensuing huge economic losses across many sectors (Hernández-Sancho et al., 2010). The dead zones are estimated to affect over 245,000 km^2 of marine ecosystems, predominantly equivalent to the total global area of coral reefs (Diaz, 2008). A broad variety of toxic pollutants from land-based sources are found in both fresh and salt waters, ranging from farming and industrial chemicals, such as organic compounds and heavy metals, to personal-care goods and pharmaceuticals (Amobonye et al., 2023; Aransiola et al., 2022). The impacts of these are extensive.

Estimates of the universal burden of water-associated human diseases, according to the World Health Organization (WHO, 2004), that some 2.2 million persons die annually from diarrheal estimate disease, 3.7 percent of all deaths, and over half of the people suffering in hospitals are from water-related diseases. Annually, out of 10.4 million deaths of children that die under the ages of five, 17 percent are ascribed to diarrheal disease. The underlying cause is unsafe water, poor hygiene, and inadequate sanitation (WHO, 2004). Therefore, there is an urgent need for wastewater treatment that is environmentally friendly and cost-effective.

2.2 *Vermifiltration Process*

In a typical vermifiltration system, there are lower beds where the liquid wastes are biodegraded, normally by means of attached biofilm growth. The filtered organic solids from the incoming wastewater in the upper beds are biodegraded with the aid of earthworms and associated microbes and form a layer of rich organic matter material known as humus. Humus production in the system increases the hydraulic conductivity and porosity of the upper beds. The particles that make up the humus create a high surface area that leads to more adsorption of contaminants (e.g., metals, nutrients, surfactants) (Hughes, 2013). The humus produced by the use of earthworms helps to overcome clogging and cake formation and thus reduces the solid fraction. It also helps to reducing the maintenance requirements of the system (Ojuolape et al., 2015; Sinha et al., 2008). The burrowing activities of worms in the wastes generate air spaces through movement and turning of the substrate, therefore producing an aerobic situation in the humus, with available oxygen to aerobic decomposer microbes, which accelerates the biological decomposition of the wastes. Earthworms are burrowing animals, and their alimentary canal acts as a bioreactor where enzymes like amylases, cellulases, lipases, proteases, and chitinases are secreted for biochemical change of the proteinaceous and cellulosic materials in the organic wastes. The accelerated oxygenation of the humus produces greater oxygen exchange with the wastewater flowing through the vermifiltration system and leads to greater removal of chemical oxygen demand (COD) from the system. Nitrification in the system also increases with the oxygenation of the incoming wastewater (Hughes, 2013).

3 Key Factors in the Vermifiltration Process

There are a number of factors that will impact the performance of a vermifiltration system. The key factors are earthworm species and biomass, hydraulic conductivity, hydraulic retention time, constituents and characteristics of the wastewater, and seasonal variation.

3.1 *Biology of Earthworm Species Used in Vermifiltration*

Earthworms belong to the *Annelida* phylum, and the class *Oligochaeta* comprises over 1800 species; most of the species belong to the *Lumbricidae* family, including the genera *Dendrobaena, Eisenia, and Lumbricus.* Some of the earthworm species involved in vermifiltration are *Lumbricus terrestris, Eudrilus eugeniae, Eisenia fetida, Libyodrilus violaceous, Eisenia andrei, Megascolex mauritii, Perionyx excavatus, Lampito mauritii, Lampito rubellus,* and *Drawida willsi* (Sinha et al., 2008).

Earthworms are cylindrical, bilaterally symmetrical, long, narrow, and segmented animals with a hydraulic skeletal system (without bones). The body is covered with delicate cuticle, dark brown, which glistens. The body weight range is between 1400 and 1500 mg during 8–10 weeks. The life span of an adult earthworm is about 3–7 years, depending upon the environmental situation and type of species. Earthworms port millions of "nitrogen-fixing" and "decomposer microbes" in their alimentary canal. They search for food with the aid of "chemoreceptors" in the gut. Their body contains 14% carbohydrates, 14% fats, 65% protein (70–80% "lysine-rich protein" on a dry weight), and 3% ash. Generally, earthworms can also exhibit high water loss tolerance by dehydration. Earthworms proliferate very speedily and are bisexual animals. After copulation, each worm produces a "cocoon" where sperm enter to fertilize the eggs. About three cocoons can be produced per worm per week. Earthworms take approximately only 4–6 weeks to become sexually mature and continue to grow throughout their life. Earthworms are grouped into three categories. Diverse worm species have dissimilar burrowing characteristics and, as a result, have different impacts on the treatment practice. The optimal worm density is one of the essential parameters for the efficient performance of a vermifiltration system (Li et al., 2008). The treatment efficiency of vermifiltration is affected positively by the quantity of worms per unit area in the vermifilter bed. A vermifiltration system ought to be started with sufficient worms to vermicompost the received wastes and generate an appropriate humus filter. It has been projected that a comparatively high quantity of at least 15,000–20,000 worms/m^3 of the vermifiltration system should be used (Sinha et al., 2008).

3.2 *Hydraulic Retention Time*

Hydraulic retention or residence time (HRT) is the normal time wastewater remains in the vermifiltration treatment system. HRT is a vital factor in vermifiltration treatment where the worms and microbes convert and stabilize nutrients, suspended materials, biochemical oxygen demand (BOD), and COD in the wastewater. The efficiency treatment of the system increases with HRT. As noticed in soil-based treatment systems, a high hydraulic loading rate (HLR) can reduce the hydraulic retention time of the system and consequently the treatment's efficiency (Sinha et al., 2008). The reason is because wastewater requires a definite contact time with the wastewater for the adsorption of contaminants, conversion of nitrogen, BOD, and COD reduction (Hughes, 2013). The greater will be the efficiency of vermiprocessing and retention of nutrients if there is longer contact of wastewater with earthworms in the system. Therefore, the flow of wastewater in the vermifilter is an essential factor, as it determines the retention rate of suspended organic matter and solids. The slower rate of wastewater discharge on the top of the vermifilter bed can result in maximum HRT and hence slower percolation into the bed. The quantity of live adult worms and performance per unit area in the vermifilter can also influence HRT. In a novel study to treat rural sewage constantly, Li et al. (2008) observed that once HLR exceeded

3 m^3/m^2/day, some worms escaped from the vermifilter. Nevertheless, an application of 1 m^3/m^2/day was accepted to be the most appropriate for more efficient and stable treatment.

3.3 *Hydraulic Loading Rate*

The amount of wastewater that a given vermifiltration (VF) system can reasonably treat at a given time is the hydraulic loading rate of the vermifilter (VF) system. HLR can thus be defined as the pace at which wastewater enters the vermifiltration system. A high HLR can reduce the HRT of the system and therefore the treatment's efficiency. This is mainly for the reason that wastewater requires a certain contact time with the humus and attached growth in vermifiltration to allow for the adsorption of contaminants, transformation of nitrogen, and reduction of COD (Hughes, 2013). It significantly depends upon the number of live adult earthworms operating per unit area in the vermifilter bed. The size and physical condition of the worms are also vital for determining the HLR. Hydraulic loading rates vary among soil types. The penetration rates depend upon the characteristics of the soil defining pore size distribution, soil morphological characteristics, and clay mineralogy (Sinha et al., 2008).

3.4 *Characteristics of Wastewater*

The alkalinity or acidity (pH) of the wastewater influences the continued existence and activity of worms. Vermifiltration systems have been found to stabilize the alkaline or acidic wastewater. It is also established that the earthworm species *E. fetida* and *E. andrei* can tolerate pH values between 6.2 and 9.7, with juvenile destruction at both upper and lesser pH levels. The composition of wastewater may significantly influence the population of earthworms and can limit the treatment process. The toxicity of diverse components and their threshold limits have not been studied extensively. In an attempt to surmount this information gap, Hughes (2013) studied the concentration of sodium and ammonium salts that reduce the performance of a vermifiltration system and found that sodium chloride is one of the toxic ionic compounds in wastewater. Redox potential of the wastewater is vital in the vermifiltration system (Morand et al., 2005). It ought to remain positive to sustain the population of worms, as worms cannot live in an environment with low availability of oxygen for a long time (Li et al., 2008).

3.5 Effects of Seasonal Variation on Vermifiltration Performance

The successful function of vermifiltration systems may depend on the existing climatic conditions, as they influence earthworm's continued existence. This is a significant consideration in places wherever the temperature drops severely, especially during winter periods, and also in areas with exceeding temperatures above 40 °C. Li et al. (2008) studied the effects of seasonal variation on the treatment efficiency of sewage in vermifiltration systems. The HLR was adjusted by the authors to correct the varying temperature by raising the loading rate throughout the summer months and lessening it during the winter period. It was established that the treatment effectiveness was not considerably affected owing to seasonal fluctuations, excluding the hottest and coldest days.

3.6 Vermifilter Design and Operation

The temperature in the laboratory should be maintained at 21.5 °C with 50% humidity. The vermifilter is made of about 30–40 kg of gravel with a layer of compost or garden soil on top, which forms the vermifilter bed. The vermifilter has provisions to collect filtered water at the bottom through a pipe fitted with a tap. A net of wire mesh lies above the chamber to allow only water to trickle down while holding the gravels above. The bottommost layer of the vermifilter is made of gravel with an aggregate size of 7.5 cm, and it fills up to the depth of 25 cm. Above this layer lies another layer of 25 cm with aggregates of 3.5–4.5 cm sizes filling up. On the top of this is the 20-cm layer of aggregates of 10–12 mm sizes mixed with sand. The topmost layer of about 10 cm consists of garden or compost soil in which the earthworms are released (Fig. 2). The worms are given around one of week settling time to acclimatize in the new environment in the soil bed. About 8–10,000 numbers of earthworms per cubic meter and in quantity (biomass) as 10 kg per cubic meter (cum) of soil for optimal function. The design parameters of vermifiltration include stocking density of earthworms, hydraulic loading rate (HLR), hydraulic retention time (HRT), and the filter media. The quantity of wastewater a vermifiltration system can reasonably treat in a given period of time is the volume of wastewater applied per unit area of the vermifilter bed per unit time, also known as the hydraulic loading rate of the vermifilter system. High hydraulic loading rate leads to reduced hydraulic retention time and possibly reduces the treatment efficiency (Sinha et al., 2008). The hydraulic retention time (HRT) was kept uniformly between 1 and 2 h in all experiments. Around 5–6 L of municipal wastewater is kept in a calibrated 10-L capacity polyvinyl chloride (PVC) drum. These drums are kept on an elevated platform just near the vermifilter kit. The PVC drums had a tap at the bottom to which an irrigation system is attached. The irrigation system consists of simple 0.5-inch polypropylene pipe with holes for trickling water that allows uniform distribution of wastewater on the soil surface (vermifilter bed). Wastewater from the drums flows through the irrigation pipes by gravity. The wastewater percolates down through various layers in the vermifilter bed, passing through the soil layer inhabited by earthworms, the sandy layer, and the gravels and, at the end, was collected in a chamber at the bottom of the vermifilter (Fig. 3). The treated wastewater is collected the next day and analyzed for the efficiency performance of the system which has been measured in terms of chemical oxygen demand (COD), biochemical oxygen demand (BOD), water quality indicator organisms like fecal coliform and total coliform, and removal performance efficiencies of pathogenic microbes like *Salmonella* spp. and *Escherichia coli* (Sinha & Herat, 2002) (Fig. 1).

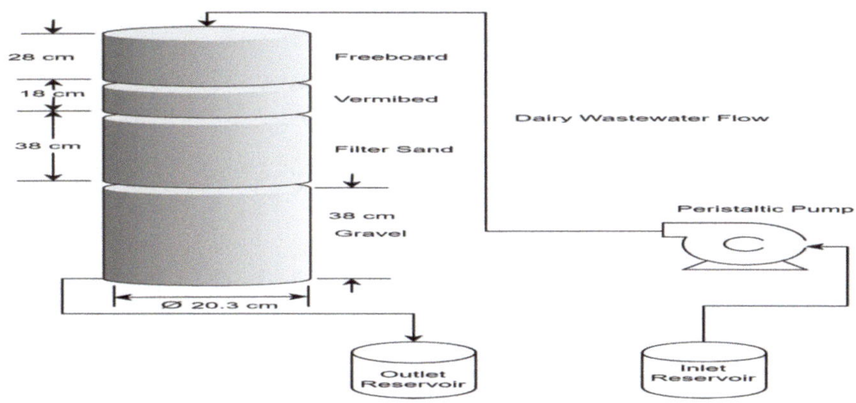

Fig. 1 Schematic setup of vermifiltration (Miito et al., 2024)

Fig. 2 Appearance of sewage before and after treatment. Bottle A: clear vermifiltered sewage water; bottle B: hazy water from the controlled kit; and bottle C: turbid and cloudy sewage water (Sinha et al., 2008)

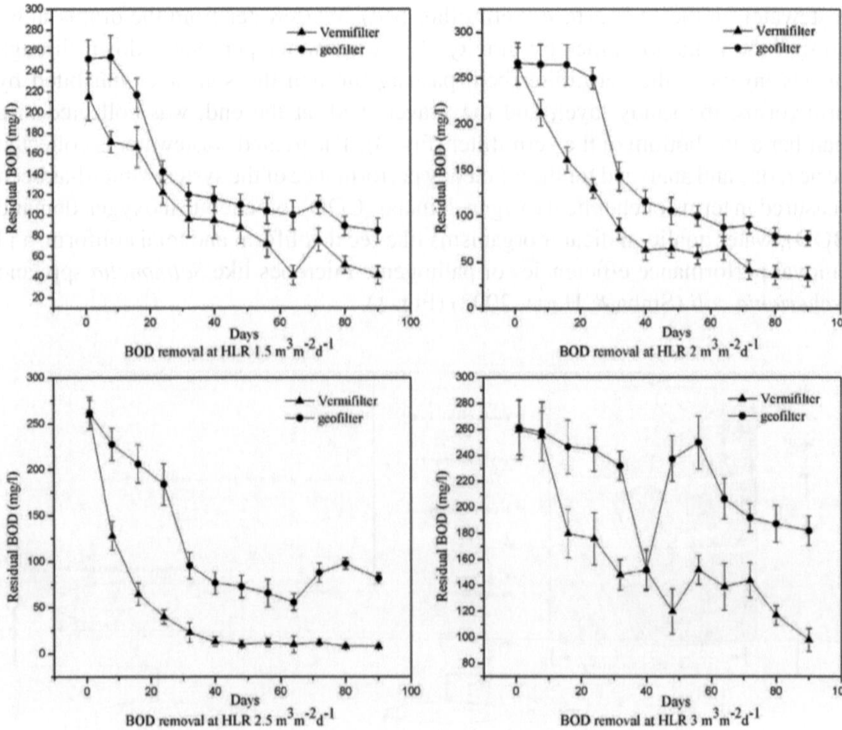

Fig. 3 BOD removal at different hydraulic loading rate (Kumar et al., 2022)

4 Significance of Vermifiltration

Vermifiltration systems are widely recognized for their ability to lower wastewater's levels of solids, organics, and nutrients. These decreases are mostly ascribed to the filtering capabilities of the vermifilters as well as the cooperative activities of earthworms and microbes. Singh et al. (2019) claimed that the primary reason for organic reduction through vermifilters is because the solids are held in the bedding's pores and capillaries, where earthworms eat them and then expel them as nutrient-rich castings that are extremely advantageous for plant growth. The accelerated mineralization of organic phosphates to inorganic orthophosphates is responsible for the observed increases in orthophosphates, especially in systems with higher earthworm densities.

4.1 Reduction of Chemical Oxygen Demand (COD) in Relation to Hydraulic Loading Rate (HLR)

Chemical oxygen demand means the chemical decomposition of organic and inorganic pollutants in wastewater that cannot be biologically removed. The COD amount removed was reported to be higher than that removed by the microbial system. This was actually because the enzymes in the alimentary canal of earthworms aid in the breakdown of those numerous chemicals that could not be degraded by microbes.

Table 1 displays the BOD/COD ratio at various HLRs. In comparison with the influent's initial characteristics, the ratio in vermifilter effluent was found to be significantly lower. This may be because earthworms' guts contain a variety of enzymes that aid in the breakdown of substances that are impossible for geomicrobial processes to break down. The lowest BOD/COD ratio value of 0.14 ± 0.1 was recorded at HLR $2.5\ m^3 m^{-2}\ d^{-1}$. BOD removal in vermifiltration was higher than COD removal in the same reactor. This might be because earthworms rely on the biodegradable portion of wastewater.

Laboratory-scale geofilter and vermifilter tests at varying hydraulic loading rates were conducted by Kumar et al. (2022). Figure 3 describes the differences in BOD removal at various HLRs. It can be shown from Fig. 1 that BOD of effluent was remarkably low both in vermifilter and geofilter. At an HLR of $2.5\ m^3\ m^{-2}\ d^{-1}$, the greatest BOD removal in a vermifilter was 96%, but in a geofilter, it was 70%. The efficiency of the vermifiltration process in comparison with the geomicrobial system is demonstrated by the symbiotic activity of earthworms and aerobic microorganisms, which speeds up and improves the decomposition of organic matter (Rajpal et al., 2012; Tomar & Suthar, 2011). When the HLR was $3.0\ m^3\ m^{-2}\ d^{-1}$, the removal was unsuccessful. This may be connected to elevated humidity and vermifilter scouring, both of which are detrimental to earthworm development and vermifiltration process efficiency (Xing et al., 2010).

In another related study, Miito et al. (2024) treated dairy wastewater in vermifilter systems and achieved reduction efficiencies ranging between 35 and 66% of the COD (Fig. 3a). Higher earthworm densities generally resulted in significantly ($p < 0.05$) greater reductions in COD compared to lower earthworm densities, with the control units (without earthworms) demonstrating the lowest COD reduction efficiency at 35% (Fig. 3b).

Table 1 BOD/COD ratio for vermifilter and geofilter at different HLRs (Kumar et al., 2022)

HLR ($m^3\ m^{-2}\ d^{-1}$)	Vermifilter[a]	Geofilter[a]
1.5	0.33 ± 0.12	0.40 ± 0.15
2.0	0.33 ± 0.17	0.49 ± 0.17
2.5	0.14 ± 0.10	0.50 ± 0.17
3.0	0.14 ± 0.13	0.53 ± 0.10

[a]Mean concentration ± standard deviation of the physicochemical parameters

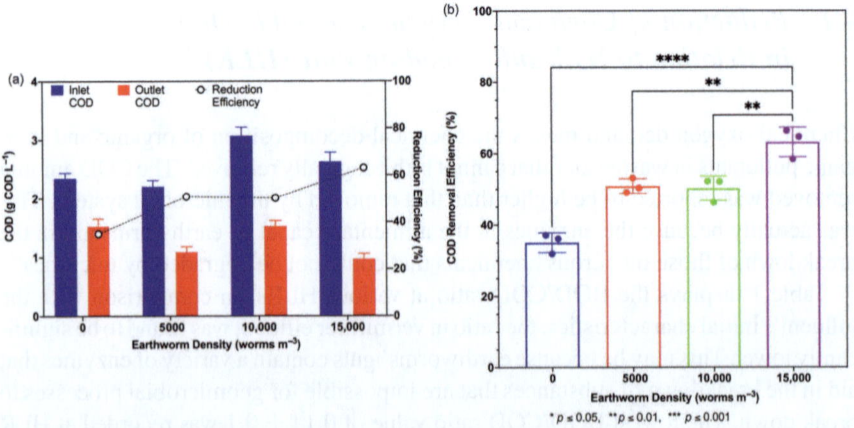

Fig. 4 The effect of earthworm population density on **a** COD reduction efficiency and **b** the significance of the various earthworm densities on COD reduction from dairy wastewater. Error bars indicate the standard deviation of the dataset (Miito et al., 2024)

Higher COD removal in vermifilter is credited to the activity of earthworms and their associated microorganisms that reduce the wastewater organics by their enzymatic activity (Kumar & Kaushal, 2022; Sinha et al., 2008). The average COD removal rate in a vermifilter is over 74%, whereas in a geofilter, it is 68%. The percentage COD removal rate is significantly higher in vermifilter as compared to geofilter. This is attributed to the enzymatic activity of the microorganisms in the earthworms. Though significant reduction of COD is achieved by the microbial–geological system in the geofilter, the formation of sludge and colonies of bacteria and fungi for longer times frequently chokes the system, and it fails to work (Li et al., 2008) (Fig. 4).

4.2 Reduction of Total Suspended and Dissolved Solids

A total suspended and dissolved solid (inorganic and organic pollutants) that are moreover suspended or dissolved in the wastewater. These solids are trapped and assemble over time as sludge and then clog the system that ceases to function properly. Nevertheless, in the vermifilter, the biosolids are ingested by the earthworms and egested as vermicompost. Therefore, there will be no choking and discontinuous functioning or stationary phase of the vermifilter bed. Sinha et al. (2008) established that total dissolved solids (TDS) and total suspended solids (TSS) removal efficiency of 90–92% and 90–95%, respectively, for dairy wastewater. Manyuchi et al. (2013) reported the removal efficiency of TDSS by 95% using a vermifilter to treat sewage wastewater, and Tomar and Suthar (2011) also found that the removal efficiency of TSS and TDS by a vermifilter was 88.6% and 99.8%, respectively,

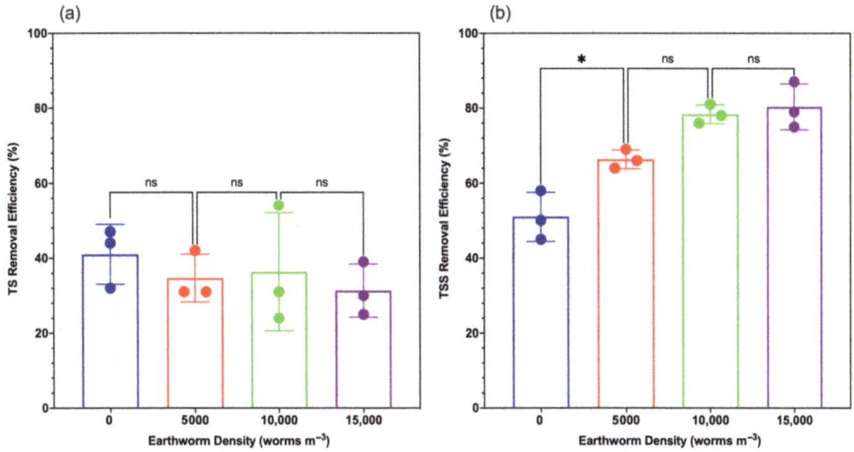

Fig. 5 The effect of earthworm population density on **a** total solids (TS) and **b** total suspended solids (TSS) reduction efficiencies (Miito et al., 2024)

for urban wastewater. Xing et al. (2010) reported 57–77.9% removal efficiency of total suspended solids (TSS) using pilot-scale vermifiltration. Kumar et al. (2011) reported that the efficiency of TDS decreased by 82% in a vermifilter for sewage wastewater treatment.

Dairy wastewater concentrations of TS and TSS ranged from 1.1 to 5.4 g L^{-1} and 0.7 to 2.6 gL^{-1}, respectively. The vermifilter units showed reduction efficiencies ranging from 24 to 54% for TS and 50% to 87% for TSS. Notably, no significant differences ($p < 0.05$) were found between the TS reduction efficiencies among all earthworm densities (Fig. 5a) and the control unit (without earthworms), as shown in Fig. 5b. The highest reduction efficiencies of suspended solids were found in the vermifilter units with 10,000 and 15,000 worms m^{-3}, which were attributed to the earthworms' increased consumption of the suspended solids (Miito et al., 2024).

4.3 Reduction of Turbidity

Higher levels of turbidity are usually associated with disease-causing microorganisms. Untreated raw sewage turbidity value of 120 nephelometric turbidity unit (NTU) was reduced to 1.5 NTU with worms (vermifiltration) and 3.6 NTU without worms (geofilter). The results specify that the average decrease in turbidity by earthworms is over 98%, while that without earthworms is also significantly high (Chaudhuri et al., 2000). It appears that the geological system too plays a very significant role in turbidity elimination by "adsorption" of suspended particles on the surface of the soil, sand, and the gravels. Turbidity of treated wastewater is affected by high retention time, and percent removal of turbidity increases with an increase in HRT.

4.4 Reduction of Nitrogen

Nitrogen compound concentrations in dairy wastewater were found to be 238–386 mg (total nitrogen [TN]) L^{-1}, 99–151 mg [NO_3^-–N] L^{-1}, and 72–184 mg [NH_4^+–N] L^{-1}. For TN, NO_3–N, and NH_4^+–N, the vermifilter systems obtained reduction values of 41%–89%, 34%–74%, and 46%–86%, respectively (Fig. 6a–c). The findings also showed that the reduction efficiencies of the three nitrogen species increased significantly in general with larger earthworm concentrations. The systems that had the highest earthworm population density (15,000 earthworms m^{-3}) also had the highest TN reduction efficiencies (86%, 74%, and 74%) among all the systems that were examined.

The vermifilter's reported increases in TN, NH_4^+–N, and NO_3–N reduction efficiencies at greater earthworm densities imply that the rates of ammonification, nitrification, and denitrification are influenced by earthworm activity. These activities take place as wastewater moves through the units' anoxic and aerobic levels. Ammonium nitrogen and/or ammonia are produced when organic sources of nitrogen undergo ammonification. Nitrogen from ammonium undergoes additional nitrification to produce nitrate nitrogen, which may subsequently undergo denitrification to produce free nitrogen gas. The reduction and transformation of nitrogen are additionally facilitated by a few more physicochemical processes, such as adsorption, filtering, and sedimentation.

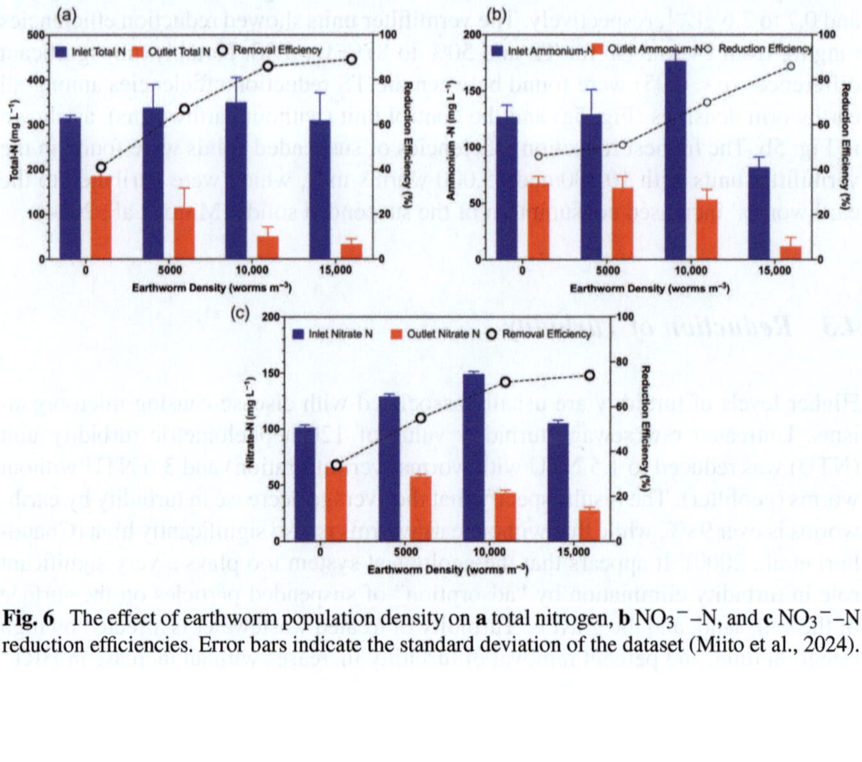

Fig. 6 The effect of earthworm population density on **a** total nitrogen, **b** NO_3^-–N, and **c** NO_3^-–N reduction efficiencies. Error bars indicate the standard deviation of the dataset (Miito et al., 2024).

Increased earthworm activity has also been shown to support aerobic nitrogen degradation, enzymatic activity ammonification/nitrification, and oxygen transfer in vermifilter systems (Singh et al., 2019; Wang et al., 2021).

4.5 Reduction of Phosphorus

The vermifiltration systems had reduction efficiencies of between 3 and 17% of total phosphorus (TP) for the total phosphorus concentration in dairy wastewater of 34–44 mg [TP] L^{-1} and 9.1–18 mg [ortho-P] L^{-1}; however, they also showed up to 38% increases in ortho-P concentrations (Fig. 7a, c). The control units had the highest TP reduction efficiencies (Fig. 7b), while the units with the highest earthworm densities (between 10,000 and 15,000 earthworms m^{-3}; Fig. 7d) had the highest increases in ortho-P. Samal et al. (2017) state that the main causes of phosphorus reduction in vermifilters are adsorption on bedding surfaces and biotransformation to soluble forms like ortho-P.

The reduced adsorptive capacity of the vermifiltration beds as a result of earthworm burrowing activity and phosphorus fixing in the earthworms' excreta is the reason for the lower TP reduction efficiency in the vermifiltration units with greater earthworm densities. However, the bioconversion of organic phosphorus and polyphosphates into ortho-P, a more soluble form of phosphorus, is primarily responsible for the rise in ortho-P concentrations with higher earthworm populations. Jiang et al. (2016) reported on the conversion of organic phosphorous and polyphosphates to ortho-P during vermifiltration, attributing this to microbiological and enzymatic activity triggered by earthworms.

The reduced adsorptive capacity of the vermifiltration beds as a result of earthworm burrowing activity and phosphorus fixing in the earthworms' excreta is the reason for the lower TP reduction efficiency in the vermifiltration units with greater earthworm densities. However, the bioconversion of organic phosphorus and polyphosphates into ortho-P, a more soluble form of phosphorus, is primarily responsible for the rise in ortho-P concentrations with higher earthworm populations. Jiang et al. (2016) reported on the conversion of organic phosphorous and polyphosphates to ortho-P during vermifiltration, attributing this to microbiological and enzymatic activity triggered by earthworms.

5 Mechanisms of Earthworm Activities in Vermitechnology

Earthworms are decomposers and adaptable waste eaters. Earthworms feed primarily on organic waste and employ different actions to degrade the waste.

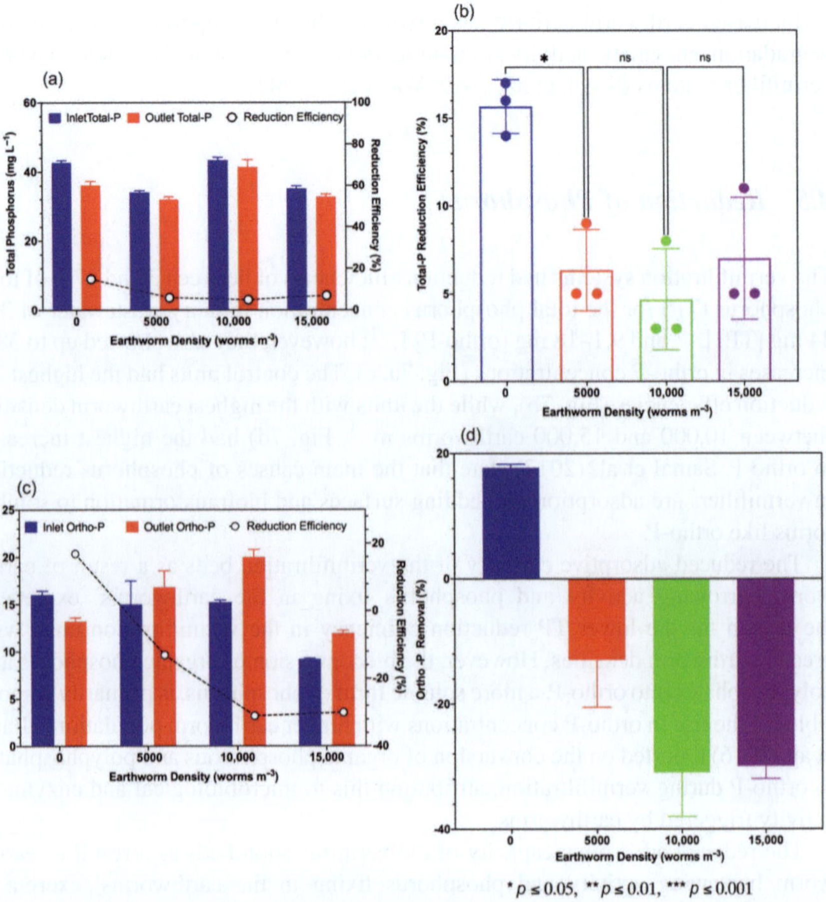

Fig. 7 The influence of **a** earthworm population density on total phosphorus reduction efficiency, **b** the significance of the various earthworm densities on total phosphorus reduction rates, **c** the effect of earthworm population density on ortho-P reduction efficiency, and **d** significance of the various earthworm densities on ortho-P reduction, during vermifiltration. Error bars indicate the standard deviation of the dataset (Miito et al., 2024).

5.1 *Grinding Action*

They act as aerators, crushers, chemical degraders, grinders, and biological stimulators, thus promoting the growth of "beneficial decomposer bacteria" (Sinha & Herat, 2002). The stones in the muscular gizzard of the earthworm aid in grinding the solid waste into fine particles measuring about 2–4 microns, which pass for enzymatic actions in the intestine. The intestine and gizzard collectively act as a "bioreactor" (Sinha & Herat, 2002). The organic loadings of wastewater in the vermifilter soil bed are strengthened by earthworms when the clay particles are granulated, thus raising the "hydraulic conductivity" of the system. They also grind sand and silt particles, hence giving a large total specific surface area, which enhances the capability to "adsorb" inorganics and organics from the wastewater. In addition, the vermicast formed as well offers excellent hydraulic conductivity of sand. This is ultimate for diluted wastewater like sewage.

5.2 Enzymatic Action

Earthworms exude enzymes, cellulases, amylases, proteases, chitinases, and lipases in their alimentary canal, which aid speedy biochemical change of the proteinous and cellulosic materials in the organics from solid waste. They consume the food substances, pick the harmful microbes, and mixed them with minerals together with beneficial microorganisms as "vermicasts" in the earth (Zhao et al., 2010). The two vital processes—

(i) vermiprocess and microbial processes—work simultaneously in the vermifiltration system.

Earthworms accommodate millions of decomposer (biodegrader) microorganisms in their alimentary canal and are excreted alongside nitrogen (N) and phosphorus (P) in their excreta (vermicast). The plant nutrients phosphorus and nitrogen are further used by microbes for reproduction and vigorous action. The microflora connected with the intestine and vermicasts of the earthworms and discovered species like *Pseudomonas, Spiroplasma, Mucor, Paenibacillus, Azoarcus, Burkholderia Alcaligenes,* and *Acidobacterium* which possess the potential to degrade numerous organic pollutants.

(ii) Suspended and dissolved (organic and inorganic) solids

They are held by adsorption and stabilized during complex biodegradation processes in the soil occupied by earthworms and aerobic microbes. Soil processes intensification and aeration by earthworms facilitate soil stabilization and filtration system to become successful and lesser in size (Sinha et al., 2008).

(iii) Digestive enzymes live in the body of earthworm such as phosphates, protease, alkaline, and cellulase. These enzymes had an important relationship with the N and P cycle and the turnover of carbon (Aira et al., 2007).

Other kinds of earthworm enzymes were the antioxidant enzymes, such as SOD and CAT, which had been often used as biomarkers of environmental stress. These enzymes could protect cells against adverse effects of reactive oxygen species. An increase in the activities of these enzymes indicated deterioration in environmental conditions.

The production of enzymes by microorganisms in earthworms is found to degrade and stabilize the organics in wastewater is significant due to their ability to decompose cellulose, proteins, starch, and sugars, which ensures the integrity of the vermifiltration system. The examination on the enzymatic activity of the isolated bacterial species would supply vital data for understanding organic matter degradation in vermifiltration (Xing et al., 2010).

(iv) Earthworms consume harmful microbes in the wastewater, thus preventing choking of the system and maintain a culture of useful biodegrader microbes to function (Sinha et al.,).

5.3 Effect of Earthworms on Microbial Activity and Sludge Treatment

Polymerase chain reaction denaturing gradient gel electrophoresis (PCR-DGGE) is widely used to analyze and measure bacterial diversity. PCR-DGGE reveals the distributions and compositions of bacterial communities in the vermifilter at the genetic level and express evidence of the biological reaction of the microbes to earthworms. The different banding patterns and intensities obtained upon analysis of samples from different depths of the filter beds in the biofilter and vermifilter systems are likely due to the presence or absence of earthworms. The available organic matter reduced when the influent passed through the filter bed; accordingly, the microbial diversity distribution along the depth followed the same pattern in the BF. However, in the VF, there is a higher bacterial diversity and richness in the film collected from the top and the bottom of the filter bed. These findings show that the burrowing activities of the earthworms led to improved aerobic conditions in the vermifilter bed, which favored the conditions for aerobic microorganisms. In addition, there are more earthworms dwelling at the top and the bottom of the filter bed, where more oxygen is available, which tally well with the greater bacterial diversity observed at the top and the bottom of the filter bed (Zhao et al., 2010).

The mucus and casts produced by earthworms have a stimulatory effect on the microorganisms. Mucus is a source of carbon for microorganisms, and the casts are often enriched with available C, N, and P (Aira et al., 2007). They contain more active microbial communities compared to the foods that earthworms consume (Suthar, 2009). In addition, it has been recognized that *Eisenia fetida* has a unique native gut-associated microflora that contributes to the growth of a diverse microbial community in vermifilter systems (Toyota & Kimura, 2000). Analysis of VF biofilm is dominated by members of the phylum Proteobacteria, and Pseudomonas sp. ascribed to the phylum Proteobacteria was exclusively detected in the VF (Table 2).

5.4 Sludge Stabilization by the Filter System

It is apparently complicated to calculate the exact total biomass quantity in the filters, as the biomass amount gradually decreases along the depth of the reactor.

Table 2 Sequences closely related to those of the denaturing gradient gel electrophoresis bands (Zhao et al., 2010)

Band name	Closely related sequences (accession no.)	Identity (%)
1	Uncultured Acidobacteriaceae bacterium clone GASP-WA1S1_E10	96
	16S ribosomal RNA gene (FJ495179)	
2	Uncultured gamma proteobacterium clone	98
	E03_SGPL02 16S gene (EF221170)	
3	Uncultured bacterium clone nbw503d06c1	94
	16S ribosomal RNA gene (GQ102089)	
4	Uncultured Xanthomonadaceae bacterium clone	96
	Amb_16S_839 16S (EF018572)	
5	Uncultured Acidobacteriaceae bacterium clone GASP-WA1S1_E10	96
	16S ribosomal RNA gene, partial sequence (EF072059)	
6	Uncultured Bacteroidetes bacterium clone MA00162B11	97
	16S ribosomal RNA gene, partial sequence (FJ532911)	
7	Pseudomonas sp. An30H-SC-S gene for 16S rRNA, partial sequence (AB267465)	97
8	Pseudomonas sp. An30H-SC-S gene for 16S rRNA, partial sequence (AB267465)	97
9	Uncultured soil bacterium clone FACE.R2.EC.B04	98
	Small subunit ribosomal RNA gene, partial sequence (FJ621004)	
10	Uncultured bacterium clone Toolik_Jun2005_Intertussock_39	93
	16S ribosomal RNA gene (DQ510149)	

Therefore, the volatile suspended solids (VSS) reduction can be determined using the mass balance between the quantity of total influent and effluent (including both the mixture in the sedimentation tank and supernatant) according to 24 h of operation using Eq. (1).

$$\text{VSS reduction}(\%) = \left(Q.C_o.a - Q.C'.b\right)/(Q.C_o.a) \tag{1}$$

where Q is the influent flow during a day period (24 h); a is the influent, ratio of VSS/ S; b is the effluent VSS/SS; C_o is the influent SS concentration; and C' is calculated by Eq. (2).

$$C' = \frac{(Q - V_1).C_1 + V_1.C_2}{Q} \tag{2}$$

where C_1 is the supernatant SS concentration, mg/L; C_2 is the SS concentration of the mixture in the sedimentation tank at the end of the period, mg/L; and V_1 is the effective volume of the sedimentation tank, L. The VSS reduction obtained by the digestion of earthworms ($R_{VSS,earthworm}$) in the VF system is defined as the amount of sludge digested by the earthworms (W_{digest}) divided by the amount of organic matter in the influent as shown in Eq. (3).

$$R_{VSS,earthworm}(\%) = \left(W_{digest}.m_{earthworm}\right)/(Q.C_o.a) \quad (3)$$

$$W_{digest} = \left(W_{ingest} - W_{cast}\right) \quad (4)$$

where $m_{earthworm}$ is the total biomass of earthworms in the vermifilter

Q, C_o, and a are defined above; W_{digest} is the digested sludge by earthworms; W_{ingest} is the sludge ingested by earthworms; and W_{cast} is the earthworm casts. An estimate for W_{ingest} was calculated by Eq. (5).

$$W_{ingest} = W_{cast}(1 - OM_1)/(1 - OM_2) \quad (5)$$

where OM_1 is the organic matter content in the cast of earthworm (%), OM_2 is the organic matter content in the raw sludge, and W_{cast} of the VF could not be obtained directly using Eq. (5).

Therefore, the excrement of the earthworms was dried at 105 °C for 2 h, after which the dry weight was measured. W_{cast} was expressed in mg of dry weight of the cast per gram of fresh weight of earthworm per day (mg g^{-1} d^{-1}). Earthworms in the VF ingest and digest sludge (Table 2). On average, the cast production per gram of earthworms and the corresponding organic matter content for the cast were 11.9 ± 1.9 (mg g^{-1} d^{-1}) and 52.6 ± 4.2%, respectively (Table 3). Therefore, the average VSS reduction due to direct digestion by the earthworms was 13.6%. Furthermore, earthworms mostly digest the organic portion of their feed. Sludge breakdown produces lesser-sized sludge particles, which leads to an enhancement in the sludge biodegradation by microbes.

The typical scanning electron microscopy (SEM) micrographs of influent and effluent samples were collected, in addition to the earthworm cast. The monographs revealed loosely filled, fluffy composition characterized by a dominance of rod-shaped cells in the raw sludge and obvious extracellular polymeric substance (EPS) medium in which the cells were entrenched. The floc structure in the biofilter effluent

Table 3 The amount of sludge ingested and digested by earthworms (Zhao et al., 2010)

Organic matter content in earthworm cast (%)	Earthworm cast production (mg g^{-1} d^{-1})	Sludge digested by earthworms (W_{digest}) (mg g^{-1} d^{-1})	Sludge ingested by earthworms (W_{ingest}) (mg g^{-1} d^{-1})
52.6 ± 4.2	11.9 ± 1.9	6.4 ± 1.0	18.3 ± 2.8

Note Organic matter content in earthworm cast and earthworm cast production were measured, while W_{digest} and W_{ingest} were calculated using Eqs. (4) and (5), respectively had similar characteristics to that of the raw influent, except that a remarkable reduction in rod-shaped cells was observed, indicating the incomplete interference of extracellular polymers in the biofilter effluent.

The new casts produced by the earthworms in the vermifilter exhibited a different physical appearance, being porous, and fragmented. The casts also showed a significant reduction of flocs. On the contrary, the surface composition of the vermifilter effluent sludge was more compacted than that of the earthworm cast, an indication of a significantly lower amount of filamentous bacteria.

6 Advantages of Vermifiltration Technology Over Conventional Systems of Sewage Treatment

There are a number of advantages of the vermifiltration system when compared with other biological wastewater treatment systems.

Vermifiltration of wastewater treatment is low-energy system with distinct advantages over all the conventional biological wastewater treatment systems, which is highly energy demanding and expensive to install and manage without any income generated. There is one hundred percent (100%) capture of organic materials in the vermifilter process, with less capital and operating costs and production of the vermifilter, a high value-added end product.

6.1 No Formation of Sludge

The hydraulic conductivity in any filtration system of the infiltrative surface is important in achieving effective treatment. Though, in the soil filter, there is a significant removal of biological oxygen demand (BOD), COD, and suspended solids, the formation of sludge and films of microorganisms that multiply under conditions of excess nutrients as slime capsules, creating a "clogging zone" that can decrease the hydraulic conductivity of the filtration medium and may cause system failure (US EPA, 2002). The vermifiltration system is convenient to operate as there is no sludge formation when compared to conventional activated sludge treatment methods (Sinha et al., 2010a, 2010b). Over 10% removal rate of nitrogen and total phosphorus with the use of vermifilters compared to activated sludge systems (Li et al., 2008). Aeration and clogging that that are limitation in constructed wetland systems and reed beds can be handled by vermifiltration.

6.2 *Vermifiltration is a Low-Energy System*

Most vermifiltration systems require no external source of energy to pump wastewater, though some energy may be required in pumping the wastewater to the vermifiltration unit if gravity flow is not adequate (Sinha et al., 2008). The vermifiltration method of sewage management is a low-energy reliant and has distinct benefits over every conventional natural wastewater management system—the "trickling filters," "rotating biological contactors," and "activated sludge process," which are extremely energy demanding, expensive to install and manage, and do not produce any income. Since the conventional methods are typically the flow technologies and have limited hydraulic retention time (HRT), it constantly results in a "left behind stream" of compound organics and heavy metals (while just the simple organics are used) in the "sludge" that requires further treatment (requiring additional energy) prior to landfill disposal. This becomes unproductive. There is 100% capture of organic materials in the vermifilter process. The assets and cost of operation are less with a high rate added end product (vermicompost) (Sinha & Herat, 2002).

6.3 *Reduction of Heavy Metals and Endocrine-Disrupting Chemicals by Earthworms*

Although some trace amounts of heavy metals are essential for living organisms for metabolism, any excess amount of these metals can be harmful to life (Aemere & Ogunlaja, 2007). Earthworms are capable of bioaccumulation of heavy metal ions by the formation of organometallic complexes in their gut, which reduces heavy metals solubility. This decreases the concentration of water-soluble heavy metals (Suthar, 2009). The decrease in mobility and bioavailability of heavy metals throughout vermicomposting by earthworms is achieved through two major forms of cellular adaptations: the binding of metals to nuclear proteins to form inclusion nuclear bodies and the cytoplasmic method concerning the production of a specific metal-binding protein, metallothionein (MT), inside the chloragogenous tissue. Earthworms have a specific ability to regulate heavy metals; particularly, trace metals, accumulation, and control mechanisms could be metal-specific (Aemere et al., 2020). Metalloprotein and metallothionein work against the ecological stress caused by a mixture of contaminants, specifically by metal (Vondel, 2006). The binding of metals to organic matter mostly more firmly bound fractions, partially reducing the accessibility of metals to earthworms. The earthworm gut could adjust the mobility of metals and support their absorption. It is well recognized that MT also acts as a part of the antioxidant system and is involved in the scavenging of free radicals and reactive oxygen species (Aemere et al., 2020). Every contaminant in the environment has a threshold, though, beyond which detoxification occurs.

These are defense mechanisms that organisms develop in the presence of the contaminants. One of such mechanisms is that concerning metallothioneins (MTs), proteins containing cysteine, which are capable of linking toxic metals (Conti, 2008). Metallothioneins are low-molecular weight cysteine-rich (up to 33% by composition) proteins expressed by organisms under stress states, mainly when induced by metals at certain levels. MTs are encoded by multigene families that differ in their responses to diverse inducers, including heavy metals, hormones, hydroxyl radicals, oxidants, strenuous exercise, superoxide, and glucocorticoids, generated by gamma radiation and cold exposure. The major roles of MTs comprise of protection against oxidative stress, detoxification of xenobiotic metals, and the homeostasis of trace metals (Zn, Cu, Mn, Fe, Pb, and Cd among others) and maintaining redox pool, metal ion transport, scavenging of radical and regulation of expression (Adebayo & Anumudu, 2015).

Earthworms normally use three ways to metabolize heavy metals:

(i) Control in fatty (chloragogen) cells of the wall of the alimentary canal
(ii) Storage of heavy metals in waste nodules produced within the body cavity
(iii) Excretion via the calciferous glands.

The gut-related processes in earthworms may also raise metal accessibility. The earthworm-modified epithelial cells, which form the chloragosomes, the eleocytes of the alimentary canal, contain constituents of ion exchange compounds—phenolic hydroxyl, carboxyl, phosphoric acid, and sulfonic acid groups—that act as cation exchange system with the ability of taking up and accumulating heavy metals. They willingly bioaccumulate metals (lead [Pb], copper [Cu], cadmium [Cd], mercury [Hg], manganese [Mn], calcium [Ca], iron [Fe], and zinc [Zn]). The transformation of highly toxic Cr(VI) to nontoxic form Cr(III) during the metabolic process in the mitochondrial and cytoplasmic fractions was reported through vermicomposting with *Eisenia fetida* (Singh & Kalamdhad, 2016). Metal toxicity will take place when the ability of these mechanisms to bind metals is exceeded.

Endocrine-disrupting chemicals (EDCs) alter the regular function of the endocrine system. Harmful effects of EDCs are more prominent in young animals, which have lesser body mass, a quicker cell division rate than adults, and are in the middle of their developmental process. EDCs that influence the endocrine system by fastening to hormone receptors alter the reproductive anatomy, decrease reproduction rates (Heindel et al., 2012), result in somatic anatomical abnormality, and change the body state of aquatic organisms. Tissues of earthworms have been reported to contain significantly high concentrations of EDCs (dibutyl phthalate dioctyl phthalate, bisphenol-A, and 17β-estradiol) in (*E fetida*). Earthworms either accumulate or degrade "organochlorine pesticide" and "polycyclic aromatic hydrocarbons" (PAHs) residues in the medium in which they feed. It has been established that earthworms can also tolerate toxic chemicals in soil environments. In 1976 in Italy, there was a chemical plant explosion at Seveso that contaminated vast areas of the soil with chemicals like 2, 3, 7, 8-tetrachlorodibenzo-p-dioxin (TCDD), which is extremely toxic; different fauna species perished except for some earthworms that survived. The earthworms were able to ingest TCDD-contaminated soils and bioaccumulate dioxin in their tissues, which was then concentrated on an average 14.5-fold. Studies also showed that *Eisenia fetida* survived 1.5% crude oil with several toxic organic pollutants (Sinha, 2014).

6.4 Reduction of Pathogens and Antibacterial Properties of Earthworms

The presence of coliforms is frequently used as an indicator of general pathogenicity in any sample. Previous studies have revealed that vermicomposting involves a relationship between earthworms and microorganisms in which microorganisms perform biochemical degradation of the waste material, while earthworms degrade and regulate sludge through muscular actions of their upper gut and add mucus to the ingested substance, thus increasing the surface area for microbial action (Aira et al., 2007; Suthar, 2009). This indicates a mutualistic association between earthworms and microorganisms. The earthworm and microorganism connections accelerate the mineralization of organic materials, which favors the breakdown of excreted polysaccharides.

Earthworms are vital macerator organisms, and they constantly devour the colonies of microorganisms in the vermifiltration process. Earthworms are capable of consuming pathogens (fungus, bacteria, protozoa, and nematodes) present in both wastewater and sludge. They also produce a coelomic fluid that possesses "antibacterial" properties and captures the formation of all microorganisms that cause decaying. Some bacteria and fungi found in the worms also generate "antibiotics" that kill pathogenic organisms in the waste biomass, making the medium virtually germ-free. The fungus *Penicillium* spp. which produces ~~antibiotics~~ antibiotic "penicillin," has been reported to be present in the intestine of earthworms. The removal rate of pathogens, *Salmonella* spp., enteric viruses, fecal coliforms (*E. coli*), and ova of helminths from sewage and sludge is much faster when they are processed by *Eisenia fetida*. Among all the pathogens, *Escherichia coli* and *Salmonella* spp. are greatly reduced (Aira et al., 2007).

Vermifiltration can decrease pathogens to harmless levels through the action of intestinal enzymes in earthworms (Table 4). In addition, the microbes that earthworms leave behind are useful to the VF as they compete with pathogenic organisms for the limited nutrients. The likely reason for pathogen removal is ascribed to the fact that these pathogens get subjected to different toxic and antibiotic secretions from the earthworms and related microflora (Sinha et al., 2008). So, there is a chance of the antibacterial activity of the microorganisms that inhibits or prevents the increase of pathogens during the treatment.

Diversity of microorganisms exists in the body of earthworms, which exhibits the potential to reduce or prevent the growth of other known pathogens (Zhao et al., 2010). Some bacteria and fungi fostered by earthworms also produce antibiotics, which kill the pathogenic organisms in the waste biomass, making the medium virtually sterile and odorless. So there is a likelihood of the resistance mechanism to inhibit the growth

Table 4 Pathogen removal performance of vermifiltration

Organisms	Influent	Effluent VF	K	Effluent GF	K
Total coliforms (MPN/100 mL)	3.5×10^8	2.5×10^5	3.15	2.0×10^6	2.24
Fecal coliforms (MPN/100 mL)	2.0×10^6	$8.3 \times 10^{2.5}$	2.88	6.5×10^4	1.49
Fecal streptococci (MPN/100 mL)	3.3×10^6	$1.9 \times 10^{2.5}$	3.74	3.0×10^4	2.04
Total heterotrophic bacteria (CFU/mL)	1.4×10^9	2.0×10^5	3.85	6.8×10^6	2.32
Total fungi (CFU/mL)	2.2×10^6	7.5×10^2	3.46	1.6×10^4	2.14
Actinomycetes (CFU/mL)	6.5×10^5	5.2×10^4	1.09	2.5×10^5	0.4
Salmonella (CFU/mL)	1.2×10^6	1.5×10^2	3.9	1.2×10^4	2.0
E. coli (CFU/mL)	1.5×10^6	1.4×10^4	2.03	1.5×10^5	1.0

All values are average; $n = 17$; and K = log removal (Arora et al., 2014).

of pathogens (Sinha et al., 2010a, 2010b). There exists a mutualistic interaction of earthworms and microorganisms—in an environment that is best appropriate for earthworms that may further improve the antibacterial activity and cause the death of pathogens in a vermifilter. This significant observation described the antibacterial property of the isolated microflora and a possible reason for the removal of pathogens.

6.5 *Vermifilter Treatment of Sewage is an Odorless System*

There is a lack of stinking smell as the earthworms arrest putrefying matters in the wastewater and sludge. By their burrowing behaviors, they also create an aerobic condition in the soil bed and the waste substrate therefore preventing the actions of anaerobic microorganisms that release stinking hydrogen sulfide and mercaptans (Sinha et al., 2008).

6.6 *Source of Fish and Animal Feeds*

Protein-rich probiotic feeds for fishery, poultry, and piggery are produced from earthworm biomass. The availability of food (organic matter and microbes) and water that provide moisture are available in wastewater for rapid multiplication of earthworms to produce a huge population (biomass) within a short period of time. Large quantities of worm biomass are available as "probiotic" feed for fish, poultry, and livestock

Fig. 8 Vermifiltration of wastewater demonstration site in Scotland (IMMOQUA, 2018)

farming. It is a good source of essential amino acids, lysine, and methionine (Hughes, 2013).

6.7 Wastewater Recycling and Reuse

Water is a limited and scarce resource. A global increase in population implies more water will be required for food production, which means placing an extra burden on accessible water provisions. The development of affordable and accessible natural-based wastewater treatment processes like vermifiltration is very vital. For example, in Scotland, an innovation sanitation system (Littlemill) uses earthworms to treat wastewater (Fig. 8). It is therefore relevant that recycling and reuse are fundamental for sustainability. Vermifiltration of restaurant wastewater could be reused for agricultural and horticultural purposes (Addy et al., 2019).

7 Conclusion

Vermifiltration is a novel technology capable of providing sustainable management of organic wastewater treatment by means of recycling water and nutrients for an eco-friendly environment. However, to bring vermifiltration to a large-scale commercial treatment protocol, more comprehensive research is essential to develop the effectiveness of vermifiltration system. The investigation of potentially harmful microbes before and after treatment is also advantageous, as a large amount of wastewater contains unsafe pathogens; therefore, it is ideal to subject them to proper analyzing before the treated water can be discharged or reused.

Vermifiltration systems rely on living organisms (earthworms and microbes) to treat wastewater. With increasing earthworm bulk densities during vermifiltration, there was larger reduction efficiencies of 41%–89% (TN), 46%–86% (NH_4^+-N), 34%–74% (NO_3^-–N), 3%–17% (TP), 18%–38% (ortho-P), 35%–66% (COD), 24%–54% (TS), and 50%–87% (TSS). In addition, their burrowing behaviors create aerobic environment, thus preventing anaerobic microorganisms from generating mercaptans and hydrogen sulfide that are accountable for foul smell. There is lack of sludge production, which is a main challenge faced in the conventional wastewater management. Most critical factors needed for optimal performance in vermifilter system include climatic conditions, constituents of wastewater, water loading capacity or rate, retention time, species of earthworm, and biomass. Studies from different scholars recommend vermifiltration for rural areas, smaller communities, and small-scale industries to produce organic wastewater as a feasible substitute to conventional wastewater treatment as it provides conditions that are favorable for the growth of earthworms.

References

Addy, J. V., Aguoru, C. U., Imandeh, N. G., Azua, E. T., & Olasan, J. O. (2019). Studies on vermifiltration of restaurant effluent and reuse in Benue state, North Central, Nigeria. *International Journal of Environment, Agriculture and Biotechnology, 4*(2), 456–461. https://doi.org/10.22161/ijeab/4.2.29

Adebayo, A., & Anumudu, C. (2015). Differential expression of metallothionein-I and cytochrome p450–2a5 (cyp2a5) in mice in response to lead acetate exposure and industrial effluents in Ibadan, Nigeria. *5*(July). https://doi.org/10.1177/0748233715594107

Adnan, M. (2019). The uniqueness of microbial diversity from the gut of earthworm and its importance, April.

Aemere, O., & Ogunlaja, O. O. (2007). Physico-chemical analysis of water sources in Ubeji community and their histological impact on organs of albino mice. *Materials and Methods.*

Aemere, O., Sharma, V., & Lin, J. (2020). *Metallothioneins in earthworms: The journey so far. 5*, 14–21.

Aguoru, C. U., Azua, E. T., & Olasan, O. J. (2015). Approaches to minimizing and overcoming current biodiversity loss. *British Journal of Environmental Sciences, 3*(3), 12–26.

Aira, M., Monroy, F., & Domı, J. (2007). *Microbial Ecology, 54*, 662–671. https://doi.org/10.1007/s00248-007-9223-4

Amobonye, A., Aruwa, C., Aransiola, S. A., John, O., Alabi, T. D., & Lalung, J. (2023). Exploring the potential of fungi in the bioremediation of pharmaceutically active compounds. *Frontiers in Microbiology, 14*, 1207792. https://doi.org/10.3389/fmicb.2023.1207792

Aransiola, S. A., Ijah, U. J. J., Abioye, O. P., Victor-Ekwebelem, M. O. (2021). ANAMMOX in wastewater treatment. In N. R. Maddela, L. C. García Cruzatty, & S. Chakraborty (Eds.), *Advances in the domain of environmental biotechnology. Environmental and microbial biotechnology*. Springer. https://doi.org/10.1007/978-981-15-8999-7_15

Aransiola, S. A., Ijah, U. J. J., Abioye, O. P., & Bala J. D. (2022). Vermicompost-assisted phytoremediation of toxic trace element-contaminated soil in Madaka, Nigeria using Melissa officinalis

L and Sida acuta. *International Journal of Environmental Science and Technology*. https://doi.org/10.1007/s13762-022-04105-y

Arora, S., Rajpal, A., Bhargava, R., Pruthi, V., Bhatia, A., & Kazmi, A. A. (2014). Antibacterial and enzymatic activity of microbial community during wastewater treatment by pilot scale vermifiltration system. *Bioresource Technology, 166*(January 2021), 132–141. https://doi.org/10.1016/j.biortech.2014.05.041

Auta H. S., Aboyeji D. O., Bala J. D., Abioye O. P., Adabara N. U., Aransiola S. A., Hassana A., & Azize A. (2022). Marine microbial enzymes—An overview. In N. R. Maddela, S. A. Aransiola, & R. Prasad (Eds.), *Ecological interplays in microbial enzymology* (1st ed.) Springer Nature Singapore Pte Ltd. ISSN: 2662-1681. https://www.springer.com/series/16324

Chaudhuri, P., Bhattacharjee, G., Vivekananda, S., Mohanpur, M., & Dey, S. (2000). Chemical changes during vermicomposting (Perionyx excavatus) of kitchen wastes, June 2016.

Cohen, M. J., Christian-smith, J., & Ross, E. N. (2010). Clearing the waters.

Conti, M. E. (2008). 1 Environmental biological monitoring. *30*, 1–23. https://doi.org/10.2495/978-1-84564-002-6/01

Diaz, R. J. (2008). Marine ecosystems, September. https://doi.org/10.1126/science.1156401

Domínguez, J., Velando, A., Aira, M., & Monroy, F. (2003). Uniparental reproduction of Eisenia fetida and E. andrei (Oligochaeta: Lumbricidae): Evidence of self-insemination. *1963*, 530–534.

Glibert, P. M., Mayorga, E., & Seitzinger, S. (2008). Prorocentrum minimum tracks anthropogenic nitrogen and phosphorus inputs on a global basis: Application of spatially explicit nutrient export models. https://doi.org/10.1016/j.hal.2008.08.023

Gopal, M., Gupta, A., Palaniswami, C., Dhanapal, R., & Thomas, G. V. (2010). Coconut leaf vermiwash: A bio-liquid from coconut leaf vermicompost for improving the crop production capacities of soil. *98*(9).

Heindel, J. J., Jobling, S., Kidd, K. A., & Zoeller, R. T. (2012). Endocrine disrupting chemicals—2012.

Hernández-sancho, F., Molinos-senante, M., & Sala-garrido, R. (2010). Economic valuation of environmental benefits from wastewater treatment processes: An empirical approach for Spain. *Science of the Total Environment, 408*, 953–955. https://doi.org/10.1016/j.scitotenv.2009.10.028

Hoornweg, D., & Pope, K. (2016). Population predictions for the world's largest cities in the 21st century. *Environment and Urbanization, 29*(1), 195–216. https://doi.org/10.1177/0956247816663557

Hughes, R. J. (2013). Vermifiltration systems for liquid waste management: A review. *12*(4), 382–396.

IINNOQUA. (2018). INNOQUA innovative wastewater treatment. https://innoqua-project.eu/. Accessed February 14, 2024.

Jayakody, P. (2014). Drivers and characteristics of wastewater agriculture in developing countries: Results from a global assessment (Issue January 2009).

Jiang, L., Liu, Y., Hu, X., Zeng, G., Wang, H., Zhou, L., Tan, X., Huang, B., Liu, S., & Liu, S. (2016). The use of microbial-earthworm ecofilters for wastewater treatment with special attention to influencing factors in performance: A review. *Bioresource Technology, 200*, 999–1007.

Kirschner, A. K., Schachner-Groehs, I., Kavka, G., Hoedl, E., Kovacs, A., & Farnleitner, A. H. (2024) Long-term impact of basin-wide wastewater management on faecal pollution levels along the entire Danube River. *Environmental Science and Pollution Research, 31*, 45697–45710. https://doi.org/10.1007/s11356-024-34190-0

Kumar, M., & Kaushal, S. (2022). Vermicomposting. *11*(5), 1505–1508. https://doi.org/10.21275/SR22518093637

Li, Y. S., Robin, P., Cluzeau, D., Bouch, M., Qiu, J. P., & Laplanche, A. (2008). *Vermifiltration as a Stage in Reuse of Swine Wastewater: Monitoring Methodology on an Experimental Farm, 2*, 301–309. https://doi.org/10.1016/j.ecoleng.2007.11.010

Lorena, C. M., Andrade-Yucailla, V. C., Alda, G. L., Oswaldo, V. R., Lorena, C. C. R., Sarah, A. C. T., Javier, F. G., Jose, B. H., Aransiola, S. A., & Maddela, N. R. (2025). Use of modified activated carbon in groundwater remediation for human consumption. *Water, 2025*(17), 207. https://doi.org/10.3390/w17020207

Manyuchi, M., & Phiri, A. (2013). Vermicomposting in solid waste management: A review. *International Journal of Scientific Engineering and Technology, 2(12), 1234-1242.*

Miito, G. J., Alege, F., Harrison, J., & Ndegwa, P. (2024). Influence of earthworm population density on the performance of vermifiltration for treating liquid dairy manure. *Journal of Environmental Quality, 53*, 1176–1187. https://doi.org/10.1002/jeq2.20626

Morand, P., Robin, P., Hamon, G., Qiu, J. P., & Bouché, M. (2005). Extensive treatments for a piggery with minimal pollution, March 2015.

Ojuolape, O. T., Aransiola, S. A., & Obideyi, O. A. (2015). Vermitechnology: A solution to environmental problems; A review. *Journal of Global Ecology and Environment, 3*(3), 127–135. http://www.ikpress.org/abstract/4516

Oliveira, J. P., Pessoa, F. L. P., Mehl, A., Alves, F. C., & Secchi, A. R. (2024). Sustainability indicators to MSW treatment assessment: The Rio de Janeiro case study. *Sustainability, 16*, 7445.

Pimentel, D. (2009). Energy inputs in food crop production in developing and developed nations. *1*, 1–24. https://doi.org/10.3390/en20100001

Rajpal, A., Bhargava, R., Sasi, S. K., & Chopra, A. K. (2012). On site domestic organic waste treatment through vermitechnology using indigenous earthworm species. *Waste Management and Research, 30*, 266–275.

Raphela, T., Manqele, N., & Erasmus, M. (2024). The impact of improper waste disposal on human health and the environment: A case of Umgungundlovu District in KwaZulu Natal Province, South Africa. *Frontiers in Sustainability, 5*, 1386047. https://doi.org/10.3389/frsus.2024.1386047

Samal, K., Dash, R. R., & Bhunia, P. (2017). Treatment of wastewater by vermifiltration integrated with macrophyte filter: A review. *Journal of Environmental Chemical Engineering, 5*(3), 2274–2289.

Singh, W. R., & Kalamdhad, A. S. (2016). Transformation of nutrients and heavy metals during vermicomposting of the invasive green weed Salvinia natans using Eisenia fetida. *International Journal of Recycling of Organic Waste in Agriculture, 5*(3), 205–220. https://doi.org/10.1007/s40093-016-0129-3

Singh, R., Samal, K., Dash, R. R., &Bhunia, P. (2019). Vermifiltration as a sustainable natural treatment technology for the treatment and reuse of wastewater: A review. *Journal of Environmental Management, 247*, 140–151.

Sinha, R. K. (2014). Sewage treatment by vermifiltration with synchronous treatment of sludge by earthworms: A low-cost sustainable technology over conventional systems with potential for decentralization, December 2008. https://doi.org/10.1007/s10669-008-9162-8

Sinha, R. K., Bharambe, Æ. G., & Chaudhari, U. (2008). Sewage treatment by vermifiltration with synchronous treatment of sludge by earthworms: A low-cost sustainable technology over conventional systems with potential for decentralization.

Sinha, R. K., & Herat, S. (2002). Vermiculture and waste management : Study of action of earthworms Elsinia foetida, Eudrilus euginae and Perionyx excavatus on biodegradation of some community wastes in India and Australia, June 2014.

Sinha, R. K., Herat, S., Karmegam, N., Chauhan, K., & Chandran, V. (2010a). Vermitechnology—The emerging 21st century bioengineering technology for sustainable development and protection of human health and environment: a review.

Sinha, R. K., Herat, S., & Valani, D. B. (2010b). Earthworms—The environmental engineers: Review of vermiculture technologies for environmental management and resource development, December 2013. https://doi.org/10.1504/IJGENVI.2010.037271

Steffen, W., Noone, K., Lambin, E., Lenton, T. M., Scheffer, M., Folke, C., Schellnhuber, H. J., Wit, C. A. De, Hughes, T., Leeuw, S. Van Der, Rodhe, H., Snyder, P. K., Costanza, R., Svedin, U., Falkenmark, M., Karlberg, L., Corell, R. W., Fabry, V. J., Hansen, J., … Foley, J. (2009). *Planetary boundaries: Exploring the safe operating space for humanity.*

Suthar, S. (2009). Vermistabilization of municipal sewage sludge amended with sugarcane trash using epigeic Eisenia fetida (Oligochaeta). *163*, 199–206. https://doi.org/10.1016/j.jhazmat.2008.06.106

Tomar, P., & Suthar, S. (2011). Urban wastewater treatment using vermibiofiltration system. *Desalination, 28*(2), 95–103.

Toyota, K., & Kimura, M. (2000). Microbial community indigenous to the earthworm Eisenia foetida, July 2022. https://doi.org/10.1007/s003740050644

UNEP. (2015). Global waste management. Global waste management outlook (no date). Outlook.

United Nations Environment Programme. (2024). Global waste management outlook 2024: Beyond an age of waste—Turning rubbish into a resource. Nairobi LXTW-HSM.SVK0.20.00.2244

US EPA. (2002). Onsite wastewater treatment systems manual, February.

Van De Vondel, S. (2006). Assessment of biological activity along a metal pollution gradient applying diatoms and macroinvertebrates.

Wang, X., Bai, J., Tian, Y., Wang, T., Zhou, X., & Zhang, C. (2021). Synergistic effects of natural ventilation and animal disturbance on oxygen transfer, pollutants removal and microbial activity in constructed wetlands. *Chemosphere, 283*, Article 131175.

Water, U. (2017). Wastewater the untapped resource.

WHO. (2004). The global burden of disease 2004.

WHO. (2019). Water, sanitation, hygiene and health.

Xing, M., Li, X., & Yang, J. (2010). Treatment performance of small-scale vermifilter for domestic wastewater and its relationship to earthworm growth, reproduction and enzymatic activity, November. https://doi.org/10.5897/AJB10.811

Zhao, L., Wang, Y., Yang, J., Xing, M., Li, X., Yi, D., & Deng, D. (2010). Earthworm—microorganism interactions : A strategy to stabilize domestic wastewater sludge. *Water Research, 44*(8), 2572–2582. https://doi.org/10.1016/j.watres.2010.01.011

Vermistabilization Through Earthworms

Srinivasan Kameswaran, Bellamkonda Ramesh, Ramesh B. Kasetti, Manjunatha Bangeppagari, P. Sudhakar Reddy, Sudhakara Gujjala, and Bhadramraju Ramu

Abstract On a global level, the administration of solid materials has grown to be a significant problem. There is an urgent need to recycle these wastes because they can be a rich source of fertilizers for plants and soil conditioning agents. Earthworms can effectively break down various organic wastes into nutrient-rich vermicompost. By utilizing several earthworm species, an effort has been made in this chapter to emphasize the vermicomposting of various organic wastes. A thorough literature search was conducted using search terms on "Google Scholar, PubMed Central, SpringerLink, Science Direct, and" acceptable studies of vermicomposting of various organic wastes were chosen. Vermicomposting may turn any type of organic waste into fertilizer. It was discovered that the waste needed to be combined with additional organic material for different types of vermicomposting (cattle dungs). The excessive amount of organic waste kills the earthworms even though the vermipit with 20–30% of litter wastes combined with 70–80% additional biologically rich material, like calf dung, may easily turn garbage into a valuable product. The present chapter illustrated how vermicomposting is an effective and superior way to recycle these wastes even

S. Kameswaran · B. Ramu
Department of Botany, Vikrama Simhapuri University College, Kavali, Andhra Pradesh, India

B. Ramesh (✉)
Department of Biochemistry and Molecular Biology, University of Nebraska Medical Center, Omaha, NE, USA
e-mail: rammygp@gmail.com

R. B. Kasetti
The North Texas Eye Research Institute, University of North Texas Health Science Center at Fort Worth, Fort Worth, TX, USA

M. Bangeppagari
Department of Cell Biology and Molecular Genetics, Sri Devaraj Urs Academy of Higher Education and Research, Kolar, Karnataka, India

P. Sudhakar Reddy
Department of Zoology, Vikrama Simhapuri University College, Kavali, Andhra Pradesh, India

S. Gujjala
Department of Biochemistry, Sri Krishnadevaraya University, Ananthapuramu, Andhra Pradesh, India

S. A. Aransiola and N. R. Maddela (eds.), *Vermitechnology: Economic, Environmental and Agricultural Sustainability*, https://doi.org/10.1007/978-3-031-96212-7_8

if many forms of organic solid waste cannot be directly digested by earthworms. A variety of crops can be grown in the fields with the help of the vermicompost that is created in this way. Also, farmers need to be persuaded to utilize vermicompost in their fields by educating them on the negative impacts of artificial pesticides and fertilizers.

Keywords Vermistabilizing · Organic waste · Vermicompost · Earthworms

1 Introduction

The Latin name for earthworms and vermicomposting is vermes. Due to its widespread use in the treatment of many types of organic waste, vermicomposting has gained increased attention in recent years (Bhat et al., 2018; Chauhan & Singh, 2013; Ghatnekar et al., 1998). Vermicomposting speeds up the natural decomposition of vegetative matter, which nonetheless occurs eventually (Benitez et al., 2005; Ojuolape et al., 2015). The organic material as well as farm waste materials of the soil can be efficiently maintained by making use of natural manure along with other waste from agriculture. Earthworms have been widely used for the recycling of a variety of organic wastes, including municipal solid wastes (Ciavatta et al., 1993), wheat straw (Bannik & Joergensen, 1993), sewage sludge (Govi et al., 1993), forestry waste (Marttinez-Inigo & Almedros, 1994), vegetable waste (Vallini & Pera, 1989), farmyard manure (Jakobsen et al., 1988), sorghum stalk, wheat straw, paddy straw (Gaur & Singh, 1995), and coir pith (Jothimani, 1994). Famous scientists Charles Darwin and Aristotle referred to earthworms as the "intestine of earth" and the "unheralded soldiers of mankind," respectively, because they could digest a variety of organic resources. Animal excretions from various species are produced all over the world, and when appropriately employed through vermicomposting, they may improve the physical and chemical properties of soil and can be utilized for crop fertilizers (Bhardwaj, 1995; Chauhan & Singh, 2015).

Vermicomposting is an effective method for reducing the quantity of natural waste that, unless correctly handled, could damage rivers and groundwater with its highly concentrated flow of nitrates, phosphates, and ammonia (Aransiola et al., 2022; Krishna et al., 2017; Ramnarain et al., 2019). As soon as food enters all earthworms' digestive tracts, the bacteria within start to break down the organic components it contains, turning them into vermicast (Singh, 2018). This vermicompost changes the soil's biological, chemical, and physical properties, which encourages plant growth (Datta et al., 2016). Vermicompost contains the following types of commonly found nutrients: organic carbon (9.5–17.98%), nitrogen (0.5–1.50%), phosphorus (0.1–0.30%), potassium (0.15–0.56%), sodium (0.06–0.30%), calcium and magnesium (22.67–47.60 meq/100 g), copper (2–9.50 mg/kg), iron (2–9.30 mg/kg), zinc (5.70–11.50 mg/kg), and sulfur (128–548 mg/kg) (Kale, 1995). In light of this, vermicomposting enables the biological conversion of trash into a useful organic fertilizer (Bozym, 2012; Kostecka, 2000). One of the key elements of an organic farming system, vermicompost, is sometimes known as "black gold" (Cresent, 2003). Comparing vermicomposting to thermophilic composting, the nitrogen loss was dramatically reduced by 10–20%. Methane and nitrous oxide emissions were both reduced by 25–36% and 22%, respectively, via vermicomposting. A higher earthworm density resulted in a 3–14% increase in carbon dioxide emissions, but a 10–35% decrease in methane emissions (Nigussie et al., 2016). Biological characteristics of the earth include the stimulation of bacterial populations that lessen the onslaught of different pathogenic microorganisms. The soil's chemical composition includes minerals like both micro- and macronutrients required for the development of plants, as well as its physical characteristics like soil texture and capacity for holding water (Bhat et al., 2018; Datta et al., 2016; Hansen, 2007; Singh, 2018).

2 Earthworm Category

Earthworms were divided into three types by Bouche (1977) based on their biological makeup and kind of feeding habitat: epigeic, anecic, and endogeic. The most popular earthworm species for vermicomposting are those found in epigeic ecological groups, like *Eisenia andrei* and *Eisenia fetida* (Aransiola et al., 2022; Dominguez, 2018). The epigeic species have a high rate of bioconversion and are surface feeders. They can withstand a variety of adverse environmental situations like reduced food availability, predator pressure, etc., and remain stable in an unstable and unpredictable ecosystem. Earthworm species can quickly develop their population for vermiculture and produce vermicompost thanks to these epigeic species' rapid rate of reproductive as well as grazing as opposed to the endogeic and anecic ecological categories (Dominguez, 2018; Julka, 2008). Several investigations have utilized anecic earthworm populations such as *Lampito mauritii* for vermicomposting (Tripathi & Bharadwaj, 2004) and endogeic earthworm populations such as *Metaphire posthuma* (Das et al., 2016; Sahariah et al., 2015).

3 Mechanism of Worm Action in Vermistabilization

Vermistabilization, a challenging physical as well as biological process for transforming sludge, employs earthworms (Fig. 1). Among other things, worms can serve as a biological stimulating, grinder, crusher, aerator, and chemical degrader (Sinha et al., 2002). Earthworms consume infections, mineralize nutrients, break down the organic portion of the sewage sludge, and consume heavy metals (bacteria, fungi, nematodes, and protozoa). They essentially serve as a "sludge digester" (Fig. 2).

1. The sludge is softened before it enters the digestive tract by the grume that the earthworms' mouths exude.

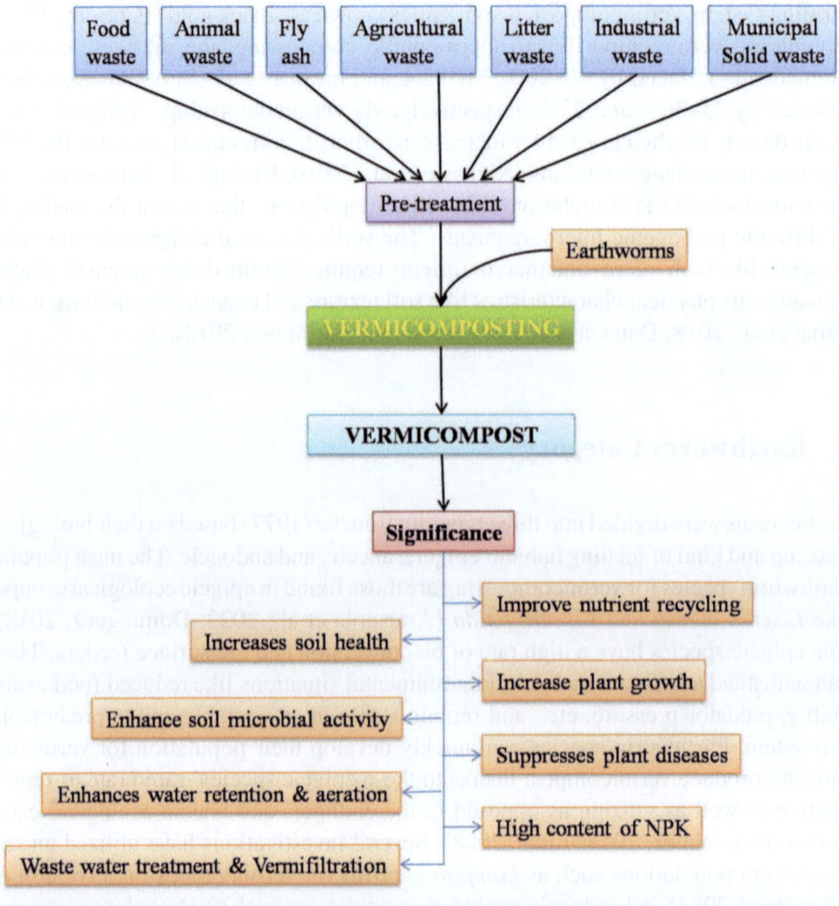

Fig. 1 Stages of vermistabilization and its significance

Fig. 2 Preparation of vermistabilization through earthworms

2. The digestive tract's inner walls secrete calcium, which neutralizes the softened sediment components in the esophagus and sends them to the stomach and gizzard for additional processing.
3. It is first delivered through the intestines over the enzymatic breakdown of food, where it is broken down into 2- to 4-m-sized granules with the aid of stones inside of the neuromuscular gizzard. A "bioreactor" is the intestine and gizzard.
4. Proteases, amylases, lipases, chitinases, and cellulases, which are secreted here and then absorbed, break down the ground and pulped sludge components in the intestine.
5. Humification, the last step in the vermiprocessing and sludge degradation process, transforms the big organic particles into intricate, amorphous colloids that include phenolic compounds. This stabilized sludge is then expelled as waste (vermicast).

4 Earthworms Boost Microbial Sludge Communities in Sewage that Work Together to Stabilize Wastes

Earthworms improve aeration, which boosts the number of microorganisms that break down waste materials (Binet et al., 1998; Dash, 1978). Additionally, they keep millions of bacteria in their guts that are decaying, excreting these bacteria in their waste with minerals like phosphate as well as nitrogen (Singleton et al. 2003). The microorganisms also utilize the nitrogen and phosphorus for increased activity and cellular reproduction. According to Edward and Fletcher (1988), ingested material contains bacteria and actinomycetes that can multiply up to 1000-fold as it travels through the stomach. A worm population of roughly 15,000 will support a billion-strong microbial community (Morgan & Burrows, 1982). In their study of the ecology of the microbes in the gastrointestinal tract and vermicasts of earthworms, Singleton et al. (2003) found that species of *Paenibacillus, Pseudomonas, Acidobacterium, Azoarcus, Mucor, Burkholderia, Alcaligenes,* and *Spiroplasma* that were capable of decomposing an extensive variety in chemical compounds. Even PCBs and Mucor spp. dieldrin can be broken down by *Alcaligenes* species. When conditions are favorable, earthworms and microorganisms collaborate to hasten and enhance the breakdown of the organic stuff found in the waste (Morgan & Burrows,1982; Xing et al., 2005).

5 Advantages of Earthworms' Vermistabilization of Wastes

When vermicompost is stabilized, its quality is noticeably better than that of conventional composting, which is thermophilic (temperatures can rise to 55 °C), killing many beneficial microbes and losing nutrients, especially nitrogen. It is also rich in important minerals and beneficial soil microbes (owing to nitrogen gassing off).

Additionally, earthworms keep their bodies well-oxygenated, and aerobic processes advance around tenfold more rapidly than anaerobic ones.

6 Almost Odorless Procedure

The worms create aerobic conditions within the waste products by their burrowing efforts, suppressing the activity of anaerobic bacteria that release unpleasant hydrogen sulfide along with mercaptans.

7 Several Vermicomposting Substrates

Vermicomposting is a well-liked and cost-effective method for the disposal of various kinds of solid waste. Numerous scientists have examined how various earthworm species perform on various types of garbage and have also reported on the earthworms' reproductive behavior based on the formation of cocoons and hatched eggs at the conclusion of each experiment.

8 Agricultural Residues

The interactions of agricultural wastes with bacteria and fungus control how quickly agricultural residues decompose in the soil. The earthworms may also turn crop residue into vermicompost. Tian et al. (1997) examined the function of the earthworms for the decomposition in five botanical residues—"*Dactyladenia barteri, Gliricidia sepium, Leucaena leucocephala,* maize, and rice straw" —by combining them to generate mulch. They discovered that the number of earthworms is larger in *Leucaena* and *Gliricidia* mulches (54%) than in controls. Cortez and Bouche (2001) conducted a study on the breakdown of fresh and composted *Nicodrilus meridionalis* leaf litter in the Mediterranean region and came to the conclusion that pre-composting the litter made it more palatable and enhanced the earthworm biomass. In contrast, earthworm biomass decreased when new litter was fed. Earthworms and oyster mushrooms could be used in a two-stage bioconversion process to turn inedible biomass into biohumus, according to Manukovsky et al. (2001). They mixed potato trash using the straw from wheat (1:3), processed it by adding mushrooms, and then added earthworms to create humus. This produced a healthy supply of mushrooms in addition to the biohumus needed for plant growth. In order to solve the issue of lignocellulosic waste degradation, the scientific viability of composting straw from wheat with biological inoculations (*Pleurotus sajor-caju, Trichoderma harzianum, Aspergillus niger,* and *Azotobacter chroococcum*), while following vermicomposting was examined by Singh and Sharma (2002). They discovered that the addition of bioinoculants and earthworms to lignocellulosic waste resulted in a faster overall development of nutrient-rich vermicompost. The concentration of N, P, and K increased significantly throughout the experiment. Chaudhuri and Bhattacharjee (2002) looked at *Perionyx excavatus* breeding and biomass production on a variety of different experimental dieting, including cattle dung (control) along with its mixture with food waste, litter wastes, and bamboo straw (1:3 dried weight of cattle dung with each other waste). In contrast to a mixture of kitchen waste, they discovered that straws of bamboo and dry leaves had the highest rates of biomass and reproduction. Manna et al. (2003) investigated the effects of epigeic earthworms belonging to tropical on the breakdown of forest waste, specifically "*Tectona grandis, Madhuca indica,* and *Butea monosperma.*" These worms included "*E. fetida, P. excavatus,* and *Dichogaster bolaui.*" They determined that the *T. grandis* waste might have proven more appropriate. They also discovered that *E. fetida* had a higher rate of reproduction in the forest litter than *P. excavatus* and *D. bolaui.* According to a study by Gajalakshmi and Abbasi (2004) on the vermicomposting of leaves from neem, earthworms (*Eudrilus eugeniae*) consumed neem leaves enthusiastically turning as much as 7% of the food they consumed as vermicomposting on a daily basis. The researchers additionally discovered that brinjal's growth and fruiting were considerably enhanced by the vermicomposts made with leaves of neem. Tripathi and Bhardwaj (2004) carried out comparisons of the decomposing capability, generation of biomass, and lifecycle of *L. mauritii* and *E. fetida* utilizing heterogeneous beds comprised of cattle dung, sawdust, kitchen waste, and wheat straw. Due to *E. fetida*'s high reproduction rate, they found that it was a more successful breeder than *L. mauritii.* Gajalakshmi et al. (2005) effectively converted the mango litter wastes into vermicomposts with a pair of vermibeds having 62.5 and 75 individual earthworms for each liter. A vermibed with 62.5 worms per liter yielded 13.6 g of vermicast per liter per day, but one having 75 worms per liter produced 14.9 g per liter per day. They also noted that the carbon/nitrogen decreased over the course of the research in comparison with the beginning, which was a result of the earthworm's gut bacteria using carbon. In the vermireactor, earthworm biomass grew by 103%, they added. According to Garg et al. (2006), *E. fetida* is effective at vermicomposting a variety of organic wastes, including agricultural waste. The amount of "total organic carbon (TOC)" dropped in crop waste as a factor of three, followed by kitchen waste by a factor of two, institutional waste at a proportion of one, and textile waste from industries at a proportion of 1.5 when compared to control.

Kurien and Ramasamy (2006) looked at the capacity of two types of earthworms, "*E. fetida* and *E. eugeniae,*" in the biological conversion of *Colocasia esculenta* in various vermibeds. They discovered that *E. fetida* was not as effective in producing vermicasts as *E. eugeniae.* By using *Perionyx sansibaricus* earthworms to study the degradation of different kinds of wastes, including farm wastes, farm manures, and urban municipality solid wastes, Suthar (2007) discovered that vegetable waste and leaf litter substrates produce the most biomass and have the highest growth rates (mg/day), the average number of cocoons, and the highest reproduction rates (cocoon/worm) when compared to other substrates. This study discovered large increases in total nitrogen from 80.8 to 42.3%, phosphorus from 33.1 to 114.6%, and potassium from 26.3 to 25.2%, while finding significant decreases in overall the amount of organic carbon from 14 to 37% and the carbon/nitrogen ratio ranges from 52.4 to 69.8%. The study revealed that the population death was highest with a combination of domestic trash and cow dung and lowest with a combination of vegetable market-place wastes and litter waste. Suthar (2010a) sought to convert vegetable waste into nutrient-rich biofertilizers by mixing it alongside wheat straw waste, cattle dung, and biogas slurry with different ratios. This was done using the earthworm *E. fetida*. He claimed that beds with quickly digesting bulked residues, for example, biogas slurry 40% with cattle dung 60%, have higher rates of waste reduction and humification. Potassium (3.2–15.3%) and organic carbon (4.8–12.7%) are significantly lower in the product, whereas total nitrogen (5.9–25.1%), accessible phosphorus (1.2–10.9%), exchangeable calcium (2.3–10.9%), and exchangeable magnesium (4.5–14%) are all higher.

Pineapple waste broke down quickly while vermicomposting at a pilot size with *E. eugeniae*, according to Mainoo et al. (2009). Vermicomposting, which may be applied to recycle degradable waste into a nutritious biofertilizer with a carbon/nitrogen ratio between 9 and 10 and up to 0.4% of total nitrogen, 0.4% total phosphorus, and 0.9% total potassium, was also promoted by them. The composting process of grass and paddy straw was studied by Ramnarain et al. (2019) using three distinctive treatments: paddy straw only, paddy straw with grass litter, and grass litter only. Researchers noticed that the lowest rates for vermicompost generation were for grass litter alone and straw from rice alone, respectively. The number of earthworms (*E. eugeniae*) and the formation of vermicomposts, on the other hand, were the subjects of Khwanchai and Kanokkorn's (2018) research. Among the waste from agriculture that they examined were watermelon peeling, soybean meal waste, coir, coffee waste, and so on. They discovered that *E. eugeniae* helped create vermicompost that had nitrogen, phosphorus, and potassium levels that were, respectively, 01.10%, 00.66%, and 01.31% larger than initial garbage and that the biomass of individual earthworms increased from 28.33 to 104.67% over time.

9 Weeds

The scientific community is very interested in weed management because of the negative effects it has on several ecosystems, including a forest, a farm, and a city (Suthar & Sharma, 2013). One of the naturally occurring organic materials that are easily available for vermicomposting is weeds like water hyacinth (*Eichhornia crassipes*). In addition to serving as a water cleanser against water pollution, water hyacinth helps several fish species by releasing their sticky eggs in its roots. However, the high concentration of this plant can be used to produce vermicompost, which is not beneficial for the water bodies (Saha et al., 2018). In their study of the ecological vermi-conversion of the water hyacinth within different beds using *E. eugeniae*, Gajalakshmi et al. (2002a) found that a vermi-bed with a high earthworm population yielded five to six times greater quantity of vermicompost than a vermi-bed with a low earthworm population. Various varieties of water hyacinths, namely fresh, dried, and chopped fresh plants, wasted weeds eliminated by the vermi-beds after isolating essential fatty acid compounds, pre-composed fresh weeds, and pre-composed spent weeds, were tested in the vermicomposting process by Gajalakshmi et al. (2002b). They discovered that the earthworms favored the pre-composed shapes as a feed the most. Additionally, they discovered that mixing cattle dung (14% of the feed volume) alongside different kinds of water hyacinth plants was a significantly positive impact on vermicast output compared to the corresponding unblended feed and that different types during spent weeds were preferred over the corresponding kinds of novel weeds. According to Gupta et al. (2007), who noticed the possibility of water hyacinths spiking with cattle dung producing vermicompost, earthworms developed and reproduced satisfactorily in a 25% WH and 75% CD diet mixture. The biomass increase, amount of eggs laid, and number of cocoons produced were all significantly impacted by a higher proportion of WH in the diet combination. Gajalakshmi et al. (2001a) underlined the importance of "two epigeic species, *E. eugeniae* (Kinberg) and *P. excavatus* (Perrier)," as well as "two anecic species, *L. mauritii* (Kinberg) and *Drawida willsi* (Michaelsen)," for the efficiency and long-term viability of water hyacinth vermicomposting. They discovered that "*E. eugeniae*, *P. excavatus*, *L. mauritii*, and *D. willsi*" were the top four producers of vermicompost. Singh and Kumar (2017) examined the vermicomposting processes of two land-invasive plants, "*Lantana camara* L. and *Parthenium hysterophorus* L.," using *E. fetida*. They used cattle dung to create five different concentrations of weed, namely 0%, 20%, 40%, 60%, and 80%. Throughout the experimental time, they also assessed the physicochemical characteristics of each concentration. According to their findings, aquatic vegetation may be utilized as an initial substrate for composting because the amount of EC, pH, and total organic carbon decreased whereas the quantities of nitrogen, phosphorous, and potassium increased at all dosages. Singh and Kalamdhad (2016) observed comparable outcomes for the physicochemical characteristics of vermicompost to be compared with starting trash while conducting composting of the aggressive green plant *Salvinia natans* using the earthworm *E. fetida*. Additionally, they reported that during the research trials, the earthworm population, as well as development, increased. Physicochemical characteristics, microbial populations, and enzyme activities were all examined by Gusain and Suthar (2020) when *Ageratum conyzoides* were vermicomposted in ratios of 25%, 50%, and 75% (v/v). They discovered that the pH and the amount of organic carbon decreased, while the levels of N, PO43-, K, and Ca improved by 69.9%, 148.7%, and 92.43%, respectively, in vermicompost. In comparison with the initial feed mixture, they also reported higher microbial populations and enzyme activity in vermicompost. They found that among the aforementioned three concentrations, vermicomposting can be effective in the range of 50–75% of *A. conyzoides* waste combined with 25–50% of cattle dung.

10 Animal Dung

According to Nasiru et al. (2013), animal dung was a valuable resource for supplying plants with a combination of macronutrients and micronutrients. When huge amounts of animal waste are generated, it can be challenging to appropriately collect and dispose of the animal excreta, even though only a small portion of the nutrients expelled are used as manure. Uncontrolled and excessive production of animal excrement led to overfertilization, disease dissemination, soil contamination, stench, water pollution, and a rise in emissions of greenhouse gases (Nasiru et al., 2013). The optimum waste management solution for this type of trash is vermicomposting. In order to evaluate the possibility of breeding *Dendrobaena veneta* at various temperatures (10, 15, 20, or 25 °C) on aerobic paper sludge or on horse manure, Fayolle et al. (1997) researched the life cycle of the species. They concluded that horse dung would not be an ideal breeding medium for this particular species after noting that the type of food can alter the *D. veneta* life cycle. Dominguez et al. (2001) studied the physiology as well as population growth of *E. eugeniae* in animal dung in miniature containers kept at 15, 20, 25, and 30 °C. They discovered that at 25 °C, growth, maturity, and biomass productions were all significantly higher than at 15, 20 and 30 °C. Gunadi and Edwards (2003) looked at growth as well as death rates of *E. fetida* using pre-composted including fresh livestock manure, fresh pig dung of different maturities, vegetable, and fruit waste (Aransiola et al., 2022). They showed how *E. fetida* could not survive with fresh cattle dung, fresh young pig dung, vegetables, and fruit wastes and that the earthworm's proliferation accelerated in partially decomposing pig dung compared with pre-composed cattle dung. For the treatment of natural wastes from cattle, vermicomposting is frequently used. When the manure of goats and cow manure were compared for their composting process capacity using *E. fetida*, Loh et al. (2005) discovered that the latter had a higher population and reproduction rates for *E. fetida* over the former, while cocoons and the embryo growth were unaffected. Borges et al. (2017) additionally looked at the progress of biological degradation alongside an upsurge in earthworms' rate of growth using an assortment of pig dung and calf manure. They concluded that, under controlled temperature and moisture conditions, animal dung is a good waste for the vermicomposting process. The method is successful at stabilizing the metals, according to Lv et al. (2016) evaluation of the consequences of the composting process on the species with the movement of toxic metals (Zinc, Lead, Chromium, and Copper) within pig manure and animal manure.

11 Industrial Waste

In response to global modernization, the industry sector has been growing daily. These industries create various products that are helpful to society, but they also create various kinds of sludge and effluent. In addition to polluting the land and groundwater, the open disposal of this toxic sludge will waste a valuable carbon resource (Singh et al., 2010). Government awareness, environmental protection laws, and landfill space reduction have brought to our attention alternative methods for retrieving a rich reservoir of vital nutrients from waste. These forms of biological waste can be recycled and broken down by earthworms to natural fertilizer, which is full of nutrients and helps to reduce pollutants in the environment (Bhat et al., 2018; Julka, 2008).

12 Paper Industry Waste

According to Tripathi (2014), while paper consumption climbed by 9% in 2012, the paper and pulp production industry only grew by 8.4%. The US paper industry recorded a record profit of $100 million in just 2009 alone (Kaur et al., 2010). The harmful paper mill sludge produced by these paper manufacturers is enormous. The quantity and composition of wastes are influenced by the initial components used, the processes used, and the characteristics of the intended paper. The two main constraints limiting the process of biodegradation are polysaccharides and low oxygen content, or below 0.05% in the sludge of pulp and paper plants (Elvira et al., 1997). The cellulose content of the paper mill waste is high, but the nitrogen amount is low. Cattle waste, which acts as an organic inoculant for bacterial populations, can be combined with pulp and paper factory waste to solve these issues, and the area can subsequently be treated with earthworms (Butt, 1993).

Gajalakshmi et al. (2001b) discovered that increasing the amount of dairy manure in the mixture used for feeding between 14.3 and 20% had an overall moderately favorable impact on earthworm survival in their investigation with regard to vermicomposting for pulp and paper waste with varied levels of cow manure. Gupta and Garg (2009) discovered that earthworm proliferation and reproduction were unaffected by paper waste up to a concentration of 30% in a vermireactor when it was mixed with cattle manure to generate vermicompost with *E. fetida*. Mohan (2017) examined the biological transformation of pulp and paper factory waste to produce high in nutrient compost using *E. fetida* and documented any variations in physical and chemical characteristics in the waste prior to and after vermicomposting. The 25% combination of sludge from paper factories and a 75% mixture of cattle manure were found to generate valuable manure with great efficiency. Amouei et al. (2017) similarly came to the conclusion that earthworms may quickly digest paper mill sludge waste. Hence, vermicomposting is an effective solution to address the issues associated with the disposal of waste from paper mills.

13 Sugar Industry Waste

A sugar refinery represents a single of the essential agricultural industries for the production of sugars and ethanol as well. According to Tewari et al. (2007), India has over 500 sugar factories and about 300 molasses-based liquor refineries. During the grinding and extraction of sugarcane, these industries produce a large amount of reusable organic residue, including press mud, sugarcane refuse, and bagasse. According to Yadav (1995), each tonne of fresh crushed sugarcane results in the production of 35 kg of press mud. Press mud production in India is thought to be around 3.6 and 3.9 million metric tonnes annually (Karwal & Kaushik, 2020; Rasappan et al., 2015). According to Bhat et al. (2014), this press mud is an excellent source of protein, catalysts, macronutrients, micronutrients, organic matter, and organic carbon. The press mud possesses a pH value of 7.1, 313 g/kg of carbon from organic matter, 24 g/kg of N, 3.6 g/kg of phosphorus, 0.86 g/kg of potassium, and 12.1 g/kg of calcium, accordingly, whereas the amounts about iron, copper, manganese, and zinc are 22,440, 870, 2008, and 1392 mg/kg. Several researches vehemently supported the usage of press mud in vermicomposting on a commercial basis.

Bhat et al. (2014) investigated the physicochemical changes and the genotoxicity potential of press mud trash before and subsequent to vermicomposting. They discovered that the nitrogen, phosphorus, sodium, electrical conductivity, and pH level contents of the final product were higher than those of their original input combinations. On another hand, values for organic matter, potassium, and the ratio of carbon to nitrogen fell from the initial feed combination to the final one. They also concluded that the percent deviation rose as waste concentration rose and vice versa. However, when vermicompost was substituted for the initial waste combination, the percentage of aberration was reduced. A pair of imported earthworms (*E. fetida* and *E. eugeniae*) together with a single indigenous earthworm (*P. excavatus*) have been successfully used by Khwairakpam and Bhargava (2009) in monocultures as well as poly-cultures to observe the modifications in press mud. Their research revealed that while monoculture and polyculture gave comparable results, earthworm polyculture produced the best outcomes. According to Sangwan et al. (2010), excellent vermicomposting material that had higher nitrogen, phosphorous, and potassium levels were produced when press mud and calf dung were mixed in a ratio of 50%. When sugary beet mud and paper pulp were mixed with cow manure for vermicomposting, Bhat et al. (2015) observed that the amount of waste present showed a big effect on how quickly the earthworms developed.

The vermicomposting of various types of waste from sugar refineries was researched by Shweta Kumar et al. (2010) with an improvement in vermicompost quality and a decrease in stabilization time. Their research showed that the composting of trash with microbes accelerated vermicomposting manure. Suthar (2010a, 2010b) tested the consistency of vermicomposting by combining press mud, distilleries wastewater, animal manure, and biogas generator slurry, with wheat stubble in different ratios. *E. fetida* was able to thrive and reproduce thanks to the combinations of industrial sludge used in the vermicomposting process. They discovered that when a certain amount of waste within the feeding combination decreased and vice versa, the rate of mineralization was raised. *Perionyx ceylanensis* may successfully convert press mud and an equivalent volume of cow manure into vermicompost, according to Prakash and Karmegam (2010).

14 Textile Industry Waste

The textile industries produce some of the most hazardous solid wastes, and their disposal at open dump sites, farming regions, near street borders, or close to railroad lines may contaminate the ground, water sources, or ground waters, endangering the health of the general population. *E. fetida* was used in many trials by Kaushik and Garg (2003, 2004) to create vermicompost from textile mill waste mixed with cow manure. According to their research, as much as 30% of the combined product of animal dung and textile industry wastewater possesses excellent vermicomposting materials. However, the earthworms' development and sexual maturation are delayed when extra textile industry wastewater can be added to the cattle manure. Garg and Kaushik (2005) also had successful results using *E. fetida* to vermi-stabilize textile industry wastewater mixed using poultry manure. As a result, using a diet that was a mixture made by combining 70% poultry manure along with 30% textile industrial waste, earthworms thrived and replicated favorably. The scientist also advised against using any chemicals from the textile sector in the sludge waste used in vermicomposting because doing so might make the worms sick.

Garg et al. (2009) tested vermicompost manure using textile industrial waste combined with both cattle manure and horse manure using *E. fetida*. Temperature fluctuations had a substantial impact on *E. fetida*'s development and fertility rate, according to their research. In contrast to increases in P and N levels, vermicomposting manure caused decreases in electrical conductivity, pH level, and carbon/nitrogen proportion, along with potassium. Bhat et al. (2013) tested the composting process of that waste product by mixing the textile dyes along with cattle manure in a number of independent proportions. The electrical conductivity, carbon/nitrogen proportion, and organic matter, along with K salt, all decreased, while the nitrogen (N), sodium, pH levels, and quantity of phosphorus all increased.

15 Food Industry Waste

Waste materials generated through the food sector, particularly effluent sewage, is a significant source of organic material. The open discharge of food industry sludge may be the primary cause of environmental contamination. *Eisenia fetida* and *E. andrei* were examined by Tajbakhsh et al. (2008) for their capacity to convert ingested mushrooms into vermicompost. They observed a decrease in pH levels from 7.23 to

6.69 with electrical conductivity (forty percent below starting garbage) throughout the research. In contrast, they also caused vermicompost's nitrogen and phosphorus contents to rise relative to the original raw material, by 42–85% and three times, respectively. Yadav and Garg (2009) also demonstrated the efficiency of using *E. fetida* for recovering minerals from food sector wastewater treatment plant sludge. The results of the studies demonstrated that while composting, pH levels, electrical conductivity, organic matter, organic carbon, carbon/nitrogen proportion, nitrogen, phosphorus, and potassium all decreased. *Eisenia fetida* has been used to vermicompost sludge from biogas plant slurry and sewage treatment plants serving the food processing industry, according to Yadav and Garg (2010). They discovered that "*E. fetida*" could not endure in an environment that included only sludge. The maturation of vermicompost is aided, however, by the addition of sludge to the slurry from between 20 and 30% biogas plant. Researchers came to the conclusion with "vermicomposting" by a combination of industry sewage waste and slurry from biogas plants could serve as a better way of reducing sludge from food processing plants.

While generating the waste materials in different ratios using cow manure, Garg et al. (2012) evaluated the efficacy of "*E. fetida* for vermicomposting" of food processing sludge mixed with a variety of biological wastes. Nitrogen, from 60 to 214%; phosphorus, from 35.8 to 69.6%; sodium, from 39 to 95%; and potassium, from 43.7 to 74.1%, all increased considerably from the beginning feed combinations, whereas the pH levels, from 8.45 to19.7%; organic carbon, from 28.4 to 36.1%; and carbon/nitrogen ratio, from 61.2 to 77.8%, all decreased. Trash can be vermicomposted into excellent-quality manure if it is combined together with organic matter in large enough amounts, according to numerous studies.

16 Alcohol Industry Waste

The distillery is a significant segment of the sugarcane industry. Since alcohol industries produce a sizable amount of sewage sludge, much like other industries, properly disposing of this solid waste was a top priority. Due to its high concentrations of vital plant nutrients like nitrogen, potassium, phosphorus, calcium, zinc, manganese, copper, and iron, distillery sludge can serve as a soil conditioner once it has undergone the proper biological processing. Suthar and Singh (2008) used the earthworm *P. excavatus* to examine the composting process of cattle manure and industrial residue from a distillery. According to their findings, pH levels, zinc, manganese, iron, organic carbon, and copper significantly decreased, whereas nitrogen, phosphorus, potassium, calcium, and magnesium significantly increased. The effective "vermicomposting" process of the industrial wastewater from distilleries boosted using cow manure was also studied by Suthar (2008) using various *E. fetida* ratios. Final manure showed large increase during nitrogen, phosphorus, and potassium but a considerable decline in pH and organic carbon. Additionally, they discovered that earthworm populations along with fertility rates had been greater in beds that contained as much as 40% reduced distillery sludge, indicating that using more industrial sewage might limit earthworms' potential. Nevertheless, the wastewater treatment that included alcohol sewage collected from 40 and 60% distilleries showed an elevated increase in the amounts of macronutrients and a reduction in the pH and the amount of organic carbon in the vermicomposting. As a result, earlier research revealed that earthworms can only break down distillery waste when it is combined in exactly the right amounts with cow manure. The amount of solid waste within the mix for feed has increased along with the rate of waste decomposition, and the contrary is also true. In Singh et al.'s (2014) investigation on the vermi-remediation recycling distilleries sewage produce the compost, *E. fetida* was used. They mixed the industrial sewage solid waste from the distillery along with cow manure in several unique percentages, for example, 0, 25, 50, 75, and 100%. They investigated their rates about survival, development, maturation, and cocoon, along with multiplication accumulation, that boosted via an increase within the proportion of cows manures.

17 Thermal Power Plant Fly Ash

Fly ashes' disposal difficulties are problematic in several nations, particularly in India (Karwal & Kaushik, 2020). The development of techniques for utilizing both small- and huge-scale volumes of fly ash is crucial because India generates 150 million metric tonnes of fly ashes each year (Gupta et al., 2005). Bhattacharya and Chattopadhyay (2006) claim that toxic metals (lead, cadmium, and chromium) and several incompatible fertilizers (iron, manganese, copper, and zinc) significantly changed during the vermicomposting of fly ash. They created various combinations of both fly ashes along with cattle manure in addition to treating with and without earthworms. The solubility of a number of trace elements has also been estimated repeatedly. According to their investigation, adding *E. fetida* into different mixtures of fly ashes with calf manure resulted in the conversion of a significant amount of micronutrients into forms that could be accessed.

Gupta et al. (2005) tested the vermicomposting of a fly ash mixture with animal manure with four different proportions, which includes 20, 40, 60, and 80%, and they discovered that 40% of the fly ash mixture produced the most earthworm hatchlings. They discovered that *E. fetida* can lessen the toxic effect of metals and that as much as 60% of fly ash plus animal manure mixtures might be successfully composted via vermi-composting. Moreover, fly ash spiked with calf manure was vermicomposted by Venkatesh and Eevera (2008) in several ratios, including 1:3, 1:1, and 3:1. In comparison with fly ash alone, cattle dung mixes treated with earthworms had increased macronutrients and micronutrient contents. The general accessibility of nutrition with the 1:3 proportion had substantially greater than it was for the other treatments. Singh et al. (2016) additionally investigated composting of cattle manure with fly ash from thermal power plants in a variety of proportions. They discovered that while pH levels, electrical conductivity, total organic carbon, and carbon/nitrogen proportions all considerably altered between the original into the final product, total K, and P increased significantly. Also, the vermicompost was discovered to have the lowest levels of heavy metals, indicating that the microflora in earthworms' guts can convert waste's toxic effects into a nontoxic state.

18 Winery and Beverage Industry Waste

The squeezing dripping, and compressing procedures needed to make wine result in large amounts of waste being produced by the wine and beverage industries, namely grape seeds, grape marc, skin, and stalks. The beverage industry, which had a market value of twenty-two billion dollars and an annual growth rate of around 19% in India during 2013, will continue to grow dramatically through the next twenty years. The production of biological waste by these businesses will then rise as a result. *E. andrei* was evaluated by Nogales et al. (2005) for its potential to bioconvert different vineyard wastes through vermicomposting into vermicompost, a valuable product. They postulated that vermicomposting raised the agricultural benefit of vineyard waste from factories via lowering the amount of carbon to nitrogen in addition to raising the humic materials, pH, and mineral content by losing between 19 and 31% of the total amount of carbon compounds contained in plant material by the end of the process. Biosludge coming from the alcoholic beverage industry that was mixed with cow dung, according to Singh et al. (2010), was vermicomposed. The findings show that 50:50 mixes could deteriorate or break down after 75 days during earthworms have been given currently a dosage of 25 g/kg of feeding mixture. Several studies have shown that beverage industry sludge can be stabilized through vermicomposting, but that it must be mixed with animal dung because pure sludge is hazardous to worms.

19 Vermicompost's Physicochemical Characteristics After Conversion from Waste

The physical and chemical characteristics have changed between garbage toward vermicompost, although the modification also becomes reliant on the waste's initial composition. One of the most crucial factors in determining the vermicompost's maturity is the pH (Esmaeili et al., 2020). Lower pH concentrations can be found in grasses (*Trapa natans, P. hysterophorus, L. camara*) as well as in cattle, goats, rabbits, pigeons, horses, diverse waste from industries, waste from agriculture, and other wastes. The generation of organic acids throughout the waste mineralization cycle may be responsible for the reduction in pH of vermicompost (Nakasaki et al., 2005). However, certain garbage additionally demonstrated an elevation in pH levels, which might have been brought on by the breakdown of organic acids, volatile fatty acids, and organic nitrogen into ammonium during the vermicomposting process. The EC content, which is also used as a criterion for deciding whether to utilize vermicompost as fertilizer, indicates the quantity of all of the dissolved salts formed throughout the course of vermicomposting (Gusain & Suthar, 2020). Several types of organic wastes have experienced increases and decreases in EC concentration. The rise in various soluble salts during the vermicomposting process may be the cause of the increase in EC (Singh et al., 2019). However, the use of particular essential salts by the microbial populations throughout the procedure of vermicomposting may cause a decrease in electrical conductivity (Chauhan & Singh, 2013; Sharma & Garg, 2018). Vermicompost that has an EC level under 4 mS/cm is suitable for use as fertilizer, according to Wong et al. (2001).

All types of organic garbage contain less carbon than they did at the beginning. According to Sharma and Garg (2018), earthworms as well as microorganisms may have contributed to the decline in organic carbon in vermicompost by converting organic molecules into CO_2. These earthworms may have contributed to the vermicompost's lower carbon content by consuming carbon for their own growth (Soobhany et al., 2017). Also, the decline in carbon demonstrated how effectively earthworms degrade garbage (Ansari & Rajpersaud, 2012). There has been an increase in nitrogen, phosphorous, and potassium concentration from the initial garbage to the finished vermicompost. The nitrogen, phosphorus, and potassium contents in vermicompost are influenced by the trash's initial concentrations of each element. The amounts of the elements nitrogen, phosphorus, and potassium may have raised into the vermicompost as a result of the accumulation of mucous, elimination materials, and catalysts produced by the earthworms (Ganiger et al., 2020).

20 Vermicompost Storage and Packaging

The gathered vermicompost needs to be kept cold and dark to prevent moisture and nutrient loss from sunshine. Additionally, rather than being stuffed into sacs, gathered vermicompost material ought to be stored in the open. Packing must be completed at the time of sale, and laminated packaging is always recommended. To keep the moisture level and beneficial microbial population stable when compost is being stored outside, water should be sprinkled on a regular basis. If its moisture level is kept at 40%, vermicompost can be kept for longer periods of time—up to one year—without losing any of its quality.

21 Benefits of Vermicompost

The recycled organic wastes mentioned above can be utilized to create vermicompost, which can be used as effective organic manure. The worms in excess can be utilized to recover vermiproteins, and the vermicompost that is thus produced can be applied to agricultural land. Vermicomposting improves soil microbial activity, soil porosity, plant development, water retention, and aeration in fields. It also suppresses plant disease. Vermicompost is a good fertilizer due to the rise in total N, P, and K content.

With a lot of waste that is discharged into the environment and the demand for chemical fertilizers, these vermicompost manures also decrease.

22 Conclusion

Vermicomposting may be a better option for the treatment of various forms of solid organic debris, according to this study, which also found that it is a cheap, efficient process. Earthworms cannot directly digest different types of organic waste, but they can mix the waste with other nutrient-rich organic materials, such as leaf litter and cow dung, to effectively destroy it. Another discovery is that when between 25 and 30% of the organic matter is combined with 70 and 75% of some other organic matter, such as cow manure, vermicompost can be created easily. Nevertheless, the high percentage or amount of organic waste results in earthworm mortality. Farmers need to be taught how to use vermicompost technology to turn cattle manure through high in nutrient vermicompost using agricultural uses in order to cut their costs in conventional agriculture. Therefore, by using this technology, companies, governments, and also farmers would additionally benefit financially in addition to contributing to the protection of the surroundings.

References

Amouei, A. I., Yousefi, Z., & Khosravi, T. (2017). Comparison of vermicompost characteristics produced from sewage sludge of wood and paper industry and household solid wastes. *Journal of Environmental Health Science and Engineering, 15*, 5. https://doi.org/10.1186/s40201-017-0269-z

Ansari, A. A., & Rajpersaud, J. (2012). Physicochemical changes during vermicomposting of Water hyacinth (*Eichhornia crassipes*) and grass clippings. *International Scholarly Research Notices*. https://doi.org/10.5402/2012/984783

Aransiola, S. A, Ijah, U. J. J., Abioye, O. P, & Bala, J. D. (2022). Vermicompost-assisted phytoremediation of toxic trace element-contaminated soil in Madaka, Nigeria using *Melissa officinalis* L and *Sida acuta*. *International Journal of Environmental Science and Technology*. https://doi.org/10.1007/s13762-022-04105-y

Bannik, C. G., & Joergensen, R. G. (1993). Change in N fractions during composting of wheat straw. *Biology and Fertility of Soils, 16*, 269–274. https://doi.org/10.1007/BF00369303

Benitez, E., Sainz, H., & Nogales, R. (2005). Hydrolytic enzyme activities of extracted humic substances during the vermicomposting of a lignocellulosic olive waste. *Bioresource Technology, 96*, 785–790. https://doi.org/10.1016/j.biortech.2004.08.010

Bhardwaj, K. K. (1995). Recycling of crop residue soil cakes and other plant products in agriculture. Recycling of crop, animal, human and industrial wastes in agriculture (pp. 9–30). Fertilizer Development and Consultation Organization.

Bhat, S. A., Singh, J., & Vig, A. P. (2013). Vermiremediation of dyeing sludge from textile mill with the help of exotic earthworm *Eisenia fetida* Savigny. *Environmental Science and Pollution Research, 20*(9), 5975–5982. https://doi.org/10.1007/s11356-013-1612-2

Bhat, S. A., Singh, J., & Vig, A. P. (2014). Genotoxic assessment and optimization of press mud with the help of exotic earthworm *Eisenia fetida*. *Environmental Science and Pollution Research, 21*, 8112–8123. https://doi.org/10.1007/s11356-014-2758-2

Bhat, S. A., Singh, J., & Vig, A. P. (2015). Vermistabilization of sugarbeet (*Beta vulgaris* L) waste produced from sugar factory using earthworm *Eisenia fetida*: Genotoxic assessment by *Alliumcepa* test. *Environmental Science and Pollution Research, 22*, 11236–11254. https://doi.org/10.1007/s11356-015-4302-4

Bhat, S. A., Singh, S., Singh, J., Kumar, S., & Vig, A. P. (2018). Bioremediation and detoxification of industrial wastes byearthworms: Vermicompost as powerful crop nutrient in sustainable agriculture. *Bioresource Technology, 252*, 172–179. https://doi.org/10.1016/j.biortech.2018.01.003

Bhattacharya, S. S., & Chattopadhyay, G. N. (2006). Effect of vermicomposting on the transformation of some trace elements in fly ash. *Nutrient Cycling in Agroecosystems, 75*, 223–231. https://doi.org/10.1007/s10705-006-9029-7

Binet, F., Fayolle, L., Pussard, M., Crawford, J. J., Traina, S. J., & Tuovinen, O. H. (1998). Significance of earthworms in stimulating soil microbial activity. *Biology and Fertility of Soils, 27*(1), 79–84.

Borges, Y. V., Alves, L., Bianchi, I., Espíndola, J. C., Oliveira, J. M., Jr., Radetski, C. M., & Somensi, C. A. (2017). Optimization of animal manure vermicomposting based on biomass production of earthworms and higher plants. *Journal of Environmental Science and Health, Part b: Pesticides, Food Contaminants, and Agricultural Wastes, 52*(11), 791–795. https://doi.org/10.1080/03601234.2017.1356162

Bouche, M. B. (1977). Strategies lombriciennes, in soil organism as component of ecosystems edited by U Lohmand Persson. *Biology Bulletin (Stockholm) 25*, 122–132.

Bozym, M. (2012). Biologiczne przetwarzanie biodegradowalnej frakcji odpadów komunalnych i osadów ściekowych w wermikulturze. *Prace Instytutu Ceramiki i Materiałów Budowlanych, 5*, 335–369.

Butt, K. R. (1993). Utilization of solid paper-mill sludge and spent brewery yeast as feed for soil-dwelling earthworms. *Bioresource Technology, 44*, 105–107. https://doi.org/10.1016/0960-8524(93)90182-B

Chaudhuri, P. S., & Bhattacharjee, G. (2002). Capacity of various experimental diets to support biomass and reproduction of *Perionyx excavatus*. *Bioresource Technology, 82*, 147–150. https://doi.org/10.1016/S0960-8524(01)00169-9

Chauhan, H. K., & Singh, K. (2015). Potency of vermiwash with neem plant parts on the infestation of *Earias vittella* (Fabricius) and productivity of Okra (*Abelmoschus esculentus* L.) Moench. *Asian Journal of Research in Pharmaceutical Sciences, 5*, 36–40. https://doi.org/10.5958/2231-5659.2015.00006.5

Chauhan, H. K., & Singh, K. (2013). Effect of tertiary combinations of animal dung with agro wastes on the growth and development of earthworm *Eiseniafetida* during organic waste management. *International Journal of Recycling of Organic Waste in Agriculture, 2*, 11. https://doi.org/10.1186/2251-7715-2-11

Ciavatta, C., Govi, M., Pasotti, L., & Sequi, P. (1993). Changes in organic matter during stabilization of compost from municipal solid waste. *Bioresource Technology, 43*, 141–145. https://doi.org/10.1016/0960-8524(93)90173-9

Cortez, J., & Bouche, M. (2001). Decomposition of Mediterranean leaf litters by *Nicodrilus meridionalis* (Lumbricidae) in laboratory and filed experiments. *Soil Biology and Biochemistry, 33*, 2023–2035. https://doi.org/10.1016/S0038-0717(01)00124-9

Cresent, T. (2003). Vermicomposting. Development alternatives (DA) sustainable livelihoods [Internet]. Available from: http://www.dainet.org/livelihoods/default.htm

Das, S., Deka, P., Goswami, L., Sahariah, B., Hussain, N., & Bhattacharya, S. S. (2016). Vermiremediation of toxic jute mill waste employing *Metaphire posthuma*. *Environmental Science and Pollution Research, 23*, 15418–15431. https://doi.org/10.1007/s11356-016-6718-x

Dash, M. C. (1978). Role of earthworms in the decomposer system. Glimpses of ecology (pp. 399–406). India International Scientific Publication. New Delhi.

Datta, S., Singh, J., Singh, S., & Singh, J. (2016). Earthworms, pesticides and sustainable agriculture: A review. *Journal of Environmental Science and Pollution Research, 23*, 8227–8243. https://doi.org/10.1007/s11356-016-6375-0

Dominguez, J. (2018). Earthworms and vermicomposting. In: *Earthworms the ecological engineers of soil*. https://doi.org/10.5772/intechopen.76088

Dominguez, J., Edwards, C. A., & Dominguez, J. (2001). The biology and population dynamics of *Eudriluseugeniae* (Kinberg) (Oligochaeta) in cattle waste solids. *Pedobiologia, 45*, 341–353. https://doi.org/10.1078/0031-4056-00091

Edwards, C. A., & Fletcher, K. E. (1988). Interactions between earthworms and microorganisms in organic-matter breakdown. *Agriculture, Ecosystems & Environment*, *24*(1–3), 235–247. https://doi.org/10.1016/0167-8809(88)90069-2

Elvira, C., Sampedro, L., Dominguez, J., & Mato, S. (1997). Vermicomposting of wastewater sludge from paper-pulp industry with nitrogen rich materials. *Soil Biology & Biochemistry, 29*, 759–762. https://doi.org/10.1016/S0038-0717(96)00202-7

Esmaeili, A., Khoram, M. R., Gholami, M., & Eslami, H. (2020). Pistachio waste management using combined composting-vermicomposting technique: Physico-chemical changes and worm growth analysis. *Journal of Cleaner Production, 242*, Article 118523. https://doi.org/10.1016/j.jclepro.2019.118523

Fayolle, L. H., Michaud, D. C., & Stawiecki, J. (1997). Influence of temperature and food source on the life cycle of the earthworm *Dendrobaena veneta* (Oligochaeta). *Soil Biology & Biochemistry, 29*, 747–750. https://doi.org/10.1016/S0038-0717(96)00023-5

Gajalakshmi, S., & Abbasi, S. A. (2004). Neem leaves as a source of fertilizer-cum-pesticide vermicompost. *Bioresource Technology, 92*, 291–296. https://doi.org/10.1016/j.biortech.2003.09.012

Gajalakshmi, S., Ramasamy, E. V., & Abbasi, S. A. (2001). Potential of two epigeic and two anecic earthworm species in vermicomposting of water hyacinth. *Bioresource Technology, 76*, 177–181. https://doi.org/10.1016/S0960-8524(00)00133-4

Gajalakshmi, S., Ramasamy, E. V., & Abbasi, S. A. (2001b). Screening off our species of detritivorous (humus–former) earthworms for sustainable vermicomposting of paperwaste. *Environmental Technology, 22*, 679–85. https://doi.org/10.1080/09593332208618240

Gajalakshmi, S., Ramasamy, E. V., & Abbasi, S. A. (2002a). High rate composting vermicomposting of water hyacinth (*Eichhornia crassipes*, Mart. Solms). *Bioresource Technology, 83*, 235–239. https://doi.org/10.1016/S0960-8524(01)00216-4

Gajalakshmi, S., Ramasamy, E. V., & Abbasi, S. A. (2002). Vermicomposting of different forms of water hyacinth by the earthworm *Eudrilus eugeniae*, Kinberg. *Bioresource Technology, 82*, 165–169. https://doi.org/10.1016/S0960-8524(01)00163-8

Gajalakshmi, S., Ramasamy, E. V., & Abbasi, S. A. (2005). Composting–vermicomposting of leaf litter ensuing from the trees of mango (*Mangifera indica*). *Bioresource Technology, 96*, 1057–1061. https://doi.org/10.1016/j.biortech.2004.09.002

Ganiger, K. S., Patil, S. R., & Biradar, P. M. (2020). Nutrient status of compost and vermicompost produced by different organic wastes. *Asian Journal of Biological Sciences, 34*, 19–24.

Garg, P., Gupta, A., & Satya, S. (2006). Vermicomposting of different types of waste using *Eisenia foetida*: A comparative study. *Bioresource Technology, 97*, 391–395. https://doi.org/10.1016/j.biortech.2005.03.009

Garg, V. K., Gupta, R., & Kaushik, P. (2009). Vermicomposting of solid textile mill sludge spiked with cow dung and horse dung: A pilot-scale study. *International Journal of Environment and Pollution, 38*, 385–396. https://doi.org/10.1504/IJEP.2009.027271

Garg, V. K., Suthar, S., & Yadav, A. (2012). Management of food industry waste employing vermicomposting technology. *Bioresource Technology, 126*, 437–443. https://doi.org/10.1016/j.biortech.2011.11.116

GargVK, K. P. (2005). Vermistabilization of textile mill sludge spiked with poultry droppings by anepigeic earthworm *Eisenia foetida*. *Bioresource Technology, 96*, 1063–1071. https://doi.org/10.1016/j.biortech.2004.09.003

Gaur, A. C., & Singh, G. (1995). Recycling of rural and urban wastes through conventional composting and vermicomposting. In H. L. S. Tandon (Ed.), *Recycling of crop, animal, human and industrial wastes in agriculture* (pp. 1–49). Fertilizer development and consultation organization.

Ghatnekar, S. D., Mahavash, F. K., & Ghatnekar, G. S. (1998). Management of solid waste though vermiculture biotechnology. In *Ecotechnology for Pollution Control and Environmental Management*. ISBN:8186421033.

Govi, M., Ciavatta, C., & Gessa, C. (1993). Evolution of organic matter in sewage sludge a study based on the use of humification parameters and analytical electro focusing. *Bioresource Technology, 44*, 175–180. https://doi.org/10.1016/j.wasman.2010.04.030

Gunadi, B., & Edwards, C. A. (2003). The effect of multiple applications of different organic wastes on the growth, fecundity and survival of *Eisenia foetida* (Savigny) (Lumbricidae). *Pedobiologia, 47*, 321–330. https://doi.org/10.1078/0031-4056-00196

Gupta, R., & Garg, V. K. (2009). Vermiremediation and nutrient recovery of non-recycleable paper waste employing *Eisenia fetida*. *Journal of Hazardous Materials, 162*, 430–439. https://doi.org/10.1016/j.jhazmat.2008.05.055

Gupta, R., Mutiyar, P. K., Rawat, N. K., Saini, M. S., & Garg, V. K. (2007). Development of a water hyacinth based vermireactors using an epigeic earthworm *Eisenia fetida*. *Bioresource Technology, 13*, 2605–2610. https://doi.org/10.1016/j.biortech.2006.09.007

Gupta, S. K., Tewari, A., Srivastava, R., Murthy, R. C., & Chandra, S. (2005). Potential of *Eisenia fetida* for sustainable and effective vermicomposting of flyash. *Water, Air, & Soil Pollution, 163*, 293–302. https://doi.org/10.1007/s11270-005-0722-y

Gusain, R., & Suthar, S. (2020). Vermicomposting of invasive weed *Ageratum conyzoids*: Assessment of nutrient mineralization, enzymatic activities, and microbial properties. *Bioresource Technology, 123537*. https://doi.org/10.1016/j.biortech.2020.123537

Hansen, D. (2007). Vermicomposting: Innovative kitchen help. [Online] Available: http://www.dnr.mo.gov/env/swmp/docs/vermicomposting.pdf

Jakobsen, S. T., Gade, N., Moller, I., & Franzen, E. (1988). Experiments on decomposition of farmyard manure at Egtved field station. *Egtved Avlsstationer for KodProdukter, 1990*, 1–48.

Jothimani, S. (1994). Organic farming in coconut. *Indian Coconut Journal* 48–49.

Julka, J. M. (2008). Know your earthworms. Foundation for Life Sciences and Business Management, Solan, 51 pp.

Kale, R. D. (1995). Vermicomposting has a bright scope. *Indian Silk, 34*, 6–9.

Karwal, M., & Kaushik, A. (2020). Co-composting and vermicomposting of coal fly-ash with press mud: Changes in nutrients, micro-nutrients and enzyme activities. *Environmental Technology & Innovation, 27*, Article 100708. https://doi.org/10.1016/j.eti.2020.100708

Kaur, A., Singh, J., Vig, A. P., Dhaliwal, S. S., & Rup, P. J. (2010). Cocomposting with and without *Eisenia fetida* for conversion of toxic paper mill sludge to a soil conditioner. *Bioresource Technology, 101*, 8192–8198. https://doi.org/10.1016/j.biortech.2010.05.041

Kaushik, P., & Garg, V. K. (2003). Vermicomposting of mixed solid textile mill sludge and cow dung with epigeic earthworm *Eisenia foetida*. *Bioresource Technology, 90*, 311–316. https://doi.org/10.1016/S0960-8524(03)00146-9

Kaushik, P., & Garg, V. K. (2004). Dynamics of biological and chemical parameters during vermicomposting of solid textile mill sludge mixed with cow dung and agricultural residues. *Bioresource Technology, 94*(2), 203–209. https://doi.org/10.1016/j.biortech.2003.10.033

Khwairakpam, M., & Bhargava, R. (2009). Bioconversion of filter mud using vermicomposting employing two exotic and one local earthworm species. *Bioresource Technology, 100*, 5846–5852. https://doi.org/10.1016/j.biortech.2009.06.038

Khwanchai, K., & Kanokkorn, S. (2018). Effect of agricultural waste on vermicompost production and earthworm biomass. *Journal of Environmental Science and Technology, 11*, 23–27. https://doi.org/10.3923/jest.2018.23.27

Kostecka, J. (2000). Badania nad wermikompostowaniem odpadów organicznych. *Zeszyty Naukowe Akademii Rolniczej w Krakowie. Rozprawy, 68*, 1–88.

Krishna, I. M., Manickam, V., Shah, A., & Davergave, N. (2017). *Environmental management: Science and engineering for industry*. Butterworth-Heinemann.

Kurien, J., & Ramasamy, E. V. (2006). Vermicomposting of Taro (*Colocasia esculenta*) with two epigeic earthworm species. *Bioresource Technology, 97*(11), 1324–1328. https://doi.org/10.1016/j.biortech.2005.05.018

Loh, T. C., Lee, Y. C., Liang, J. B., & Tan, D. (2005). Vermicomposting of cattle and goat manures by *Eisenia foetida* and their growth and reproduction performance. *Bioresource Technology, 96*, 111–114. https://doi.org/10.1016/j.biortech.2003.03.001

Lv, B., Xing, M., & Yang, J. (2016). Speciation and transformation of heavy metals during vermicomposting of animal manure. *Bioresource Technology, 209*, 397–401. https://doi.org/10.1016/j.biortech.2016.03.015

Mainoo, N. O. K., Barrington, S., Whalen, J. K., & Sampedro, L. (2009). Pilot-scale vermicomposting of pineapple wastes with earthworms native to Accra, Ghana. *Bioresource Technology, 100*, 5872–5875. https://doi.org/10.1016/j.biortech.2009.06.058

Manna, M. C., Jha, S., Ghosh, P. K., & Acharya, C. L. (2003). Comparative efficacy of three epigeic earthworm sunder different deciduous forest litters decomposition. *Bioresource Technology, 88*, 197–206. https://doi.org/10.1016/S0960-8524(02)00318-8

Manukovsky, N. S., Kovalev, V. S., & Gribovskaya, I. V. (2001). Two stage of biohumus production from inedible potato biomass. *Bioresource Technology, 78*, 273–275. https://doi.org/10.1016/S0960-8524(01)00022-0

Marttinez-Inigo, M. J., & Almedros, G. (1994). Kinetic study of the composting of evergreen oak forestry waste. *Waste Management and Research, 12*, 305–314. https://doi.org/10.1177/0734242X9401200403

Mohan, S. M. (2017). Vermicomposting of papermill sludge with *Eisenia fetida* for its conversion to nutrient using different seed materials. *Journal of The Institution of Engineers (India): Series A, 98*, 545–553. https://doi.org/10.1007/s40030-017-0253-8

Morgan, M., & Burrows, I. (1982). Earthworms/microorganisms interactions. Rothamsted Experimental Station Reports, 104.

Nakasaki, K., Nag, K., & Karita, S. (2005). Microbial succession associated with organic matter decomposition during thermophilic composting of organic waste. *Waste Management & Research, 23*, 48–56. https://doi.org/10.1177/0734242X05049771

Nasiru, A., Ismail, N., & Ibrahim, M. H. (2013). Vermicomposting: Tool for sustainable ruminant manure management. *Journal of Waste Management*. https://doi.org/10.1155/2013/732759

Nigussie, A., Kuyper, T. W., Bruun, S., & de Neergaard, A. (2016). Vermicomposting as a technology for reducing nitrogen losses and greenhouse gas emissions from small-scale composting. *Journal of Cleaner Production, 139*, 429–439.

Nogales, R., Cifuentes, C., & Benitez, E. (2005). Vermicomposting of winery wastes: A laboratory study. *Journal of Environmental Science and Health, 40*, 659–673. https://doi.org/10.1081/PFC-200061595

Ojuolape, O. T., Aransiola, S. A., & Obideyi, O. A. (2015). Vermitechnology: A solution to environmental problems; A review. *Journal of Global Ecology and Environment, 3*(3), 2454–2644, 127–135. http://www.ikpress.org/abstract/4516.

Prakash, M., & Karmegam, N. (2010). Vermistabilization of press mud using *Perionyx ceylanensis* Mich. *Bioresource Technology, 101*, 8464–8468. https://doi.org/10.1016/j.biortech.2010.06.002

Ramnarain, Y. I., Ansari, A. A., & Ori, L. (2019). Vermicomposting of different organic materials using the epigeic earthworm *Eisenia foetida. International Journal of Recycling of Organic Waste in Agriculture, 8*, 23–36. https://doi.org/10.1007/s40093-018-0225-7

Rasappan, K., Kumar, A., & Santhosh, P. (2015) Studies on sugarcane press mud and distillery wastes as bio fertilizer through biocomposting. *Journal of Chemical Sciences, 13*, 1333–1344.

Saha, B., Devi, C., Khwairakpam, M., & Kalamdhad, A. S. (2018). Vermicomposting and anaerobic digestion–viable alternative options for terrestrial weed management–A review. *Biotechnology Report, 17*, 70–76. https://doi.org/10.1016/j.btre.2017.11.005

Sahariah, B., Goswami, L., Kim, K. H., Bhattacharyya, P., & Bhattacharya, S. S. (2015). Metal remediation and biodegradation potential of earthworm species on municipal solidwaste: A parallel analysis between *Metaphire posthuma* and *Eisenia fetida. Bioresource Technology, 180*, 230–236. https://doi.org/10.1016/j.biortech.2014.12.062

Sangwan, P., Kaushik, C. P., & Garg, V. K. (2010). Vermicomposting of sugar industry waste (press mud) mixed with cow dung employing an epigeic earthworm *Eisenia foetida. Waste Management Resource, 28*, 71–75. https://doi.org/10.1177/0734242X09336315

Sharma, K., & Garg, V. K. (2018). Comparative analysis of vermicompost quality produced from rice straw and paper waste employing earthworm *Eisenia fetida* (Sav.). *Bioresource Technology, 250*, 708–715. https://doi.org/10.1016/j.biortech.2017.11.101

Shweta Kumar, R., Singh, B. L., & Verma, D. (2010). Integrating microbial composting and vermicomposting for Effective utilization of by-products of sugarcane-processing industries. *Bioremediation Journal, 14*, 158–167. https://doi.org/10.1080/10889868.2011.554357

Singh, A., & Sharma, S. (2002). Composting of a crop residue through treatment with microorganisms and subsequent vermicomposting. *Bioresource Technology, 85*, 107–111. https://doi.org/10.1016/S0960-8524(02)00095-0

Singh, C. K., & Kumar, A. (2017). Vermicomposting of terrestrial weeds *Lantana camara* L. and *Parthenium hysterophorus* L.: Agriculture solid waste. *Ecological Questions, 28*, 63–69. https://doi.org/10.12775/EQ.2017.040

Singh, J. (2018). Role of earthworm in sustainable agriculture. In C. M. Galanakis (Ed.), *Sustainable food systems from agriculture to industry* (pp. 83–122). Academic Press. https://doi.org/10.1016/B978-0-12-811935-8.00003-2

Singh, J., Kaur, A., & Vig, A. P. (2014). Bioremediation of distillery sludge into soil-enriching material through vermicomposting with the help of *Eisenia fetida. Applied Biochemistry and Biotechnology, 174*(4), 1403–1419. https://doi.org/10.1007/s12010-014-1116-7

Singh, J., Kaur, A., Vig, A. P., & Rup, P. J. (2010). Role of *Eiseniafetida* in rapid recycling of nutrients from biosludge of beverage industry. *Ecotoxicology and Environmental Safety, 73*(3), 430–435. https://doi.org/10.1016/j.ecoenv.2009.08.019

Singh, S., Bhat, S. A., Singh, J., Kaur, R., & Vig, A. P. (2016). Vermistabilization of thermal power plant fly ash using *Eisenia fetida. Journal of Industrial Pollution Control, 32*, 554–561. ISSN (0970-2083).

Singh, S., Singh, J., Kaur, A., Kaur, J., Vig, A. P., & Bhat, S. A. (2019). Nutrient recovery from pigeon dropping by using exotic earthworm *Eisenia fetida. Sustainable Chemistry and Pharmacy, 12*, Article 100126. https://doi.org/10.1016/j.scp.2019.01.003

Singh, W. R., & Kalamdhad, A. S. (2016). Transformation of nutrients and heavy metals during vermicomposting of the invasive green weed *Salvinia natans* using *Eisenia fetida. International Journal of Recycling of Organic Waste in Agriculture, 5*(3), 205–220. https://doi.org/10.1007/s40093-016-0129-3

Singleton, D. R., Hendrix, P. F., Coleman, D. C., & Whitman, W. B. (2003). Identification of uncultured bacteria tightly associated with the intestine of the earthworm Lumbricus rubellus (Lumbricidae; Oligochaeta). *Soil Biology and Biochemistry, 35*(12), 1547–1555. https://doi.org/10.1016/S0038-0717(03)00244-X

Sinha, R. K., Heart, S., Agarwal, S., Asadi, R., & Carretero, E. (2002). Vermiculture technology for environmental management: Study of the action of earthworms Eisenia foetida, Eudrilus eugeniae and Perionyx excavatus on biodegradation of some community wastes in India and Australia. *The Environmentalist, 22*, 261–268. https://doi.org/10.5281/zenodo.13899379

Soobhany, N., Gunasee, S., Rago, Y. P., Joyram, H., Raghoo, P., & MoheeR, GargVK. (2017). Spectroscopic, thermogravimetric and structural characterization analyses for comparing municipal solid waste composts and vermicomposts stability and maturity. *Bioresource Technology, 236*, 11–19. https://doi.org/10.1016/j.biortech.2017.03.161

Suthar, S. (2007). Vermicomposting potential of *Perionyx sansibaricus*(Perrier) in different waste materials. *Bioresource Technology, 98*, 1231–1237. https://doi.org/10.1016/j.biortech.2006.05.008

Suthar, S. (2008). Bioremediation of aerobically treated distillery sludge mixed with cow dung by using an epigeic earthworm *Eisenia fetida. The Environmentalist, 28*, 76–84. https://doi.org/10.1007/s10669-007-9031-x

Suthar, S. (2010). Recycling of agro-industrial sludge through vermitechnology. *Ecological Engineering, 36*, 1028–1036. https://doi.org/10.1016/j.ecoleng.2010.04.015

Suthar, S. (2010). Pilot-scale vermireactors for sewage sludge stabilization and metal remediation process: Comparison with small-scale vermireactors. *Ecological Engineering, 36*, 703–712. https://doi.org/10.1016/j.ecoleng.2009.12.016

Suthar, S., & Sharma, P. (2013). Vermicomposting of toxicweed—*Lantana camara* biomass: Chemical and microbial properties changes and assessment of toxicity of end product using seed bioassay. *Ecotoxicology and Environmental Safety, 95*, 179–187. https://doi.org/10.1016/j.ecoenv.2013.05.034

Suthar, S., & Singh, S. (2008). Feasibility of vermicomposting in biostabilization of sludge from a distillery industry. *Science of the Total Environment, 394*, 237–243. https://doi.org/10.1016/j.scitotenv.2008.02.005

Tajbakhsh, J., Abdoli, M. A., Mohammadi Goltapeh, E., Alahdadi, I., & Malakouti, M. J. (2008). Trend of physico-chemical properties change in recycling spent mushroom compost through vermicomposting by epigeic earthworms *Eisenia fetida* and *Eisenia andrei. Journal of Agricultural Technology, 4*, 185–198.

Tewari, P. K., Batra, V. S., & Balakrishnan, M. (2007). Water management initiatives in sugarcane molasses-based distilleries in India. *Resources, Conservation & Recycling, 26*, 351–367. https://doi.org/10.1016/j.resconrec.2007.05.003

Tian, G., Kang, B. T., & Brussaard, L. (1997). Effect of mulch quality on earthworm activity and nutrient supply in humidtropics. *Soil Biology and Biochemistry, 29*, 369–373. https://doi.org/10.1016/S0038-0717(96)00099-5

Tripathi, G., & Bhardwaj, P. (2004). Comparative studies on biomass production, life cycles and composting efficiency of *Eiseniafoetida* (Savigny) and *Lampitomauritii* (Kinberg). *Bioresource Technology, 92*, 275–278. https://doi.org/10.1016/j.biortech.2003.09.005

Tripathi, J. G. (2014). Study of India's paper industry-potential and growth in 21st century. *Indian Journal of Applied Research, 4*, 112–115. https://doi.org/10.36106/ijar

Vallini, G., & Pera, A. (1989). Green compost production from vegetable waste separately collected in metropolitan garden produce markets. *Biological Wastes, 29*, 33–41. https://doi.org/10.1016/0269-7483(89)90101-8

Venkatesh, R. M., & Eevera, T. (2008). Mass reduction and recovery of nutrients through vermicomposting of fly ash. *Applied Ecology and Environmental Research, 6*(1), 77–84. ISSN: 15891623.

Wong, J. W. C., Mak, K. F., Chan, N. W., Lam, A., Fang, M., Zhou, L. X., Wu, Q. T., & Liao, X. D. (2001). Co-composting of soybean residues and leaves in Hong Kong. *Bioresource Technology, 76*, 99–106. https://doi.org/10.1016/S0960-8524(00)00103-6

Xing, M., Yang, J., & Lu, Z. (2005, May). Microorganism-earthworm integrated biological treatment process—a sewage treatment option for rural settlements. In *ICID 21st European regional conference*, Frankfurt (pp. 15– 19).

Yadav, A., & Garg, V. K. (2009). Feasibility of nutrient recovery from industrial sludge by vermicomposting technology. *Journal of Hazardous Materials, 168*, 262–268. https://doi.org/10.1016/j.jhazmat.2009.02.035

Yadav, A., & Garg, V. K. (2010). Bioconversion of food industry sludge into value-added product (vermicompost) using epigeic earthworm *Eisenia fetida. World Review of Science, Technology and Sustainable Development, 7*, 225–238. https://doi.org/10.1504/WRSTSD.2010.032526

Yadav, D. V. (1995). Recycling of sugar factory press mud in agriculture. In H. L. S. Tandon (Ed.), *Recycling of crop, animal, human and industrial wastes in agriculture* (pp. 91–108). Fertilizer Development and Consultation Organization Publication.

Agro-industrial Sludge and Vermitechnology

Yemisi Tosin Aluko, Labake Agunbiade, and Ifekristi Benson

Abstract Improvements in agricultural methods and technology, spurred by the Green Revolution and the rapidly increasing human population, have resulted in the rise of agro-industries over the past five decades. This, in turn, led to the production of high amounts of waste, such as husks, bagasse, and pulp residues of fruits. These agro-industrial wastes and residues have the potential for conversion into new products with increased value while reducing environmental pollution. Vermicomposting technology has emerged over the past few decades as a sustainable and cost-effective treatment method for effectively utilizing wastes from a variety of agro-industrial processes and turning them into products with additional value for use in land restoration techniques. Vermicompost, the end result of the process, is a humus-like, finely ground material that can be used as a fertilizer source to reintegrate organic matter into agricultural soils. This makes the process an attractive and sustainable treatment option among researchers and organic farmers. The present chapter is intended to provide information on the processes involved in vermicomposting, the use of vermicompost for agricultural soil restoration and increment in crop yield, and its economic importance.

Keywords Wastes · Sludge · Vermitechnology · Vermicompost · Agro-industrial

Y. T. Aluko (✉)
Department of Environmental Management, Pan African University Institute of Life and Earth Sciences (Including Health and Agriculture), University of Ibadan, Ibadan, Nigeria
e-mail: aweyemi4u@gmail.com

L. Agunbiade
Department of Soil, Water and Ecosystem Sciences, University of Florida, Gainesville, USA

I. Benson
Department of Agronomy, University of Florida, Gainesville, USA

S. A. Aransiola and N. R. Maddela (eds.), *Vermitechnology: Economic, Environmental and Agricultural Sustainability*, https://doi.org/10.1007/978-3-031-96212-7_9

1 Introduction to Agro-industrial Sludge

Agro-industries contribute immensely to the global economy with a measured production mean value of 23.7 million tonnes of food every day, of which a larger percentage is from developing countries that often depend on small-scale agriculture and income from formal as well as informal agro-industries (Amor et al., 2019). Improvement in agricultural practices and technology, which was initiated by Green Revolution and the rapidly increasing human population, has led to the expansion of agro-industries over the past five decades (Hernández et al., 2013).

Agro-industries produce high amounts of waste, such as husks, bagasse, and pulp residues of fruits. These residues have the potential for reuse and for obtaining new products with increased value with a possible reduction of environmental pollution and sustainability (Ojuolape et al., 2015; Pelizer et al., 2007). Agro-industrial wastes include peanut, wheat, corn, potato, and rice by-products. Sugar-processing waste and fermentation liquor are also examples of wastes generated from agro-industries that are very suitable for microbial growth (Siddeeg et al., 2019). Agricultural wastes generated from oil extraction industry or other agro-industries are usually in large quantities. These wastes are indiscriminately dumped or buried in landfills which results in environmental and socioeconomic issues (González-Moreno et al., 2022). On a global scale, they contribute majorly to the issue of environmental contamination and pollution. It cannot be overemphasized that continuous utilization of natural resources, rapidly increasing population growth, and rising individual energy consumption will lead to an increase in waste generated from agro-industries (Sekeri et al., 2020).

1.1 *Composition of Agro-industrial Sludge*

Agro-industrial sludge produced from agro-industries includes dairy, vinasse, cheese whey, potato, coffee, cassava, beverage, palm oil, olive oil, pulp and paper mill, and slaughterhouse wastewaters. The composition of the sludge is greatly influenced by the raw materials' source, the nature of the products, the system type, the methods of operation, and the stages involved in processing (Prazeres et al., 2012; Teh et al., 2014). Cheese whey is an essential waste stream of cheese factories. The distillery factories generate vinasse as a wastewater, which consists of a wide range of organic components that may vary in composition depending on the feedstock and the distillation methods adopted (Robles González et al., 2012). Likewise, coffee-processing wastewater from coffee-producing agro-industries contains organic substances like pectin, sugars, and proteins (Von Enden, 2002). The physical and chemical composition of agro-industrial sludge is diverse; it usually consists of high concentrations of inorganic and organic pollutants, nutrients such as nitrogen, phosphorus, and potassium, pesticides, and toxic compounds including heavy metals that may affect water, soil, and air quality (Abubakar et al., 2016; Leh-Togi Zobeashia et al., 2018). Agro-industrial residues can also contain some beneficial nutrients. Pardo et al. (2014) discovered that agro-industrial pineapple residues had higher fiber content than the edible portion or pulp, which was dominant in the leaf bracts, shell, and core residues. Other research has also shown some beneficial uses of agro-industrial pineapple residues, which include their use as a source of protein supplement (Díaz-Vela et al., 2017; Mensah & Twumasi, 2017), bromelain extraction (Chaurasiya et al., 2015; Manosroiet al., 2014), vinegar manufacture (Madurai et al., 2016), enzyme production (Aransiola et al., 2023a; Arun & Sivashanmugam, 2015; Selvakumar & Sivashanmugam, 2017), biofuel production (Aransiola et al., 2023b; Aworanti et al., 2017; Ogunleye et al., 2016; Shamsul et al., 2017), among others.

1.2 Environmental Effects of Improper Disposal of Agro-industrial Sludge

The uncontrolled release of untreated agro-industrial effluents into the environment can enhance eutrophication processes and cause instability in the ecosystem leading to several harmful effects on humans, animals, and plants. These effluents can contaminate shallow and groundwater aquifers (Amor et al., 2019), which renders the water bodies unsuitable for other utilization. They harbor large amounts of organic compounds that have the potential to cause environmental pollution (Abubakar et al., 2016). These organic compounds with high chemical oxygen demand (COD) and color values, when released into the ecosystem, affect aquatic organisms and human health and cause visual contrast in the natural characteristics of landscapes (Chen et al., 2020). Other adverse effects caused by agro-industrial sludge include changes in soil physicochemical properties, increased salt content in soil, soil microbial population, increased heavy metal concentrations in the soil, foul odors, depletion of water resources, outbreaks of endemic/native/indigenous diseases, and a rise in the depletion of dissolved oxygen in water (Yaqoob et al., 2021). Thus, when released without being properly treated, it poses a serious risk to the environment (Chen et al., 2020). The numerous adverse effects of improper disposal of agro-industrial sludge indicate that effluents should be properly treated to significantly reduce the pollutant load and volume before disposal. For instance, Roldi et al. (2013) reported that in the northeastern region of Paraná, agro-industrial wastes from cassava, sugar, and ethanol production are all easily available, and proper disposal is necessary as these waste products could become harmful to the environment. Agro-industrial sludge is usually disposed of on agricultural fields and utilized as a source of fertilizer. If it is not properly treated and is applied in its raw state, it poses various environmental hazards such as pathogen infestation and groundwater pollution with nitrate and organic pollutants (Chiochettaet al., 2014; Düring & Gäth, 2002; Huguier et al., 2015). Therefore, before and after soil application, the benefit over the risk must be carefully considered (Fig. 1).

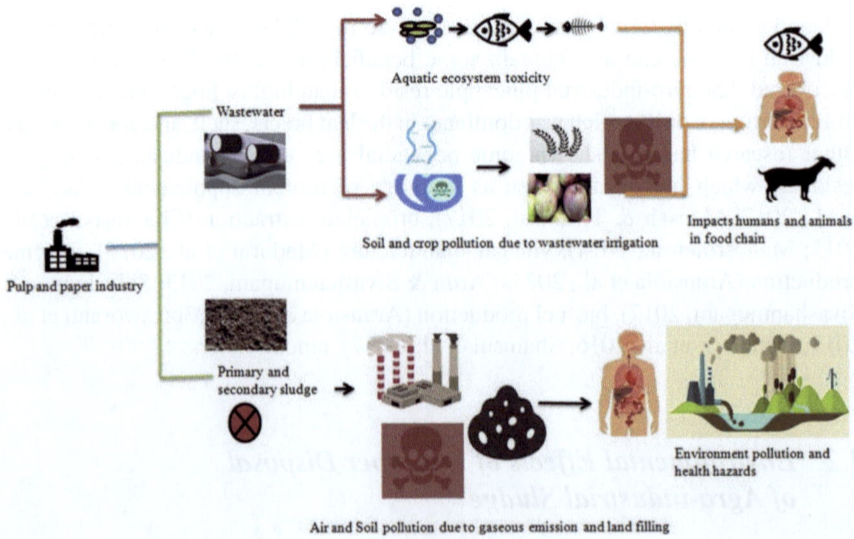

Fig. 1 Flow chart illustrating the emission pathways of wastewater and sludge from the pulp and paper industry and their environmental and health Impacts. (adapted from Gupta et al., 2019)

2 Application of Vermitechnology to Agro-industrial Sludge

As discussed above, agro-industries generate many by-products and wastes that will most likely have a deleterious impact on the environment if they are not treated appropriately. Included among these by-products is agro-industrial sludge, which has an enormous impact on the environment; hence, it is of necessity to find the right technology for mitigating these impacts. One of such technologies that will not only reduce the waste or make it less harmful to the environment, but also make the by-product a very useful material, is vermitechnology. Vermitechnology is the study which involves the application of technologies that utilize earthworms to break down waste organic materials for sanitation and convert them for agricultural reuse (Kumar, 2005; Ojuolape et al., 2015). It has three main branches: vermifiltration, which is a process for purifying industrial and agricultural effluents; vermicomposting, which is a process that depends on earthworms for composting organic waste material; and vermiculture, which is the rearing of earthworms on a larger scale to be utilized for other processes. Vermicomposting is the method needed for composting agro-industrial sludge. It is an alternative option in solid waste disposal with greater socioeconomic benefits (Aransiola et al., 2022; Sharma et al., 2022).

2.1 Vermicomposting

Vermicomposting involves a simple, low-cost biotechnology for composting that utilizes a particular species of earthworms (epigeic worms such as *Eisenia fetida*) to enhance the process of waste conversion and produce a more valuable end product called vermifertilizer, vermicompost, or vermicast (Ahmad et al., 2021; Aransiola et al., 2022; Vavouraki & Kornaros, 2023). It is a process of synergetic interaction between earthworms and microorganisms to biooxidize and stabilize organic waste (Dume et al., 2022; Sharma et al., 2022). It is considered the best technique to manage organic waste because it is environmentally friendly, economically viable, simple, practicable, and socially acceptable (Ahmad et al., 2021; Sharma & Garg, 2018). Vermicompost can be considered an organic fertilizer source and a biological control agent, which helps in protecting plants from pests and pathogens (Thakur et al., 2021). The end product of vermicomposting is also referred to as black gold (Ahmad et al., 2021); thus, it can be said to be a simple waste-to-wealth technology.

In composting, organic matter must first undergo an accelerated biooxidation process known as thermophilic stage (45 °C to about 65 °C), during which microbes, primarily bacteria, fungi, and actinomycetes, release heat, CO_2, and water. With turning or aeration, the heterogeneous organic material is converted into a homogeneous and stabilized humus-like product. While vermicomposting is also a biooxidation and stabilization process of organic material, it involves the joint action of earthworms and microorganisms and does not involve a thermophilic stage. The turning, fragmentation, and aeration processes are carried out by the earthworms (Ahmad et al., 2021). Vermicomposting is, in other words, a mesophilic process that makes use of microorganisms and earthworms that are agile at 10–32 °C. This procedure takes less time than composting because the waste goes via the gut of the earthworm, where a considerable but little understood transformation occurs, producing earthworm castings (vermifertilizer) that are rich in microbial activity, plant growth regulators, and pest-repelling properties (Crescent, 2003; Nagavallemma et al., 2004).

The soil harbors over 3000 species of earthworms around the world. The body of an earthworm contains about 60–70% protein with a high level of essential amino acids like methionine and lysine, which are higher than that of livestock and fish (Rostami, 2011). The body of an earthworm consists of 2–3% minerals, 6–11% fat, 5–21% carbohydrates, niacin, and vitamin B12 in particular (Edwards, 1985). Pathogens are adversely affected by earthworm activity on organic waste, and some studies have found that vermicompost is a healthier organic fertilizer than compost and manure (Asgharnia, 2003). Of the over 3000 species of earthworms, *Eudrilus eugeniae, E. fetida, Eisenia andrei*, and a few others are the most desirable for vermicomposting (Ahmad et al., 2021; Rostami, 2011). These worm species are epigeic which prefer to inhabit the top surface of soil and feed on organic materials such as compost, organic bedding, vegetable waste, and other products that are naturally rich in nutrients as compared with species that feed on plain soil. Unlike other species, these species have the capacity to eat as much as half their body weight daily; hence, they are able to adequately break down and decay organic waste, leading to production of high-quality organic compost. Furthermore, they are more resistant to unfavorable environmental conditions caused by fluctuations in moisture and temperature regimes. They are very active throughout the year, able to decompose organic matter very quickly and produce vermicompost in the shortest possible time (Ahmad et al., 2021). Earthworm tissues accumulate significant amounts of potentially toxic elements (PTEs), which suggests that vermicomposting could be an appropriate alternative technology for reducing toxic materials being released into the environment (Dume et al., 2022). The movement of earthworms aerates, mixes, and grinds the substrate, which creates a favorable condition for microbes present in both the waste and the worm's intestinal tracts, which are responsible for the biochemical degradation of organic material (Pizzanelli et al., 2023). The presence and activities of earthworms have been proven to accelerate cow dung decomposition, resulting in the mineralization of organic compounds (Vavouraki & Komaros, 2023).

Among the epigeic species, *E. fetida,* also termed as banded worms, is the most utilized all over the globe due to their high efficiency, breeding capacity, and adaptability (Ali et al., 2015), while *E. eugeniae* is very common in tropical and subtropical nations (Kumar, 2005). *Eisenia fetida* has been widely used in different types of agro-industrial sludge such as textile mill sludge (Garg & Kaushik, 2005), vegetable market waste and wheat straw (Suthar, 2009), tannery sludge (Viget al., 2011), food industry sludge (Yadav & Garg, 2010), and so on (Fig. 2).

Fig. 2 A close-up image of earthworms on rich, moist soil during vermicomposting

2.2 *Processes Involved in Vermicomposting of Agro-industrial Sludge*

The essential steps in vermicomposting are adapted from Ahmad et al. (2021) but modified (Fig. 1):

(a) Choosing a suitable site: Factors to consider for a proper site for setting up vermicomposting include safe and secure space, availability of dung or manure, favorable environment for earthworms, and access to water.

(b) Preparing the site for the chosen vermicomposting system: This means constructing the bed/windrow system of choice. For example, construct tanks made with normal bricks with dimensions of 4.5 m in length, 1.5 m in breadth, and 0.9 m in height. However, the size could be decreased or increased to dimensions suitable for operation. Provide partition walls with small outlets for smooth movement of earthworms from one tank to another. Moisten the surface of the soil thoroughly by sprinkling it with water. Subsequently, apply bedding materials (such as wood chips, dry leaves, straw, or grass) of about 20 mm thickness into the tank as the first layer. Water is once again sprinkled over the layer of bedding materials. Lay out a generous amount, about 0.3–0.5 m thick, of farmyard manure or cow dung evenly over the bedding materials and apply water in a sprinkling fashion to make it adequately moist.

It should be noted that the cow dung used should not be too fresh or too old. It should be an estimated 10–15 days old at least, since the heat produced from fresh cow dung can kill earthworms. Similarly, cow dung that is too old will be depleted of nutrients, as it would have decomposed, and earthworms will not get sufficient food from it.

(c) Adding the agro-industrial sludge: Add the agro-industrial sludge (such as sludge from sugar, palm oil, winery, dairy, and meat processing). Again, spread about 0.3–0.5 m layer of cow dung evenly and sprinkle an adequate quantity of water.

It is worthy of note that sprinkling water after each layer is necessary in order to moisten the materials as this initiates the activities of microorganisms for the initial decomposition of the material. If bedding materials are too dry, it is advised to first soak it in water before applying to the bed. This enhances easy moistening of materials and prevents moistening of piles.

(d) Introducing earthworms: Lay out about 1 kg of vermiculture (with earthworm population of about 800–1000) over the layer of cow dung. Again, apply evenly a 2- to 3-in. layer of green leaves, and sprinkle water on it. Next, cover the vermicompost bed with jute/gunny bags or other materials that can provide cooling and prevent entry of sunlight. Grass straws, broad leaves, and reeds can also be used. To achieve optimum moisture and temperature conditions in the vermicompost bed, sprinkle water regularly over the gunny or jute bags.

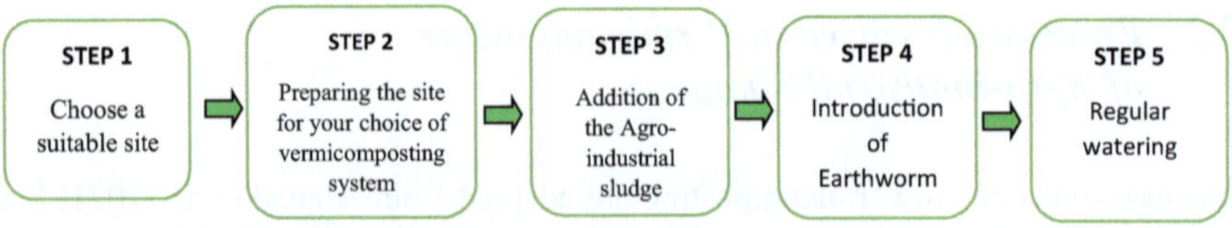

Fig. 3 Flow chart illustrating the steps for setting up a vermicomposting system

(e) Watering regularly: To sustain optimum conditions for earthworm growth and functioning, maintain a moisture content of 35–40% and temperature of 15–30 °C in the bed. This can be achieved by regular watering of the bed at least two times a week, and turning the bed once or twice a month if necessary.

Following the steps above, vermicompost is ready in about 8–12 weeks. Watering should be stopped before harvesting to allow drying of the top part of the vermicompost. On maturity, vermicompost appears dark brown in color, very porous, granulated, and free of any foul odor as observed by Ahmad et al. (2021). The vermicompost is screened to remove undecomposed materials and worms. The vermicast is allowed to dry for few days, weighed, and stored (Fig. 3).

2.3 *Composition of Vermicompost from Agricultural Sludge*

Vermicomposting is regarded as an effective process for getting rid of organic waste from agro-industrial sludge (Rupani et al., 2023), and the resulting vermicompost has more exchangeable plant nutrients than the traditional/garden compost. Also, because of the different makeup of the sludge from which the vermicompost is made, the nutrient content of the vermicompost tends to be different. Several researchers have reported increased nutrient content in sludge after vermicomposting. This includes Yadav & Garg (2011), Kumar et al. (2010), and Sen and Chandra (2007), who observed an increase in nutrients in sugar sludge after vermicomposting. Nogales et al. (2005), Gómez-Brandón et al. (2022), and Fernández-Bayo et al. (2008) also studied the effect of composting with earthworms on winery sludge and observed an increase in plant nutrients in the sludge after vermicomposting compared with its nutrient content before vermicomposting. Yadav & Garg (2011) found that the C:N ratio and organic matter content in sugar sludge increased after vermicomposting. Vermicompost from agro-industrial sludge (such as sugar, winery, dairy, and meat processing) has a lot of plant nutrients; hence, it is employed as a soil amendment (Yadav & Garg, 2011). They provide important plant nutrients which help to enrich the soil. Aside from this, they are also rich in organic carbon, sugar, protein, and enzyme. Although farmers are hesitant to feed it directly to the soil because of its smell, significant degradation and fermentation have been used to make sure they are suitable for agricultural use (Sen & Chandra, 2007). In addition to the nutrients included in these agro-industrial sludge, the composting process used to create a sludge also adds nutrients, making the finished product more densely nutrient-rich for plants (Rupani et al., 2023).

Press mud is an example of an agro-industrial product that is generated after sugar has been extracted in the sugar industry (Yadav & Garg, 2011). Organic leftovers after the sugarcane juice was clarified had a pH of 7.1, 313 g/kg of organic carbon, 24 g/kg of nitrogen, 3.6 g/kg of phosphorus, 0.86 g/kg of potassium, 12.1 g/kg of calcium, a C:N ratio of 13,0, 870 mg/kg of copper, 22,440 mg/kg of iron, 1392 mg/kg of zinc, and 2008 mg/kg of manganese (Sangwan et al., 2008). These nutrients were reported to be more than those contained in the pressmud before vermicomposting. Sen and Chandra (2007) also found a 2.0-fold higher nitrogen concentration than in compost after vermicomposting.

The impact of vermicomposting on the C:N ratio and nutrient content of wine sludge, as well as the micro- and macronutrient composition of the final product, aligned with the quality criteria for high-quality vermicompost, as studied by Gómez-Brandón et al. (2022). They reported a significant increase in the nutrient content of the wine sludge after vermicomposting. Singh et al. (2014) investigated how vermicomposting with *E. fetida* affected distillery sludge and found that the amounts of the tested micro- and macronutrients increased. They also noted that this combination improved physicochemical properties and aided faster stabilization (low C:N and greater electrical conductivity [EC]). This evidence made vermicomposting a superior choice to traditional composting.

All micro- and macronutrients, including nitrogen (N), phosphorus (P), potassium (K), iron (Fe), zinc (Zn), copper (Cu), manganese (Mn), growth promoters and regulator hormones like auxin and gibberellin, enzymes like protease, lipase, and chitinase, and helpful bacteria like *Bacillus subtilis* are abundant in vermifertilizer (Ahmad et al., 2021; Vavouraki & Kornaros, 2023). Vermicompost typically contains between 32 and 66% moisture and has a pH of 7.0. As shown in Table 1, vermicompost has a higher percentage of micro- and macronutrients than conventional compost. These nutrients are readily available for plant uptake.

2.4 *Examples of Vermicomposting of Agro-industrial Sludge*

2.4.1 Vermicomposting of Sludge from Pulp and Paper Industry

There has been a steady growth of the pulp and paper industry in recent years, owing to the importance of pulp production in many new markets. For instance, total global paper consumption in 2009 was 371 million tonnes, and by 2012, global paper and paperboard production was over 390 million tonnes, with 490 million tonnes predicted for 2020 (Bajpai, 2013). Every year, the paper and pulp industries consume large amounts of resources, such as wood and water, while also producing huge volumes of solid waste and wastewater that must be treated (Gopal et al., 2019). According to Abdullah et al. (2015), daily production of solid waste from pulp and paper mills increased from 16,200 to 19,100 tons between 2001 and 2005. The amount

Table 1 Nutrient composition of vermicompost and garden compost

Nutrient element	Vermicompost (%)	Traditional compost (%)
Organic carbon	9.8–1313.4	12.222
Nitrogen	0.51–1.61	0.8
Phosphorus	0.19–1.02	0.35
Potassium	0.15–0.73	0.48
Calcium	1.18–7.61	2.27
Magnesium	0.093–0.568	0.57
Sodium	0.058–0.158	< 0.01
Zinc	0.0042–0.110	0.0012
Copper	0.0026–0.0048	0.0017
Iron	0.2050–1.3313	1.1690
Manganese	0.0105–0.2038	0.0414

Source Nagavallemma et al. (2004)

of industrial sludge produced by the pulp and paper sectors is clearly increasing. The industry has found it difficult to manage and properly dispose of sludge because of strict environmental rules on solid waste disposal. It is therefore expedient to use suitable methodologies to treat and utilize this waste in an environmentally friendly manner.

Sludge from the pulp and paper industry is often composed of water, carbohydrates, micro- and macronutrients, trace metals, and wood fibers such as cellulose, hemicellulose, and lignin. The presence of structural polysaccharides and a low nitrogen level (< 0.5%) in sludge can cause difficulty in biodegradation. However, the difficulties could be mitigated by adding nitrogen-rich substances that act as natural inoculants for microbial populations in the sludge. Quintern (2014) proposed that pulp mill sludge be subjected to the vermicomposting process with the addition of nutrient-rich municipal biosolids. Fernández-Gómez et al. (2013) investigated the possibility of employing vermicomposting to break down waste from a paper mill that was combined with tomato plant debris in various ratios. They concluded that the best feed ratio for promoting *E. fetida* growth and reproduction throughout the vermicomposting process was a 1:2 mixed ratio of paper mill sludge and tomato plant waste. Moreover, vermicompost made from a mixture containing a greater percentage of tomato plant waste displayed a higher level of humic acid. Negi and Suthar (2018) reported that paper mill sludge mixed with cow dung in 50–75% proportion yielded good results.

Likewise, in another study by Yadav and Madan (2013), paper mill sludge was combined with various forms of waste (i.e., mixture of agricultural, municipal, and poultry waste) for vermicomposting using *E. fetida*. Different combinations of paper mill sludge (PMS) and different wastes (containing agricultural waste, municipal solid waste, and poultry waste) were used in three proportions for composting and vermicomposting: 1:1, 1:3, and 3:1. Different chemical parameters were measured every 15 days during the 60-day experiment, and it was discovered that the total Kjeldahl nitrogen, available phosphorus, and total potassium increased, while the organic carbon decreased, as the composting and vermicomposting processes progressed. It was concluded that among the three treatment units, paper mill sludge and different waste mixture in the ratio of 1:3 showed the best results. This implies that vermicomposting of paper mill sludge will yield a better result when combined with a higher percentage of nutrient-rich materials. An additional benefit of vermicomposting of paper mill sludge to the environment is that it may be a useful method to remove heavy metals, which could have posed toxicological risks, from the sludge. Suthar et al. (2014) tested the vermicomposting of effluent from paper mills that had been mixed with cow dung and reported a noticeable decrease in the amount of heavy metals present in the final products after 60 days. Lead (Pb) showed the greatest reduction (95.3–97.5%), followed by Cu (68.8–88.4%), Cr (47.3–80.9%), and Cd (32–37%). Vermicomposting, they concluded, might be a promising technique for bioremediation of heavy metals from industrial sludge. The viability of using *Perionyx excavatus* to vermicompost paper mill sludge that was combined with cow manure and food processing waste in various ratios was examined by Sonowal et al. (2013). They found that the total phosphorus and total nitrogen had increased by 76.1 and 58.7%, respectively, whereas the total organic carbon had reduced by 74.5% when using the feedstock of sludge, cow dung, and food processing waste in equal ratios. They concluded that vermicomposting was a superior method for handling and getting rid of the sludge generated by pulp and paper mills.

2.4.2 Vermicomposting of Sludge from Sugar Industry

The sugar industry is an important agro-industry around the world (Martinez-Burgos et al., 2021). It produces sugar as well as numerous sludges or wastes during the manufacturing process. Sugarcane bagasse, pressmud, molasses, sugar beet mud, and pulp are all by-products of the industrial processing of sugarcane and sugar beets. Every year, more than 30 million tons of sludge from sugar industries are produced worldwide, with India accounting for 12 million tons. These sludges are often disposed of in open fields, contaminating and degrading the land and water of that area. Disposal problems of sugar industrial sludge include prohibitive costs, environmental contamination, offensive odors, and prolonged natural decomposition times (Bhat et al., 2014).

According to Bhat and Vig (2019), vermistabilization using earthworms is one alternative approach for utilizing sugar industry sludges to produce value-added organic manure. It is a low-cost bioconversion method (Muthukrishnan & Swaminathan, 2017) and is also effective in converting toxic chemicals in sludge into harmless and usable forms. Vasanthi et al. (2014) successfully turned pressmud sludge mixed with cow manure and Jeevamirtham *Azospirillum* using *E. eugeniae* into fertilizer. The organic fertilizer generated from the vermicomposting process was odorless, more mature, and nutrient rich. It contained higher nitrogen, phosphorus, and potassium but lower organic carbon and C:N ratio. Bhat et al. (2014) biotransformed pressmud sludge mixed with cow dung in various compositions using *Eudrilus fetida*. At the conclusion of the vermicomposting process, they observed that potassium and the C:N ratio had dropped while nitrogen, phosphorus, sodium, electrical conductivity, and pH had increased. They also observed that pressmud sludge's genotoxicity was decreased through vermicomposting, as evidenced by the final vermicompost. Recent studies that utilize sludge from sugar production to produce nutrient-rich organic fertilizer by combining it with other substrates include Namli et al. (2020) and Sharma et al. (2022).

2.4.3 Vermicomposting of Sludge from Palm Oil Processing Industry

One of the most important vegetable oils produced worldwide is palm oil, which plays a significant role in the global oils and fats market. Its production generates solid wastes and effluents, the sustainable treatment of which is vital for the growth of the oil chain in palm oil-producing countries (Choudhary & Grover, 2019; Koura et al., 2017). Palm oil mill effluent (POME) is regarded as the most hazardous waste from palm oil processing industry if released untreated into the environment. The effluent from palm oil mills is a viscous brownish liquid containing high solids, oil and grease, COD, and biological oxygen demand (BOD) values. Despite being organic in nature, POME is difficult to disintegrate under natural conditions. Owing to the fact that earthworms can digest POME and produce valuable products, effective treatment of POME using vermicomposting technique is recommended as a good alternative sustainable management approach that is cost-effective (Chozhavendhan et al., 2022; Rupani et al., 2022; Vyas et al., 2022).

Research conducted by Lim et al. (2014) in a laboratory-scale experiment utilized *E. eugeniae* to biodegrade palm oil mill effluent (POME). POME was absorbed in varied ratios into amendments (soil or rice straw) for the earthworm. The pH, electrical conductivity, and nutritional content all increased significantly in the presence of earthworms, whereas the C:N ratio, soluble chemical oxygen demand, and volatile solids all decreased significantly from 0.687 to 75.8%, from 19.7 to 87.9%, and from 0.687 to 52.7%, respectively. They reported a decrease in earthworm development toward the end of the experiment. With more nutrients and a lower C:N ratio in the vermicompost, rice straw performed better as an amendment and absorbent than soil. It also reduced soluble chemical oxygen demand more than soil. The highest quality, nutrient-dense vermicompost was produced by the treatment using rice straw and POME in a 1:3 ratio.

In a similar laboratory-scale experiment conducted by Bidattul Syirat et al. (2013), changes in the physicochemical properties of vermicompost resulted in a decrease in pH (1.8%), C (1.9%), C:N ratio (12.86%), P (37.5%), Fe (48.5%), Cu (24.7%), Zn (10%), and Mn (11.6%) and an increase in available micro- and macronutrients such as K (3.8%), N (11.2%), Ca (5.9%), and Mg (15.4%) compared to those of the initial substrate. These two findings proved that palm oil mill sludge could be recycled to produce vermicompost that is rich in micro- and macronutrients, which are essential

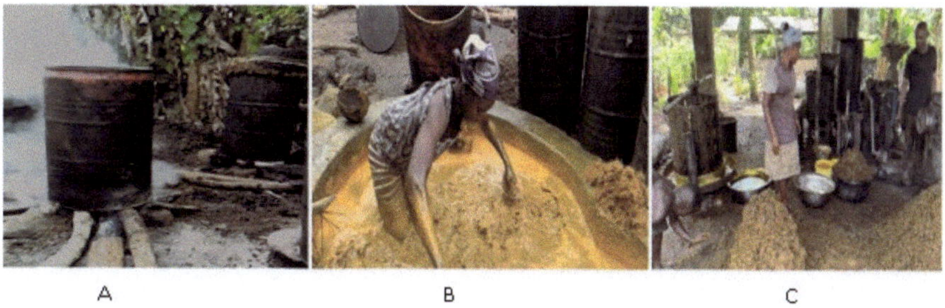

Fig. 4 Images showing traditional palm oil processing in Nigeria. **A** Boiling of oil palm fruit in a large drum until softened; **B** Manual extraction of oil from the softened fruit and separation of sludge; **c** use of local machines to separate oil from sludge

for plant growth, has good physical properties, a low C:N ratio, and optimal stability and maturity (Fig. 4).

2.4.4 Vermicomposting of Sludge from Dairy Industry

The dairy industry is one of the world's greatest producers of wastewater, generating massive amounts of dairy processing sludge (DPS) (Shi et al., 2021), particularly the milk processing industry (Lee et al., 2018). India, the world's largest milk producer, has been producing 94.5 million tonnes of milk annually, and in 2015, that number rose to 155 million tonnes and this will keep increasing (Lee et al., 2018; Suthar, 2012). Dairy processing sludge is classified into two main types: lime-treated dissolved air flotation sludge and biochemically treated activated sludge. Traditional disposal practices of the sludge like landfilling and disposal into surface water have caused pollution and damage to soil, groundwater, and surface water.

Suthar (2012) conducted an experiment using the composting earthworm *E. fetida* to stabilize wastewater sludge from milk processing industry that was combined with cow dung as an amendment. He concluded that the mixture of wastewater sludge and amendment in a ratio of 3:2 had a greater rate of mineralization and had much lower levels of pH, organic carbon, and C:N ratio. The waste mixture also contained more total nitrogen, available phosphorus, exchangeable cations (K^+ and Ca^{2+}), and extractable trace metals (Fe, Mn, and Zn). When a higher concentration of wastewater sludge was used, significant earthworm mortality was recorded, indicating that the formation of various chemicals, such as nitrogen oxide, ammonia,and organic acids, may be detrimental to earthworms. However, Suthar et al. (2012) also investigated the possibility of using vermicomposting on wastewater sludge from dairy industry, combined with cow manure and plant waste (such as wheat straw and sugarcane trash) in various ratios. The study found that the optimal feed condition for earthworm growth and activity was a waste mixture that contained wastewater sludge, cow dung, and plant waste in proportions of 60%, 10% and 30%, respectively. These results validate that earthworm (*E. fetida*) is suitable for the conversion of noxious

Fig. 5 Images of milk production stages. **A** Extraction of milk from cows; **B** processing of the collected milk; **C** generation of wastewater/sludge from milk production

wastewater sludge from the dairy industry into enriched products for land restoration programs. Recent studies on efficiently using wastewater from the dairy industry include Liu et al. (2021) and Sharma et al. (2022). A study to assess the synergistic effect of biochar amended in dairy industry sludge mixed with cattle dung resulted in vermicompost with physicochemical properties suitable for vermiremediation (Dutta et al., 2023) (Fig. 5).

2.5 *Environmental and Economic Importance of Vermicompost of Agro-industrial Sludge*

Several researchers have reported increased nutrient content in sludge after vermicomposting. This includes Yadav & Garg (2011), Kumar et al. (2010), and Sen and Chandra (2007), who observed an increase in nutrients in sugar sludge after vermicomposting. Nogales et al. (2005), Gómez-Brandón et al. (2022), and Fernández-Bayo et al. (2008) also studied the effect of composting with earthworms on winery sludge and observed an increase in plant nutrients contained in the sludge after vermicomposting compared to the nutrient content of the sludge before vermicomposting. Yadav & Garg (2011) also reported an increase in the C:N ratio and organic matter of sugar after vermicomposting. These advantages ensure plant growth-promoting activity and improved crop yield.

Generally, many studies have reported an increased crop yield after the application of vermicompost, for example, increased yield of rice stalks (Jeyabal & Kuppuswamy, 2001), improved tomato growth (Federico et al., 2007), and enhanced maize growth performance (Abd Karim et al., 2022).

Hence, it can be said that the use of vermicompost benefits farmers, industries, the environment, and the broader national economy.

To farmers: The use of fewer inorganic nutrients or other forms of input nutrients saves production costs. Improved soil quality leads to increased soil productivity, which enhances agricultural production quality and quantity. It also provides an extra source of income.

To industries: Vermicomposting is a technology for reducing pollutants at a low cost.

To the environment: Vermicompost reduces pollution from agro-industrial sludge because it is used as a raw material to improve soil fertility. Vermicomposting is also a method for removing heavy metals.

To the national economy: Vermicomposting boosts the rural economy by providing additional source of income. It leads to savings in purchased input in agriculture. There is also less wasteland formation.

3 Conclusion

Vermicomposting is becoming a more popular and cost-effective treatment option for a variety of agro-industries due to the pollution and environmental difficulties associated with the incorrect disposal of most agro-industrial waste. Vermicompost frequently offers a better overall nutritional profile than regular compost. Vermicompost treatment improves soil structure and aggregation while increasing organic matter content, nutrient status, cation exchange ability, microbial activity, microbial biomass carbon, and enzyme activities. As a result, soil health improves and plant growth increases. The practice of vermicomposting of agro-industrial waste should be promoted globally.

References

Abdullah, R., Ishak, C. F., Kadir, W. R., & Bakar, R. A. (2015). Characterization and feasibility assessment of recycled paper mill sludges for land application in relation to the environment. *International Journal of Environmental Research and Public Health, 12*, 9314–9329.

Abd Karim, K. N., Isa, I. M., Fauzi, M., Ramlan, A. M., Khairuddin, M. N., & Rizal, M. (2022). Influence of Palm Oil Mills Effluent (POME) sludge vermicomposting on soil physicochemical properties and Zea mays growth performances.

Abubakar, S., Lawal, I., Hassan, I., & Jagaba, A. (2016). Quality water analysis of public and private boreholes (a case study of Azare Town, Bauchi, Nigeria). *American Journal of Engineering Research, 5*(2), 204–208.

Ahmad, A., Aslam, Z., Bellitürk, K., Iqbal, N., Naeem, S., Idrees, M., et al. (2021). Vermicomposting methods from different wastes: An environment friendly, economically viable and socially acceptable approach for crop nutrition: A review. *International Journal of Food Science and Agriculture, 5*(1), 58–68.

Ali, U., Sajid, N., Khalid, A., Riaz, L., Rabbani, M. M., Syed, J. H., & Malik, R. N. (2015). A review on vermicomposting of organic wastes. *Environmental Progress & Sustainable Energy, 34*(4), 1050–1062.

Amor, C., Marchão, L., Lucas, M. S., & Peres, J. A. (2019). Application of advanced oxidation processes for the treatment of recalcitrant agro-industrial wastewater: A review. *Water, 11*(2), 205. [CrossRef]

Aransiola, S. A., Afolabi, F., Joseph, F., & Maddela, N. R. (2023a). Soil enzymes: Distribution, interactions and influencing factors. In P. Jeschke & E. B. Starikov (Eds.), *Agricultural biocatalysis: Enzymes in agriculture and industry* (1st ed., pp. 303–333). ISBN 9789814968478, Jenny Stanford Publishing. https://www.routledge.com/Agricultural-Biocatalysis-Enzymes-in-Agriculture-and-Industry/Jeschke-Starikov/p/book/9789814968478

Aransiola, S. A, Ijah, U. J. J. Abioye, O. P., & Bala J. D. (2022). Vermicompost-assisted phytoremediation of toxic trace element-contaminated soil in Madaka, Nigeria using *Melissa officinalis* L and *Sida acuta. International Journal of Environmental Science and Technology*. https://doi.org/10.1007/s13762-022-04105-y

Aransiola, S. A., Victor-Ekwebelem, M. O., Leh-Togi Zobeashia, S. S., & Maddela, N. R. (2023b). Sources and techniques for biofuel generation. In M. P. Shah (Ed.), *Green approach to alternative fuel for a sustainable future*. Elsevier Inc. https://doi.org/10.1016/B978-0-12-824318-3.00026-6

Arun, C., & Sivashanmugam, P. (2015). Solubilization of waste activated sludge using a garbage enzyme produced from different pre-consumer organic waste. *RSC Advances, 5*(63), 51421–51427. https://doi.org/10.1039/C5RA07959D

Asgharnia, H. (2003). Comparison of aerobic compost and vermicompost in the view of maturation time and microbial and chemical quality. In *The 6th National Environmental Health Congress, Mazandaran.*

Aworanti, O. A., Agarry, S. E., & Ogunleye, O. O. (2017). Biomethanization of the mixture of cattle manure, pig manure and poultry manure in co-digestion with waste peels of pineapple fruit and content of chicken-gizzard—part ii: Optimization of process variables. *The Open Biotechnology Journal, 11*, 54–71. https://doi.org/10.2174/1874070701711010054

Bajpai, P. (2013). *Recycling and deinking of recovered paper*. Elsevier.

Bhat, S. A., & Vig, A. P. (2019). Vermistabilization and detoxification of sugar industry sludges by earthworms. In *Industrial and municipal sludge* (pp. 61–81). Butterworth-Heinemann.

Bhat, S. A., Singh, J., & Vig, A. P. (2014). Genotoxic assessment and optimization of pressmud with the help of exotic earthworm *Eisenia fetida. Environmental Science and Pollution Research, 21*, 8112–8123.

Bidattul Syirat, Z., Ibrahim, M. H., & Astimar, A. A. (2013). Studies on vermicompost production of palm oil mill effluent sludge using *Eudrilus eugeniae. Online International Interdisciplinary Research Journal, 3*(5), 42–50.

Chaurasiya, R. S., Sakhare, P. Z., Bhaskar, N., & Hebbar, H. U. (2015). Efficacy of reverse micellar extracted fruit bromelain in meat tenderization. *Journal of Food Science and Technology, 52*(6), 3870–3880. https://doi.org/10.1007/s13197-014-1454-z

Chen, W., Oldfield, T. L., Patsios, S. I., & Holden, N. M. (2020). Hybrid life cycle assessment of agro-industrial wastewater valorisation. *Water Research, 170*, Article 115275.

Chiochetta, C. G., Goetten, L. C., Almeida, S. M., Quaranta, G., Cotelle, S., & Radetski, C. M. (2014). Leachates from solid wastes: Chemical and eco (geno) toxicological differences between leachates obtained from fresh and stabilized industrial organic sludge. *Environmental Science and Pollution Research, 21*, 1090–1098.

Choudhary, M., & Grover, K. (2019). *Palm (Elaeis guineensis jacq.) oil. In Fruit oils: Chemistry and functionality* (pp. 789–802). Cham: Springer International Publishing

Chozhavendhan, S., Karthigadevi, G., Bharathiraja, B., Kumar, R. P., Abo, L. D., Prabhu, S. V.,... & Jayakumar, M. (2022). Current and prognostic overview on the strategic exploitation of anaerobic digestion and digestate: A review. *Environmental Research*, 114526.

Crescent, T. (2003). *Vermicomposting. Development alternatives (DA) sustainable livelihoods.*

Díaz-Vela, J., Totosaus, A., Escalona-Buendía, H. B., & Pérez-Chabela, M. L. (2017). Influence of the fiber from agro-industrial co-products as functional food ingredient on the acceptance, neophobia and sensory characteristics of cooked sausages. *Journal of Food Science and Technology, 54*, 379–385. https://doi.org/10.1007/s13197-016-2473-8

Dume, B., Hanc, A., Svehla, P., Michal, P., Chane, A. D., & Nigussie, A. (2022). Vermicomposting technology as a process able to reduce the content of potentially toxic elements in sewage sludge. *Agronomy, 12*(9), 2049.

Düring, R.-A., & Gäth, S. (2002). Utilization of municipal organic wastes in agriculture: Where do we stand, where will we go? *Journal of Plant Nutrition and Soil Science, 165*(4), 544–556.

Dutta, R., Angmo, D., Singh, J., Chowdhary, A. B., Quadar, J., Singh, S., & Vig, A. P. (2023). Synergistic effect of biochar amendment in milk processing industry sludge and cattle dung during the vermiremediation. *Bioresource Technology, 371*, 128612.

Edwards, C. A. (1985). Production of feed protein from animal waste by earthworms. *Biological Sciences, 310*(1144), 299–307.

Federico, J. S., Borraz, J. A., Molina, M., Nafate, C., Archila, C., & Oliva, L. M. (2007). Vermicompost as a soil supplement to improve growth, yield and fruit quality of tomato (*Lycopersicum esculentum*). *Bioresource Technology, 98*(15), 2781–2786.

Fernández-Bayo, J. D., Romero, E., Schnitzler, F., & Burauel, P. (2008). Assessment of pesticide availability in soil fractions after the incorporation of winery-distillery vermicomposts. *Environmental Pollution, 154*(2), 330–337.

Fernández-Gómez, M. J., Díaz-Ravina, M., Romero, E., & Nogales, R. (2013). Recycling of environmentally problematic plant wastes generated from greenhouse tomato crops through vermicomposting. *International Journal of Environmental Science and Technology, 10*, 697–708.

Garg, V. K., & Kaushik, P. (2005). Vermistabilization of textile mill sludge spiked with poultry droppings by an epigeic earthworm Eisenia foetida. *Bioresource Technology, 96*, 1063–1071.

Gómez-Brandón, M., Fornasier, F., de Andrade, N., & Domínguez, J. (2022). Influence of earthworms on the microbial properties and extracellular enzyme activities during vermicomposting of raw and distilled grape marc. *Journal of Environmental Management, 319*, Article 115654.

González-Moreno, M. Á., García Gracianteparaluceta, B., Marcelino Sádaba, S., Prieto Cobo, E., & Seco Meneses, A. (2022). Vermicomposting of lavender waste: A biological laboratory investigation. *Agronomy, 12*(12), 2957.

Gopal, P. M., Sivaram, N. M., & Barik, D. (2019). Paper industry wastes and energy generation from wastes. In *Energy from toxic organic waste for heat and power generation* (pp. 83–97). Woodhead Publishing.

Gupta, G. K., Liu, H., & Shukla, P. (2019). Pulp and paper industry–based pollutants, their health hazards and environmental risks. *Current Opinion in Environmental Science & Health, 12*, 48–56.

Hernández, D., Riaño, B., Coca, M., & García-González, M. C. (2013). Treatment of agro-industrial wastewater using microalgae–bacteria consortium combined with anaerobic digestion of the produced biomass. *Bioresource Technology, 135*, 598–603. [CrossRef] [PubMed]

Huguier, P., Manier, N., Chabot, L., Bauda, P., & Pandard, P. (2015). Ecotoxicological assessment of organic wastes spread on land: Towards a proposal of a suitable test battery. *Ecotoxicology and Environmental Safety, 113*, 103–111.

Jeyabal, G., & Kuppuswamy. (2001). Recycling of organic wastes for the production of vermicompost and its response in rice-legume cropping system and soil fertility. *European Journal of Agronomy, 15*(3), 153–170.

Koura, T. W., Dagbenonbakin, G. D., Kindomihou, V. M., & Sinsin, B. A. (2017). Palm oil mill solid waste generation and uses in rural area in Benin Republic: Retrospection and future outlook. In *Solid Waste Management in Rural Areas*. IntechOpen.

Kumar, A. (2005). *Verms & vermitechnology*. APH Publishing.

Kumar, R., Verma, D., Singh, B. L., & Kumar, U. (2010). Composting of sugar-cane waste by-products through treatment with microorganisms and subsequent vermicomposting. *Bioresource Technology, 101*(17), 6707–6711.

Lee, L. H., Wu, T. Y., Shak, K. P. Y., Lim, S. L., Ng, K. Y., Nguyen, M. N., & Teoh, W. H. (2018). Sustainable approach to biotransform industrial sludge into organic fertilizer via vermicomposting: A mini-review. *Journal of Chemical Technology & Biotechnology, 93*(4), 925–935.

Leh-Togi Zobeashia, S. S., Aransiola, S. A., Ijah, U. J. J., & Abioye, O. P. (2018). Anaerobic digestion and agricultural application of organic wastes. *Advances in Environmental Research, 7*(2), 73–85. http://www.techno-press.org/content/?page=article&journal=aer&volume=7&num=2&ordernum=1

Lim, S. L., Wu, T. Y., & Clarke, C. (2014). Treatment and biotransformation of highly polluted agro-industrial wastewater from a palm oil mill into vermicompost using earthworms. *Journal of Agricultural and Food Chemistry, 62*(3), 691–698.

Liu, X., Geng, B., Zhu, C., Li, L., & Francis, F. (2021). An improved vermicomposting system provides more efficient wastewater use of dairy farms using *Eisenia fetida*. *Agronomy, 11*(5), 833.

Madurai, R., John, S., & Bhavani, I. L. G. (2016). Study on preparation, nutrient analysis and shelf life of biovinegar and its formulations. *Biosciences Biotechnology Research Asia, 7*(2), 849–855.

Manosroi, A., Chankhampan, C., Pattamapun, K., Manosroi, W., & Manosroi, J. (2014). Antioxidant and gelatinolytic activities of papain from papaya latex and bromelain from pineapple fruits. *Chiang Mai J. of Sci, 41*(3), 635–648.

Martinez-Burgos, W. J., Sydney, E. B., Medeiros, A. B. P., Magalhães, A. I., de Carvalho, J. C., Karp, S. G., et al. (2021). Agro-industrial wastewater in a circular economy: Characteristics, impacts and applications for bioenergy and biochemicals. *Bioresource Technology, 341*, Article 125795.

Mensah, J. K., & Twumasi, P. (2017). Use of pineapple waste for single cell protein (SCP) production and the effect of substrate concentration on the yield. *Journal of Food Process Engineering, 40*(3), Article e12478. https://doi.org/10.1111/jfpe.12478

Muthukrishnan, B., & Swaminathan, P. (2017). Sugar Industry Waste Recycling through Vermicompost by Eisenia Foetida (Savigny).

Nagavallemma, K. P., Wani, S. P., Stephane Lacroix, V. V., Padmaja, C., Vineela, M., Rao, B., & Sahrawat, K. L. (2004). Vermicomposting: Recycling wastes into valuable organic fertilizer. Global Theme on Agroecosystems Report no. 8.

Namli, A., Akça, H., & Akça, M. O. (2020). Vermicomposting of agro-industrial waste by-product of the sugar industry. *Eurasian Journal of Soil Science, 9*(4), 292–297.

Negi, R., & Suthar, S. (2018). Degradation of paper mill wastewater sludge and cow dung by brown-rot fungi Oligoporus placenta and earthworm (*Eisenia fetida*) during vermicomposting. *Journal of Cleaner Production, 201*, 842–852.

Nogales, R., Cifuentes, C., & Benitez, E. (2005). Vermicomposting of winery wastes: A laboratory study. *Journal of Environmental Science and Health Part B, 40*(4), 659–673.

Ogunleye, O. O., Aworanti, O. A., Agarry, S. E., & Aremu, M. O. (2016). Enhancement of animal waste biomethanation using fruit waste as co-substrate and chicken rumen as inoculums. *Energy Sources, Part A: Recovery, Utilization, and Environmental Effects, 38*(11), 1653–1660. https://doi.org/10.1080/15567036.2014.933286

Ojuolape, O. T., Aransiola, S. A., & Obideyi, O. A. (2015). Vermitechnology: A solution to environmental problems; A review. *Journal of Global Ecology and Environment, 3*(3), 2454–2644: 127–135. http://www.ikpress.org/abstract/4516

Pardo, M. E. S., Cassellis, M. E. R., Escobedo, R. M., & García, E. J. (2014). Chemical characterisation of the industrial residues of the pineapple (*Ananas comosus*). *Journal of Agricultural Chemistry and Environment, 3*(2), 53–56.

Pelizer, L. H., Pontieri, M. H., & Moraes, I. de O. (2007). Utilização de Resíduos Agro-Industriais em Processos Biotecnológicos como Perspectiva de Redução do Impacto Ambiental. *Journal of Technology Management & Innovation, 2*(1), 118–127.

Pizzanelli, S., Calucci, L., Forte, C., & Borsacchi, S. (2023). Studies of organic matter in composting, vermicomposting, and anaerobic digestion by 13C solid-state NMR spectroscopy. *Applied Sciences, 13*(5), 2900.

Prazeres, A. R., Carvalho, F., & Rivas, J. (2012). Cheese whey management: A review. *Journal of Environmental Management, 110*, 48–68.

Quintern, M. (2014). Full scare vermicomposting and land utilization of pulpmill solids in combination with municipal biosolids (sewage sludge). *Journal of Ecological Environment, 18*, 65–76.

Robles-González, V., Galíndez-Mayer, J., Rinderknecht-Seijas, N., & Poggi-Varaldo, H. M. (2012). Treatment of mezcal vinasses: A review. *Journal of Biotechnology, 157*(4), 524–546.

Roldi, M., Dias-Arieira, C. R., Abe, V. H. F., Mattei, D., Severino, J. J., Rodrigues, D. B., & Felix, J. C. (2013). Agro industrial waste ans sewage sludge can control Pratylenchus brachyurus in maize. *Acta Agriculturae Scandinavica, Section B-Soil & Plant Science, 63*(3), 283–287.

Rostami, R. (2011). Vermicomposting. In *Integrated waste management-volume II*. IntechOpen.

Rupani, P. F., Embrandiri, A., Garg, V. K., Abbaspour, M., Dewil, R., & Appels, L. (2023). Vermicomposting of green organic wastes using *Eisenia fetida* under field conditions: A case study of a green campus. *Waste and Biomass Valorization*, 1–12.

Rupani, P. F., Zabed, H. M., & Domínguez, J. (2022). Vermicomposting and bioconversion approaches towards the sustainable utilization of palm oil mill waste. In *Advanced organic waste management* (pp. 193–205). Elsevier.

Sangwan, P., Kaushik, C. P., & Garg, V. K. (2008). Vermiconversion of industrial sludge for recycling the nutrients. *Bioresource Technology, 99*(18), 8699–8704.

Sekeri, S. H., Ibrahim, M. N. M., Umar, K., Yaqoob, A. A., Azmi, M. N., Hussin, M. H., Othman, M. B. H., & Malik, M. F. I. A. (2020). Preparation and characterization of nanosized lignin from oil palm (*Elaeis guineensis*) biomass as a novel emulsifying agent. *International Journal of Biological Macromolecules, 164*, 3114–3124. [CrossRef] [PubMed].

Selvakumar, P., & Sivashanmugam, P. (2017). Optimization of lipase production from organic solid waste by anaerobic digestion and its application in biodiesel production. *Fuel Processing Technology, 165*, 1–8. https://doi.org/10.1016/j.fuproc.2017.04.020

Sen, B., & Chandra, T. S. (2007). Chemolytic and solid-state spectroscopic evaluation of organic matter transformation during vermicomposting of sugar industry wastes. *Bioresource Technol, 98*, 1680–1683.

Shamsul, N. S., Kamarudin, S. K., Kofli, N. T., & Rahman, N. A. (2017). Optimization of biomethanol production from goat manure in single stage bio-reactor. *International Journal of Hydrogen Energy, 42*(14), 9031–9043. https://doi.org/10.1016/j.ijhydene.2016.05.228

Sharma, D., Prasad, R., Patel, B., & Parashar, C. K. (2022). Biotransformation of sludges from dairy and sugarcane industries through vermicomposting using the epigeic earthworm *Eisenia fetida. International Journal of Recycling Organic Waste in Agriculture, 11*(2), 165–175.

Sharma, K., & Garg, V. K. (2018). Vermicomposting: A green technology for organic waste management. *Waste to Wealth*, 199–235.

Shi, W., Healy, M. G., Ashekuzzaman, S. M., Daly, K., Leahy, J. J., & Fenton, O. (2021). Dairy processing sludge and co-products: A review of present and future re-use pathways in agriculture. *Journal of Cleaner Production, 314*, Article 128035.

Siddeeg, S. M., Tahoon, M. A., & Rebah, F. B. (2019). Agro-industrial waste materials and wastewater as growth media for microbial bioflocculants production: A review. *Materials Research Express, 7*(1), Article 012001.

Singh, J., Kaur, A., & Vig, A. P. (2014). Bioremediation of distillery sludge into soil-enriching material through vermicomposting with the help of Eisenia fetida. *Applied Biochemistry and Biotechnology, 174*, 1403–1419.

Sonowal, P. K. D., Khwairkpam, M., & Kalamdhad A. S. (2013). Feasibility of vermicomposting dewatered sludge from paper mills using *Perionyx excavatus. European Journal of Environmental Science, 3*, 17–26.

Suthar, S. (2009). Vermistabilization of municipal sewage sludge amended with sugarcane trash using epigeic *Eisenia fetida* (Oligochaeta). *Journal of Hazardous Materials, 163*, 199–206.

Suthar, S. (2012). Vermistabilization of wastewater sludge from milk processing industry. *Ecological Engineering, 47*, 115–119.

Suthar, S., Mutiyar, P. K., & Singh, S. (2012). Vermicomposting of milk processing industry sludge spiked with plant wastes. *Bioresource Technology, 116*, 214–219.

Suthar, S., Sajwan, P., & Kumar, K. (2014). Vermiremediation of heavy metals in wastewater sludge from paper and pulp industry using earthworm *Eisenia fetida*. *Ecotoxicology and Environmental Safety, 109*, 177–184.

Teh, C. Y., Wu, T. Y., & Juan, J. C. (2014). Optimization of agro-industrial wastewater treatment using unmodified rice starch as a natural coagulant. *Industrial Crops and Products, 56*, 17–26. [CrossRef].

Thakur, A., Kumar, A., Kumar, C. V., Kiran, B. S., Kumar, S., & Athokpam, V. (2021). A review on vermicomposting: By-products and its importance. *Plant Cell Biotechnology and Molecular Biology, 22*, 156–164.

Vasanthi, K., Chairman, K., Ranjit Singh, A. J. A. (2014). Sugar factory waste (vermicomposting with an epigeic earthworm, *Eudrilus eugeniae*). *American Journal of Drug Discovery and Development, 4*, 22–31.

Vavouraki, A. I., & Kornaros, M. (2023). Vermi-conversion of anaerobic sludges by *Eisenia fetida* earthworms. *Fermentation, 9*(6), 512.

Vig, A. P., Singh, J., Wani, S. H., & Dhaliwal, S. S. (2011). Vermicomposting of tannery sludge mixed with cattle dung into valuable manure using earthworm *Eisenia fetida* (Savigny). *Bioresource Technology, 102*(17), 7941–7945.

Von Enden, J. C. (2002). Best practices at wet processing pay financial benefits to farmers and processors. GTZ-PPP Project on Improvement of coffee quality and sustainability of coffee production in Vietnam.

Vyas, P., Sharma, S., & Gupta, J. (2022). Vermicomposting with microbial amendment: Implications for bioremediation of industrial and agricultural waste. *BioTechnologia, 103*(2), 203–215.

Yadav, A., & Garg, V. K. (2010). Bioconversion of food industry sludge into value-added product (vermicompost) using epigeic earthworm *Eisenia fetida*. *World Review of Science, Technology and Sustainable Development, 7*(3), 225–238.

Yadav, A., & Garg, V. K. (2011). Industrial wastes and sludges management by vermicomposting. *Reviews in Environmental Science and Bio/technology, 10*(3), 243–276.

Yadav, A., & Madan, S. (2013). Nutrient status of vermicompost of Paper Mill Sludge with different wastes by using *Eisenia fetida*. *European Journal of Applied Sciences, 5*(2), 62–66.

Yaqoob, A. A., Noor, N. H. B. M., Umar, K., Adnan, R., Ibrahim, M. N. M., & Rashid, M. (2021). Graphene oxide–ZnO nanocomposite: An efficient visible light photocatalyst for degradation of rhodamine B. *Applied Nanoscience, 11*, 1291–1302.

Biorestoration of Polycyclic Aromatic Hydrocarbon- and Heavy Metal-Contaminated Soil: The Role of Vermitechnology

Joshua Ibukun Adebomi, Babafemi Raphael Babaniyi, Bukola Rukayat Olowoyeye, and Oluwatosin Emmanuel Daramola

Abstract This chapter is about the biorestoration of soil contaminated with polycyclic aromatic hydrocarbons and heavy metals through the process of vermitechnology. Polycyclic aromatic hydrocarbons are chemical compounds made up of two or more aromatic rings that are fused and arranged in a linear or clustered pattern. They are highly hydrophobic and resistant to environmental degradation. The pollution of soil by heavy metals poses a huge risk to food security and the environment due to the increase in human population and anthropogenic activities. Biorestoration is a process by which microorganisms and their products are employed to remove contaminants from the soil. Different sources of heavy metals and polycyclic aromatic hydrocarbons in the environment have been identified, including, but not limited to, the application of chemicals in agricultural practices, industrial activities, and naturally occurring forms. Vermitechnology is a process that uses earthworms and their gut-related organisms to accelerate the decomposition of various types of biomass into fertilizer. The formation of the drilosphere, which improves soil aeration, is enhanced through the action of earthworms, leading to the enhancement of plant development and yield via complex mechanical and biochemical reactions between soil abiotic and biotic factors. Vermitechnology is an eco-friendly method of degrading or transforming solid waste into useful or easily degradable products, and it presents an opportunity for resource recovery and recycling. Bacteria are the most widespread and minute microscopic organisms in the soil and are responsible for the degradation of organic matter, nitrogen and sulfur transformation, and nitrogen fixation. Fungi, such as those in the family Actinomycetes and Basidiomycetes (mostly

J. I. Adebomi
Chemical Engineering Department, University of Saskatchewan, Saskatoon, Canada

B. R. Babaniyi (✉)
Bioresources Development Centre, National Biotechnology Development Agency, Abuja, Nigeria
e-mail: babafemiraphael@gmail.com

B. R. Olowoyeye
Department of Microbiology, Federal University of Technology, Akure, Nigeria

O. E. Daramola
Department of Chemistry and Biochemistry, Texas Tech University, Lubbock, USA

S. A. Aransiola and N. R. Maddela (eds.), *Vermitechnology: Economic, Environmental and Agricultural Sustainability*, https://doi.org/10.1007/978-3-031-96212-7_10

mushrooms), also aid in the decomposition of organic matter. The text explains the different methods of soil biorestoration, such as ex situ biorestoration, in situ techniques, natural attenuation, projected bioremediation, and composting. Therefore, vermitechnology can be described as a biological method suitable for the cleanup of contaminated soil.

Keywords Vermitechnology · Vermifiltration · Biorestoration · Earthworm

1 Introduction: Biorestoration of Soil

1.1 Soil

Soil can be defined as the loose surface material that covers the land, consisting of both organic and inorganic matter. It serves as a fundamental resource in agriculture, providing essential nutrients and water to plants, as well as structural support to their root systems (Agriculture Victoria, 2021). The composition and characteristics of soil vary widely due to differences in its physical and chemical properties. These variations are influenced by several natural processes, including weathering, microbial activity, and leaching (Agriculture Victoria, 2021). Weathering, which involves the breakdown of rocks and minerals over time, contributes to soil formation and nutrient availability. Microbial activity plays a crucial role in decomposing organic matter, releasing nutrients that plants can absorb. Leaching, on the other hand, involves the movement of dissolved substances through the soil profile, which can impact soil fertility and nutrient distribution (Abdalqadir et al., 2024; Mir et al., 2024).

1.2 Soil Organisms

Several organisms can be found in the soil, namely bacteria, fungi, and macroscopic soil animals such as earthworms. The microorganisms commonly found in the soil include:

(i) Fungi
(ii) Bacteria
(iii) Algae
(iv) Nematodes
(v) Protozoa

Plant growth can be enhanced and hindered depending on the diversity of soil organisms. Beneficial activities of soil organisms (both micro and macro) include:

(i) Transformation of the form of essential elements
(ii) Organic matter decomposition

(iii) Nitrogen fixation
(iv) Improvement of the structure of soil via aggregation
(v) Improvement of aeration and drainage.

It has been established that some soil organisms can be in competition with plants for inorganic nutrients under some conditions. Earthworms, by means of their channels, help in improving aeration and drainage in the soil. They also integrate organic matter into the soil.

Bacteria are ubiquitous in the soil and are the most connected to the decomposition of organic matter, change of nitrogen and sulfur, and nitrogen fixation. Fungi, especially actinomycetes and basidiomycetes, help in organic matter decomposition.

1.3 Physical Features of Soil

The physical characteristics of soil include:

(i) Texture
(ii) Structure
(iii) Color
(iv) Depth
(v) Stone content
(vi) Porosity (the space between the particles).

The structure of soil plays a crucial role in determining plant health, as it directly influences key factors such as aeration, water retention, and drainage within the soil profile. Good soil structure facilitates the movement of air and water, ensuring that plant roots receive adequate oxygen and moisture for optimal growth and development. Conversely, poor soil structure can lead to issues such as waterlogging, compaction, and restricted root penetration, which can negatively impact plant health and productivity. Different soil types exhibit varying degrees of structural quality, with some naturally possessing better aggregation and porosity than others. However, with appropriate management practices, the physical characteristics of soil can be significantly improved to enhance its suitability for plant growth (ref). Techniques such as organic matter addition, proper irrigation management, crop rotation, and reduced tillage can help enhance soil structure by promoting the formation of stable aggregates and improving porosity. To effectively assess and manage soil conditions, it is essential to carefully examine its physical properties, including texture, structure, porosity, and compaction levels. Understanding these features allows farmers and land managers to implement strategies that support sustainable soil health and long-term agricultural productivity (Manning et al., 2024). Regular monitoring and soil testing can provide valuable insights into the physical state of the soil, enabling informed decisions to optimize its structure and function.

1.4 Chemical Characteristics of Soil

The chemical properties of soil include:

(i) Organic matter
(ii) Inorganic matter which consists of iron, aluminum, and silicon
(iii) Colloidal properties
(iv) Acidity and basicity.

1.5 Soil Texture

Soil texture is the measurement or determination of the percentage amounts of sand, silt, and clay particles in the fine earth fraction (UBC Wiki, 2020). It also includes the quantity of organic matter, which determines the texture of soil. The percentage quantity of sand, silt, organic matter, and clay present determines the grade of the texture. Fractions such as clay, silt, and organic matter make up the solid part of the soil that forms aggregates. Aggregates are held together by organic matter and clay particles. The major cementing agent of soil aggregates is organic matter (Fig. 1).

The structure of soil impacts plant growth by affecting aeration, percolation, and availability of nutrients to plants. Soil texture is a function of the comparative fractions of inorganic matter of varying sizes.

The inorganic fraction of soils in Australia are described using the sizes below (Agriculture Victoria, 2021):

(i) Particles less than 0.002 mm in diameter—clay
(ii) Particles between 0.002 and 0.02 mm in diameter—silt
(iii) Particles between 0.02 and 0.2 mm in diameter—fine sand
(iv) Particles between 0.2 and 2 mm in diameter—coarse sand
(v) Particles greater than 2 mm in diameter—gravel.

Fig. 1 Soil aggregate (Agriculture Victoria, 2021)

Table 1 Microbiological classification of soil

Type of soil	Microorganism type	Agriculture types
Disease suppressors	*Penicillium* sp. *Aspergillus* sp.	Ecological
Disease Inducers	*Fusarium* sp. *Pythium* sp. *Botrytis* sp.	Modern
Nutrient synthesizers	*Saccharomyces* sp. *Rhizobium* sp. *Azospirillum* sp. *Azotobacter* sp.	Natural
Zymogenous	*Trichoderma* sp. *Lactobacillus* sp.	Organic

Source WilliamUsher (2021)

Clay

Clay is known for its very large surface area which makes it chemically active and able to bind nutrients on its surface. These nutrients dissolve into soil water to be utilized by plants. The factors that distinguish clay from silt and sand are its sticky nature, swelling capacity, and ability to form shape (ref).

Sand

The predominant mineral in the sand proportion of most soils is quartz. The particles of sand have poor water retention capacity, low chemical activity, and reduced surface area per unit weight compared with silt and clay (Mureithi et al., 2024).

Silt

Silt has little chemical activity and a limited surface area. Soils that contain a high proportion of silt tend to compact when exposed to heavy traffic which in turn affects aeration and percolation.

1.6 Microbiological Classification of Soils

Table 1 shows the microbiological classification of soils.

1.7 Important Functions of Microbes in the Soil

The important functions of microbes in the soil include:

(i) Mineral solubilization
(ii) Mineralization of organic matter

(iii) Nitrogen fixation
(iv) Chelation of minerals
(v) Root growth and morphology
(vi) Absorption and translocation of minerals
(vii) Enzyme and vitamin production
(viii) Production of phytohormones
(ix) Soil aggregation and stability.

1.8 Soil Contamination

Soil contamination is defined as the accumulation of chemical compounds, radioactive waste, salts, and persistent harmful substances or pathogens that can negatively affect biological systems (Okrent, 1999; Mareddy, 2017). The increased accumulation of pollutants such as heavy metals, pesticides, and petroleum derivatives increases the levels of toxic compounds in the soil, which negatively affects the ecosystems and, consequently, human health (Aransiola et al., 2021; Palansooriya et al., 2020; Leena et al., 2023). In the soil, pollutants can be removed, washed away by wind and runoff, adsorbed, or leached by infiltrating water that passes through the lower layers to the groundwater (CETESB 2020).

1.8.1 Main Sources of Soil Contamination

The main sources of soil contamination include:

(i) Petroleum products (containing pollutants such as hydrocarbons, heavy metals, radioactive metals, and non-hydrocarbons)
(ii) Urban waste
(iii) Pesticides
(iv) Herbicides
(v) Industrial effluents
(vi) Chemical and biological warfare (Ashraf et al., 2020) (Table 2).

2 Biorestoration

Biorestoration is a process that employs microorganisms and/or their products to rid soil of contaminants (USEPA 2012). It can also be described as the application of microorganisms to degrade or immobilize waste materials (Shanahan, 2004). Biorestoration technology takes advantage of microorganisms to remediate, degrade, destroy, contain, or transform benign contaminants in air, water, sediments, and soils. The use of microbes in modern biorestoration is credited, in part, to George Robinson

Table 2 Sources of soil contamination

Petroleum derivatives	Chemical war	Urban source	Agrochemical source	Biological warfare
Exploration, production, refining, transport, and consumption	Contaminants, toxic chemical compounds, and arms; soil contamination from Cold War army operations	Energy generation emissions; soil pollution by transportation and manufacture; soil contamination by residues and sludge from wastewater treatment	Insecticides, herbicides, fungicides, and fuel spills on farms	Bacteria, viruses, fungi, and toxins

(US Microbics, 2003). He pioneered the use of microorganisms in the elimination of contaminants in the late 1960s, along the coast of Santa Barbara, California, demonstrating with an oil spill. Since the 1980s, there has been an increase in the application and research in the area of biorestoration (Shannon & Unterman, 1993). Very importantly, natural soil microorganisms play a vital role in soil biorestoration as biogeochemical tools to degrade polymers into monomers/simple inorganic compounds or into their constituent elements (ref). This process is known as mineralization. By the mechanism of ionic exchange, the microorganisms are attached to soil particles. By mineralization, transformation, or alteration, the harmful chemicals are targeted using detoxification process (Shannon & Unterman, 1993). The use of biorestoration for treatment of wastewater is well established in different civilizations, but its application in the degradation and elimination of hazardous contaminants has been receiving more attention in recent time.

The classification of soil remediation methods includes:

(i) Physical
(ii) Chemical
(iii) Biological methods (carried out either in [in situ] or outside [ex situ] of the polluted area).

Biological processes such as biorestoration and phytoremediation are eco-friendly, have the ability to remove diverse contaminants, and are more cost-effective compared to preexisting techniques (Soleimani, 2014). Due to these features, they are being studied intensively.

2.1 Types of Biorestoration

Biorestoration can be classified into two, in situ or ex situ, depending on the site of application (Fig. 2). Ex situ biorestoration is expensive because it involves transportation and excavation, but it is suitable for the removal of a higher number of

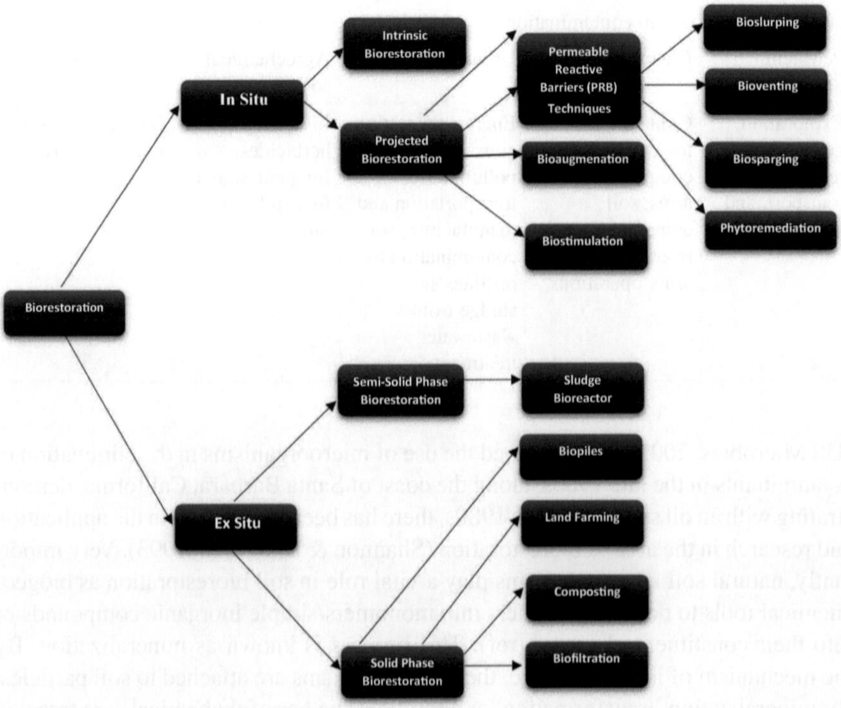

Fig. 2 Different types of biorestoration

contaminants under a monitored condition(s). In situ biorestoration does not involve excavation but has the drawbacks of the inability to see and carry out effective monitoring below the surface of the contaminated site and the high cost of equipment. These factors sometimes make in situ methods unfeasible. Consequently, cost is not a factor that determines the method of biorestoration to use, but the type of contaminant(s) present at a site.

2.2 *In Situ Techniques*

This type of biorestoration takes place at the site of contamination without the excavation of contaminated materials. It can be classified into two types, namely, intrinsic and projected bioremediation.

2.2.1 Natural Attenuation/Passive/Intrinsic Biorestoration

This is a degradation process that is natural and employs the use of microorganisms that are indigenous to the site to destroy contaminants without enhancing the process with any artificial means (Kumar et al., 2018). It is the cheapest in situ biorestoration technique because the process does not involve any external factors. However, continuous monitoring is necessary for biorestoration to progress and be sustainable. Before the application of intrinsic biorestoration, a risk evaluation should be carried out to ensure that there is a relatively short time for the contaminant to undergo complete biorestoration to avoid exposure to humans and animals.

2.2.2 Projected Biorestoration

There are different types of projected biorestoration, namely:

(i) Improved techniques
(ii) Permeable reactive barriers
(iii) Bioaugmentation
(iv) Biostimulation.

Permeable Reactive Barriers

This method is used to remediate contaminated groundwater polluted by heavy metals and chlorinated hydrocarbons. The reactive barriers are made up of iron that is buried in the contaminated groundwater stream. Polluted water flows naturally through the iron barrier; the pollutants are trapped and react, releasing purified water. Permeable reactive barriers (PRBs) are reactive to captured pollutants, passive with little energy consumption, cheap, and permeable to allow water to flow. The performance of PRBs depends on the type of medium, which further depends on factors such as environment, type of contaminant, effects on health, system steadiness, hydrogeology, biogeochemistry, and financial implications.

Improved Techniques

Bioslurping/Multiphase Extraction

Bioslurping integrates bioventing, soil vapor extraction, and enhanced vacuum pumping for restoration of soil and groundwater using indirect oxygen supply to improve the degradation of contaminants (Azubuike et al., 2016). This technique is effective in removing groundwater contaminants that have low density and are water insoluble such as light nonaqueous phase liquids (LNAPLs), which float on the water table. It transports pollutants upward with the aid of a tube that sucks LNAPLS from a tank. This technique should only be employed when the pollutants are at 7 m or less below the soil surface. This is because the performance of the vacuum pump is poor at greater depth. Bioventing, on the other hand, is used to biodegrade water and air contaminants after their removal from the surface. The principal cons of its usage are oxygen transfer rate, microbial activity, and the high moisture content of the soil which reduces air permeability. Moreover, bioslurping is not an ideal method for soil that is low in permeability.

Bioventing

Bioventing entails enhancing biorestoration via the improvement of microbial activity through increased oxygen supply and controlled stimulation of airflow. It involves improving restoration by the addition of nutrients and humidity to effect the degradation of contaminants to products that will be harmless to the environment. This method has been successfully applied in the restoration of soils contaminated by hydrocarbons.

Biosparging

Biosparging is a process that involves the introduction of air into the soil to elevate the microorganisms' degradation capacity. As opposed to bioventing, air is blown into the covered, area and this causes upward movement of unstable contaminants. The success of biosparging depends on the permeability of the soil, which in turn controls the accessibility of microorganisms to pollutants and the biodegradability of pollutants.

Phytoremediation

This method involves the application of flora in a contaminated area to enhance biological, chemical, biochemical, physical, and microbiological interactions to reduce contaminant toxicity. This process occurs via diverse means such as biodegradation, filtration, vaporization, and so on, depending on the contaminant types. Most times, radioactive elements and heavy metals are the pollutants that are normally removed, degraded, and sequestered, while organic pollutants are extracted using biodegradation, vaporization/stabilization, and rhizodegradation (Aransiola et al., 2013). Plants interact with pollutants in many different ways such as phytoextraction, phytovolatilization, phytostimulation, phytotransformation or phytodegradation, phytostabilization, rhizodegradation, and rhizofiltration.

Bioaugmentation

In this process, previously isolated and selected native or genetically modified microbial species are added to the natural microflora of the polluted site so as to elevate oil degradation. This method is known to be very effective when the natural microflora of the contaminated site are incapable of degrading the pollutants. A lot of work is still to be done on bioaugmentation for its application to achieve successful and desirable results.

Biostimulation

Biostimulation is a process that entails the addition of growth-enhancing factors such as nutrients like nitrogen and phosphorus to stimulate native microorganisms. If it does not produce effective restoration of the polluted site, oxidizing agents or oxygen can be added. The stimulating agents/nutrients are usually applied underground using injection wells.

2.3 *Ex Situ Techniques*

This is a method that entails the excavation of polluted material to be degraded in a special facility at a different site. The factors to consider before choosing this technique are contaminant types, depth and extent of contamination, operating costs, location, and geological characteristics of the polluted area (Azubuike et al., 2016). In ex situ methods, better rate of degradation is achieved because of a more robust control of environmental conditions, compared to in situ treatment techniques. Furthermore, there is uniformity, and shorter time is required because of the capability to homogenize the excavated polluted material. However, it has the disadvantage of being expensive due to excavation costs and site restoration. Also, the excavation of soil increases the mobility of pollutants and consequently exposure to them. It is, therefore, necessary that the site be preadapted by installing coating systems in the affected site in order to prevent the leakage of pollutants (Azubuike et al., 2016). Ex situ techniques are classified into two types, namely semisolid-phase restoration and solid-phase restoration.

2.3.1 Semisolid-Phase Biorestoration

Bioreactors

Bioreactors are defined as manufactured facilities or equipment that aid a bioactive system. Sludge bioreactors are safely and easily used in the treatment of hydrocarbon contaminants. The remediation can include the use of oxygen (aerobic) or the absence of oxygen (anaerobic). They are cylindrical in shape and range in capacity from a few liters to cubic meters, and they are usually made of stainless steel and tough glass. The polluted substances usually end as suspension or dried material are delivered (ref).

The polluted material can be delivered to the reactor as a suspension or dry substance which is an advantage. Other benefits include the ability to effectively treat pesticides, heavy metals, volatile organic compounds, etc. and to satisfactorily control variables such as pH, temperature, aeration rate, mixing intensity, substrate concentration, and inoculum level. Bioreactors are one of the most effective ways to treat contaminated soil due to the fact that the conditions of production can be monitored/controlled, consequently leading to an increase in the ability of microorganisms to biodegrade.

2.3.2 Solid-Phase Biorestoration

This type of biorestoration involves four methods, namely:

(i) Biopiles
(ii) Landfarming
(iii) Composting

(iv) Biofiltration.

Biopiles

Biopiles involve the piling of polluted soil, which is then aerated so as to improve microbial biodegradation activity. They are built on concrete slabs and membranes that are waterproof in nature to reduce spreading of contaminants to the surface and to prevent the diffusion/dispersion of contaminants around the polluted site as a result through wind and rain. Biopiles are effective in the biorestoration of most hydrocarbons. Oil fractions with low molecular weight, such as gasoline, are partially reduced through evaporation and released during aeration, while the medium-weight products, such as diesel and kerosene, are majorly reduced by the process of biodegradation.

Landfarming

This is a soil biorestoration method that entails the mixing of polluted soil with hydrocarbons to enhance physical, biological, and chemical processes of the soil for biodegradation. It is a basic technique with a low cost and low footprint. Depending on where the treatment of contaminated soil is carried out, landfarming can be classified as either ex situ or in situ technology.

Some of the drawbacks of using this method are poor efficiency in the removal of inorganic pollutants, low microbial activity due to hostile environmental conditions, additional cost incurred due to excavation, and the need for a large work site. A major drawback of the technique is the release of volatile organic compounds into the environment. However, this method requires low rainfall (274 mm), climates with high rate of evaporation (annual evaporation of 2700 mm), and large expanses of land to produce desirable results.

Composting

This is an ex situ method that involves the decomposition of organic waste using thermophilic biological agents under aerobic conditions to achieve a humic transformation known as compost. The compost is used as a soil fertilizer. In order to achieve this, a temperature range of 40–70 °C, high presence of oxygen, and nutrient pH in the region of neutral are key to obtaining extensive biodegradation. Composting is not just used to recycle organic waste but also to bioremediate polluted soil or sludge. During the process, microbial activity degrades hazardous organic compounds and reduces the bioavailability of metals. Microorganisms from the soil become part of the process when the compost is mixed with soil. Of recent, this method of bioremediation has enjoyed more attention chiefly because it has been established that it has high efficiency with respect to degradation of petroleum hydrocarbons, pesticides, chlorophenols, etc.

3 Polycyclic Aromatic Hydrocarbons (PAHs) Contaminated Soil

Polycyclic aromatic hydrocarbons (PAHs) are micropollutants with the potential to cause cancer, and they persist in the environment as a result of their highly hydrophobic (properties that repels water) nature. PAHs are chemical compounds that comprise many intertwined aromatic rings arranged in a clustered or linear manner. Naturally, they contain only hydrogen (H) and carbon (C) atoms, but oxygen (O), nitrogen (N), and sulfur (S) atoms may readily substitute in the benzene ring to form heterocyclic aromatic compounds.

3.1 Categorization of Polycyclic Aromatic Hydrocarbons (PAHs) Sources in the Environment

Pollution of soil with PAHs is categorized into three aspects, namely:

(i) Heavily polluted (PAH > 1000 ng g^{-1})
(ii) Unpolluted ($\sum$PAH < 200 ng g^{-1})
(iii) Weakly polluted (PAH 200–600 ng g^{-1}) (Wu et al., 2019).

Many (several hundreds) of PAHs have been identified over the years, but just 28, which have been established to be hazardous, are presented in Table 3 by the US Environmental Protection Agency (EPA) (USEPA, 2008).

The main sources of polycyclic aromatic hydrocarbon (PAHs) contamination are:

(i) Natural
(ii) Anthropogenic activities (Mojiri et al., 2019).

3.1.1 Natural Emission Sources

The natural emission sources include forest infernos, volcanic eruptions, and moorland fires (they are of less importance) induced by lightning flashes (Abdel-Shafy & Mansour, 2016; Ravindra et al., 2008; Srogi, 2007).

3.1.2 Anthropogenic Activities

Human activities are the major factors in PAH pollution. They are categorized into four types, namely:

(i) Mobile
(ii) Industrial
(iii) Domestic
(iv) Agricultural emission sources (Ravindra et al., 2008).

Table 3 Priority PAHs as listed by US EPA (USEPA, 2008)

Number*	PAH compound	Number of rings
1	Benzo(a)anthracene	4
2	Chrysene	4
3	Benzo(a)pyrene	5
4	Benzo(b)fluoranthene	5
5	Benzo(j)fluoranthene	5
6	Benzo(k)fluoranthene	5
7	Fluoranthene	4
8	Benzo(r,s,t)pentaphene	6
9	Dibenz(a,h)acridine	5
10	Dibenz(a,j)acridine	5
11	Dibenzo(a,h)anthracene	5
12	Dibenzo(a,e)fluoranthene	6
13	Dibenzo(a,e)pyrene	6
14	Dibenzo(a,h)pyrene	6
15	Dibenzo(a,l)pyrene	6
16	7H-Dibenzo(c,g)carbazole	5
17	7,12-Dimethylbenz(a)anthracene	4
18	Indeno(1,2,3-cd)pyrene	6
19	3-Methylcholanthrene	4
20	5-Methylchrysene	4
21	1-Nitropyrene	4
22	Acenaphthene	3
23	Acenaphthylene	3
24	Anthracene	3
25	Benzo(g,h,i)perylene	6
26	Fluorene	3
27	Phenanthrene	3
28	Pyrene	4

*Compounds numbered 1–21 are listed on the Toxic Release Inventory reported by the US EPA National Waste Minimization Program, while those numbered 22–28 are listed on the US EPA Priority Chemical List

3.1.3 Industrial Sources

Some industrial emission sources include large machineries powered by gasoline, gasification of coal, oxygen furnace, diesel engine, and electric arc furnace (Ravindra et al., 2008; Srogi, 2007). Other sources like iron and steel production, insecticide and fungicide production, exhaust from refineries, waste incineration, aluminum production, dye manufacturing, cement production, coal-tar pitch manufacturing, rubber tire production, and asphalt-producing industries are the primary sources of PAH emissions as a result of incomplete combustion (Abdel-Shafy & Mansour, 2016; Gupte et al., 2016; Mojiri et al., 2019; Ravindra et al., 2008; Srogi, 2007).

3.1.4 Mobile Sources

These include exhaust from many automobiles and large means of transportation such as trains, aircrafts, ships, and off-road lightweight and heavyweight vehicles (Ravindra et al., 2008; Srogi, 2007).

3.1.5 Domestic Sources

Sources from this type of anthropogenic activities include household activities like cooking on gas burners, wood burning, coal coking, garbage burning, kerosene stoves, and so on (Gupte et al., 2016; Johnsen & Karlson, 2007; Ravindra et al., 2008).

3.1.6 Agricultural Sources

These sources occur when there is burning of agricultural waste and open biomass under conditions that cause incomplete combustion (Ravindra et al., 2008). In urban areas, high PAH pollution is mainly due to industrial, mobile, and domestic sources, while in rural areas it is due to domestic and agricultural sources. The extent of PAH contamination peaks in winter, then spring, autumn, and summer, in that order. PAH contamination is high in winter and spring because of poor diffusion due to atmospheric conditions like gentle winds and low temperatures, increased residential heating, high rate of incomplete combustion of fossil fuel, and lower photodegradation (Miura et al., 2019). Different types of PAH emission sources are shown in Fig. 3.

Atmospheric PAHs in their gaseous state, like aerosols, get deposited in the environment (soil, plants, and water) in the particulate phase through processes such as dry or wet deposition (Abdel-Shafy & Mansour, 2016). As a result of high hydrophobicity and low vapor pressure, polycyclic aromatic hydrocarbons with three and above aromatic rings have very strong adsorption to soil particles (Abdel-Shafy & Mansour, 2016). Subsequent pollution of plants, groundwater, and food is caused by accumulation of PAHs in the soil. Roots of plants absorb PAHs from contaminated soil and translocate them to other parts of the plant. Exposure to PAHs can occur in three ways, namely, ingestion, inhalation, and skin contact (Burchiel & Luster, 2001). In addition, exposure can occur through more than one route, that is, more than one way; for example, inhalation and skin exposure to contaminated soil (Abdel-Shafy & Mansour, 2016; Rengarajan et al., 2015).

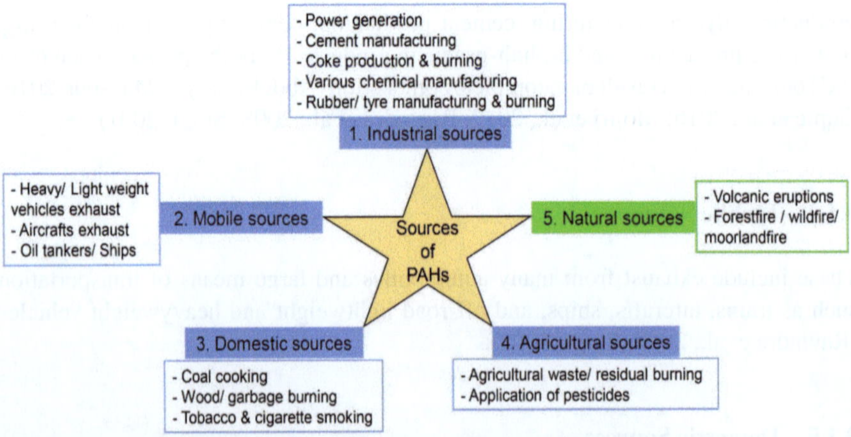

Fig. 3 Different types of PAH emission sources

The most common place of exposure for many people is their offices or places of work. For instance, police officers are exposed via inhalation of vehicle exhaust gases and road dust containing polycyclic aromatic hydrocarbons during traffic. Also, workers in factories where coke is produced are exposed (Lee & Vu, 2010). Other ways of exposure include consuming foods that are smoked or grilled, drinking polluted water, smoking, etc. (Suman et al., 2016). Smoking one cigarette exposes the smoker to 20–40 ng of benzo(a)pyrene (Skupinska et al., 2004). For nonsmokers, their diet can have up to 70% of PAH exposure (Skupinska et al., 2004).

Furthermore, the quantity of PAHs in drinking water is between 1 ng/L and 11 μg/L, and the WHO's highest acceptable quantity of benzo(a)pyrene is 0.7 μg/L (Skupinska et al., 2004). Foods that are processed domestically and industrially by toasting, roasting, drying, frying, barbecuing, and baking always contain PAHs (Rose et al., 2015). Vegetables and fruits can become contaminated with PAHs through polluted soil and atmospheric deposition (Zelinkova & Wenzl, 2015). Furthermore, tea and coffee which are the most commonly consumed beverages in the world get contaminated with PAHs via heating during production, industrial drying and roasting, and atmospheric disposition (Duedahl-Olesen et al., 2015).

In 2015, Duedahl-Olesen et al. carried out an investigation to check for the presence of PAH4 (benzo(b)fluoranthene, chrysene, benzo(a)anthracene, and benzo(a)pyrene) in some brands of coffee and tea. The results show that black tea and coffee have high PAH4 of about 25–115 μg/kg and 2.2–5.1 μg/kg, respectively. According to German Environment Agency (GEA), products that are used daily such as mouse pads, bicycle handles, toys, bathing shoes, sports items, and so on contain PAHs (Brandt & Einhenkel-Arle, 2016). Also, industrial soot, tar oils, and extender oils used in the industry to achieve desired elasticity, softness, and flexibility of rubber-made products contain extended PAHs. In 2015, Paschke et al. detected benzo(a)pyrene, pyrene, phenanthrene, and naphthalene in inks used in newspapers, measuring up to 52, 553, 778, and 283 μg/kg, respectively. Regardless of their toxicity, PAHs are unavoidable in daily life due to their use in the manufacture of domestic products.

3.2 Polycyclic Aromatic Hydrocarbons (PAHs) in Soil

The accumulation of PAHs in the soil organic components is fast because of its stable chemical structure and hydrophobic nature (Abdel-Shafy & Mansour, 2016). The major sources of PAHs in soil environments are limitless including the use of commercial composts for horticulture, vehicular exhaust emissions, industrial pyrolysis, and fossil fuel combustion. (Ambade & Sethi, 2021; Ambade et al., 2020; Arora & Reddy, 2013; Bosetti et al., 2007; Kumar et al., 2020; Zhang et al., 2015). The contamination of soils with PAHs has the potential to affect ecological and human health. In addition, PAHs pose an ecotoxicological risk to air quality, aquatic life, plants, and soil functions and biomes. Chemical properties and environmental conditions are the factors that decide the movement and occurrence/persistence of PAHs in the environment (CCME, 2010; Neff et al., 2005; Wilcke, 2000). The natural/initial makeup of PAH mixtures which can be altered by post-emission transport, transformation, and other processes is determined by soil properties, compound properties, vegetation, and other ambient conditions (Katsoyiannis et al., 2011; Keyte et al., 2013; Tobiszewski & Namiesnik, 2012; Wilcke et al., 2014).

In the developed world, research activities are being carried out to determine the characteristics of PAHs in soils and the hazards they pose to plants and animals (Arp et al., 2014; Bandowe et al., 2019; Cai et al., 2008; CCME, 2010; Davie-Martin et al., 2017; Desaules et al., 2008; Sun et al., 2018; Wilcke, 2007). Procedures, regulatory guidelines, and rules are being set in different countries so as to safeguard and assess the risks PAHs pose to humans and ecosystems. In addition, they are used to identify soils that require remediation (CCME, 2010; Desaules et al., 2008).

3.3 Hazards of PAHs to Humans and Other Animals

Polyaromatic hydrocarbons pose health risks to humans, wildlife, and livestock (CCME, 2010; Douben, 2003; IARC, 2010). They are recognized as persistent organic pollutants due to the many fused aromatic rings in their structure. They have mutagenic properties and are carcinogenic (U.S. EPA, 2003). Either directly or indirectly, the human body gets exposed to PAHs intended for the soil (Ambade et al., 2021a, 2021b, 2021c, 2021d; Roy et al., 2017; Tarafdar et al., 2018). Furthermore, in human organs, the development of cancerous tumors is a result of long-term exposure to PAHs (IARC, 2010). The evaluation of their potential to be mutagenic or carcinogenic requires reliable data on exposure which is near impossible to collate in practice (White & Claxton, 2004). Despite this difficulty, nations like Denmark, Canada, and the Netherlands have developed guidelines for the cleanup of soil polluted with PAHs (Chung et al., 2006).

4 Heavy Metals Contaminated Soil

A group of elements with metallic properties such as actinides, lanthanides, metalloids, and transition metals are known as heavy metals. (Singh et al., 2011). They can be classified as essential (with different enzymes) and nonessential (Aransiola et al., 2023; Theron et al., 2012). Based on this classification, essential heavy metals include trace elements such as copper, iron, zinc, manganese, cobalt, selenium, and molybdenum. They interact with the body by binding to proteins, consequently affecting cell functions and making their concentrations closely regulated (Babaniyi et al., 2023; Theron et al., 2012).

The pollution of soil by heavy metals is a major source of concern for food security and the environment as a result of increase in human population and anthropogenic activities, astronomical growth of industry and agricultural activities, and the disturbance of natural ecosystems by humans (Sarwar et al., 2017). Anthropogenic activities, such as the use of materials containing metals in homes and agriculture and the increase in the number of industries and mining activities, are causes of environmental pollution and increase in human exposure to heavy metals (Tchounwou et al., 2012). Globally, about five million sites are estimated to be contaminated with heavy metals, and they are above the allowed levels (Li et al., 2019). Heavy metal contamination affects humans and ecosystems and causes issues pertaining to nonavailability of land for agricultural production, food quality, and food chain safety, which leads to land tenure and food security problems (Wuana & Okieimen, 2011). Nonessential heavy metals are lead, mercury, cadmium, arsenic, tungsten, plutonium, and vanadium (Abioye et al., 2013). They can penetrate the cells/tissues because of their physicochemical characteristics such as ionic charge and potent toxins (Duce & Bush, 2010; Johri et al., 2010). Vanadium requires special attention because it is established as the major threat to all life forms as a result of its high toxicity at low levels of exposure. In addition, nonessential heavy metals serve no essential function in living organisms (Atobatele & Olutona, 2015).

Mercury, lead, cadmium, and arsenic are among the metals with the most toxicity in the environment according to the Environmental Protection Agency (EPA) (Goyer, 2004). Over 20 heavy metals with acute toxicity are listed by the US Agency for Toxic Substances and Disease Registry (ATSDR), but mercury (Hg), arsenic (As), lead (Pb), and cadmium (Cd) pose a risk to human health. Arsenic is ranked number 1 on ATSDR's list, followed by lead, and cadmium ranks seventh. Furthermore, arsenic is the most common cause of acute heavy metal poisoning (Fay & Mumtaz, 1996; Flora et al., 2011). Arsenic (As) is classified as a metalloid because of its metallic and nonmetallic properties: it is a naturally abundant element in the Earth's crust (Kesici, 2016; Nriagu et al., 2007). It is in inorganic (high toxicity) and organic forms in the soil (Shrivastava et al., 2015). In well-drained (toxic) surface soils, arsenic occurs predominantly as oxyanions of As^{5+} (arsenate), whereas As^{3+} (arsenite) species are more abundant in reducing environments, such as waterlogged soils (Roberts et al., 2010).

Cadmium (Cd) is a major pollutant because it is soluble in water and highly toxic (Benavides et al., 2005; Pinto et al., 2004). With inorganic and organic ligands, it tends to form stable dissolved complexes, thus inhibiting its sorption and precipitation (Kubier et al., 2019). In addition, it can inhibit the DNA-mediated transformation in microorganisms and interrupt enzyme activities; its main sources are from human activities such as agriculture, industry, and mining (Kabata-Pendias, 2010; Kubier et al., 2019).

Lead (Pb) is a widely available toxic heavy metal; it negatively affects living organisms at different levels: biochemically, physiologically, and morphologically. It has a high mobility and persistency in soil and water. Also, it accumulates in the upper part of the ground (Pourrut et al., 2011; Tangahu et al., 2011; Zeng et al., 2007). Even though lead is a naturally occurring element, human activities such as manufacturing, mining, and fossil fuel usage all play a huge role in increasing its concentrations in soils (Tchounwou et al., 2012).

4.1 Sources of Heavy Metals in Contaminated Soils

Soil is a major and important part of the ecosystem. Soil is prone to contamination from different sources because of its absorbing and emitting capabilities. Soil is our habitat and where we engage in different forms of agriculture, but it is not safe from heavy metals. Heavy metals enter the soil through human activities and parent materials (lithosphere). Factors such as climatic conditions, composition of parent rock, extent of weathering, biological, physical, and chemical features of the soil, etc. affect the presence and division of metals in the soil. Human activities are the major cause and are gradually increasing day by day, which results in the deterioration of the environment (Xu et al., 2019).

Heavy metals exist in the air in the form of aerosols and reach the soil by rainfall and natural processes of sedimentation. They enter the soil through many sources which include: wastewater, metal mining and milling processes, fertilizers, pesticides, biosolids (sewage sludge) and manures, industrial wastes, and air-borne sources. They enter the environment through dust produced by transportation, dirt, gas, metallurgy, energy, and construction. Industrial effluents are known to be harmful chemicals which are usually released to water bodies or an open area. Nowadays, people make use of a huge number of pesticides and fertilizers for agricultural production. The incessant use of these substances leads to soil degradation.

Fertilizers are reported to contain pollutants like heavy metals. Fertilizers containing nitrogen and potash have low amounts of toxic heavy metals while phosphoric fertilizers have high amounts of such heavy metals. High concentration of heavy metals is found in vegetables and crops grown close to industrial areas, municipal and agricultural wastewater, and busy roads (Xu et al., 2019). Many types of industries release different kinds of toxic heavy metals which end up in the soil. Some of them are lead from paints industries, cadmium from metal smelters, and nickel from steel industries. Exhaust gases released from vehicles are the major sources of lead contamination in plants, air, and soil (Xu et al., 2019). (Look for more references for this section).

4.2 *Heavy Metal Contamination in Soil*

Globally, contamination of soil by heavy metals from the industries, agricultural chemicals, volcanoes, etc. is a huge concern (Hinojosa et al., 2004). Pollution by heavy metals has significant effects on microbial activity, plant quality, and yield (Yao et al., 2003). As a result of this, heavy metals are seen as one of the main soil pollutants. Soil properties such as pH, clay content, and organic matter play a huge role in how long metals can affect biological and biochemical properties and activities of plants (Speira et al., 1999).

Enzyme activities of the soil are indirectly affected by heavy metals; this happens by having negative effects on the microorganisms which synthesize the enzymes. Additionally, in Karaca et al. (2010) concluded that different metals influence activities of enzymes in various ways because of differing chemical affinities of enzymes in the soil system. On the contrary, the longer the effect of heavy metal, the higher the bacterial and fungal community tolerance to their presence. A very good example is the arbuscular mycorrhizal (AM) fungus, which is key in the restoration of polluted ecosystems (Mora et al., 2005). In Chen et al. (2010) reported that heavy metals triggered an increase in soil actinomycetes while there was a decrease in bacterial biomass, species, and richness.

Activity of soil microbes and their diversity is responsible for the maintenance of soil structure, recycling of plant nutrients, plant growth communities, control of plant pests, and detoxification of toxic chemicals. They are key determinants of soil quality. Sustained pollution by heavy metals can have adverse effects on soil, by causing changes in the population, diversity, and activities of microbes in that soil.

Cadmium (Cd) has lower affinity and greater mobility for soil colloids, which is responsible for its higher toxicity to enzymes compared to lead (Pb). In the soil, chromium appears as Cr(III) and Cr(VI); they are known for their toxicities and different chemical properties. Cr(VI) is a powerful oxidizing agent and is highly toxic, whereas Cr(III) is a micronutrient and not known to be harmful; it is 10–100 times less toxic than Cr(VI). At high concentrations, Cr(VI) is known to have damaging effects on microbial cell metabolism and to shift the diversity of soil microbial populations (Rehman et al., 2023). Heavy metals are harmful to soil microorganisms, leading to changes in the number, diversity, and interactions of soil microbial communities (Ashraf & Ali, 2007).

It is therefore safe to conclude that soil pollution affects soil microbial properties such as enzyme activity, microbial population, diversity, and replication rate, which in turn serve as useful indicators of soil pollution.

5 Role of Vermitechnology

All over the globe, the increase in the rate of waste generation and deposition is a subject of concern (Anand & Sinha, 2019; Dada et al., 2021). Soil is exposed to contamination by heavy metals and polycyclic aromatic hydrocarbons (PAHs) which pose a lot of threats to the biotic community due to their nondegradability and recalcitrancy in the environment (Wei et al., 2020). Metals at high concentrations are toxic to plants and animals (Singh et al., 2022). Bioremediation is established as a safe and efficient method of restoring polluted sites compared to other technologies and methodologies of waste management (chemical, physicochemical, and thermal processes) available. In the search for a cost-effective, eco-friendly method that can be used directly on land for bioremediation of polluted sites, global attention seems to be turning toward vermitechnology.

5.1 Vermitechnology

Vermitechnology is the application of earthworms to degrade organic materials for environmental cleanup and possible use of end products in agricultural manuring. It has become a global biological technology which employs earthworms to enhance degradation and simultaneously removing contaminants in polluted environments. This approach of bioremediation involves using earthworms in the bioaccumulation of contaminants for enzyme-mediated biodegradation and biotransformation into safe products (Almutairi, 2019). Vermitechnology is an emerging technology and an eco-friendly approach to bioremediation. The technology serves as a means of recovering nutrients from organic waste to replenish soil fertility (Samal et al., 2019). Vermitechnology involves vermifiltration (purifying effluents using earthworms), vermiculture (commercial rearing of earthworms), and vermicomposting (composting organic wastes using earthworms) (Aransiola et al., 2022).

Aristotle, about 2000 years ago referred to earthworms as "the intestines of the earth." Earthworms are important and are a significant part of the soil biomass, not least because they perform indispensable role in nutrient cycling in the soil. Earthworms are detritus feeders and have been recognized for the vital role they play in the degradation of organic matter/wastes and the sensitive position they occupy in the food chain (Aemere et al., 2020; Dada et al., 2021). This confirms the statement of Darwin that "no other creature has contributed to the building of the earth as earthworm." Species of earthworms utilized in vermitechnology are the epigeic species because they carry out the process more rapidly and produce a more suitable end product. Examples are *Eisenia fetida, Perionyx excavatus*, and *Eudrilus eugeniae* (Singh et al., 2018). Earthworms are long, cylindrical, and bilaterally symmetrical annelids. They are terrestrial invertebrates with over 1800 species identified globally. The guts of earthworms host millions of enzymes and biodegrading microbes, which facilitate the bioconversion of organic matters in the soil (Dada et al., 2021) (Fig. 4).

Fig. 4 Vermicomposting facilities

Vermitechnology has found wide application in aquaculture, wastewater treatment, organic fertilizer production, anticancer drug production, and solid waste management. Vermitechnology employed in aquaculture is vermiculture. This involves the commercial growing or culturing of earthworms, cocoons, and/or vermiwash, which can be used as or incorporated into fish feed.

5.2 *Vermifiltration*

Vermifiltration is vermitechnology applied in the treatment of wastewater. In this technique, earthworm-inoculated compost or soil is used as filter bed through which the wastewater is passed and filtered. It is a biotreatment technology of wastewater using earthworms for greater utilization and recycling of used water. The vermifiltered water, also referred to as "vermiaqua," is sterile, highly nutritive, odor free, pathogen free, chemical free, and neutral in pH. Vermiaqua can be used for toilet use, domestic laundry, irrigation in farms and gardens, as well as industrial usage.

Here, earthworms function as bioreactors and biofilters, which adsorb pollutants from wastewater and reduce chemical oxygen demand (COD) and biochemical oxygen demand (BOD) to acceptable levels (Sinha et al., 2012). Earthworms are also employed in the stabilization of sewage sludge and in the purification of water in trickling filters of sewage plants in the United States and the United Kingdom (Rai, 2019). Vermifiltration is a cost-effective and eco-friendly method of wastewater treatment.

The important determinants of the efficiency of vermifiltration are the species/population of the worms, content/type and features of the wastewater, hydraulic retention time, and hydraulic conductivity (Krishnasamy et al., 2013).

5.3 *Vermiculture*

In the 1950s, the Canadian earthworm collectors started operations in vermiculture in order to meet the market demand for fishing baits by fishers. In 1970, it had spread to England, Mexico, Japan, the United States, Southeast Asia, and now to all parts of the globe (Rai, 2019). Vermiculture is the commercial rearing of earthworms for the production of protein-rich materials for fishery, poultry, and dairy industries. It is the scientific method of breeding and rearing earthworms. The goal of vermiculture is to increase the population of worms so as to ensure a sustainable harvest. It forms the basis for vermifiltration and vermicomposting. Vermiculture can also be done on a small scale at home. Cement tanks, earthen pots, wooden boxes, plastic trays, and tubs can be used for vermiculture bed with holes at the bottom for the purpose of aeration (Rai, 2019).

5.4 *Vermicomposting*

Vermicomposting has been defined by Singh et al. (2022) as a biooxidative process that involves the interaction of earthworms, microorganisms, and other degradable components of the soil environment. In vermicomposting, earthworms and their gut microbes work hand in hand to biodegrade organic waste. Through their muscular actions, earthworms convert larger organic material into smaller ones, which are further degraded by the gut microbes (Samal et al., 2019).

They find application in industrial and other human activities that cause accumulation of waste residues (Abdellah et al., 2022; Wei et al., 2020). Singh et al. (2022) reported vermicomposting as a cost-effective and eco-friendly method that could transform allopathic sludge into a safe, nutrient-rich compost that could be used for agricultural purposes. *Eudrilus eugeniae* is more suitable for tropical climates because such areas are less prone to temperature fluctuations.

5.5 *Advantages of Vermitechnology*

1. Compost produced via vermicomposting is completely organic. It produces nontoxic materials with good structure. Vermicompost is not harmful to soil, plants, or the environment.
2. There is increase in the level of nutrients available to plants. Vermicompost is higher in nitrate content, which is a better form of nitrogen available for plant utilization and growth.
3. Vermicompost acts as a soil conditioner with the ability to stimulate plant growth. They improve seed germination, seedling growth, and development. It also improves soil aeration and texture.

4. The level of beneficial microorganisms is increased.
5. Protection of plants against phytopathogens: A high level of beneficial microbes present in the soil outcompetes potential disease-causing pathogens. They also colonize plant soil-root area, preventing attachment of pathogens to plant roots.
6. It prevents unnecessary disposal of organic or vegetative food wastes. Vegetative food wastes that would have constituted environmental nuisance are converted into useful products such as agro-fertilizers by vermicomposting.

5.6 Roles of Vermitechnology in Biodegradation of PAHs

Polycyclic aromatic hydrocarbons (PAHs) are recalcitrant pollutants with many benzene rings and are known to pose a serious threat to the biotic community and the environment (Kosnar et al., 2019). Earthworms speed up the removal of PAHs as they burrow through the soil, making contaminants more available for degradation by microbes.

Kosnar et al. (2019) established through an experiment that soil biochemical activity is enhanced by adding earthworms to treat organic wastes to produce usable organic amendments. Ekperusi and Aigbodion (2015), in a study on the bioremediation of soil polluted with diesel using the earthworm *E. eugeniae*, reported that during a 90-day period of study, there was a decrease in pH, total nitrogen, chloride, electrical conductivity, total organic carbon, calcium, sodium, potassium, phosphate, nitrate, sulfate, zinc, magnesium, manganese, copper, nickel, cadmium, vanadium, chromium, lead, mercury, arsenic, and total petroleum hydrocarbons (TPH) due to the activities of earthworm. They reported a decline in total petroleum hydrocarbons, toluene, benzene, ethylbenzene, and xylene concentrations in diesel-contaminated soil treated with *E. eugeniae.* According to Soobhany et al. (2015), when *E. eugeniae* was inoculated in waste containing heavy metals, it was effective in diminishing toxic heavy metals.

5.7 Roles of Vermitechnology in Biodegradation of Heavy Metals

The soil environment is a known reservoir of heavy metals. Earthworms are capable of surviving in a heavy metal-contaminated soil through the collection and accumulation of heavy metals in their tissues. Biosorption and accumulation of heavy metals in earthworms result in striking a sense of balance between uptake and excretion of wastes, determining their survival in a heavy metal-contaminated environment (Singh et al., 2018). According to Malley et al., earthworms are biological indicators of the presence of toxic heavy metals, a matter which indicates a potential environmental hazard (Paoletti et al., 1991). Earthworm chloraogocyte cells and bacteria in their guts have the capability to reduce the genotoxic effects of harmful pollutants and heavy metals through detoxification (Singh et al., 2022). This is owing to the fact that earthworms are capable of accumulating and tolerating high levels of toxic metals without experiencing significant damage (Aemere et al., 2020).

Earthworms have been proven to be both productive and protective for the environment and society. Earthworms may be manipulated to increase their intestinal uptake of pollutants through food limitation (Dada et al., 2021). They are naturally resistant and adapt quickly to environmental changes. Wei et al. (2020) reported that adsorption efficiency of humin (an activator that can trigger the biosorption of heavy metals by improving biodiversity and biomass of heavy metal-resistant bacteria) was increased when waste residues contaminated with heavy metals were subjected to vermicomposting.

It has been argued that metallothionein-induced bioaccumulation and sequestration could be responsible for the levels of metal reduction in contaminated sites (Dada et al., 2021). Metallothioneins are heat-stable, low molecular weight, cysteine-rich ubiquitous proteins expressed by living organisms under environmental stress conditions induced by the presence of metals at certain concentration levels (Aemere et al., 2020). They are a group of metal-binding proteins which may be responsible for the detoxification of heavy metals in earthworms. They can also be used as biomarkers to assess the ecotoxicological impact of pollutants in the environment.

5.8 Factors Influencing Vermitechnology

Growth rate, onset of maturity, rate of reproduction, and population buildup of earthworms during vermicomposting are dependent on temperature, moisture content, pH, and other physicochemical properties of the environment or substrate. Earthworms have the ability to thrive well even under unfavorable environmental conditions. They can grow within a wide temperature range of 5–43 °C. Growth and maturation of earthworms have been found to be best at 20–25 °C. However, increased temperature up to 30 °C hastens the rate of growth and reduces the time to attain sexual maturity. Their reproduction and biomass production are pH dependent (Dominguez et al., 2001). Neutral or near-neutral pH is the best for the growth of earthworms (Kaplan et al., 1980; Singh et al., 2018). Activities of earthworms usually result in declines in pH, moisture content, and electrical conductivity (Ekperusi & Aigbodion, 2015). A moisture content of 80–85% is best for the growth of earthworms in a vermicompost.

Earthworms present a promising solution to combat the global problem of environmental pollution as a result of their ubiquity, high biodegradation capacity, and the millions of enzymes and microbes they harbor in their guts. For a successful vermiculture and vermicomposting, vermicompost pit should be protected from direct sunlight, ants, rats, birds, and excessive rain to ensure a high survival rate of worms.

6 Conclusion

The high levels of hazardous compounds in the soil, especially due to pesticides, heavy metals, and petroleum derivatives, negatively affect the ecosystem balance and human health. Activities such as agriculture and, by extension, food safety have been greatly compromised due to the pollution of water and soil by heavy metals and polycyclic aromatic hydrocarbons (PAHs). Over the years, several methods have been employed to deal with the issue of soil, environmental, and water pollution by different pollutants with varying degrees of success. Vermitechnology, which is the application of earthworms to degrade organic materials for environmental cleanup and the possible use of end products in agricultural manuring, has recently attracted the attention of many researchers for being the most eco-friendly method available. It is also attractive because its products are suitable for a lot of purposes, especially agriculture. Vermitechnology presents a promising solution to combat the global problem of environmental pollution. It is a sustainable and cost-effective method that can be used to restore soil health and improve food safety. By giving more attention to vermitechnology, researchers and governments all over the world can make the environment safer for humans, animals, and plants. Let us all work together to promote the use of vermitechnology and protect our planet for future generations.

References

Abdalqadir, M., Hughes, D., Rezaei Gomari, S., & Rafiq, U. (2024). A state of the art of review on factors affecting the enhanced weathering in agricultural soil: Strategies for carbon sequestration and climate mitigation. *Environmental Science and Pollution Research, 31*(13), 19047–19070.

Abdellah, Y., Shi, Z., Luo, Y., Hou, W., Yang, X., & Wang, R. (2022). Effects of different additives and aerobic composting on heavy metal bioavailability reduction and compost parameters: A meta-analysis. *Environmental Pollution, 307*(1), Article 119549.

Abdel-Shafy, H. I., & Mansour, M. S. M. (2016). A review on polycyclic aromatic hydrocarbons: Source, environmental impact, effect on human health and remediation. *Egyptian Journal of Petroleum, 25*, 107–123. https://doi.org/10.1016/j.ejpe.2015.03.011

Abioye, O. P., Ijah, U. J. J., & Aransiola S. A. (2013). Remediation mechanisms of tropical plants for lead-contaminated environment. In D. K. Gupta (Ed.), *Plant-based remediation processes, soil biology* (p. 35). Springer-Verlag Berlin Heidelberg. https://doi.org/10.1007/978-3-642-35564-6_4. https://www.springer.com/gp/book/9783642355639

Aemere, O., Sharma, V., & Lin, J. (2020). Metallothioneins in earthworms: The journey so far. *Open Journal of Environmental Biology, 5*(1), 14–21.

Agriculture Victoria. (2021). *Farm management*. https://agriculture.vic.gov.au/farm-management/soil/what-is-soil. Retrieved on September 2022.

Almutairi, M. (2019). Vermiremediation strategy for remediation of Kuwaiti oil contaminated soil. *SN Applied Sciences, 1*, 1312.

Ambade, B., Kumar, A., Kumar, A., & Sahu, L. K. (2021a). Temporal variability of atmospheric particulate-bound polycyclic aromatic hydrocarbons (PAHs) over central East India: Sources and carcinogenic risk assessment. *Air Quality, Atmosphere and Health, 14*, 1–16. https://doi.org/10.1007/s11869-021-01089-5

Ambade, B., Kumar, A., Kumar, A., & Sethi, S. S. (2020). Characterization of PAHs and N-alkanes in atmospheric aerosol of Jamshedpur City, India. *Journal of Hazardous, Toxic, and Radioactive Waste, 24*, 4. https://doi.org/10.1061/(ASCE)HZ.2153-5515.0000490

Ambade, B., Kumar, A., & Sahu, L. K. (2021b). Characterization and health risk assessment of particulate bound polycyclic aromatic hydrocarbons (PAHs) in indoor and outdoor atmosphere of Central East India. *Environmental Science and Pollution Research, 28*(40), 56269–56280. https://doi.org/10.1007/s11356-021-14606-x

Ambade, B., Kurwadkar, S., Sankar, T. K., & Kumar, A. (2021c). Emission reduction of black carbon and polycyclic aromatic hydrocarbons during COVID-19 pandemic lockdown. *Air Quality, Atmosphere and Health, 14*, 1081–1095. https://doi.org/10.1007/s11869-021-01004-y

Ambade, B., & Sethi, S. S. (2021). Health risk assessment and characterization of polycyclic aromatic hydrocarbon from hydrosphere. *Journal of Hazardous, Toxic, and Radioactive Waste, 25*, 2. https://doi.org/10.1061/(ASCE)HZ.2153-5515.0000586

Ambade, B., Sethi, S. S., Kurwadkar, S., Kumar, A., & Sankar, T. K. (2021d). Toxicity and health risk assessment of polycyclic aromatic hydrocarbons in surface water, sediments and groundwater vulnerability in Damodar River Basin. *Groundwater for Sustainable Development, 13*, Article 100553. https://doi.org/10.1016/j.gsd.2021.100553

Anand, K., & Sinha, P. B. (2019). Vermitechnology: A solution for agricultural waste.

Aransiola, S. A., Afolabi, F., Joseph, F., & Maddela, N. R. (2023). Soil enzymes: Distribution, interactions and influencing factors. In P. Jeschke, & E. B. Starikov (Eds.), *Agricultural biocatalysis: Enzymes in agriculture and industry* (1st ed., pp. 303–333). Jenny Stanford Publishing. ISBN 9789814968478. https://www.routledge.com/Agricultural-Biocatalysis-Enzymes-in-Agriculture-and-Industry/Jeschke-Starikov/p/book/9789814968478

Aransiola, S. A., Ijah, U. J. J., & Abioye, O. P. (2013). Phytoremediation of lead polluted soil by glycine max L. *Applied and Environmental Soil Science, 2013*, 631619. https://doi.org/10.1155/2013/631619. https://www.hindawi.com/journals/aess/2013/631619/abs/

Aransiola, S. A., Ijah, U. J. J., Abioye, O. P., & Bala, J. D. (2021). Microbial and heavy metal determination of contaminated soil using Melissa officinalis L. *International Journal of Environmental Planning and Management, 7*(3), 102–107. http://www.aiscience.org/journal/paperinfo/ijepm?paperid=5369

Aransiola, S. A, Ijah, U. J. J., Abioye, O. P., & Bala J. D. (2022). Vermicompost-assisted phytoremediation of toxic trace element-contaminated soil in Madaka, Nigeria using Melissa officinalis L and Sida acuta. *International Journal of Environmental Science and Technology*. https://doi.org/10.1007/s13762-022-04105-y

Arora, A. S., & Reddy, A. S. (2013). Multivariate analysis for assessing the quality of stormwater from different urban surfaces of the Patiala City, Punjab (India). *Urban Water Journal, 10*(6), 422–433. https://doi.org/10.1080/1573062X.2012.739629

Arp, H. P. H., Lundstedt, S., Josefsson, S., Cornelissen, G., Enell, A., Allard, A.-S., & Kleja, D. B. (2014). Native oxy-PAHs, N-PACs, and PAHs in historically contaminated soils from Sweden, Belgium, and France: Their soil-porewater partitioning behavior, bioaccumulation in *Enchytraeus crypticus*, and bioavailability. *Environmental Science and Technology, 48*, 11187–11195.

Ashraf, M. A., Maah, M. J., & Yusoff, I. (2014). Soil contamination, risk assessment and remediation. In *Environmental risk assessment of soil contamination* (pp. 1–56). IntechOpen. Available online: https://www.intechopen.com/books/environmental-risk-assessment-of-soil-contamination. Accessed on October 14, 2022.

Ashraf, R., & Ali, T. A. (2007). Effect of heavy metals on soil microbial community and mung beans seed germination. *Pakistan Journals of Botany, 39*(2), 629–636.

Atobatele, O. E., & Olutona, G. O. (2015). Distribution of three non-essential trace metals (cadmium, mercury and lead) in the organs of fish from Aiba reservoir, Iwo, Nigeria. *Toxicology Reports, 2*, 896–903. https://doi.org/10.1016/j.toxrep.2015.06.003

Azubuike, C. C., Chikere, C. B., & Okpokwasili, G. C. (2016). Bioremediation techniques–classification based on site of application: Principles, advantages, limitations and prospects. *World Journal of Microbiology & Biotechnology, 32*, Article e180.

Babaniyi, G. G., Olagoke, O. E., & Aransiola, S. A. (2023). Extracellular enzymatic activity of bacteria in aquatic ecosystems. In N. R. Maddela, L. K. W. Eller, & R. Prasad (Eds.), *Microbiology for cleaner production and environmental sustainability*. CRC Press. ISBN: 9781032496061. https://www.routledge.com/Microbiology-for-Cleaner-Production-and-Environmental-Sustainability/Maddela-Eller-Prasad/p/book/9781032496061

Bandowe, B. A. M., Wei, C., Han, Y. M., Cao, J. J., Zhan, C., & Wilcke, W. (2019). Polycyclic aromatic compounds (PAHs, oxygenated PAHs, nitrated PAHs and azaarenes) in soils from China and their relationship with geographic location, land use and soil carbon fractions. *Science of the Total Environment, 690*, 1268–1276.

Benavides, M. P., Gallego, S. M., & Tomaro, M. L. (2005). Cadmium toxicity in plants. *Brazilian Journal of Plant Physiology, 17*(1), 21–34. https://doi.org/10.1590/s1677-04202005000100003

Bosetti, C., Boffetta, P., & La Vecchia, C. (2007). Occupational exposures to polycyclic aromatic hydrocarbons, and respiratory and urinary tract cancers: A quantitative review to 2005. *Annals of Oncology, 18*, 431–446. https://doi.org/10.1093/annonc/mdl172

Brandt, M., & Einhenkel-Arle, D. (2016). Polycyclic aromatic hydrocarbons harmful to the environment! Toxic! inevitable? *German Environment Agency* 1–24.

Burchiel, S. W., & Luster, M. I. (2001). Signaling by environmental polycyclic aromatic hydrocarbons in human lymphocytes. *Clinical Immunology, 98*, 2–10. https://doi.org/10.1006/clim.2000.4934

Cai, Q. Y., Mo, C. H., Wu, Q. T., Katsoyiannis, A., & Zeng, Q. Y. (2008). The status of soil contamination by semi-volatile organic compounds (SVOCs) in China: A review. *Science of the Total Environment, 389*, 209–224.

CCME (Canadian Council of Ministers of the Environment). (2010). *Canadian soil quality guidelines for carcinogenic and other polycyclic aromatic hydrocarbons (Environmental and Human Health Effects)*. Scientific Criteria Document (revised) PN 1445, Quebec, Canada. https://www.ccme.ca/files/Resources/supporting_scientific_documents/pah_soqg_scd_1445.pdf. Accessed January 2020.

CETESB. (2020). *Qualidade do Solo*. Available online: https://cetesb.sp.gov.br/solo/poluicao/. Accessed on October 14, 2022.

Chen, G. Q., Chen, Y., Zeng, G. M., Zhang, J. C., Chen, Y. N., Wang, L., & Zhang, W. J. (2010). Speciation of cadmium and changes in bacterial communities in Red Soil following application of cadmium-polluted compost. *Environmental Engineering Science, 27*(12), 1019–1026.

Chung, M. K., Hu, R., Cheung, K. C., & Wong, M. H. (2006). Pollutants in Hong Kong soils: Polycyclic aromatic hydrocarbons. *Chemosphere, 67*, 464–473.

Dada, E. O., Akinola, M. O., Owa, S. O., Dedeke, G. A., Aladesida, A. A., Owagboraiye, F. O., & Oludipe, E. O. (2021). *Journal of Health Pollution, 11*(29), Article 210302.

Davie-Martin, C. L., Stratton, K. G., Teeguarden, J. G., Waters, K. M., & Simonich, S. L. M. (2017). Implications of bioremediation for polycyclic aromatic hydrocarbon contaminated soils for human health and cancer risk. *Environmental Science and Technology, 51*, 9458–9468.

Desaules, A., Ammann, S., Blum, F., Brändli, R. C., Bucheli, T. D., & Keller, A. (2008). PAHs and PCBs in soils of Switzerland—Status and critical review. *Journal of Environmental Monitoring, 10*, 1265–1277.

Dominguez, J., Edwards, C. A., & Ashby, J. (2001). The biology and population dynamics of *Eudrilus eugeniae* (Kinberg) (oligochaeta) in cattle waste solids. *Pedobiologia, 45*, 341–353.

Douben, P. E. T. (2003). *PAHs: An ecotoxicological perspective*. Wiley. https://doi.org/10.1002/0470867132

Duce, J. A., & Bush, A. I. (2010). Biological metals and Alzheimer's disease: Implications for therapeutics and diagnostics. *Progress in Neurobiology, 92*(1), 1–18. https://doi.org/10.1016/j.pneurobio.2010.04.003

Duedahl-Olesen, L., Navaratnam, M. A., Jewula, J., & Jensen, A. (2015). PAH in some brands of tea and coffee. *Polycyclic Aromatic Compounds, 35*, 74–90. https://doi.org/10.1080/10406638.2014.918554

Ekperusi, O. A., & Aigbodion, I.F. (2015). Bioremediation of heavy metals and petroleum hydrocarbons in diesel contaminated soil with the earthworm: *Eudrilus eugeniae. SpringerPlus, 4*(540)

Fay, R. M., & Mumtaz, M. M. (1996). Development of a priority list of chemical mixtures occurring at 1188 hazardous waste sites, using the HazDat database. *Food and Chemical Toxicology, 34*(11–12), 1163–1165. https://doi.org/10.1016/s0278-6915(97)00090-2

Flora, S. J. S., Pachauri, V., & Saxena, G. (2011). *Arsenic, cadmium and lead. Reproductive and developmental toxicology*. Academic Press.

Goyer, R. (2004). *Issue paper on the human health effects of metals*. US Environmental Protection Agency.

Gupte, A., Tripathi, A., Patel, H., Rudakiya, D., & Gupte, S. (2016). Bioremediation of polycyclic aromatic hydrocarbon (PAHs): A perspective. *Open Biotechnology Journal, 10*, 363–378. https://doi.org/10.2174/1874070701610010363

Hinojosa, M. B., Carreira, J. A., Ruiz, R. G., & Dick, R. P. (2004). Soil moisture pre-treatment effects on enzyme activities as indicators of heavy metal-contaminated and reclaimed soils. *Soil Biology & Biochemistry, 36*, 1559–1568.

http://www.bugsatwork.com/USMX/BUGS%20Report%20PRINT%20(07-13-04)%20Hawaii%20(paginate%201-8).pdf

IARC. (2010). Some non-heterocyclic polycyclic aromatic hydrocarbons and some related exposures. *IARC Monographs on the Evaluation of Carcinogenic Risks to Humans, 92*, 1–853.

Johnsen, A. R., & Karlson, U. (2007). Diffuse PAH contamination of surface soils: Environmental occurrence, bioavailability, and microbial degradation. *Applied Microbiology and Biotechnology, 76*, 533–543.

Johri, N., Jacquillet, G., & Unwin, R. (2010). Heavy metal poisoning: The effects of cadmium on the kidney. *BioMetals, 23*(5), 783–792. https://doi.org/10.1007/s10534-010-9328-y

Kabata-Pendias, A. (2010). *Trace elements in soils and plants*. CRC Press.

Kaplan, D. L., Hartenstein, R., Neuhauser, E. F., & Malecki, M. R. (1980). Physicochemical requirements in the environment of the earthworm *Eisenia fetida* and their growth and reproduction performance. *Soil Biology and Biochemistry, 12*, 347–352.

Karaca, A., Cetin, S. C., Turgay, O. C., & Kizilkaya R. (2010). Effects of heavy metals on soil enzyme activities. In I. Sherameti, & A. Varma (Ed.), *Soil heavy metals, soil biology* (Vol. 19, pp. 237–265). Heidelberg.

Katsoyiannis, A., Sweetman, A. J., & Jones, K. C. (2011). PAH molecular diagnostic ratios applied to atmospheric sources: A critical evaluation using two decades of source inventory and air concentration data from the UK. *Environmental Science and Technology, 45*, 8897–8906.

Kesici, G. G. (2016). Arsenic ototoxicity. *Journal of Otology, 11*(1), 13–17. https://doi.org/10.1016/j.joto.2016.03.001

Keyte, I. J., Harrison, R. M., & Lammel, G. (2013). Chemical reactivity and long-range transport potential of polycyclic aromatic hydrocarbons—A review. *Chemical Society Reviews, 42*, 9333–9391.

Kosnar, Z., Wiesnerova, L., Hrebeckova, T., & Vondrackova, S. (2019). Bioremediation of polycyclic aromatic hydrocarbons (PAHs) present in biomass fly ash by composting and vermicomposting. *Journal of Hazardous Materials, 369*, 79–86.

Krishnasamy, K., Nair, J., & Hughes, R. J. (2013). Vermifiltration system for liquid waste management: A review. *International Journal of Environment and Waste Management, 12*(4), 382–396.

Kubier, A., Wilkin, R. T., & Pichler, T. (2019). Cadmium in soils and groundwater: A review. *Applied Geochemistry, 108*, Article 104388. https://doi.org/10.1016/j.apgeochem.2019.104388

Kumar, A., Sankar, T. K., Sethi, S. S., & Ambade, B. (2020). Characteristics, toxicity, source identification and seasonal variation of atmospheric polycyclic aromatic hydrocarbons over

East India. *Environmental Science and Pollution Research, 27*(1), 678–690. https://doi.org/10.1007/s11356-019-06882-5

Kumar, V., Shahi, S. K., & Singh, S. (2018). Bioremediation: An eco-sustainable approach for restoration of contaminated sites. In *Microbial bioprospecting for sustainable development* (pp. 115–136). Springer. ISBN 9789811300530.

Lee, B.-K., & Vu, V. T. (2010). Sources, distribution and toxicity of polyaromatic hydrocarbons (PAHs) in particulate matter. In *Air pollution* (pp. 99–122). IntechOpen.

Li, C., Zhou, K., Qin, W., Tian, C., Qi, M., Yan, X., et al. (2019). A review on heavy metals contamination in soil: Effects, sources, and remediation techniques. *Soil Sediment Contamination: An International Journal, 28*(4), 380–394. https://doi.org/10.1080/15320383.2019.1592108

Manning, D. A., de Azevedo, A. C., Zani, C. F., & Barneze, A. S. (2024). Soil carbon management and enhanced rock weathering: The separate fates of organic and inorganic carbon. *European Journal of Soil Science, 75*(4), Article e13534.

Mareddy, A. R. (2017). Environmental impact assessment. In *Theory and practice; technology in EIA*. Butterworth-Heinemann. ISBN 9780128111390.

Mir, I. A., Goreau, T. J., Campe, J., & Jerden, J. (2024). India's biogeochemical capacity to attain food security and remediate climate. *Environmental Geochemistry and Health, 46*(1), 17.

Miura, K., Shimada, K., Sugiyama, T., Sato, K., Takami, A., Chan, C. K., et al. (2019). Seasonal and annual changes in PAH concentrations in a remote site in the Pacific Ocean. *Science and Reports, 9*, 1–10.

Mojiri, A., Zhou, J. L., Ohashi, A., Ozaki, N., & Kindaichi, T. (2019). Comprehensive review of polycyclic aromatic hydrocarbons in water sources, their effects and treatments. *Science of the Total Environment, 2019*, Article 133971. https://doi.org/10.1016/j.scitotenv.2019.133971

Mora, A. P., Calvo, J. J. O., Cabrera, F., & Madejon, E. (2005). Changes in enzyme activities and microbial biomass after "in situ" remediation of a heavy metal-contaminated soil. *Applied Soil Ecology, 28*, 125–137.

Mureithi, S. M., Mwendwa, S., & Neina, D. (2024). Soil types, formation processes, and characteristics in the Global South. In *Sustainable soil systems in Global South* (pp. 3–47). Springer Nature Singapore.

Neff, J. M., Stout, S. A., & Gunstert, D. G. (2005). Ecological risk assessment of polycyclic aromatic hydrocarbons in sediments: Identifying sources and ecological hazard. *Integrated Environmental Assessment and Management, 1*, 23–33.

Nriagu, J. O., Bhattacharya, P., Mukherjee, A. B., Bundschuh, J., Zevenhoven, R., & Loeppert, R. H. (2007). Arsenic in soil and groundwater: an overview. In *Trace metals and other contaminants in the environment* (Vol. 9, pp. 3–60).https://doi.org/10.1016/s1875-1121(06)09001-8

Okrent, D. (1999). On intergenerational equity and its clash with intragenerational equity and on the need for policiesto guide the regulation of disposal of wastes and other activities posing very long-term risks. *Risk Analysis, 19*, 877–901.

Palansooriya, K. N., Shaheen, S. M., Chen, S. S., Tsang, D. C. W., Hashimoto, Y., Hou, D., Bolan, N. S., Rinklebe, J., & Ok, Y. S. (2020). Soil amendments for immobilization of potentially toxic elements in contaminated soils: A critical review. *Environment International, 134*, Article 105046.

Paoletti, M. G.., Favretto, M. R., Stinner, B. R., Purrington, F. F., & Bater, J. E. (1991). Invertebrates as bioindicators of soil use. *Agriculture, Ecosystems and Environment, 34*, 341–362.

Pinto, A. P., Mota, A. M. D., De Varennes, A., & Pinto, F. C. (2004). Influence of organic matter on the uptake of cadmium, zinc, copper and iron by sorghum plants. *Science of the total Environment, 326*(1–3), 239–247.https://doi.org/10.1016/j.scitotenv.2004.01.004

Pourrut, B., Shahid, M., Dumat, C., Winterton, P., & Pinelli, E. (2011). Lead uptake, toxicity, and detoxification in plants. *Reviews of Environmental Contamination and Toxicology, 213*, 113–136. https://doi.org/10.1007/978-1-4419-9860-6_4

Rai, S. N. (2019). Vermiculture and vermicomposting: A boon for sustainable agriculture in Fiji Islands. *Haya: The Saudi Journal of Life Sciences, 4*(2), 93–102.

Ravindra, K., Sokhi, R., & Van Grieken, R. (2008). Atmospheric polycyclic aromatic hydrocarbons: Source attribution, emission factors and regulation. *Atmospheric Environment, 42*, 2895–2921. https://doi.org/10.1016/j.atmosenv.2007.12.010

Rehman, S. U., De Castro, F., Marini, P., Aprile, A., Benedetti, M., & Fanizzi, F. P. (2023). Vermibiochar: A novel approach for reducing the environmental impact of heavy metals contamination in agricultural land. *Sustainability, 15*, 9380.https://doi.org/10.3390/su15129380

Rengarajan, T., Rajendran, P., Nandakumar, N., Lokeshkumar, B., Rajendran, P., & Nishigaki, I. (2015). Exposure to polycyclic aromatic hydrocarbons with special focus on cancer. *Asian Pacific Journal of Tropical Biomedicine, 5*, 182–189.

Shakya, L., Babu, R., Singh, A., Dhiman, S., & Dhiman, S. K. (2023). Vermitechnology: A sustainable approach to manage organic waste in urban areas. *IJCSRR, 06*, 2581–2834. https://doi.org/10.47191/ijcsrr/V6-i8-05

Roberts, L. C., Hug, S. J., Dittmar, J., Voegelin, A., Kretzschmar, R., Wehrli, B., et al. (2010). Arsenic release from paddy soils during monsoon flooding. *Nature Geoscience, 3*(1), 53–59. https://doi.org/10.1038/ngeo723

Rose, M., Holland, J., Dowding, A., Petch, S. R., White, S., Fernandes, A., et al. (2015). Investigation into the formation of PAHs in foods prepared in the home to determine the effects of frying, grilling, barbecuing, toasting and roasting. *Food and Chemical Toxicology, 78*, 1–9. https://doi.org/10.1016/j.fct.2014.12.018

Roy, D., Seo, Y.-C., Sinha, S., SinghBhattacharya, G., & Biswas, P. K. (2017). Human health risk exposure with respect to particulate-bound polycyclic aromatic hydrocarbons at mine fire-affected coal mining complex. *Environmental Science and Pollution Research, 26*, 19119–19135. https://doi.org/10.1007/s11356-017-9202-3

Samal, K., Mohan, A. R., Chaudhary, N., & Moulick, S. (2019). Application of vermitechnology in waste management: A review on mechanism and performance. *Journal of Environmental Chemical Engineering, 7*(5), Article 103392.

Sarwar, N., Imran, M., Shaheen, M. R., Ishaque, W., Kamran, M. A., Matloob, A., et al. (2017). Phytoremediation strategies for soils contaminated with heavy metals: Modifications and future perspectives. *Chemosphere, 171*, 710–721. https://doi.org/10.1016/j.chemosphere.2016.12.116

Shanahan, P. (2004) *Bioremediation.* Waste Containment and Remediation Technology, Spring, Massachusetts Institute of Technology, MIT OpenCourseWare. http://ocw.mit.edu/NR/rdonlyres/Civil-and-Environmental-Engineering/1-34Spring2004/335613D5-6D6F-413F-9098-453 E8AC20BC2/0/lecture12.pdf

Shannon, M. J., & Unterman, R. (1993). Evaluating bioremediation: Distinguishing fact from fiction. *Annual Review of Microbiology, 47*(Annual 1993), 715(24).

Shrivastava, A., Ghosh, D., Dash, A., & Bose, S. (2015). Arsenic contamination in soil and sediment in India: Sources, effects, and remediation. *Current Pollution Reports, 1*(1), 35–46. https://doi.org/10.1007/s40726-015-0004-2

Singh, J., Singh, S., Vig, A. P., & Kaur, A. (2018). Environmental influence of soil toward effective vermicomposting. In S. Ray (Ed.), *Earthworms—The ecological engineers of soil.* Intech Open.

Singh, R., Gautam, N., Mishra, A., & Gupta, R. (2011). Heavy metals and living systems: An overview. *Indian Journal of Pharmacology, 43*(3), 246. https://doi.org/10.4103/0253-7613.81505

Singh, S. I., Singh, W. R., Bhat, S. A., Sohal, B., Khann, N., Vig, A. P., Ameen, F., & Jones, S. (2022). Vermiremediation of allopathic pharmaceutical industry sludge amended with cattle dung employing *Eisenia fetida. Environmental Research, 214*(1), 1113766.

Sinha, R. K., Chandran, V., Soni, B. K., Patel, U., & Ghosh, A. (2012). Earthworms: Nature's chemical managers and detoxifying agents in the environment: An innovative study on the treatment of toxic wastewaters from petroleum industry by vermifiltration technology. *The Environmentalist, 32*(4), 445–452.

Skupinska, K., Misiewicz, I., & Kasprzycka-Guttman, T. (2004). Polycyclic aromatic hydrocarbons: Physicochemical properties, environmental appearance and impact on living organisms. *Acta Poloniae Pharmaceutica, 61*, 233–240.

Soleimani, M. (2014.) Comparison of biological and thermal remediation methods in decontamination of oil polluted soils. *Journal of Bioremediation and Biodegradation.*

Soobhany, N., Mohee, R., & Garg, V. K. (2015). Comparative assessment of heavy metals content during the composting and vermicomposting of municipal solid waste employing *Eudrilus eugeniae. Waste Management, 39*, 130–145.

Speira, T. W., Kettlesb, H. A., Percivalc, H. J., & Parshotam, A. (1999). Is soil acidification the cause of biochemical responses when soils are amended with heavy metal salts? *Soil Biology and Biochemistry, 31*, 1953–1961.

Srogi, K. (2007). Monitoring of environmental exposure to polycyclic aromatic hydrocarbons: A review. *Environmental Chemistry Letters, 5*, 169–195. https://doi.org/10.1007/s10311-007-0095-0

Suman, S., Sinha, A., & Tarafdar, A. (2016). Polycyclic aromatic hydrocarbons (PAHs) concentration levels, pattern, source identification and soil toxicity assessment in urban traffic soil of Dhanbad India. *Science of the Total Environment, 545*, 353–360. https://doi.org/10.1016/j.scitotenv.2015.12.061

Sun, J., Pan, L., Tsang, D. C. W., Zhan, Y., Zhu, L., & Li, X. (2018). Organic contamination and remediation in the agricultural soils of China: A critical review. *Science of the Total Environment, 615*, 724–740.

Tangahu, B. V., Abdullah, S., Rozaimah, S., Basri, H., Idris, M., Anuar, N., et al. (2011). A review on heavy metals (As, Pb, and Hg) uptake by plants through phytoremediation. *International Journal of Chemical Engineering, 2011*, 1687–1806. https://doi.org/10.1155/2011/939161

Tarafdar, A., Chawda, S., & Sinha, A. (2018). Health risk assessment from polycyclic aromatic hydrocarbons (PAHs) present in dietary components: A meta-analysis on a global scale. *Polycyclic Aromatic Compound, 40*, 850–861. https://doi.org/10.1080/10406638.2018.1492426

Tchounwou, P. B., Yedjou, C. G., Patlolla, A. K., & Sutton, D. J. (2012). Heavy metal toxicity and the environment. *Experimental Supplement, 101*, 133–164. https://doi.org/10.1007/978-3-7643-8340-4_6

Theron, A. J., Tintinger, G. R., & Anderson, R. (2012). Harmful interactions of non-essential heavy metals with cells of the innate immune system

Tobiszewski, M., & Namiesnik, J. (2012). PAH diagnostic ratios for the identification of pollution emission sources. *Environmental Pollution, 162*, 110–119.

U.S. EPA. (2003). *Integrated risk information system.* Available at: https://www.epa.gov/iris. Accessed October 10, 2022.

UBC Wiki. (2020, June 16). Retrieved September, 2022, from https://wiki.ubc.ca/Determining_Soil_Texture

United States Environmental Protection Agency. (2012). *Solid waste and emergency response, EPA 542-R-92–011. Number 45 Oct. 2000.* United States Environmental Protection Agency Office of Research and Development National Risk Management Research Laboratory.

United States Environmental Protection Agency, Polycyclic Aromatic Hydrocarbons (PAHs). (2008). http://www.epa.gov/osw/hazard/wastemin/priority.htm

USMicrobics. (2003). *Annual report FY-2003.*

Wei, Y., Zhao, Y., Zhao, X., Gao, X., Zheng, Y., & Wei, Z. (2020). Roles of different humin and heavy-metal resistant bacteria from composting on heavy metal removal.

White, P. A., & Claxton, L. D. (2004). Mutagens in contaminated soil: A review. *Mutation Research, 567*, 227–345.

Wilcke, W. (2000). Polycyclic aromatic hydrocarbons in soil—A review. *Journal of Plant Nutrition and Soil Science, 163*, 229–248.

Wilcke, W. (2007). Global patterns of polycyclic aromatic hydrocarbons (PAHs) in soil. *Geoderma, 141*, 157–166.

Wilcke, W., Kiesewetter, M., & Bandowe, B. A. M. (2014). Microbial formation and degradation of oxygen-containing polycyclic aromatic hydrocarbons (OPAHs) in soil during short-term incubation. *Environmental Pollution, 184*, 385–390.

William Usher. (2021). Soil bio-renovation of heavily depleted agricultural soils. *Slide 11.*

Wu, H., Sun, B., & Li, J. (2019). Polycyclic aromatic hydrocarbons in sediments/soils of the rapidly urbanized lower reaches of the River Chaohu, China. *International Journal of Environmental Research and Public Health, 16*, 2302. https://doi.org/10.3390/ijerph16132302

Wuana, R. A., & Okieimen, F. E. (2011). Heavy metals in contaminated soils: A review of sources, chemistry, risks and best available strategies for remediation. *International Scholarly Research Notices, 2011*, 2090–4614. https://doi.org/10.5402/2011/402647

Xu, J., Liu, C., Hsu, P. C., Zhao, J., Wu, T., Tang, J., et al. (2019). Remediation of heavy metal contaminated soil by asymmetrical alternating current electrochemistry. *Nature Communications, 10*(1), 1–8.

Yao, H., Xu, J., & Huang, C. (2003). Substrate utilization pattern, biomass and activity of microbial communities in a sequence of heavy metal-polluted paddy soils. *Geoderma, 115*, 139–148.

Zelinkova, Z., & Wenzl, T. (2015). The occurrence of 16 EPA PAHs in food- a review. *Polycyclic Aromatic Compounds, 35*, 248–284. https://doi.org/10.1080/10406638.2014.918550

Zeng, L. S., Liao, M., Chen, C. L., & Huang, C. Y. (2007). Effects of lead contamination on soil enzymatic activities, microbial biomass, and rice physiological indices in soil-lead-rice (Oryza sativa L.) system. *Ecotoxicology and Environmental Safety, 67*(1), 67–74. https://doi.org/10.1016/j.ecoenv.2006.05.001

Zhang, J., Fan, S., Du, X., Yang, J., Wang, W., & Hou, H. (2015). Accumulation, allocation, and risk assessment of polycyclic aromatic hydrocarbons (PAHs) in soil-*Brassica chinensis* system. *PLoS ONE, 10*, e0115863-e115916. https://doi.org/10.1371/journal.pone.0115863

Sustainable Management of Agro-industrial Waste Using Vermitechnology

Rahil Dutta, Anu Bala Chowdhary, Surbhi Sharma, Jahangeer Quadar, Raman Tikoria, Jaswinder Singh, and Adarsh Pal Vig

Abstract Agricultural, agro-industrial, and cellulose-based industrial wastes often remain underutilized, posing environmental challenges such as noxious odor emissions, extensive land occupation, and contamination of groundwater and surface water. These waste materials harbor the potential to serve as a sustainable and scientifically managed source of renewable energy. Over the past few decades, vermicomposting technology has emerged as an environmentally sound approach for efficiently harnessing agro-industrial processing waste, transforming it into value-added products conducive to land restoration efforts. The output of the vermicomposting process, known as vermicompost, possesses properties akin to humus, characterized by fine granulation and friability. Vermicompost can be effectively employed as a soil conditioner, reintroducing organic matter into agricultural soils. The vermicomposting process enhances the concentrations of essential nutrients such as nitrogen, potassium, and phosphorus in the compost, concurrently mitigating the risk of phytotoxicity by reducing the levels of toxic elements within the compost. The efficacy of the vermicomposting process is contingent upon several variables, including the nature of the raw materials and various process parameters such as pH, temperature, moisture levels, aeration, the type of vermicomposting system employed, and the

R. Dutta · A. B. Chowdhary · S. Sharma · J. Quadar · A. P. Vig (✉)
Department of Botanical and Environmental Sciences, Guru Nanak Dev University, Amritsar, Punjab, India
e-mail: dr.adarshpalvig@gmail.com

R. Dutta
e-mail: rahildutta02@gmail.com

R. Dutta
Department of Environmental Sciences, GDC Reasi J&K, Reasi, India

R. Tikoria
Department of Zoology, Guru Nanak Dev University, Amritsar, Punjab, India

J. Singh
Post Graduate Department of Zoology, Khalsa College, Amritsar, Punjab, India

A. P. Vig
Punjab Pollution Control Board (PPCB), Vatavaran Bhawan, Patiala, Punjab, India

S. A. Aransiola and N. R. Maddela (eds.), *Vermitechnology: Economic, Environmental and Agricultural Sustainability*, https://doi.org/10.1007/978-3-031-96212-7_11

species of earthworms utilized. This chapter provides a succinct overview of vermicomposting process technology and presents the state of research in vermicomposting for the treatment of agro-industrial processing wastes.

Keywords Earthworm · Vermicomposting · Agro-industrial waste · Sustainable development

1 Introduction

In the present era, waste management has become a serious environmental issue due to increasing quantities of waste caused by fast urbanization, industrialization, and population expansion (Ahmed et al., 2023; Bhat et al., 2018). According to a report published in 2018, the yearly output of municipal solid garbage throughout the world was roughly 2.01 billion tonnes, with over 33% not being handled in a way that was safe for the environment. Additionally, it is predicted that the amount of global waste will increase to 3.4 billion tons by 2050 (Kaza et al., 2018). This highlights the urgent need for effective waste management strategies, including recycling and composting, to mitigate the adverse effects of municipal solid waste on public health and the environment.

The management of agro-industrial waste is a critical challenge confronting numerous countries, particularly those with significant agricultural and industrial activities. These wastes can come from agricultural activities such as crop residues, animal manure, and agroprocessing waste such as rice husks, sugarcane bagasse, and sawdust. Improper management of these wastes can lead to environmental pollution, soil degradation, and health hazards. In addition, agro-industrial waste can also contribute to greenhouse gas emissions, contributing to climate change (Devi et al., 2022; Leh-Togi Zobeashia et al., 2018). Effective sustainable waste mitigation and management strategies such as anaerobic digestion, composting, and utilization of waste for energy production can help reduce the environmental effect of agro-industrial waste and encourage sustainable agriculture practices (Koul et al., 2022). Governments, private sector, and other stakeholders need to work together to develop and implement effective waste management policies and practices to mitigate the negative effects of agro-industrial waste.

Vermicomposting is a type of composting method in which earthworms are utilized to convert organic materials into a high-quality final organic manure. The aerobic mesophilic biooxidation of the organic waste during vermicomposting involves the coaction of earthworms and microbes that transforms waste into valuable manure (vermicast) (Aransiola et al., 2024; Chen et al., 2023; Swati & Hait, 2018). During the process of vermicomposting, mineralization of waste results in liberation of CO_2, heat, and water (Fig. 1). In addition to biodegradation process, the earthworm and microbial respiration contribute to the release of CO_2. Earthworm castings are rich in plant growth-promoting hormones and nutrients (Ahmed & Al-Mutairi, 2022; Aransiola et al., 2022).

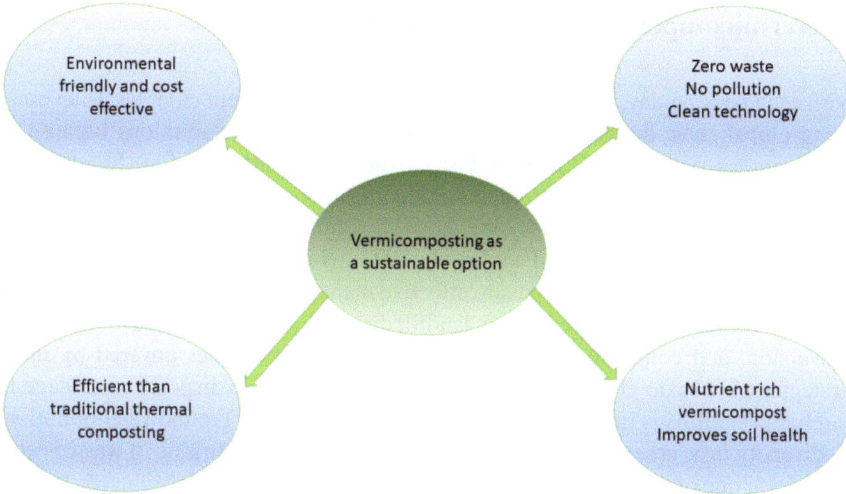

Fig. 1 Vermicomposting process as a sustainable option for solid waste management

As part of the vermicomposting process, earthworms play a vital role in breaking down and aerating the substrate, thereby increasing the available surface area for microbial activity (Bhat et al., 2018). During the vermicomposting process, the synergistic interaction between earthworms and microbes leads to the secretion of a diverse range of enzymes, including amylases, chitinases, proteases, lipases, cellulases, and more. This enzymatic activity facilitates the conversion of organic materials, thereby making previously unavailable micro- and macronutrients accessible to plants in forms that can be readily absorbed and utilized (Babaniyi et al., 2023; Bhat et al., 2018; Dutta et al., 2023). Earthworms contribute to the production of coelomic fluids, which possess antimicrobial properties capable of eliminating parasites and harmful bacteria found within the waste material and producing pathogen- and odor-free vermicompost (Aransiola et al., 2023; Ojuolape et al., 2015; Sinha et al., 2010). Vermicomposting has been proven to be a superior approach to conventional composting. Vermicomposting is faster than traditional composting because it uses the ability of earthworms and microorganisms to metabolize organic components (Thakur et al., 2021). Additionally, the collaborative action of earthworms and microbes in the vermicomposting process leads to an accelerated rate of mineralization and humification. As a result, the final product, vermicompost, exhibits significantly higher nutritional value compared to conventional thermal composting methods (Bhat et al., 2018). Different agro-industrial residues are found to be viable feedstocks for earthworms (Garg & Gupta, 2009). Different studies have proved the potential of vermiremediation in treating waste from various industries (Maharjan et al., 2022). As a result of its low cost, simplicity, and improved nutrient availability, vermitechnology is one of the most appropriate technologies for sustainable and efficient management of all types of organic waste.

2 Vermicomposting and Vermiculture

With each passing day, the severity of solid waste management emerges as an escalating global issue demanding urgent attention and effective solutions because of the enormous rise in population, industrialization, and urbanization (Singh et al., 2011). Several environmental problems, including unpleasant odors and groundwater contamination, are brought on by the massive volumes of organic waste produced globally (Chia et al., 2020; El-Fadel et al., 1997). Numerous approaches have been suggested and implemented to ensure effective management of solid waste. Material recovery, recycling, source reduction, landfill disposal, incineration, waste-to-energy conversion, and composting are a few of the many techniques covered by these strategies (Abubakar et al., 2022; Ojuolape et al., 2015). The disposal of garbage in unreliable landfills is linked to a number of issues, including groundwater pollution, biodiversity loss, greenhouse gas emissions, etc., while open burning of waste is also considered one of the major problems of solid waste management (Mohan & Joseph, 2021).

The productivity of soil can be maintained with the help of various micro- and macroflora species, which have the ability to convert waste material into usable plant nutrients. Several types of remedial technologies are being employed to decompose waste material. Among the multitude of sustainable and environmentally friendly practices today, vermicomposting stands out as a remarkable process (Ravindran et al., 2021; Sharma et al., 2019). The technique of vermicomposting employs the biodegradation activity of earthworms to balance the nutrient flow from one system to another (Sim & Wu, 2010). The detailed process of vermicomposting has been illustrated in Fig. 2a, b.

2.1 Earthworms and Types of Earthworms Used in Vermicomposting

Earthworms are classified as cylindrical-shaped, simple, and segmented coelomates (Table 1). They are known to have rounded, elongated bodies, and sharply shaped heads. Moreover, their distinguishing characteristic lies in the absence of both cartilage and bones. Earthworms are devoid of appendages, but they have various hooks that match the chaetae for catching the substrate. Earthworms exhibit remarkable flexibility and locomotion abilities, owing to the presence of circular rings encircling their delicate and moist bodies. By utilizing the motion of their setae, these creatures efficiently navigate their surroundings, propelling themselves forward through a rhythmic back-and-forth movement (Dutta et al., 2021; Prasad & Kashyap, 1989).

Vermicomposting is a biotic process that involves the participation of various microorganisms, including bacteria, fungi, and oligochaetes, in the biochemical decomposition of organic substrates (Domínguez et al., 2010). The process commences with the inoculation of the organic substrate into a container filled with

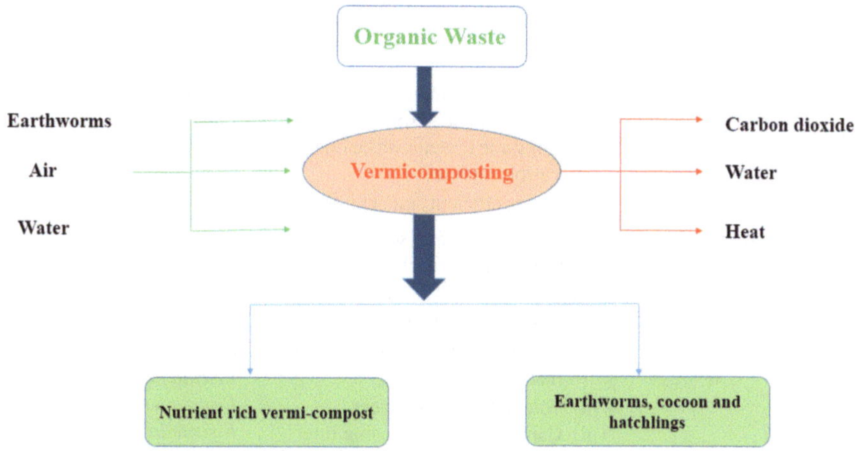

(a)

(b)

Fig. 2 Mechanism (**a**) and process of vermicomposting (**b**)

bedding material, which acts as a habitat for the earthworms. As the earthworms feed on the organic substrate, they generate excreta in the form of castings. Castings exhibit elevated concentrations of essential plant macronutrients, such as phosphorus, nitrogen, potassium, as well as enriched microbial communities that facilitate the breakdown of the organic matter (Vuković et al., 2021). The vermicomposting process is modulated by the physicochemical parameters of the environment, such as moisture content, pH, and temperature. Thus, the proficient management of these

Table 1 Classification and characteristics of different types of earthworms

S. No	Classification	Characteristic features	Examples
1	Anecic	(i) Reside in permanent, vertical burrows in deep soil (ii) Consume organic waste on the topsoil surface (iii) Act as efficient soil aerators	*Lampito mauritii* *Lumbricus terrestris*
2	Endogeic	(i) Reside in mineral soil layers and horizontally branched burrows (ii) Feed on mineral soil particles and decaying organic matter	*Metaphire posthuma* *Pontoscolex corethrurus* *Megascolex konkanensis*
3	Epigeic	(i) Capable of biodegrading organic matter (ii) Tolerant to disturbances (iii) Capable of decomposing litter	*Eudrilus eugeniae* *Eisenia andrei* *Eisenia fetida* *Eiseniella tetraedra*

Adapted from Ali et al. (2015), Dutta et al. (2021)

factors is fundamental to the achievement of successful vermicomposting (Loehr et al., 1985). Vermicomposting represents a sustainable and eco-friendly means of waste management, offering the benefits of waste reduction and the production of an effective soil amendment for use in agricultural and horticultural contexts.

Earthworms have been referred to as the "unheralded soldiers of mankind" because they are known to digest a variety of organic materials. The density of earthworms is a representative of soil degradation, as earthworm density decreases with the enhancement of soil degradation (Pathma & Sakthivel, 2012). Earthworms involve themselves in various mechanisms of soil formation and humus accumulation (Blouin et al., 2013). The internal system of earthworms contains an array of microorganisms, certain hydrolytic enzymes (cellulase, amylase, lipase, protease, and urease), and hormones that prove helpful in the fast decomposition of wastes which further transforms the complex organic matter into final product called "vermicompost" in a comparatively shorter duration than traditional composting mechanisms (Dutta et al., 2021; Singh et al., 2016).

Depending upon their feeding behavior, physiological traits, morphological characteristics, and burrowing mechanisms, they are classified as anecic, endogeic, and epigeic. Table 2 presents a comprehensive compilation of various earthworm species, showcasing their distinct characteristics and features.

3 Factors Affecting Vermicomposting

The effectiveness of the vermicomposting process depends on a number of control variables viz., moisture, pH, temperature, carbon-to-nitrogen (C:N) ratio, and aeration, which regulate the continuity of the vermicomposting process (Ali et al., 2015;

Table 2 Vermiculture properties of some earthworm species

Earthworm species	Temperature (°C)	Cocoon generation age (weeks)	Upper limit of soil temperature tolerance	Vermistabilization time (weeks)	No. of young cocoons	Incubation period (weeks)	Average size (g)
Eisenia fetida	18–25	5–9	25	6–8	2–4	3–4	0.5
Perionyx excavatus	20–25	7–10	30	3–4	2–3	4	1
Eudrilus eugeniae	25–30	15–18	30	4–5	1	4	1

Adapted from Garg and Gupta (2009), Ali et al. (2015)

Kumar, 2011; Vuković et al., 2021) (Table 3). The detailed explanation of the factors affecting vermicomposting is discussed below:

(a) **Moisture**

The optimal moisture condition in which the earthworms can thrive efficiently is ranged between 60 and 70%. Various factors that affect the moisture content in vermicomposting include the physical characteristics of the waste material used, its porosity, and the system in which the process of vermicomposting is carried out. By periodically sprinkling water, the moisture content is maintained throughout the process of vermicomposting (Ali et al., 2015; Garg & Gupta, 2009).

(b) **pH**

The efficiency of earthworms is maximum in the pH range of 5–9. Furthermore, a range of 7.5–8 is considered to be optimal for the proper functioning of earthworms. The increase in the pH leads to the release of high amount of ammonia gas that results in unpleasant odors from the feed mixtures. The pH values are first lowered by the creation of CO_2 and organic acids; as the process moves forward, the pH values rise due to the breakdown of proteins (Suthar & Singh, 2008).

(c) **Temperature**

Scientific observations have unequivocally demonstrated that temperature serves as the paramount factor influencing the development, metabolism, and growth of earthworms. Extensive research reveals that a vast majority of earthworm species exhibit optimal performance within a temperature range of 25–35 °C for the vermicomposting process. When the temperature goes beyond the optimal temperature,

Table 3 Factors affecting vermicomposting

Factor	Description
Temperature	The optimal temperature range for vermicomposting is 25–35 °C. Higher or lower temperatures can negatively affect worm activity and the decomposition of organic matter
Moisture	The moisture content of the composting material should be 60–70%. Too much moisture can lead to anaerobic conditions and odors, while too little moisture can slow down the decomposition process
pH	The pH range for vermicomposting should be 6.0–8.0. Acidity or alkalinity outside of this range can negatively affect worm activity and the decomposition of organic matter
Carbon-to-nitrogen ratio	The optimal carbon-to-nitrogen (C:N) ratio for vermicomposting is between 20:1 and 30:1. Too much carbon can lead to slow decomposition, while too much nitrogen can result in odors and a buildup of ammonia
Oxygen	Vermicomposting requires adequate oxygen for the worms and microorganisms to breathe. Proper aeration is essential to prevent anaerobic conditions and to promote decomposition

metabolic activities of earthworms become unstable that ultimately leads to the death of the earthworm species. Different earthworm species respond to temperature in different ways. *Dendrobaena veneta*, for example, exhibits superior growth performance at lower temperatures and exhibits a lower tolerance for high temperatures compared to *Eisenia fetida*. *Eisenia fetida* demonstrates optimal growth at 25 °C and exhibits a wider temperature tolerance range of 0–35 °C. Although *Eudrilus eugeniae* and *Perionyx excavatus* similarly demonstrate optimal development at around 25 °C, within a range of temperatures typically spanning from 9 to 35 °C, these earthworms exhibit their characteristic tolerance for environmental conditions (Ali et al., 2015; Samal et al., 2019).

(d) **Carbon-to-Nitrogen Ratio (C:N)**

The C:N ratio is important for a number of processes in earthworms, including cell formation, growth, and metabolism. Carbon and nitrogen should be provided as substrates in the right proportions for healthy nutrition (Ndegwa & Thompson, 2000). The C:N ratio, one of the most crucial factors in waste stabilization, determines the index for compost maturity. When the substrate's initial C:N ratio is 25, the increased compost maturity is evident with a C:N ratio of less than 20. Microbial respiration, driven by rapid mineralization and the breakdown of organic matter, results in the release of carbon dioxide, leading to the loss of carbon from the system. Earthworms also increase nitrogen levels by producing mucus, which lowers the ratio of carbon to nitrogen overall. However, the initial nitrogen levels in the substrate have a major role in determining the vermicompost's ultimate N content and the overall level of breakdown. Since nitrogen is lost as volatile ammonia at high pH, pH drop also has a significant impact on nitrogen retention (Suthar, 2009a, 2009b).

(e) **Aeration**

An adequate oxygen supply is necessary for a vermicomposting system to function. Worms have high rates of death due to a combination of factors, including oxygen deprivation and the production of poisonous chemicals (such as ammonia and other phytotoxic metabolites) (Garg & Gupta, 2009). By physically turning the substrate biomass on a regular basis or by mechanically mixing it, proper aeration might be accomplished.

4 Vermicompost as Organic Manure

Vermicompost, a unique organic fertilizer, is meticulously created through the controlled decomposition of organic waste by the collaborative efforts of earthworms, microorganisms, and other soil fauna. Vermicomposting process is efficient in breaking down a wide spectrum of organic substrates, including green waste, food waste, and agricultural residues, into nutrient-rich manure. The physical properties of vermicompost include a dark brown, crumbly texture with a pleasant earthy aroma.

It is a well-aerated material with good water-holding capacity, which helps improve soil structure and fertility. Vermicompost has a high nutrient content, with a good mix of macronutrients (phosphorus, nitrogen, and potassium) and micronutrients (iron, sulfur, calcium, magnesium, zinc, and copper). The nutrient composition of vermicompost exhibits variability, influenced by both the specific feedstock utilized and the prevailing composting conditions. However, in general, vermicompost has been reported to have higher nutrient concentrations than traditional compost. The nutrients in vermicompost are more readily available to plants, as they are present in soluble forms and are released slowly over time, thereby reducing the risk of leaching and nutrient runoff.

Vermicompost is also a rich source of beneficial microbes, such as bacteria, fungi, actinomycetes, and protozoa. These microorganisms play a crucial role in soil health by decomposing organic matter, releasing plant-available nutrients, suppressing plant pathogens, and improving soil structure. The presence of these microbes in vermicompost enhances soil fertility, stimulates plant growth, and improves plant resistance to stress. The application of vermicompost to soil has numerous benefits for plant growth and soil health. The organic matter in vermicompost helps improve soil porosity, structure, and water-holding capacity, leading to better root growth and nutrient absorption by plants. The slow release of nutrients from vermicompost also ensures a steady supply of plant nutrients over an extended period, which reduces the need for additional fertilizer applications. The beneficial microorganisms in vermicompost help suppress soil-borne diseases, improve plant resistance to pests and environmental stresses, and enhance the overall health and productivity of the soil–plant system (Aransiola et al., 2022).

5 Vermicomposting of Agro-industrial Wastes

Vermicomposting, a unique process that harnesses the power of earthworms to transform organic waste into a valuable compost enriched with nutrients, has gained attention as a sustainable solution for agro-industrial waste management. Agro-industrial waste is a potential source of organic matter and plant nutrients that can be effectively processed through vermicomposting (Table 4). This approach not only reduces waste volumes and associated environmental impacts but also generates a valuable end product that enhances crop productivity and soil fertility. Vermicomposting thus presents a cost-effective and environmentally friendly option for agro-industrial waste management, promoting circular economy principles in agriculture.

Sangwan et al. (2010) conducted a study utilizing the epigeic earthworm *E. fetida* to explore the potential of vermicomposting for converting waste materials from the sugar industry, specifically pressmud (PM) in combination with cow dung (CD), into vermicompost. The results showed that while 100% CD produced the greatest earthworm growth, a 1:1 PM and CD ratio still allowed sustained growth and reproduction. Higher PM percentages in reactors negatively affected earthworm growth

Table 4 Agro-industrial processing wastes tested for vermicomposting

S. No.	Agro-industrial waste	Organic amendment	Earthworm species
1	Biosolid vinasse and vine shoots	Vermicompost	*E. fetida*
2	Empty fruit bunches from palm oil mill	Cow dung	*E. eugeniae*
3	Filter cake from sugarcane factory		*E. fetida*
4	Grape marc from winery industry	Mature vermicompost	*E. andrei*
5	Pressmud from sugarcane industry	Cow dung	*E. fetida*
6	Dairy sludge	(a) Cereal straw, (b) wood shavings	*E. andrei*
7	Lignocellulosic wastes		*E. fetida*
8	Paper-pulp mill sludge	Cattle manure	*E. andrei*
9	Paper-pulp mill sludge	(a) Pig slurry, (b) poultry slurry, (c) sewage sludge	*E. andrei*
10	Paper-pulp mill sludge	Brewery yeast	*L. terrestris*

Adopted from Garg and Gupta (2009), Lim et al. (2016)

and fecundity. The process of vermicomposting led to a reduction in carbon concentration while increasing the concentrations of phosphorus, nitrogen, and calcium in the resulting vermicompost. This study demonstrates that vermicomposting is an environmentally friendly and practical solution for effectively managing pressmud (PM) waste and generating valuable fertilizer material. It is important to note that the optimal mixture for successful vermicomposting involves a maximum of 50% pressmud (PM) combined with cow dung (CD).

In a study conducted by Bhat et al. (2016), the potential of vermicomposting using the earthworm *E. fetida* to reduce genotoxicity in postvermicompost feed mixtures of bagasse waste was investigated. Different proportions of bagasse waste mixed with cattle dung (B0, B25, B50, B75, and B100) were prepared on a dry-weight basis. The genotoxicity of the initial bagasse extracts and the postvermicompost extracts was evaluated using the *Allium cepa* root chromosomal aberration assay. The results demonstrated that the postvermicompost extracts exhibited enhanced root length and mitotic index compared to the initial bagasse waste. Among the mixtures, the B50 extract showed the highest increase in root length (96.60%), while the B100 mixture exhibited the highest mitotic index (14.20 ± 0.60) after 6 h of treatment, which was similar to the control group. Furthermore, the genotoxicity analysis of the postvermicompost extracts revealed a significant reduction (21–44%) in aberration frequencies compared to the initial extracts. The highest reduction in genotoxicity was observed in the B75 extract (44.50%). These findings indicate that the vermicomposting process utilizing *E. fetida* effectively reduces the genotoxicity of bagasse waste. Therefore, vermicomposting can serve as a potential alternative for managing and mitigating the harmful effects of bagasse waste.

In a 90-day pilot-scale vermicomposting study conducted by Kumar et al. (2012), *Eudrilus eugeniae* worms were utilized along with various agro-industrial wastes as substrates. The researchers observed significant changes in different parameters when vermicomposting beds were prepared using a mixture of spent wash and pressmud (PS), combined with cow dung (CD) at different ratios. The vermicomposting process led to notable reductions in pH levels (ranging from 11.4 to 14.8%) and organic carbon content (ranging from 4.2 to 30.5%). Conversely, there were increases in total nitrogen content (ranging from 6 to 29%), available phosphorus content (ranging from 5 to 29%), exchangeable potash content (ranging from 6 to 21%), and turnover rate (ranging from 52 to 66%). Worm mortality was found to be highest when 100% pressmud was used as the substrate, indicating its adverse impact on the worms. On the other hand, the vermicompost produced from a ratio of 75% pressmud to 25% cow dung (PS:CD) exhibited the highest quality. Based on the results, it can be concluded that vermicomposting effectively decomposes pressmud when mixed with spent wash and enhances the quality of the resulting vermicompost.

In a 60-day research, Ganguly and Chakraborty (2021) assessed the efficiency of vermicomposting using *E. fetida* in handling primary waste (VCP) and secondary waste (VCS) from paper mills. To find the ideal circumstances, several mixtures of straw, cow dung, and sun-dried paper mill waste were investigated. The VCP1 and VCS1 experiment sets, which used paper mill waste, cow dung, and straw in a ratio of 5:4:1, were successful in preserving physiochemical parameters and supporting earthworm populations. The study demonstrated that vermicomposting led to enzymatic enrichment of the resulting vermicompost, with 19 enzymes aiding in the maintenance of the compost's carbon-to-nitrogen (C:N) ratio. Both VCP1 (Zn: 40% > Pb: 36% > Cr: 28% > Cu: 25%) and VCS1 (Zn: 44% > Pb: 41% > Cu: 19% > Cr: 13%) significantly reduced their heavy metal contents as a consequence of the vermicomposting process, with a clear link between the two being shown in the decline of the C:N ratio. The bioremediation process in the earthworm populations of VCP1 (Zn: 0.40 > Cu: 0.23 > Cr: 0.18 > Pb: 0.11) and VCS1 (Zn: 0.47 > Cu: 0.25 > Cr: 0.21 > Pb: 0.14) was further validated by monitoring the bioaccumulation factor (BAF). Based on its findings, the study hypothesized that adding straw, cow dung, and paper mill waste in a ratio of 5:4:1 might significantly improve the vermicomposting of various harmful organic paper mill wastes. Through vermicomposting, this strategy manages and enriches such waste products while lowering their heavy metal concentration and fostering bioremediation.

Badhwar et al. (2020) conducted a study focusing on vermicomposting paper mill sludge (PMS) and tea waste (TW) using *E. fetida* earthworms in combination with cow dung (CD). The main objective of the study was to employ an eco-friendly vermiconversion technique to address the waste disposal issue associated with PMS. Physicochemical parameters were carefully monitored at 30-day intervals throughout a 90-day period. The final pH values recorded for all vermicomposting units fell within the range of 6.09–6.95, indicating a suitable environment for the process. Following the vermicomposting process, total nitrogen (TN), total phosphorus (TP), and total potassium (TK) contents increased by factors ranging from 0.30 to 0.87, 0.53 to 3.23, and 0.33 to 0.63, respectively. This increase was accompanied by elevated electrical conductivity (EC) and ash content. The treatment containing the highest proportion of PMS exhibited the most significant reduction in total organic carbon (TOC), with a decrease of 23.91%. This reduction was attributed to the activity of the earthworms. Furthermore, all treatment combinations resulted in a noteworthy decrease in the carbon-to-nitrogen (C:N) ratio, ranging from 38.63 to 54.05%. Based on the study findings, it was concluded that a combination of paper mill sludge, tea waste, and cow dung can be effectively converted into valuable manure through vermicomposting facilitated by earthworms. This approach offers a sustainable solution for managing these waste materials while producing beneficial organic fertilizers.

In a study conducted by Dutta et al. (2023), the potential of vermiremediation was evaluated utilizing bovine dung (BD) and milk-processing industry sludge (MS) in various diet mixes. The synergistic impact of adding paddy straw biochar (BC) at a 10% application rate during vermicomposting was also studied by the researchers. The findings showed that in terms of earthworm quantity, biomass, cocoon formation, and hatchlings, the feed combinations MS25 and MS25BC10 (25:75 + 10% BC) displayed the lowest earthworm mortality and the maximum growth. Notably, compared to nonbiochar treatments, all feed mixes with biochar amendments demonstrated improved earthworm growth and reproduction. The pH, total organic carbon (TOC), and carbon-to-nitrogen (C:N) ratio all decreased in the final vermicompost generated from all feed mixes. Additionally, following vermicomposting, all feed mixes showed an increase in electrical conductivity (EC), total Kjeldahl nitrogen (TKN), total accessible phosphorus (TAP), total potassium (TK), total sodium (TNa), and ash content. All feed combinations had much lower heavy metal concentrations as a result of the biochar addition. According to the study's findings, adding biochar might improve mineralization, boost nutrient concentration, and lower heavy metal level, producing vermicompost of high quality that is appropriate for agricultural use.

The vermicomposting capacity of postharvest wastes from regional commodities such as wheat, millets, and pulses was examined in a study by Suthar (2009a, 2009b). Three different types of vermibeds were created by combining crop remnants with animal dung, and a fourth vermibed was created using manure from cow sheds. Vermicomposting resulted in a decrease in the amount of organic carbon and an increase in total nitrogen, accessible phosphorus, exchangeable potassium, and exchangeable calcium. Among the several vermibeds, the one containing cow stall dung had the highest growth rate, biomass gain, and cocoon formation. The potential of vermicomposting as a method for producing value-added agricultural waste products, such as vermicompost and worm biomass, which can support sustainable crop production, was emphasized in the paper.

6 Conclusion

The sustainable management of agro-industrial waste using vermitechnology holds great promise for addressing the dual challenges of waste reduction and soil enrichment. This approach not only contributes to environmental preservation but also supports the transition toward more sustainable and resilient agricultural systems. Furthermore, the use of earthworms in waste management aligns with the principles of circular economy and sustainability. By converting waste materials into a valued resource (vermicompost), vermitechnology contributes to closing the nutrient loop and reducing the reliance on synthetic fertilizers, which often have negative environmental impacts. As we look to the future, it is imperative to further explore and harness the potential of vermitechnology while considering its broader implications for agricultural sustainability and environmental well-being.

References

Abubakar, I. R., Maniruzzaman, K. M., Dano, U. L., AlShihri, F. S., AlShammari, M. S., Ahmed, S. M. S., & Alrawaf, T. I. (2022). Environmental sustainability impacts of solid waste management practices in the global South. *International Journal of Environmental Research and Public Health, 19*(19), 12717.

Ahmed, F., Hasan, S., Rana, M. S., & Sharmin, N. (2023). A conceptual framework for zero waste management in Bangladesh. *International Journal of Environmental Science and Technology, 20*(2), 1887–1904.

Ahmed, N., & Al-Mutairi, K. A. (2022). Earthworms effect on microbial population and soil fertility as well as their interaction with agriculture practices. *Sustainability, 14*(13), 7803.

Ali, U., Sajid, N., Khalid, A., Riaz, L., Rabbani, M. M., Syed, J. H., & Malik, R. N. (2015). A review on vermicomposting of organic wastes. *Environmental Progress & Sustainable Energy, 34*(4), 1050–1062.

Aransiola, S. A., Afolabi, F., Joseph, F., & Maddela, N. R. (2023). Soil enzymes: Distribution, interactions and influencing factors. In P. Jeschke, & E. B. Starikov (Eds.), *Agricultural biocatalysis: Enzymes in agriculture and industry* (1st ed., pp. 303–333). Jenny Stanford Publishing. ISBN 9789814968478. https://www.routledge.com/Agricultural-Biocatalysis-Enzymes-in-Agriculture-and-Industry/Jeschke-Starikov/p/book/9789814968478

Aransiola, S. A., Ijah, U. J. J., Abioye, O. P., & Bala, J. D. (2022). Vermicompost-assisted phytoremediation of toxic trace element-contaminated soil in Madaka, Nigeria using *Melissa officinalis* L and *Sida acuta*. *International Journal of Environmental Science and Technology*. https://doi.org/10.1007/s13762-022-04105-y

Aransiola, S. A., Ijah, U. J. J., Abioye, O. P., Bala, J. D., Rivadeneira-Mendoza, R. F., Luque, R., Rodríguez-Díaz, J. M., & Maddela, N. R. (2024). Micro and vermicompost assisted remediation of heavy metal contaminated soils using phytoextractors. *Case Studies in Chemical and Environmental Engineering, 9*(2024), Article 100755. https://doi.org/10.1016/j.cscee.2024.100755

Babaniyi, G. G., Olagoke, O. E., & Aransiola, S. A. (2023). Extracellular enzymatic activity of bacteria in aquatic ecosystems. In N. R. Maddela, L. Kretli, W. Eller, R. Prasad (Eds.), *Microbiology for cleaner production and environmental sustainability*. CRC Press. ISBN: 9781032496061. https://www.routledge.com/Microbiology-for-Cleaner-Production-and-Environmental-Sustainability/Maddela-Eller-Prasad/p/book/9781032496061

Badhwar, V. K., Singh, S., & Singh, B. (2020). Biotransformation of paper mill sludge and tea waste with cow dung using vermicomposting. *Bioresource Technology, 318*, Article 124097.

Bhat, S. A., Singh, J., & Vig, A. P. (2016). Genotoxicity reduction in bagasse waste of sugar industry by earthworm technology. *Springerplus, 5*, 1–9.

Bhat, S. A., Singh, S., Singh, J., Kumar, S., & Vig, A. P. (2018). Bioremediation and detoxification of industrial wastes by earthworms: Vermicompost as powerful crop nutrient in sustainable agriculture. *Bioresource Technology, 252*, 172–179.

Blouin, M., Hodson, M. E., Delgado, E. A., Baker, G., Brussaard, L., Butt, K. R., & Brun, J. J. (2013). A review of earthworm impact on soil function and ecosystem services. *European Journal of Soil Science, 64*(2), 161–182.

Chen, Y., Zhang, Y., Shi, X., Shi, E., Zhao, T., Zhang, Y., & Xu, L. (2023). The contribution of earthworms to carbon mineralization during vermicomposting of maize stover and cow dung. *Bioresource Technology, 368*, Article 128283.

Chia, W. Y., Chew, K. W., Le, C. F., Lam, S. S., Chee, C. S. C., Ooi, M. S. L., & Show, P. L. (2020). Sustainable utilization of biowaste compost for renewable energy and soil amendments. *Environmental Pollution, 267*, Article 115662.

Devi, M. K., Manikandan, S., Oviyapriya, M., Selvaraj, M., Assiri, M. A., Vickram, S., Subbaiya, R., Karmegam, N., Ravindran, B., Chang, S. W., & Awasthi, M. K. (2022). Recent advances in biogas production using agro-industrial waste: A comprehensive review outlook of techno-economic analysis. *Bioresource Technology, 363*, 127871.

Domínguez, J., Aira, M., & Gómez-Brandón, M. (2010). Vermicomposting: Earthworms enhance the work of microbes. In *Microbes at work: From wastes to resources* (pp. 93–114).

Dutta, R., Angmo, D., Singh, J., Chowdhary, A. B., Quadar, J., Singh, S., & Vig, A. P. (2023). Synergistic effect of biochar amendment in milk processing industry sludge and cattle dung during the vermiremediation. *Bioresource Technology*, 128612.

Dutta, R., Chowdhary, A. B., Angmo, D., Quadar, J., Singh, J., & Vig, A. P. (2021). Vermicomposting of different organic wastes into organic manure: A review. *Journal Punjab Academy of Sciences, 21*(1), 01–14.

El-Fadel, M., Findikakis, A. N., & Leckie, J. O. (1997). Environmental impacts of solid waste landfilling. *Journal of Environmental Management, 50*(1), 1–25.

Ganguly, R. K., & Chakraborty, S. K. (2021). Valorisation of toxic paper mill waste through vermicomposting: An insight towards cleaner engineering through alleviation of wastes. *Cleaner Engineering and Technology, 2*, Article 100070.

Garg, V. K., & Gupta, R. (2009). Vermicomposting of agro-industrial processing waste. In *Biotechnology for agro-industrial residues utilization: Utilisation of agro-residues* (pp. 431–456). Springer.

Kaza, S., Yao, L., Bhada-Tata, P., & Van Woerden, F. (2018). *What a waste 2.0: a global snapshot of solid waste management to 2050*. World Bank Publications.

Koul, B., Yakoob, M., & Shah, M. P. (2022). Agricultural waste management strategies for environmental sustainability. *Environmental Research, 206*, Article 112285.

Kumar, S. (2011). Composting of municipal solid waste. *Critical Reviews in Biotechnology, 31*(2), 112–136.

Kumar, V. V., Shanmugaprakash, M., Aravind, J., & Namasivayam, S. K. R. (2012). Pilot-scale study of efficient vermicomposting of agro-industrial wastes. *Environmental Technology, 33*(9), 975–981.

Leh-Togi Zobeashia, S. S., Aransiola, S. A., Ijah, U. J. J., & Abioye, O. P. (2018). Anaerobic digestion and agricultural application of organic wastes. *Advances in Environmental Research, 7*(2), 73–85. http://www.techno-press.org/content/?page=article&journal=aer&volume=7&num=2&ordernum=1

Lim, S. L., Lee, L. H., & Wu, T. Y. (2016). Sustainability of using composting and vermicomposting technologies for organic solid waste biotransformation: Recent overview, greenhouse gases emissions and economic analysis. *Journal of Cleaner Production, 111*, 262–278.

Loehr, R. C., Neuhauser, E. F., & Malecki, M. R. (1985). Factors affecting the vermistabilization process: Temperature, moisture content and polyculture. *Water Research, 19*(10), 1311–1317.

Maharjan, K. K., Noppradit, P., & Techato, K. (2022). Suitability of vermicomposting for different varieties of organic waste: a systematic literature review (2012–2021). *Organic Agriculture*, 1–22.

Mohan, S., & Joseph, C. P. (2021). Potential hazards due to municipal solid waste open dumping in India. *Journal of the Indian Institute of Science, 101*(4), 523–536.

Ndegwa, P. M., & Thompson, S. A. (2000). Effects of C-to-N ratio on vermicomposting of biosolids. *Bioresource Technology, 75*(1), 7–12.

Ojuolape, O. T., Aransiola, S. A., & Obideyi, O. A. (2015). Vermitechnology: A solution to environmental problems; A review. *Journal of Global Ecology and Environment, 2454–2644*(3), 127–135. http://www.ikpress.org/abstract/4516

Pathma, J., & Sakthivel, N. (2012). Microbial diversity of vermicompost bacteria that exhibit useful agricultural traits and waste management potential. *Springerplus, 1*(1), 1–19.

Prasad, S. N., & Kashyap, V. (1989). *A textbook of vertebrate zoology*. New Age International.

Ravindran, B., Karmegam, N., Yuvaraj, A., Thangaraj, R., Chang, S. W., Zhang, Z., & Awasthi, M. K. (2021). Cleaner production of agriculturally valuable benignant materials from industry generated bio-wastes: A review. *Bioresource Technology, 320*, Article 124281.

Samal, K., Mohan, A. R., Chaudhary, N., & Moulick, S. (2019). Application of vermitechnology in waste management: A review on mechanism and performance. *Journal of Environmental Chemical Engineering, 7*(5), Article 103392.

Sangwan, P., Kaushik, C. P., & Garg, V. K. (2010). Vermicomposting of sugar industry waste (press mud) mixed with cow dung employing an epigeic earthworm *Eisenia fetida*. *Waste Management & Research, 28*(1), 71–75.

Sharma, B., Vaish, B., Singh, U. K., Singh, P., & Singh, R. P. (2019). Recycling of organic wastes in agriculture: An environmental perspective. *International Journal of Environmental Research, 13*(2), 409–429.

Sim, E. Y. S., & Wu, T. Y. (2010). The potential reuse of biodegradable municipal solid wastes (MSW) as feedstocks in vermicomposting. *Journal of the Science of Food and Agriculture, 90*(13), 2153–2162.

Singh, A., Tiwari, R., Sharma, A., Adak, A., Singh, S., Prasanna, R., & Singh, R. V. (2016). Taxonomic and functional diversity of the culturable microbiomes of epigeic earthworms and their prospects in agriculture. *Journal of Basic Microbiology, 56*(9), 1009–1020.

Singh, R. P., Singh, P., Araujo, A. S., Ibrahim, M. H., & Sulaiman, O. (2011). Management of urban solid waste: Vermicomposting a sustainable option. *Resources, Conservation and Recycling, 55*(7), 719–729.

Sinha, R. K., Herat, S., Bharambe, G., & Brahambhatt, A. (2010). Vermistabilization of sewage sludge (biosolids) by earthworms: Converting a potential biohazard destined for landfill disposal into a pathogen-free, nutritive and safe biofertilizer for farms. *Waste Management & Research, 28*(10), 872–881.

Suthar, S. (2009a). Bioremediation of agricultural wastes through vermicomposting. *Bioremediation Journal, 13*(1), 21–28.

Suthar, S. (2009b). Vermicomposting of vegetable-market solid waste using *Eisenia fetida*: Impact of bulking material on earthworm growth and decomposition rate. *Ecological Engineering, 35*(5), 914–920.

Suthar, S., & Singh, S. (2008). Vermicomposting of domestic waste by using two epigeic earthworms (*Perionyx excavatus* and *Perionyx sansibaricus*). *International Journal of Environmental Science & Technology, 5*(1), 99–106.

Swati, A., & Hait, S. (2018). Greenhouse gas emission during composting and vermicomposting of organic wastes–a review. *CLEAN–Soil, Air, Water, 46*(6), 1700042.

Thakur, A., Kumar, A., Kumar, C. V., Kiran, B. S., Kumar, S., & Athokpam, V. (2021). A review on vermicomposting: By-products and its importance. *Plant Cell Biotechnology and Molecular Biology, 22*, 156–164.

Vuković, A., Velki, M., Ečimović, S., Vuković, R., Štolfa Čamagajevac, I., & Lončarić, Z. (2021a). Vermicomposting—Facts, benefits and knowledge gaps. *Agronomy, 11*(10), 1952.

Vermitechnology—Application in Agricultural Sustainability

Phytoremediation and Vermicomposting

Ayodele Omotayo Eniola, Oluwasanmi Anuoluwapo Adeyemi, Josephine Amerley Well Tetteh, Shakirat Yewande Biyaosi, Tomisin Oyawoye, Oluwaseyi Peter Adewale, Semiratu Wakaso Abdullahi, and Ruth Makanjuola

Abstract Phytoremediation and vermicomposting are sustainable, eco-friendly, and cost-effective alternatives to conventional approaches for remediating both contaminants and waste from the environment. Remedying air, water, and soil from contaminants and wastes remains a major challenge worldwide. Phytoremediation removes, immobilizes, and/or degrades organic and inorganic contaminants in the environment using plants and associated microorganisms. On the other hand, vermicomposting recycles solid organic wastes and materials into nutrient-rich vermicompost using earthworms. With both phytoremediation and vermicomposting, the presence of hazardous contaminants and waste in the environment is greatly minimized. Consequently, the environmental and health implications of these contaminants and waste are curbed, and the environment is remediated.

Keywords Phytoremediation · Vermicomposting · Earthworm · Bioremediation · Wastes · Hazardous

A. O. Eniola (✉) · O. P. Adewale
Department of Microbiology, Atiba University, Oyo, Nigeria
e-mail: ayodeleomotayoeniola@gmail.com

O. A. Adeyemi
Department of Microbiology and Biotechnology, Ajayi Crowther University, Oyo, Nigeria

J. A. W. Tetteh
Department of Biological Sciences, University for Development School, Navrongo, Ghana

S. Y. Biyaosi
Department of Biological Sciences, Fountain University, Osogbo, Nigeria

T. Oyawoye
Department of Microbiology, University of Ilorin, Ilorin, Nigeria

S. W. Abdullahi
Federal Ministry of Health, Epidemiology Division, Abuja, Nigeria

R. Makanjuola
Department of Social Care, Health and Wellbeing, University of Bolton, Bolton, UK

S. A. Aransiola and N. R. Maddela (eds.), *Vermitechnology: Economic, Environmental and Agricultural Sustainability*, https://doi.org/10.1007/978-3-031-96212-7_12

1 Introduction

Owing to changeable natural processes and rapid anthropogenic activities, contaminants and wastes have accumulated in the environment (Khan et al., 2023a; Wani et al., 2023). Among others, the use of phosphate fertilizers, drilling of metals and amalgamation (Maqbool et al., 2022), pesticide application (Tarla et al., 2020), as well as food processing (Tripathi et al., 2020) are sources from which contaminants and wastes invade our soil, water, and air. The high accumulation of contaminants in the environment imposes significant environmental and health implications on plants, animals, and humans (Sabreena et al., 2022; Yan et al., 2020).

The presence of contaminants within agricultural soil impairs plant growth, reduces food production, and interferes with food supply (Alengebawy et al., 2021; Rashid et al., 2023). Also, metals can be part of the food we eat through absorption by crops and may pile up in the body over time, causing diseases and even deaths (Sabreena et al., 2022; Yan et al., 2020; Zaynab et al., 2022). The accumulation of chromium (Cr) in the human body causes cancer (Chen et al., 2022). The accumulation of zinc (Zn) results in kidney dysfunction (Maywald & Rink, 2022). Similarly, chronic exposure to lead (Pb) damages the nervous system and causes cardiovascular problems (Collin et al., 2022).

Phytoremediation and vermicomposting play huge roles in environmental cleanup. Contrasting to conventional remediation methods such as encapsulation and electrokinetics, phytoremediation and vermicomposting offer promising, economically feasible, and environmentally friendly technologies (Kafle et al., 2022; Wani et al., 2023; Yan et al., 2020). In addition to remediation benefits, the recovery of contaminants from plant tissues as a result of phytoremediation presents commercialization privileges (Alsafran et al., 2023; Zarull et al., 2002). Among others, there is a high possibility for the reuse of remediated (recovered) metals in phytomining.

Along with the above, vermicomposting of organic residuals does not just nullify the necessity of disposing of residuals; there are also the benefits of high-nutrient vermicompost (Ratnasari et al., 2023). Increasingly, vermicompost is being accepted as a substitute for chemical fertilizers in agriculture and horticulture (Rehman et al., 2023).

In this chapter, we:

- offer an overview of phytoremediation of organic and inorganic contaminants
- provide an overview of vermicomposting of organic wastes
- discuss mechanisms, benefits, and limitations of both phytoremediation and vermicomposting
- examine the combined application of phytoremediation and vermicomposting.

1.1 *Phytoremediation*

Phytoremediation is a plant-based approach that involves the capitalization of green plants to draw out elementary contaminants and/or reduce the bioavailability of contaminants within the environment (Khan et al., 2023a). Phytoremediation techniques are based on the abilities of some green plants to degrade, immobilize, bioaccumulate, and/or dissipate contaminants from the environment (Kafle et al., 2022; Sabreena et al., 2022). Pertinently, ornamental plants are becoming increasingly relevant in phytoremediation (Kelechi & Njoku, 2022; Rocha et al., 2022; Sadasivam, 2022). Largely, trees with high transpiration rates and large roots such as willow (*Salix* sp.), hemp (*Cannabis* sp.), and poplar (*Populus* sp.) are eminent in phytoremediation (Sadasivam, 2022).

Generally, plants with the potential for high metabolism (absorption capacity) and high biomass production are fundamental in phytoremediation (Hostyn et al., 2022; Khan et al., 2022). Also, plants with high tolerance to toxic effects of contaminants, hardy and less prone to herbivore infestations, are relevant (Hostyn et al., 2022). The appropriate selection of plants is pivotal to maximizing the benefits of phytoremediation (Kafle et al., 2022). Plants' effectiveness can be enhanced by amending the soil with ethylenediaminetetraacetic acid (EDTA) (Saman et al., 2022), biochar (Zhao et al., 2022), and endophytic microorganisms (Wani et al., 2023; Yan et al., 2020), among other soil amendments (Saman et al., 2022). Additionally, plants can be genetically engineered to enhance the effectiveness of phytoremediation (Yan et al., 2020).

Kafle et al. (2022) and Sharma et al. (2023) documented the following contaminants that can be effectively remediated during the phytoremediation process:

- Metalloids
- Metals
- Agrochemicals
- Chlorinated solvents
- Petroleum hydrocarbons
- Radiochemical elements
- Explosives
- Inorganic substances
- Organic waste.

1.1.1 Mechanisms of Phytoremediation

Tables 1 and 2 and Fig. 1 show mechanisms that plants adopt in sequestering, containing, and/or detoxifying organic and inorganic contaminants in phytoremediation. Basically, the mechanisms involve uptake, transformation, translocation, compartmentalization, and occasionally mineralization of contaminants from the environment. Different plants utilize different mechanisms, either in solitary or in combination, depending on the forms and types of contaminants and the environment. For this chapter, mechanisms that are specific to the soil will be elaborated (Table 1).

1.1.1.1 Phytodegradation

This is a degradation technique that involves the mineralization of organic contaminants by certain enzymes found within plant cells (Asante-Badu et al., 2020; Esposito et al., 2021). Through sunlight-driven chemical reactions, enzymes mineralize contaminants into simpler compounds needed for plant growth (Verma, 2022). Agrochemicals and chlorinated solvents are notable examples of organic contaminants that can be degraded through this mechanism (Table 3). Bhandari et al. (2021) reported the effective degradation of chlorinated solvents and pesticides by dehalogenase enzymes. Similarly, aniline and nitroaromatic compounds were effectively decomposed by volatile enzymes and nitroreductase enzymes, respectively (Bhandari et al., 2021).

Table 1 Mechanisms of phytoremediation by process (Sharma et al., 2023)

Process	Mechanisms	Types of contaminants
Degradation	• Phytodegradation • Rhizodegradation	• Organic • Organic
Accumulation	• Rhizofiltration • Phytoextraction	• Organic and inorganic • Inorganic
Dissipation	• Phytovolatilization	• Organic and inorganic
Immobilization	• Biological hydraulic containment • Phytostabilization	• Organic and inorganic • Organic and inorganic

Table 2 Mechanisms of phytoremediation with consideration to medium (Kafle et al., 2022)

Medium	Mechanisms
Soil, sludge, or sediment	• Phytodegradation • Rhizodegradation • Phytoextraction • Phytovolatilization • Phytostabilization
Wastewater and surface water	• Phytodegradation • Rhizodegradation • Rhizofiltration
Groundwater	• Phytodegradation • Rhizodegradation • Rhizofiltration • Phytovolatilization • Biological hydraulic containment

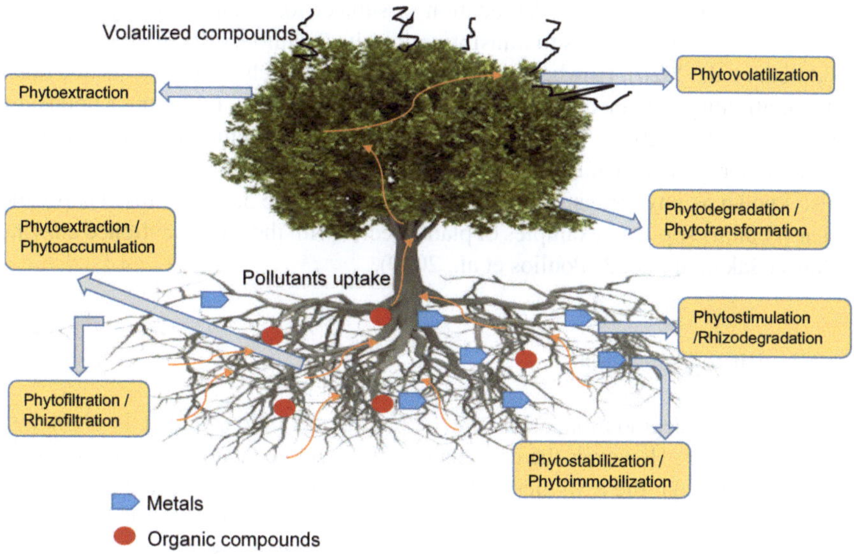

Fig. 1 Schematic representation of phytoremediation of inorganic and organic contaminants

Table 3 Examples of plants known to effectively utilize phytodegradation, rhizodegradation, and phytostabilization during phytoremediation

Mechanism	Plants		Contaminants	References
	Common name	Scientific name		
Phytodegradation	Horseradish Sage Poplar Four o'clocks French marigold	*Armoracia rusticana* *Salvia officinalis* *Populus* spp. *Mirabilis jalapa* *Tagetes patula*	Benzophenone Agrochemicals (such as herbicides) Saturated hydrocarbons	Maqbool et al. (2022) Poulios et al. (2020), Bak et al. (2023) Abdullah et al. (2020)
Rhizodegradation	Maize Sesbania Cucurbits European blackberry Bermuda grass	*Zea mays* *Sesbania cannabina* *Cucurbita* sp. *Rubus fruticosus* *Cynodon dactylon*	Petroleum hydrocarbons Polycyclic aromatic hydrocarbons Total petroleum hydrocarbons	Sivaram et al. (2020) Kafle et al. (2022) Njoku et al. (2022)
Phytostabilization	Sword lily Lovegrass Alyssum Haumaniastrum Black nightshade Willow	*Gladiolus* spp. *Eragrostis* spp. *Alyssum* spp. *Haumaniastrum* sp. *Solanum nigrum* *Salix* spp.	Metals (Ni, Cu, Cr, Zn, and Pb, among others)	Wani et al. (2023), Lacalle et al. (2023), Sharma et al. (2023), Fernández-Braña et al. (2023)

The effectiveness of phytodegradation is subject to a plant's ability to absorb and translocate contaminants to transformation sites, without effectuating cell death (Abdullah et al., 2020). Howbeit, studies have noted that chelating agents and other soil conditionings enhance contaminants' solubility, thereby aiding their uptake by plants (Yang et al., 2022). In addition, other remediation techniques are sometimes used in synergy with plant transformation (Dong et al., 2023). Phytodegradation is also known as phytotransformation. As shown in Table 3, sage (*Salvia* spp.) and poplar (*Populus* spp.) are examples of plant species with the aforementioned enzyme systems (Bak et al., 2023; Poulios et al., 2020).

1.1.1.2 Rhizodegradation

This is another degradation technique in phytoremediation, and it is similar to phytodegradation. However, the degradation of organic contaminants occurs in the rhizosphere through the activities of free-living symbiotic microorganisms that are sustained by root exudates (Sivaram et al., 2020; Wani et al., 2023). In a recent study, microbial association within the roots of Bermuda grass (*Cynodon dactylon*) resulted in 81% degradation of total petroleum hydrocarbons (TPHs) (Njoku et al., 2022). Similarly, polycyclic aromatic hydrocarbons (PAHs) were degraded by the red mangrove (*Rhizophora mangle*) in association with *Bacillus* sp. and *Pseudomonas aeruginosa* (Wu et al., 2023).

1.1.1.3 Phytostabilization

This is an immobilization technique in which plants absorb and precipitate contaminants (such as metals) found within plant roots or the rhizosphere, thereby limiting their mobility and bioavailability within the environment (Bakshe & Jugade, 2023; Sharma et al., 2023; Wani et al., 2023). The mechanism does not reduce contaminants' toxicity (Bakshe & Jugade, 2023; Sharma et al., 2023). Rather, through precipitation in the rhizosphere and/or accumulation by the roots, contaminants' movement in plants' unsaturated zone is restricted or prohibited (Mocek-Płóciniak et al., 2023).

Hence, the disadvantage of phytostabilization lies in the continuous existence of contaminants in the system (Sharma et al., 2023). Also, the need to regularly carry out surveillance to ascertain the continued immobility and reduced bioavailability of contaminants is a major drawback of phytostabilization (Sharma et al., 2023). Nevertheless, its potential to minimize soil erosion, decrease runoff, and inhibit the leaching of metals into groundwater are remarkable advantages (Sarkodie et al., 2022). Generally, phytostabilization improves soil fertility (Wani et al., 2023) and restores ecosystem functioning (Wani et al., 2023). Phytostabilization does not produce secondary wastes (Wani et al., 2023).

Phytostabilization comes in useful where phytoextraction is neither possible nor desirable (Wani et al., 2023). For instance, phytostabilization could be useful in limiting the off-site movement of metals from a barren contaminated site. Several studies have documented the effectiveness of phytostabilization in decreasing the mobility and bioavailability of metals and metalloids (Fernández-Braña et al., 2023; Lacalle et al., 2023; Sharma et al., 2023; Wani et al., 2023;). Sword lily (*Gladiolus* sp.) and alyssum (*Alyssum* sp.) are examples of metal-tolerant plants that have proven successful in phytostabilization (Table 3) (Wani et al., 2023). These plants have the ability to produce chelating agents (Wani et al., 2023). Phytostabilization is also known as phytoimmobilization.

1.1.1.4 Phytovolatilization

With this dissipation technique, plants absorb contaminants via their roots, transforming them into significantly diluted and less harmful forms, and subsequently releasing them into the atmosphere via their leaves (Yan et al., 2020). Naturally occurring plants capable of volatilizing metals into harmless organic compounds are very limited (Wani et al., 2023). Plants are mostly genetically modified for phytovolatilization (Wani et al., 2023). Cultivated tobacco (*Nicotiana tabacum*) and thale cress (*Arabidopsis thaliana*) are examples of genetically modified plants that have demonstrated effectiveness in phytovolatilization (Niedbała et al., 2021; Nilsson, 2022). Some other examples of plants used for phytovolatilization are listed in Table 4.

A limitation of this mechanism is that volatile compounds released into the atmosphere can undergo precipitation and consequently be redeposited into the environment (Khan et al., 2023b). However, as Meagher (2000) rightly posits, these compounds are adulterated, thereby inducing little or no environmental hazards. It has been noted that phytovolatilization undergoes photodegradation (Asante-Badu et al., 2020). Generally, phytovolatilization is considered a beneficial remediation technique (Asante-Badu et al., 2020). It is devoid of plant harvesting and biomass disposal (Yan et al., 2020). Phytovolatilization has been reported to eliminate arsenic (As), selenium (Se), mercury (Hg), and numerous noxious contaminants (Niedbała et al., 2021; Nilsson, 2022; Yan et al., 2020).

1.1.1.5 Phytoextraction

Phytoextraction, also referred to as phytoaccumulation, involves the absorption of contaminants by plant roots and their translocation and concentration aboveground in plant tissues (Asante-Badu et al., 2020). Typically, plants with a high tolerance for heavy contaminants, the hyperaccumulators, are planted on contaminated sites. After a while, the plants are harvested and incinerated, and the ash is used in landfills. To achieve a significant cleanup, numerous cycles (cultivating, harvesting, and incinerating) are used.

Poor biomass production and the slow growth rate of most hyperaccumulators are major limitations of phytoextraction (Bian et al., 2020; Yan et al., 2020). However,

Table 4 Examples of plants known to effectively utilize phytovolatilization and phytoextraction during phytoremediation

Mechanism	Plants		Contaminants	References
	Common name	Scientific name		
Phytovolatilization	Rabbitsfoot grass Cultivated tobacco Thale cress Tulip tree Two-grooved milkvetch Prince's plume Perennial reed grass	*Polypogon monspeliensis* *Nicotiana tabacum* *Arabidopsis thaliana* *Liriodendron tulipifera* *Astragalus bisulcatus* *Stanleya pinnata* *Phragmites australis*	Metalloids and Metals (especially As, Se, and Hg) Organochlorines (e.g., 1,4-dichlorobenzene)	Nilsson (2022) Niedbała et al. (2021)
Phytoextraction	Soybean Sunflower Alfalfa Tomato Chinese brake fern Alyssum Alpine penny-cress Shiny Elsholtzia Lettuce Perennial grass	*Glycine max* *Helianthus annuus* *Medicago sativa* *Solanum lycopersicum* *Pteris vittata* *Alyssum bertolonii* *Thlaspi caerulescens* *Elsholtzia splendens* *Lactuca sativa* *Lolium perenne*	Agrochemicals Explosives (Trinitrotoluene) Metals Organic compounds	Shah et al. (2023) Babu et al. (2021) Li et al. (2020)

studies have shown that phytoremediation can be assisted or induced using genetically modified plants, microorganisms, and/or synthetic and organic amendments (Wani et al., 2023). You et al. (2022) reported that fast-growing and high biomass-producing nonaccumulator plants such as willows, when used in conjunction with chelating agents, are effective in phytoextraction. Importantly, the study showed that chelating agents improve the bioaccumulation of metals without negatively affecting plant growth (You et al., 2022). Some examples of plants with evidence of effectiveness in phytoextraction are shown in Table 4.

1.1.2 Assisted Phytoremediation

As earlier mentioned, phytoremediation can be enhanced using genetically modified plants and microorganisms, as well as natural and synthetic amendments (Kafle et al., 2022; Wani et al., 2023; Yan et al., 2020).

1.1.2.1 Genetic Engineering Enhancement

With genetic engineering, desirable genes for enhanced uptake, translocation, sequestration, and detoxification of contaminants can be transferred to plants (Kafle et al., 2022). The resulting transgenic plants are characterized as fast-growing, high biomass-producing, and metal-tolerating (Kafle et al., 2022). Tussipkan and Manabayeva (2022) documented that the combination of transgenic plants with bacterial genes degraded simazine (herbicide) into diverse nonhazardous simpler forms. Several studies have reported the effectiveness of genetic engineering in in situ phytoremediation (Venegas-Rioseco et al., 2021; Zarull et al., 2002). More studies are required to fully understand its effectiveness in a real-world situation.

1.1.2.2 Microbial Enhancement

Through bioaugmentation and biostimulation, microorganisms enhance the uptake of heavy metals and metalloids, as well as the degradation of other contaminants (Kafle et al., 2022). Biostimulation is the introduction of additional nutrients such as phosphorous (P) and nitrogen (N) for the stimulation of existing microorganisms in contaminated media (e.g., soil) (Janati et al., 2021). On the other hand, bioaugmentation involves the introduction of additional microorganisms (genetically modified or natural) to a contaminated environment (Janati et al., 2021).

Zeng et al. (2021) reported that heavy metal removal in soil increased by 52, 44, and 32% for manganese (Mn), zinc (Zn), and arsenic (As), respectively, following the addition of *Burkholderia* spp. Similarly, Sarkodie et al. (2022) reported that Cd uptake increased by 43% using *Sedum plumbizincicola.* Also, Lacalle et al. (2023) noted that the introduction of actinobacterial consortium with organic matter resulted in the most efficacious transformation of hexavalent chromium to trivalent chromium (Lacalle et al., 2023). Evidently, plant growth-promoting endophytic bacteria (PGPE) and plant growth-promoting rhizobacteria (PGPR) enhance metal accumulation and plant growth through the secretion of indole acetic acid (IAA), siderophore, auxins, and other compounds (Wang et al., 2022).

Along with the above, fungi also play a huge role in phytoremediation. Arbuscular mycorrhizal fungi (AMF) enhance the bioavailability of contaminants through direct and indirect interaction with contaminants (Bhantana et al., 2021). AMF produce phytohormones (Pons et al., 2020) and enhance plant roots' absorptive surface area (Gao et al., 2023). Bioaugmentation of red clover (*Trifolium pratense)* with AMF resulted in higher Zn accumulation.

1.1.2.3 Natural and Synthetic Chelate Amendments

The treatment of soil with natural and synthetic amendments enhances the bioavailability, uptake, and translocation of metals in the phytoextraction process (Garbowski et al., 2023). Additionally, natural and synthetic amendments enhance plant growth

Table 5 Natural and synthetic chelating agents utilized in phytoremediation

Amendments	Plants		Contaminants	References
	Common name	Scientific name		
Natural organic amendments				
Sugar beet residue (SBR)	White clover Rock samphire	*Trifolium repens* *Crithmum maritimum*	Ni, Cd, Cr, Fe, and Zn	Rashmi et al. (2023), Lashen et al. (2022)
Paper beet residue (PBR)	Pine tree Red oak	*Pinus nigra* *Quercus rubra*	Pb and Cd	Mittal et al. (2021)
Wood biochar	Maize	*Zea mays*	Cu, Cd, As, and Pb	Qiu et al. (2022)
Bamboo and rice straw biochar	Sedum	*Sedum plumbizincicola*	Cu and Pb	Sakhiya et al. (2022)
Chemical amendments				
EDTA	Sedum Broad bean	*Sedum alfredii* *Vicia faba*	Pb Pb	Dong et al. (2023), Saman et al. (2022)
Ethylene glycol tetraacetic acid (EGTA)	Four o'clocks Chickpea	*Mirabilis jalapa* *Cicer arietinum*	Cd Pb	Mohrazi et al. (2023), Zhang et al. (2022)
Sodium dodecyl sulfate (SDS)	Hollyhocks Pot marigold White poplar	*Althaea rosea* *Calendula officinalis* *Populus alba*	Cd Cd Zn	Chakraborty et al. (2021), Correia et al. (2022), Khoshdast (2021)

(Rashmi et al., 2023). Several researchers have reported the efficacy of diverse chelating agents in the increased accumulation of metals and metalloids (Gul et al., 2021; Yang et al., 2023; Zulkernain et al., 2023). Table 5 shows some examples of natural and synthetic amendments.

EDTA is posited as the most efficacious chelating agent (Saman et al., 2022). According to Saman et al. (2022), EDTA increases metal uptake from contaminated soil to over 1% of shoot dry biomass. However, possibilities of long persistency in the environment, leaching in groundwater, and adverse effects on microorganisms have limited the acceptance of EDTA (Yang et al., 2022). Hence, the use of other chelating agents has been recommended by several researchers (Gul et al., 2021; Yang et al., 2022; Zulkernain et al., 2023).

1.1.3 Benefits and Limitations of Phytoremediation

The benefits of phytoremediation outweigh its limitations. Overall, phytoremediation improves the environment, leaving no significant secondary contaminants. Also, phytoremediation is highly cost-effective. The result of a study showed that while phytoremediation of one acre of soil costs between USD 60,000 to 1,000,000, soil

Table 6 Benefits and limitations of phytoremediation application

Benefits	Limitations
Application is possible ex situ and in situ (Alsafran et al., 2023; Zarull et al., 2002)	Physical accessibility of contaminants to plant roots is necessary for in situ applicability. In situ is not effective for highly hydrophobic substances. Current phytoremediation knowledge is from research carried out in situ. There is a need for numerous field research to ascertain effectiveness (Kafle et al., 2022)
Phytoremediation can be adapted to cover a substantial area and can be easily predisposed (Tan et al., 2023). Diverse mechanisms can remediate diverse contaminants (Alsafran et al., 2023; Zarull et al., 2002)	Choosing mechanisms that can remediate multiple contaminants at once remains a tough decision (Sadasivam, 2022)
It is effective even at both high and low concentrations of contaminants (Alsafran et al., 2023; Zarull et al., 2002)	Sustaining plant growth in highly contaminated environments could be difficult. Thus, applicability could be limited by high-level toxicity contaminants (Alsafran et al., 2023; Kafle et al., 2022)
Diverse plants are available for use (Alsafran et al., 2023; Zarull et al., 2002)	Some plants may only proliferate under certain environmental conditions and seasons. Also, the phytoremediation capacity of some plants may be affected by diseases and pests (Kafle et al., 2022). Additionally, the choice of plants for the remediation of different contaminants could be difficult (Alsafran et al., 2023; Zarull et al., 2002). Hence, numerous research is needed in addressing the aforementioned concerns
It is an autotrophic system that derives its source of energy from the sun and can therefore be easily managed (Yan et al., 2020)	

excavation can cost up to USD 4,000,000 (Riaz et al., 2022). A comparison of the benefits and limitations of phytoremediation is made in Tables 6, 7, and 8.

1.2 Vermicomposting

Vermicomposting is the process of recycling diverse biodegradable wastes such as domestic residue, animal manure, and agro-waste into nutrient-rich vermicompost using earthworms (Ahmad et al., 2022; Kaur, 2020). Vermicompost is a high-nutrient biofertilizer, mainly comprising decayed organic matter, microorganisms, and worm casts (Rehman et al., 2023). Vermicomposting is similar to conventional composting techniques, except for the use of earthworms (Kaur, 2020). Vermiculture, which is the science of raising and breeding earthworms, is key to vermicomposting.

Table 7 Benefits and limitations of phytoremediation with consideration to cost and time factors

Benefits	Limitations
Cost-effectiveness	
Highly cost-effective. Operation and labor costs are cheaper, in comparison to conventional remediation techniques (Alsafran et al., 2023; Zarull et al., 2002)	Mechanisms could be slow and time-consuming. Also, when plant deaths occur, there could be an increase in the cost of the process (Alsafran et al., 2023; Zarull et al., 2002)
The recovery of contaminants from plant tissues presents commercialization privileges (Alsafran et al., 2023; Zarull et al., 2002). High possibility for the reuse of remediated (recovered) metals in phytomining. High biomass-producing plants such as willow may be further utilized during incineration and construction, among others (Kalak, 2023)	To avoid accidents, additional efforts are directed (Alsafran et al., 2023; Zarull et al., 2002)
Time factor	
Though it takes years to fully remediate a contaminated site, the result always beats time. Also, assisted phytoremediation reduces the time required for cleanup (remediation) (Wani et al., 2023)	Phytoremediation is time-consuming, compared to other conventional methods (Wani et al., 2023)

Earthworms belong to the phylum Annelida and class Oligochaeta, and over 4000 species have been identified worldwide (Baturina et al., 2020). Sadly, seldom are they used in the recycling of organic wastes (Baturina et al., 2020). Earthworms can be broadly categorized into burrowing or nonburrowing earthworms (Baturina et al., 2020). Table 9 summarizes the key differences between the two categories of earthworms.

Earthworms weigh about 0.5–0.6 g, consume organic wastes equivalent to their body weight, and produce 40–60% of consumed waste as vermicast (Kaur, 2020). Earthworm gut microorganisms secrete enzymes capable of degrading consumed organic wastes into various nutrients needed for soil fertility and plant growth. Ergo, vermicompost is significantly higher in macro- and micronutrients than traditional compost (Emendu et al., 2022). Table 10 compares the nutrient contents of traditional compost with that of vermicompost prepared from cattle dung.

1.2.1 Vermicomposting Modes and Methods

Vermicomposting can be categorized into two modes: the continuous and batch modes (Kaur, 2020). In the batch mode, both substrates and earthworms are added at the beginning, without any further addition throughout composting (Kaur, 2020). On the other hand, the continuous mode involves the continuous addition of organic wastes, subject to earthworms' consumption rate (Kaur, 2020).

Table 8 Other benefits and limitations of phytoremediation

Benefits	Limitations
Efficacy	
Highly effective in the remediation of organic and inorganic contaminants (Kafle et al., 2022; Wani et al., 2023; Yan et al., 2020)	Efficacy is highly dependent on environmental conditions and seasonal factors, plant physiology, and the physicochemical condition of the media (soil) (He & Matthews, 2023; Kafle et al., 2022). Hence, success could be site-specific (Kafle et al., 2022)
Has high efficacy in remediating sites contaminated with multiple contaminants (Alsafran et al., 2023; Zarull et al., 2002)	The effects of the different contaminants on plant physiology and performance should be considered (Alsafran et al., 2023; Zarull et al., 2002)
Assisted phytoremediation using genetic engineering, microorganisms, and chelating agents enhance effectiveness (Wani et al., 2023; Yan et al., 2020)	As a result of genetic engineering of plants, undesirable plant species might ensue. Also, the use of chelating agents could increase the concentration of metals in soil, consequently increasing the possibility of leaching (Yang et al., 2022)
Environmental and human impact	
Environmentally friendly. Overall, phytoremediation improves the environment, leaving no significant secondary contaminants. It is highly conservative of natural resources, highly biologically active, nonintrusive, and nondestructive (Alsafran et al., 2023; Zarull et al., 2002). It helps check erosion and leaching of contaminants (Yan et al., 2020)	Also, there is a high risk of bioaccumulation of contaminants in the food chain (Alengebawy et al., 2021). Thus, necessary precautions must be taken, and proper biomass disposal is required (Alsafran et al., 2023; Zarull et al., 2002)
Implication on plant growth and agriculture	
Phytoremediation boosts soil fertility and health. It aids plant phytochemicals and growth rate (Bhat et al., 2022; Yan et al., 2020)	There is still the possibility of leaching (Alengebawy et al., 2021)
Contaminated soil can be remediated for agricultural purposes (Alsafran et al., 2023; Zarull et al., 2002)	

Both modes have their limitations and strengths. However, there is the need to stockpile wastes before commencing batch vermicomposting. Howbeit, wastes get composted at about the same time in batch mode. On this account, the risk of contamination that is associated with the subsequent addition of wastes in continuous mode is nullified in batch vermicomposting (Kaur, 2020). Also, space could be saved from the upward mounding of wastes in batch mode. The addition of all waste at a time is posited to increase the risk of generating too much heat for earthworms. Numerous methods fall under both modes.

Table 9 Differences between nonburrowing and burrowing earthworms (Baturina et al., 2020)

Characteristics	Nonburrowing	Burrowing
Length	10–15 cm	20–30 cm
Color	Purple or red	Pale in color
Life span	28 months	15 years
Habitat	Soil upper layers	Deep in the soil
Vermicomposting property	Have a faster rate of converting organic wastes to vermicompost	Have a slower rate of vermicomposting
Examples	African nightcrawler (*Eudrilus eugeniae*) and red worm (*Eisenia fetida*)	Asian earthworms (*Pheretima posthuma* and *Pheretima elongata*)

Table 10 Comparison of the macro- and micronutrient content of traditional compost and vermicompost with weight in mg/Kg (Emendu et al., 2022)

Nutrient content	Traditional compost	Vermicompost
pH	8.30–8.50	8.83–9.01
Carbon-to-nitrogen ratio	42.68–45.92	14.89–16.03
Total nitrogen (%)	0.79–1.27	1.20–3.60
Total organic carbon (%)	44.39–46.41	37.01–37.23
Total sodium (%)	0.51–0.91	1.03–1.79
Total phosphorous (%)	0.62–1.22	0.68–2.30
Total potassium (%)	2.80–5.21	– 0.18–3.98
Electrical conductivity (mS/cm)	3.20–3.24	2.79–2.85
Mn	11.25–14.79	37.13–38.89
Fe	618.44–621.64	588.52–591.56
Cu	7.81–8.27	8.83–9.23
Zn	9.48–10.22	11.17–11.91

1.2.1.1 Tower Method of Vermicomposting

This is highly domesticating, simple, and fast. Wastes are added to earthworms contained in polyvinyl chloride (PVC) pipes that are permanently and vertically erected in dug holes (Kaur, 2020). Wastes can be periodically added. Thus, there is no need for stockpiling.

1.2.1.2 Flow-Through Reactor Method

The term "flow-through" denotes the nondisturbance of earthworms in the reactors (Kaur, 2020). As described by Kumari and Mohan (2021), earthworms are placed in a common rectangular-shaped elevated box of about 3 m in width. Organic materials

Fig. 2 Windrow vermicomposting (vermicompost is made in the form of rectangular heaped ridges)

are subsequently added from the topmost layer. Using a hydraulically driven bar, casts are removed from the reactor via the bottom grid.

1.2.1.3 Windrow Vermicomposting

Windrows are commonly used by commercial farmers (Kaur, 2020; Liu & Wang, 2020). They are rows about 3 feet wide, 3 feet high, and 100 feet long in structure (Fig. 2) (Kaur, 2020; Liu & Wang, 2020). Windrows are seeded with earthworms, and fresh organic matter feeds are added to their edges to lure out earthworms (Kaur, 2020). Static piles, top feed, and wedges are the commonest ways of carrying out the windrow method of vermicomposting (Kaur, 2020; Liu & Wang, 2020). Batch mode is used in static pile windrow, whereas the continuous mode is utilized in top feed and wedges (Kaur, 2020; Liu & Wang, 2020).

Static pile windrows are elongated, rectangular, or square-shaped piles of mixed bedding and feed (Kaur, 2020). Before settling, the pile's height should not surpass 1 m (Kaur, 2020; Liu & Wang, 2020). To ensure ventilation, piles should be produced outdoors. Also, to prevent rain, they should be covered with corrugated iron (Liu & Wang, 2020).

Top-fed windrows follow a static pile approach, except that a layer of new bedding material about 10 cm is continuously fed into the system (Kaur, 2020; Liu & Wang, 2020). Also, the lining is utilized in preventing worms' escape (Liu & Wang, 2020). However, the lining has been criticized for limiting air movement, constituting anaerobic conditions, and causing the death of some earthworm populations, thereby limiting the effectiveness of vermicomposting.

Wedge windrow also adopts the continuous mode, as well as the use of lining (Kaur, 2020; Liu & Wang, 2020). However, there is an engrossing modification to wedges (Kaur, 2020; Liu & Wang, 2020). An initial earthworm-bedded stock is added inside an approximately 1 m tall multisided corral-type structure (Kaur, 2020; Liu & Wang, 2020). Straw, hay, bales, wood, and concrete are examples of materials utilized for the sides of the corral-type structure (Kaur, 2020; Liu & Wang, 2020). Periodically, fresh materials are added via the open side of the corral-type structure (Kaur, 2020; Liu & Wang, 2020). With each addition, worms are lured by the fresh food, leaving cast behind (Kaur, 2020). Following the filling of the corral, it is scooped for casting (Kaur, 2020; Liu & Wang, 2020).

1.2.1.4 Bin or Bed Vermicomposting

This is the easiest method of vermicomposting (Kaur, 2020). Bins and beds can be made from untreated, nonaromatic wood (Fig. 3), plastics, and plant protectors, among others (Kaur, 2020). Following the addition of decayed leaves, composted animal manure or shredded paper, a handful of soil, some water for moisture, and earthworms should be added to the bin or bed (Kaur, 2020). Primarily, added soil offers a burrowing platform for burrowing earthworms such as red earthworms (Kaur, 2020). Red earthworms remain the best worms for bin vermicomposting (Kaur, 2020).

The addition of eggshells to the bed reduces the acidity level of the bed. Additionally, egg shells enrich earthworms with calcium (Kaur, 2020). Fish, meat, and other fatty foods should not be bin-composted. Otherwise, there will be the production of a foul smell (Kaur, 2020). Stacked bins and top-fed beds are types of bin or bed methods of vermicomposting. While the top feed uses the continuous mode, the stacked bin adopts both modes.

Top-fed beds are similar to top-fed windrows, except that the bed is a floor contained within four walls (Kaur, 2020). Also, top-fed beds are commonly built with straws, bales, or insulated sides, which could be utilized in insulating them (Kaur, 2020). The stacked bin uses a vertical dimension in addressing the issues of space (Kaur, 2020). However, the stacked bin must be small in size, to enable easy lifting by forklift or by hand.

1.2.1.5 Trench or Pit Vermicomposting

Following the lining of dug pits with lining materials such as canvas feed bags, earthworms and organic materials or wastes are buried in the pit (Kaur, 2020). Pits or trenches are periodically watered to maintain moisture. Also, earthworms are subsequently added at the time of watering. Watering could be carried out weekly. Figure 4 shows a picture of a pit or trench.

Fig. 3 Bin or bed vermicomposting with wood protector (vermicompost is prepared in bins or bed-shaped containers protected by wood or other suitable materials)

1.2.1.6 Trough Vermicomposting

This is another continuous mode of vermicomposting, where a minimum of one-week-old manure is periodically spread across a trough, followed by the addition of earthworms (Kaur, 2020). More layers of manure are added every 10 days until the trough is filled.

1.2.2 Benefits and Limitations of Vermicomposting

Vermicomposting, just like phytoremediation, offers important environmental benefits. Studies have documented the effectiveness of vermicomposting in recycling diverse organic wastes into vermicompost (Eremeeva et al., 2023; Hajam et al., 2023; Ratnasari et al., 2023; Vuković et al., 2021; Vyas et al., 2022). The application of

Fig. 4 Pit or trench vermicomposting (organic materials are placed in dug-out trenches or pits)

vermicompost in agriculture enhances soil physicochemical and biological quality (Kaur, 2020). Also, vermicompost enhances plant growth and productivity (Kaur, 2020). Specific benefits of vermicompost in soil and plants are shown in Table 11.

In addition, vermiwash is a high-quality organic fertilizer. Vermiwash, just like vermicompost, is significantly rich in micro- and macronutrients, plant growth hormones, and enzymes (Kaur, 2020). Vermiwash is the clear, transparent, pale yellow liquid (fluids) extracted from the watery washing of vermicompost (Kaur, 2020). Vermiwashing is carried out by passing water into columns of earthworm actions. It takes about 40–50 days to get the clear, transparent pale yellow fluid from the vermiwash unit (Kaur, 2020).

While vermicomposting plays a pivotal role in environmental cleanup and agriculture, its application is associated with some limitations. The separation of earthworms from vermicompost requires adequate care and attention (Eremeeva et al., 2023). Also, earthworms proliferate at 60–70% moisture level, mesophilic temperature, and a neutral pH (Eremeeva et al., 2023). Earthworms' ingestion system and the resulting cast are subject to feed substrates, topography, moisture content, and

Table 11 Benefits of vermicompost on soil and plants (Kaur, 2020)

Soil	Plant
Its richness in humic acid balances soil pH	Supplies plants with essential aids and nutrients needed for the suppression of plant diseases
Allures soil indigenous deep-burrowing earthworms	Eases mineral absorbing by plants
Enhances soil nutrient recycling	Enhances root growth and plant structure of the plant in porous soil
Increases infiltration, water-holding capacity, aeration, and porosity	Ameliorates plant growth, crop germination, and yield
Offers the uttermost microbial population to the soil	Offers uttermost microbial population to plants
Increases available water-holding and water retention capacities	
Reduces leaching	
Improves soil texture and plant roots' anchoring	
Increases enzyme activities	
Soil can be amended with finished vermicompost during phytoremediation	

temperature, among other factors (Xiao et al., 2022). Similarly, microbial growth and dispersal during vermicomposting directly correlate to available nutrients, temperature, moisture, as well as pH (Xiao et al., 2022). It is important that all vermicomposting methods be protected from direct sunlight. Also, shade is crucial to the maintenance of moisture content.

1.3 Combined Application of Phytoremediation and Vermicomposting

Although phytoremediation and vermicomposting are distinct in their techniques, their combined application could offer intriguing and efficacious outcomes. Nedjimi (2021) reported that the integration of tiger worm (*Eisenia andrei*) vermicompost with black oat (*Avena strigosa*) enhanced the removal of Pb, Cr, and Cd from the soil. Also, several studies have reported the significant roles that earthworms play in the enhancement of phytoremediation of heavy metals (Aransiola et al., 2021; Babu et al., 2021; Gudeta et al., 2023; Liu et al., 2023; Nedjimi, 2021; Sharma et al., 2023).

Among the aforementioned, Kaur (2020) revealed that the integration of red earthworm (*Eisenia fetida*) with mustard greens (*Brassica juncea*) significantly enhanced the phytoremediation of Cd. Also, Nedjimi (2021) documented that the decrease in soil pH as a result of earthworms' secreted organic acids (such as humic and fulvic acids) enhanced heavy metal bioavailability in the rhizosphere of plants during phytoremediation of the metal-contaminated site.

Their combined application, especially in sites contaminated by diverse contaminants and wastes, will result in faster and more sustainable remediation. While vermicomposting offers a sustainable solution to organic wastes, phytoremediation remediates both organic and inorganic contaminants. Also, vermicompost and vermiwash are rich sources of nutrients, growth hormones, and enzymes for phytoremediation plants. Additionally, vermiwash could be applied as a biopesticide.

2 Conclusion

The dramatic rise in the accumulation of contaminants and wastes in the environment calls for sustainable environmentally friendly approaches such as phytoremediation and vermicomposting. Vermicomposting remains an effective waste management technique. With earthworms, toxic substances are significantly decomposed from waste, leaving vermicompost. Vermicompost is a high-nutrient organic fertilizer. Vermiwash, which is extracted from vermicompost, is another high-nutrient biofertilizer. Vermicompost and vermiwash are rich in nutrients, plant growth hormones, and enzymes.

Phytoremediation mechanisms are effective at remediating diverse organic and inorganic contaminants. Numerous plants have the capability to degrade, immobilize, bioaccumulate, and/or dissipate organic and inorganic contaminants. Also, plants can be genetically modified for better performance. Phytoremediation can be equally enhanced through microorganisms and a good number of soil amendments. Numerous studies have been carried out on phytoremediation and vermicomposting. Their application, separately and/or combined, is important.

References

Abdullah, S. R. S., Al-Baldawi, I. A., Almansoory, A. F., Purwanti, I. F., Al-Sbani, N. H., & Sharuddin, S. S. N. (2020). Plant-assisted remediation of hydrocarbons in water and soil: Application, mechanisms, challenges and opportunities. *Chemosphere, 247*, Article 125932. https://doi.org/10.1016/j.chemosphere.2020.125932

Ahmad, A., Aslam, Z., Bellitürk, K., Ullah, E., Raza, A., & Asif, M. (2022). Vermicomposting by bio-recycling of animal and plant waste: A review on the miracle of nature. *Journal of Innovative Sciences, 8*(2). https://doi.org/10.17582/journal.jis/2022/8.2.175.187

Alengebawy, A., Abdelkhalek, S. T., Qureshi, S. R., & Wang, M.-Q. (2021). Heavy metals and pesticides toxicity in agricultural soil and plants: Ecological risks and human health implications. *Toxics, 9*(3), 42. https://doi.org/10.3390/toxics9030042

Alsafran, M., Saleem, M. H., Al Jabri, H., Rizwan, M., & Usman, K. (2023). Principles and applicability of integrated remediation strategies for heavy metal removal/recovery from contaminated environments. *Journal of Plant Growth Regulation, 42*(6), 3419–3440. https://doi.org/10.1007/s00344-022-10803-1

Aransiola, S. A., Ijah, U. J. J., Abioye, O. P., & Bala, J. D. (2021). Vermicompost assisted phytoremediation of heavy metal contaminated soil in Madaka District, Nigeria Using *Melissa officinalis* L (Lemon balm) and *Sida acuta* (Stubborn weed). https://doi.org/10.21203/rs.3.rs-276415/v1

Asante-Badu, B., Kgorutla, L. E., Li, S. S., Danso, P. O., Xue, Z., & Qiang, G. (2020). Phytoremediation of organic and inorganic compounds in a natural and an agricultural environment: A review. *Applied Ecology and Environmental Research, 18*(5), 6875–6904. https://doi.org/10.15666/aeer/1805_68756904

Babu, S. M. O. F., Hossain, M. B., Rahman, M. S., Rahman, M., Ahmed, A. S. S., Hasan, M. M., Rakib, A., Emran, T. B., Xiao, J., & Simal-Gandara, J. (2021). Phytoremediation of toxic metals: A sustainable green solution for clean environment. *Applied Sciences (Basel, Switzerland), 11*(21), 10348.https://doi.org/10.3390/app112110348

Bak, K. H., Bauer, S., & Bauer, F. (2023). Effect of different genotypes and harvest times of sage (Salvia spp. Labiatae) on lipid oxidation of cooked meat. *Antioxidants (Basel, Switzerland), 12*(3). https://doi.org/10.3390/antiox12030616

Bakshe, P., & Jugade, R. (2023). Phytostabilization and rhizofiltration of toxic heavy metals by heavy metal accumulator plants for sustainable management of contaminated industrial sites: A comprehensive review. *Journal of Hazardous Materials Advances, 10*, Article 100293. https://doi.org/10.1016/j.hazadv.2023.100293

Baturina, M. A., Kaygorodova, I. A., & Loskutova, O. A. (2020). New data on species diversity of Annelida (Oligochaeta, Hirudinea) in the Kharbey lakes system, Bolshezemelskaya tundra (Russia). *ZooKeys, 910*, 43–78. https://doi.org/10.3897/zookeys.910.48486

Bhandari, S., Poudel, D. K., Marahatha, R., Dawadi, S., Khadayat, K., Phuyal, S., Shrestha, S., Gaire, S., Basnet, K., Khadka, U., & Parajuli, N. (2021). Microbial enzymes used in bioremediation. *Journal of Chemistry, 2021*, 1–17.https://doi.org/10.1155/2021/8849512

Bhantana, P., Rana, M. S., Sun, X.-c, Moussa, M. G., Saleem, M. H., Syaifudin, M., Shah, A., Poudel, A., Pun, A. B., & Bhat, M. A. (2021). Arbuscular mycorrhizal fungi and its major role in plant growth, zinc nutrition, phosphorous regulation and phytoremediation. *Symbiosis, 84*, 19–37.

Bhat, S. A., Bashir, O., Ul Haq, S. A., Amin, T., Rafiq, A., Ali, M., Américo-Pinheiro, J. H. P., & Sher, F. (2022). Phytoremediation of heavy metals in soil and water: An eco-friendly, sustainable and multidisciplinary approach. *Chemosphere, 303*, 134788. https://doi.org/10.1016/j.chemosphere.2022.134788

Bian, F., Zhong, Z., Zhang, X., Yang, C., & Gai, X. (2020). Bamboo–An untapped plant resource for the phytoremediation of heavy metal contaminated soils. *Chemosphere, 246*, Article 125750.

Chakraborty, I., Bhowmick, G. D., Nath, D., Khuman, C. N., Dubey, B. K., & Ghangrekar, M. M. (2021). Removal of sodium dodecyl sulphate from wastewater and its effect on anodic biofilm and performance of microbial fuel cell. *International Biodeterioration & Biodegradation, 156*(105108), Article 105108. https://doi.org/10.1016/j.ibiod.2020.105108

Chen, F., Ma, J., Akhtar, S., Khan, Z. I., Ahmad, K., Ashfaq, A., Nawaz, H., & Nadeem, M. (2022). Assessment of chromium toxicity and potential health implications of agriculturally diversely irrigated food crops in the semi-arid regions of South Asia. *Agricultural Water Management, 272*, 107833. https://doi.org/10.1016/j.agwat.2022.107833

Collin, M. S., Venkatraman, S. K., Vijayakumar, N., Kanimozhi, V., Arbaaz, S. M., Stacey, R. G. S., Anusha, J., Choudhary, R., Lvov, V., Tovar, G. I., Senatov, F., Koppala, S., & Swamiappan, S. (2022). Bioaccumulation of lead (Pb) and its effects on human: A review. *Journal of Hazardous Materials Advances, 7*, 100094. https://doi.org/10.1016/j.hazadv.2022.100094

Correia, E. L., Brown, N., Ervin, A., Papavassiliou, D. V., & Razavi, S. (2022). Contamination in sodium dodecyl sulfate solutions: Insights from the measurements of surface tension and surface rheology. *Langmuir: The ACS Journal of Surfaces and Colloids, 38*(23), 7179–7189. https://doi.org/10.1021/acs.langmuir.2c00460

Dong, W., Wang, R., Li, H., Yang, X., Li, J., Wang, H., Jiang, C., & Wang, Z. (2023). Effects of chelating agents' addition on Ryegrass extraction of cadmium and lead in artificially contaminated soil. *Water, 15*(10), 1929.https://doi.org/10.3390/w15101929

Emendu, E., Arinze, C., & Emendu, R. (2022). Analysis of micro and macro nutrient levels in compost and vermicompost fertilizer formulated from selected agro-waste and comparative assessment of the fertilizer efficiencies. *5*.

Eremeeva, N. A., Savoskina, O. A., Poddymkina, L. M., Abdulmazhidov, K. A., & Gamidov, A. G. (2023). Analysis of anthropogenic impact on the environment, measures to reduce it, and waste management. *Frontiers in Bioengineering and Biotechnology, 11*.https://doi.org/10.3389/fbioe.2023.1114422

Esposito, B., Riminucci, F., Di Marco, S., Metruccio, E. G., Osti, F., Sangiorgi, S., & Ferri, E. N. (2021). A simple device for the on-site photodegradation of pesticide mixes remnants to avoid environmental point pollution. *Applied Sciences (Basel, Switzerland), 11*(8), 3593. https://doi.org/10.3390/app11083593

Fernández-Braña, A., Salgado, L., Gallego, J. L. R., Afif, E., Boente, C., & Forján, R. (2023). Phytoremediation potential depends on the degree of soil pollution: A case study in an urban brownfield. *Environmental Science and Pollution Research International, 30*(25), 67708–67719. https://doi.org/10.1007/s11356-023-26968-5

Gao, X., Liu, Y., Liu, C., Guo, C., Zhang, Y., Ma, C., & Duan, X. (2023). Individual and combined effects of arbuscular mycorrhizal fungi and phytohormones on the growth and physiobiochemical characteristics of tea cutting seedlings. *Frontiers in Plant Science, 14*. https://doi.org/10.3389/fpls.2023.1140267

Garbowski, T., Bar-Michalczyk, D., Charazińska, S., Grabowska-Polanowska, B., Kowalczyk, A., & Lochyński, P. (2023). An overview of natural soil amendments in agriculture. *Soil & Tillage Research, 225*, Article 105462. https://doi.org/10.1016/j.still.2022.105462

Gudeta, K., Kumar, V., Bhagat, A., Julka, J. M., Bhat, S. A., Ameen, F., Qadri, H., Singh, S., & Amarowicz, R. (2023). Ecological adaptation of earthworms for coping with plant polyphenols, heavy metals, and microplastics in the soil: A review. *Heliyon, 9*(3), e14572. https://doi.org/10.1016/j.heliyon.2023.e14572

Gul, I., Manzoor, M., Hashim, N., Shah, G. M., Waani, S. P. T., Shahid, M., Antoniadis, V., Rinklebe, J., & Arshad, M. (2021). Challenges in microbially and chelate-assisted phytoextraction of cadmium and lead—A review. *Environmental Pollution (Barking, Essex: 1987), 287*, 117667. https://doi.org/10.1016/j.envpol.2021.117667

Hajam, Y. A., Kumar, R., & Kumar, A. (2023). Environmental waste management strategies and vermi transformation for sustainable development. *Environmental Challenges, 13*, Article 100747. https://doi.org/10.1016/j.envc.2023.100747

He, Y., & Matthews, M. L. (2023). Seasonal climate conditions impact the effectiveness of improving photosynthesis to increase soybean yield. *Field Crops Research, 296*, Article 108907. https://doi.org/10.1016/j.fcr.2023.108907

Hostyn, G., Schwartz, C., Côme, J.-M., & Ouvrard, S. (2022). Assessment for combined phytoremediation and biomass production on a moderately contaminated soil. *Environmental Science and Pollution Research International, 29*(39), 59736–59750. https://doi.org/10.1007/s11356-022-19963-9

Janati, W., Benmrid, B., Elhaissoufi, W., Zeroual, Y., Nasielski, J., & Bargaz, A. (2021). Will phosphate bio-solubilization stimulate biological nitrogen fixation in grain legumes? *Frontiers in Agronomy, 3*. https://doi.org/10.3389/fagro.2021.637196

Kafle, A., Timilsina, A., Gautam, A., Adhikari, K., Bhattarai, A., & Aryal, N. (2022). Phytoremediation: Mechanisms, plant selection and enhancement by natural and synthetic agents. *Environmental Advances, 8*, Article 100203. https://doi.org/10.1016/j.envadv.2022.100203

Kalak, T. (2023). Potential use of industrial biomass waste as a sustainable energy source in the future. *Energies, 16*(4), 1783. https://doi.org/10.3390/en16041783

Kaur, T. (2020). Vermicomposting: An effective option for recycling organic wastes. In *Organic agriculture*. https://doi.org/10.5772/intechopen.91892

Kelechi, L., & Njoku, S. O. (2022). Phytoremediation of heavy metals contaminated soil samples obtained from mechanic workshop and dumpsite using *Amaranthus spinosus*. *Scientific African, 17*.https://doi.org/10.1016/j.sciaf.2022.e01278

Khan, A. H. A., Kiyani, A., Santiago-Herrera, M., Ibáñez, J., Yousaf, S., Iqbal, M., Martel-Martín, S., & Barros, R. (2023). Sustainability of phytoremediation: Post-harvest stratagems and economic opportunities for the produced metals contaminated biomass. *Journal of Environmental Management, 326*, 116700. https://doi.org/10.1016/j.jenvman.2022.116700

Khan, A. U., Khan, A. N., Waris, A., Ilyas, M., & Zamel, D. (2022). Phytoremediation of pollutants from wastewater: A concise review. *Open Life Sciences, 17*(1), 488–496. https://doi.org/10.1515/biol-2022-0056

Khan, S., Masoodi, T. H., Pala, N. A., Murtaza, S., Mugloo, J. A., Sofi, P. A., Zaman, M. U., Kumar, R., & Kumar, A. (2023b). Phytoremediation prospects for restoration of contamination in the natural ecosystems. *Water, 15*(8), 1498.https://doi.org/10.3390/w15081498

Khoshdast, R. B. S. V. S. H. (2021). Removal of some cationic contaminants from aqueous solutions using sodium dodecyl sulfate-modified coal tailings. *Iranian Journal of Chemistry and Chemical Engineering, 40*, 1105–1120. https://doi.org/10.30492/ijcce.2020.111834.3682

Kumari, N., & Mohan, C. (2021). Basics of clay minerals and their characteristic properties. In *Clay and clay minerals*. https://doi.org/10.5772/intechopen.97672

Lacalle, R. G., Bernal, M. P., Álvarez-Robles, M. J., & Clemente, R. (2023). Phytostabilization of soils contaminated with As, Cd, Cu, Pb and Zn: Physicochemical, toxicological and biological evaluations. *Soil & Environmental Health, 1*(2), Article 100014. https://doi.org/10.1016/j.seh.2023.100014

Lashen, Z. M., Shams, M. S., El-Sheshtawy, H. S., Slaný, M., Antoniadis, V., Yang, X., Sharma, G., Rinklebe, J., Shaheen, S. M., & Elmahdy, S. M. (2022). Remediation of Cd and Cu contaminated water and soil using novel nanomaterials derived from sugar beet processing- and clay brick factory-solid wastes. *Journal of Hazardous Materials, 428*, 128205. https://doi.org/10.1016/j.jhazmat.2021.128205

Li, F., Yang, F., Chen, Y., Jin, H., Leng, Y., & Wang, J. (2020). Chemical reagent-assisted phytoextraction of heavy metals by Bryophyllum laetivirens from garden soil made of sludge. *Chemosphere, 253*(126574), Article 126574. https://doi.org/10.1016/j.chemosphere.2020.126574

Liu, P., Song, Y., Wei, J., Mao, W., Ju, J., Zheng, S., & Zhao, H. (2023). Synergistic effects of earthworms and plants on chromium removal from acidic and alkaline soils: Biological responses and implications. *Biology, 12*(6), 831. https://doi.org/10.3390/biology12060831

Liu, Z., & Wang, X. (2020). Manure treatment and utilization in production systems. In *Animal agriculture* (pp. 455–467). https://doi.org/10.1016/b978-0-12-817052-6.00026-4

Maqbool, A., Rizwan, M., Yasmeen, T., Arif, M. S., Hussain, A., Mansha, A., Ali, S., Alshaya, H., & Okla, M. K. (2022). Phosphorus fertilizers enhance the phytoextraction of cadmium through *Solanum nigrum* L. *Plants, 11*(3), 236. https://doi.org/10.3390/plants11030236

Maywald, M., & Rink, L. (2022). Zinc in human health and infectious diseases. *Biomolecules, 12*(12), 1748. https://doi.org/10.3390/biom12121748

Meagher, R. B. (2000). Phytoremediation of toxic elemental and organic pollutants. *Current Opinion in Plant Biology, 3*(2), 153–162.

Mittal, A., Singh, R., Chakma, S., Goel, G., & Birke, V. (2021). An integrated permeable reactive barrier and photobioreactor approach for simultaneous removal of nitrate, phosphate and hexavalent chromium: A combined batch and continuous flow study. *Bioresource Technology, 333*, Article 125201. https://doi.org/10.1016/j.biortech.2021.125201

Mocek-Płóciniak, A., Mencel, J., Zakrzewski, W., & Roszkowski, S. (2023). Phytoremediation as an effective remedy for removing trace elements from ecosystems. *Plants, 12*(8). https://doi.org/10.3390/plants12081653

Mohrazi, A., Ghasemi-Fasaei, R., Mojiri, A., & Safarzadeh Shirazi, S. (2023). Investigating an electro-bio-chemical phytoremediation of multi-metal polluted soil by maize and sunflower using RSM-based optimization methodology. *Environmental and Experimental Botany, 211*, Article 105352. https://doi.org/10.1016/j.envexpbot.2023.105352

Nedjimi, B. (2021). Phytoremediation: a sustainable environmental technology for heavy metals decontamination. *SN Applied Sciences, 3*(3). https://doi.org/10.1007/s42452-021-04301-4

Niedbała, G., Niazian, M., & Sabbatini, P. (2021). Modeling agrobacterium-mediated gene transformation of tobacco (*Nicotiana tabacum*)—A model plant for gene transformation studies. *Frontiers in Plant Science, 12*.https://doi.org/10.3389/fpls.2021.695110

Nilsson, L. (2022). *Genetically modified tobacco (Nicotiana tabacum) plants for an increased production of wax esters*. First cycle, G2E.

Njoku, K. L., Ude, E. O., Jegede, T. O., Adeyanju, O. Z., & Iheme, P. O. (2022). Characterization of hydrocarbon degrading microorganisms from Glycine max and Zea mays phytoremediated crude oil contaminated soil. *Environmental Analysis, Health and Toxicology, 37*(2), Article e2022008. https://doi.org/10.5620/eaht.2022008

Pons, S., Fournier, S., Chervin, C., Bécard, G., Rochange, S., Frei Dit Frey, N., & Puech Pagès, V. (2020). Phytohormone production by the arbuscular mycorrhizal fungus *Rhizophagus irregularis. PLoS ONE, 15*(10), e0240886. https://doi.org/10.1371/journal.pone.0240886

Poulios, E., Giaginis, C., & Vasios, G. K. (2020). Current state of the art on the antioxidant activity of sage (Salvia spp.) and its bioactive components. *Planta Medica, 86*(04), 224–238. https://doi.org/10.1055/a-1087-8276

Qiu, M., Liu, L., Ling, Q., Cai, Y., Yu, S., Wang, S., Fu, D., Hu, B., & Wang, X. (2022). Biochar for the removal of contaminants from soil and water: A review. *Biochar*, *4*(1). https://doi.org/10.1007/s42773-022-00146-1

Rashid, A., Schutte, B. J., Ulery, A., Deyholos, M. K., Sanogo, S., Lehnhoff, E. A., & Beck, L. (2023). Heavy metal contamination in agricultural soil: Environmental pollutants affecting crop health. *Agronomy (Basel, Switzerland), 13*(6), 1521. https://doi.org/10.3390/agronomy13061521

Rashmi, I., Kumawat, A., Munawery, A., Sreekumar Karthika, K., Kumar Sharma, G., Kala, S., & Pal, R. (2023). Soil amendments: An ecofriendly approach for soil health improvement and sustainable oilseed production. *IntechOpen*. https://doi.org/10.5772/intechopen.106606

Ratnasari, A., Syafiuddin, A., Mehmood, M. A., & Boopathy, R. (2023). A review of the vermicomposting process of organic and inorganic waste in soils: Additives effects, bioconversion process, and recommendations. *Bioresource Technology Reports, 21*, Article 101332. https://doi.org/10.1016/j.biteb.2023.101332

Rehman, S. U., De Castro, F., Aprile, A., Benedetti, M., & Fanizzi, F. P. (2023). Vermicompost: Enhancing plant growth and combating abiotic and biotic stress. *Agronomy (Basel, Switzerland), 13*(4), 1134. https://doi.org/10.3390/agronomy13041134

Riaz, U., Athar, T., Mustafa, U., & Iqbal, R. (2022). Economic feasibility of phytoremediation. In *Phytoremediation* (pp. 481–502). https://doi.org/10.1016/b978-0-323-89874-4.00025-x

Rocha, C. S., Rocha, D. C., Kochi, L. Y., Carneiro, D. N. M., dos Reis, M. V., & Gomes, M. P. (2022). Phytoremediation by ornamental plants: A beautiful and ecological alternative. *Environmental Science and Pollution Research International, 29*(3), 3336–3354. https://doi.org/10.1007/s11356-021-17307-7

Sabreena, Hassan, S., Bhat, S. A., Kumar, V., Ganai, B. A., & Ameen, F. (2022). Phytoremediation of heavy metals: An indispensable contrivance in green remediation technology. *Plants, 11*(9), 1255.https://doi.org/10.3390/plants11091255

Sadasivam, S. (2022). *Chemical science review and letters article cs205311542 478 role of ornamental plants in phytoremediation—An overview*. https://doi.org/10.37273/chesci.cs205309530

Sakhiya, A. K., Vijay, V. K., & Kaushal, P. (2022). Efficacy of rice straw derived biochar for removal of Pb^{+2} and Zn^{+2} from aqueous: Adsorption, thermodynamic and cost analysis. *Bioresource Technology Reports, 17*, Article 100920. https://doi.org/10.1016/j.biteb.2021.100920

Saman, R. U., Shahbaz, M., Maqsood, M. F., Lili, N., Zulfiqar, U., Haider, F. U., Naz, N., & Shahzad, B. (2022). Foliar application of ethylenediamine tetraacetic acid (EDTA) improves the growth and yield of brown mustard (*Brassica juncea*) by modulating photosynthetic pigments, antioxidant defense, and osmolyte production under lead (Pb) stress. *Plants, 12*(1), 115.https://doi.org/10.3390/plants12010115

Sarkodie, E. K., Jiang, L., Li, K., Yang, J., Guo, Z., Shi, J., Deng, Y., Liu, H., Jiang, H., Liang, Y., Yin, H., & Liu, X. (2022). A review on the bioleaching of toxic metal(loid)s from contaminated soil: Insight into the mechanism of action and the role of influencing factors. *Frontiers in Microbiology, 13*.https://doi.org/10.3389/fmicb.2022.1049277

Shah, N., Irshad, M., Hussain, A., Qadir, M., Murad, W., Khan, A., Awais, M., Alrefaei, A. F., & Ali, S. (2023). EDTA and IAA ameliorates phytoextraction potential and growth of sunflower by mitigating Cu-induced morphological and biochemical injuries. *Life (Basel, Switzerland), 13*(3), 759.https://doi.org/10.3390/life13030759

Sharma, J. K., Kumar, N., Singh, N. P., & Santal, A. R. (2023). Phytoremediation technologies and their mechanism for removal of heavy metal from contaminated soil: An approach for a sustainable environment. *Frontiers in Plant Science, 14*.https://doi.org/10.3389/fpls.2023.1076876

Sivaram, A. K., Subashchandrabose, S. R., Logeshwaran, P., Lockington, R., Naidu, R., & Megharaj, M. (2020). Rhizodegradation of PAHs differentially altered by C3 and C4 plants. *Scientific Reports, 10*(1), 16109. https://doi.org/10.1038/s41598-020-72844-4

Tan, H. W., Pang, Y. L., Lim, S., & Chong, W. C. (2023). A state-of-the-art of phytoremediation approach for sustainable management of heavy metals recovery. *Environmental Technology & Innovation, 30*, Article 103043. https://doi.org/10.1016/j.eti.2023.103043

Tarla, D. N., Erickson, L. E., Hettiarachchi, G. M., Amadi, S. I., Galkaduwa, M., Davis, L. C., Nurzhanova, A., & Pidlisnyuk, V. (2020). Phytoremediation and bioremediation of pesticide-contaminated soil. *Applied Sciences (Basel, Switzerland), 10*(4), 1217.https://doi.org/10.3390/app10041217

Tripathi, S., Singh, V. K., Srivastava, P., Singh, R., Devi, R. S., Kumar, A., & Bhadouria, R. (2020). Phytoremediation of organic pollutants: Current status and future directions. In: P. Singh et al (Eds.), *Abatement of environmental pollutants* (pp. 81–105). Elsevier.

Tussipkan, D., & Manabayeva, S. A. (2022). Alfalfa (*Medicago sativa* L.): Genotypic diversity and transgenic alfalfa for phytoremediation. *Frontiers in Environmental Science, 10*. https://doi.org/10.3389/fenvs.2022.828257

Venegas-Rioseco, J., Ginocchio, R., & Ortiz-Calderón, C. (2021). Increase in phytoextraction potential by genome editing and transformation: A review. *Plants, 11*(1), 86. https://doi.org/10.3390/plants11010086

Verma, A. (2022). Bioremediation techniques for soil pollution: An introduction. In *Biodegradation technology of organic and inorganic pollutants*.https://doi.org/10.5772/intechopen.99028

Vuković, A., Velki, M., Ečimović, S., Vuković, R., Štolfa Čamagajevac, I., & Lončarić, Z. (2021). Vermicomposting—Facts, benefits and knowledge gaps. *Agronomy (Basel, Switzerland), 11*(10), 1952. https://doi.org/10.3390/agronomy11101952

Vyas, P., Sharma, S., & Gupta, J. (2022). Vermicomposting with microbial amendment: Implications for bioremediation of industrial and agricultural waste. *Biotechnologia, 103*(2), 203–215. https://doi.org/10.5114/bta.2022.116213

Wang, Y., Narayanan, M., Shi, X., Chen, X., Li, Z., Natarajan, D., & Ma, Y. (2022). Plant growth-promoting bacteria in metal-contaminated soil: Current perspectives on remediation mechanisms. *Frontiers in Microbiology, 13*.https://doi.org/10.3389/fmicb.2022.966226

Wani, Z. A., Ahmad, Z., Asgher, M., Bhat, J. A., Sharma, M., Kumar, A., Sharma, V., Kumar, A., Pant, S., Lukatkin, A. S., & Anjum, N. A. (2023). Phytoremediation of potentially toxic elements: Role, status and concerns. *Plants, 12*(3), 429. https://doi.org/10.3390/plants12030429

Wu, B., Xiu, J., Yu, L., Huang, L., Yi, L., & Ma, Y. (2023). Degradation of crude oil in a co-culture system of Bacillus subtilis and *Pseudomonas aeruginosa. Frontiers in Microbiology, 14*, 1132831. https://doi.org/10.3389/fmicb.2023.1132831

Xiao, R., Ali, A., Xu, Y., Abdelrahman, H., Li, R., Lin, Y., Bolan, N., Shaheen, S. M., Rinklebe, J., & Zhang, Z. (2022). Earthworms as candidates for remediation of potentially toxic elements contaminated soils and mitigating the environmental and human health risks: A review. *Environment International, 158*, 106924. https://doi.org/10.1016/j.envint.2021.106924

Yan, A., Wang, Y., Tan, S. N., Mohd Yusof, M. L., Ghosh, S., & Chen, Z. (2020). Phytoremediation: A promising approach for revegetation of heavy metal-polluted land. *Frontiers in Plant Science, 11*.https://doi.org/10.3389/fpls.2020.00359

Yang, K. M., Poolpak, T., & Pokethitiyook, P. (2023). Rhizodegradation: The plant root exudate and microbial community relationship. In *Phytoremediation* (pp. 209–229). https://doi.org/10.1007/978-3-031-17988-4_11

Yang, Y., Liao, J., Chen, Y., Tian, Y., Chen, Q., Gao, S., Luo, Z., Yu, X., Lei, T., & Jiang, M. (2022). Efficiency of heterogeneous chelating agents on the phytoremediation potential and growth of *Sasa argenteostriata* (Regel) E.G. Camus on Pb-contaminated soil. *Ecotoxicology and Environmental Safety, 238*, 113603. https://doi.org/10.1016/j.ecoenv.2022.113603

You, Y., Dou, J., Xue, Y., Jin, N., & Yang, K. (2022). Chelating agents in assisting phytoremediation of uranium-contaminated soils: A review. *Sustainability, 14*(10), 6379. https://doi.org/10.3390/su14106379

Zarull, M. A., Hartig, J. H., & Krantzberg, G. (2002). Ecological benefits of contaminated sediment remediation. In *Reviews of environmental contamination and toxicology* (pp. 1–18). https://doi.org/10.1007/978-1-4757-4260-2_1

Zaynab, M., Al-Yahyai, R., Ameen, A., Sharif, Y., Ali, L., Fatima, M., Khan, K., & Li, S. (2022). Health and environmental effects of heavy metals. *Journal of King Saud University. Science, 34*(1), Article 101653. https://doi.org/10.1016/j.jksus.2021.101653

Zeng, G., Qiao, S., Wang, X., Sheng, M., Wei, M., Chen, Q., Xu, H., & Xu, F. (2021). Immobilization of cadmium by Burkholderia sp. QY14 through modified microbially induced phosphate precipitation. *Journal of Hazardous Materials, 412*, 125156. https://doi.org/10.1016/j.jhazmat.2021.125156

Zhang, H., Xu, Y., Kanyerere, T., Wang, Y.-S., & Sun, M. (2022). Washing reagents for remediating heavy-metal-contaminated soil: A review. *Frontiers in Earth Science, 10*.https://doi.org/10.3389/feart.2022.901570

Zhao, Y., Li, X., Li, Y., Bao, H., Xing, J., Zhu, Y., Nan, J., & Xu, G. (2022). Biochar acts as an emerging soil amendment and its potential ecological risks: A review. *Energies, 16*(1), 410. https://doi.org/10.3390/en16010410

Zulkernain, N. H., Uvarajan, T., & Ng, C. C. (2023). Roles and significance of chelating agents for potentially toxic elements (PTEs) phytoremediation in soil: A review. *Journal of Environmental Management, 341*(117926), Article 117926. https://doi.org/10.1016/j.jenvman.2023.117926

Symbiotic Relationship of Microbial Communities and Vermicompost

Aisha Bisola Bello, Idris Abdullahi Dabban, Ibrahim Mohammed Hussaini, Wuna Muhammad Muhammad, Abdulsamad Omotayo Aiyelabegan, and Adejoh Suleiman Ocholi

Abstract Vermicomposting is a process where organic wastes are degraded through the earthworm–microorganisms synergistic relationship. Although vermicomposting effectively reduces organic biomass and in turn generates quality bio-fertilizers required for plant growth and soil improvement, there is sparse knowledge about the microbial communities involved in the decomposition process. On the basis of habitats, sizes, and feeding habits, vermicompost earthworms are divided into epigeic, anecic, and endogeic, with the epigeic earthworms being a better option for vermicomposting as a result of their rapid conversion of organic matter to vermicast. The vermicast of earthworms is a high-quality nutrient-rich fertilizer containing nitrogen-fixing bacteria, auxin, gibberellin, micronutrients, beneficial microbes, phosphate-solubilizing bacteria, humus, and nitrogen, phosphorus, and potassium (NPK). Microorganisms involved in vermicompost are bacteria, fungi, and actinomycetes, with soil improvement and plant protection capacities. The positive impact of the symbiotic relationship between earthworms and microorganisms has been recognized. The earthworm's ability to degrade organic matter is ascribed to the microbial communities residing in the gut of the earthworms and in turn establishes a suitable environment through aeration, burrowing, secretion, and cast for an enhanced microbial activity. In view of these, this chapter highlights the biotic and

A. B. Bello (✉) · I. A. Dabban
Department of Biological Sciences, Federal Polytechnic Bida, Bida, Nigeria
e-mail: aisha.bisolabello@fedpolybida.edu.ng

I. M. Hussaini
Department of Microbiology, Faculty of Life Sciences, Ahmadu Bello University Zaria, Zaria, Nigeria

W. Muhammad Muhammad
Department of Microbiology, Faculty of Life Sciences, Federal University of Technology Minna, Minna, Nigeria

A. O. Aiyelabegan
Helix Biogen Institute, Ogbomoso, Nigeria

A. S. Ocholi
Department of Biology, Kogi State College of Education, Kabba, Nigeria

S. A. Aransiola and N. R. Maddela (eds.), *Vermitechnology: Economic, Environmental and Agricultural Sustainability*, https://doi.org/10.1007/978-3-031-96212-7_13

abiotic components of vermicompost, phases involved in the vermicompost process, and the symbiotic relationship of earthworms and some groups of microorganisms.

Keywords Bacteria · Earthworms · Fungi · Vermicompost · Vermicast

1 Introduction

Vermicomposting is a biotechnological, non-thermophilic, bio-oxidative process that involves earthworms (*Eisenia fetida*) and associated microbes (Abdelsattar et al., 2023; Pathma & Sakthivel, 2012). It is a process of waste biodegradation through the synergistic actions of earthworms and microbial communities through their guts, resulting to cast also known as earthworm manure (Ali et al., 2015; Domínguez et al., 2019; Villar et al., 2017). However, understanding vermicomposting of crop residue and manure in combination with nutrient-rich sources in the soil is required (Raza et al., 2022).

The essential factors on which the vermicomposting process depends are abiotic and biotic. The abiotic factors include pH, temperature, aeration, moisture content, C/N ratio, and many others (Kaur, 2020; Yadav et al., 2022), while the biotic factors include mainly earthworms and microorganism, although there are also abundant protozoa and many animals of varying sizes, including nematodes, and microarthropods. Earthworms are macroscopic clitellate oligochaete annelids that live in soil. They are major components of the soil fauna in a wide variety of soils and climates and are involved directly or indirectly in biodegradation of organic matter such as accelerating decomposition and recycling elements in association microbes (Munnoli, 2015; Zhang et al., 2023). Based on feeding habitat, earthworm species have been classified into epigeic, anecic, and endogeic (Bhat et al., 2017; Medina-Sauza et al., 2019; Singh et al., 2020). Among which epigeic species have been reported to have better vermicomposting potential with *E. fetida* been the most important earthworm species for vermicomposting due to its high rates of consumption, digestion and assimilation of organic matter, tolerance to a wide range of environmental factors, short life cycles, high reproductive rates, and endurance and resistance to handling (Dohaish, 2020).

Though microorganisms are principally responsible for the biodegradation process in vermicomposting or vermiremediation because of their presence in the gut earthworms, their activity and population are stimulated by earthworms through their feeding, aeration, and the excretion of casts (Aransiola et al., 2022; Gómez-Brandón et al., 2012; Thakur et al., 2022).

The biotic interactions between microbial decomposers and the soil fauna including earthworms include competition, mutualism, predation, and facilitation (Iven et al., 2023; Sampedro & Domınguez, 2008). These biotic components are found in a range of trophic levels; some feed primarily on microorganisms (microbial feeders), organic matter (detritivores), or a combination of organic matter and microorganisms (microbial detritivores), while others feed on animals (carnivores). The diverse microbial community in vermicomposting serves as the primary producers; they are numerically abundant and are responsible for breaking down and mineralization of organic matter, while the secondary and higher level consumers existing alongside microbes, feeding on and dispersing them throughout the organic matter.

The vermicomposting process takes place in two different phases based on the activity of earthworms; this include the active phase, during which earthworms process organic matter, altering its physical state and microbial composition; the duration of this phase mostly depends on the species, density of the main drivers of the vermicomposting process, and the rates at which they ingest the organic matter. The active phase is then followed by the maturation phase characterized by the displacement of the earthworms towards fresher layers of undigested waste, during which the microbial decomposers takeover the decomposition of the earthworm's processed waste (Aira & Domınguez, 2008; Ducasse et al., 2022).

Vermicomposting systems sustain a complex food web, in which earthworms interact intensively with microorganisms and other fauna within the decomposer community, accelerating the stabilization of organic matter and greatly modifying its physical and biochemical properties (Domínguez et al., 2010; Munnoli, 2015; Munnoli et al., 2010). These microorganisms and earthworms act symbiotically to accelerate and enhance the decomposition of organic matter resulting in faster mineralization and humification (Chowdhury et al., 2022; Emperor & Kumar, 2015). The earthworms augment the symbiotic gut microflora with secreted mucus and water to increase their degradation of ingested organic matter (Edwards & Arancon, 2022; Pramanik et al., 2007). The activities of the earthworm increase the surface area available for microbial activity (Ahmed & Al-Mtairi, 2022; Gómez-Brandón et al., 2012). Such activities enhance the turnover rate and productivity of microbial communities, thereby increasing the rate of decomposition. The symbiosis between earthworms and microorganisms in vermicomposting has been reported to hasten the decomposition process to a significant extent when compared to traditional composting methods that utilize only microorganisms (Bhat et al., 2017). Earthworms indirectly alter the dynamics of chemical processes in organic matters, via comminution and influencing the activity of the microflora (Pathma & Sakthivel, 2012; Yadav & Singh, 2023). Microbes serve as sources of protein-rich food for earthworms, and they work synergistically to increase organic waste decomposition (Emperor & Kumar, 2015). Vermicomposting has been reported to modify the microbial community of organic matter in diverse ways (Chen et al., 2022; Pathma & Sakthivel, 2012). For instance, a study on the vermicomposting of municipal solid wastes with a high organic component was conducted. The amount of total organic carbon was found to have significantly decreased, while the variety and evenness of the bacterial community had significantly increased. The findings showed that the composition and structure of the bacterial population were significantly influenced by earthworms (Srivastava et al., 2021). Another study by Bianco et al. (2022) examined microbial populations in the pretreatment stage and vermicomposting stage of brewer's spent grain, which is particularly rich in cellulose, hemicellulose, lignin, and nitrogen. The stabilization of organic matter was indicated by a decline in the total organic carbon accompanied by an increase in the total nitrogen. In comparison with the initial materials, the Chao1 and Shannon index of bacteria and fungus clearly increased in the final vermicompost. Similarly, Gómez-Brandón et al. (2010) discovered that during the vermicomposting of grape marc, the earthworms' presence altered the organization of the microbial community and led to a decreased abundance of bacteria (except for Gram-negative bacteria) and fungi. Nechitaylo et al. (2010) also reported on the effect of earthworms on bacterial diversity in soil and observed that the bacterial counts in gut/vermicompost were higher than those in the surrounding soil.

Earthworm activity profoundly affects the physical, chemical, and biological properties of organic substrates. These worms voraciously feed on such substrates, which are grounded into fine powder in the gizzard before been acted upon by digestive enzymes, microorganisms, and other fermenting substances further aiding their breakdown within the gut. The earthworms only utilizing a small portion of it for their growth and excrete a large proportion in a half-digested form 'casts' which can be easily decomposed into vermicompost by a wide range of microorganisms from the guts of earthworms (Pathma & Sakthivel, 2012). The decomposition of organic matter during vermicomposting process through fragmentation and ingestion is due to the earthworm's gut-associated processes, which includes all the modifications that the organic matter and the microorganisms undergo during transit in the gut of worms (Gómez-Brandón et al., 2022) (Fig. 1). Such modifications include the addition of sugars and other substances, modification of the microbial diversity and activity, modification of the micro-fauna populations, homogenization, and the intrinsic processes of digestion, assimilation, and production of mucus and excretory substances like ammonia and urea, which serve as nutrients for microorganisms. These endosymbiotic microorganisms enhance decomposition through the production of various extracellular enzymes that degrade various compounds in the organic matter. Other physical modifications of the substrate caused by the burrowing and tunnelling activities of the earthworms aerate the substrate and enable water, nutrients, oxygen, and microbes to move through it, further enhancing decomposition (Aira et al., 2008; Das & Paul, 2023).

Various bacteria and fungi species have been reported in vermicomposts produced by different earthworm species. Yasir et al. (2009) reported *Proteobacteria, Bacteroidetes,* and *Actinobacteria* in vermicompost containing *E. fetida.* Similarly, in vermicompost containing *E. fetida,* Satpathy et al. (2020) reported *Actinomyces israelii, Azotobacter, Micrococcus luteus, Bacillus cereus, Bacillus subtilis, Pseudomonas aeruginosa,* and *Enterobacter* species. *Azotobacter, autotrophic Nitrosomonas,* and *Nitrobacter* were observed in *Eudrilus* species vermicomposts by Gopal et al. (2009). Emperor and Kumar (2015) isolated *Klebsiella pneumoniae, P. aeruginosa, Enterobacter aerogenes, Morganella morganii, Proteus vulgaris, Escherichia coli, Enterococcus faecium, B. subtilis,* and *B. cereus* in vermicompost of *Eudrilus eugeniae* and *K. pneumoniae, P. aeruginosa, Enterobacter acrogens, Citrobacter diversus,* and *B. subtilis* in vermicompost of *E. fetida.* Their study also reported the following fungi species in vermicompost *Rhizopus nigricans, Chaetomium globosum, Aspergillus flavus, Aspergillus niger, Aspergillus nidulans,*

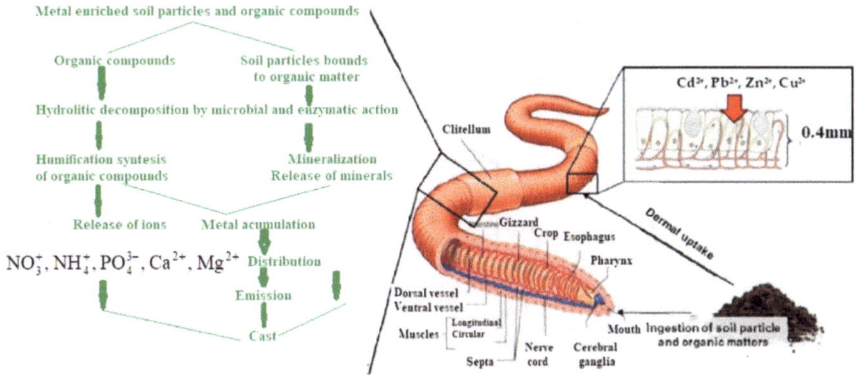

Fig. 1 The process of vermicomposting by earthworms (Rusănescu et al., 2022)

Cladosporium herbarum, Fusarium oxysporum, Fusarium moniliforme, and *Penicillium citrinum.* All these microbes found in the vermicompost were reported to be present in the gut of the earthworm species. According to Singleton et al. (2003), the gut of earthworms contains 'nitrogen-fixing' and 'decomposer microbes' which are released along with nutrients in their final excreta. Bhat et al. (2017) reported that ingested material while passing through the gut of earthworms increases the number of microbes up to 1000-fold. Karanpantzou et al. (2023) also reported that ingested waste material yields a decreased C/N ratio, increased total nitrogen, neutral pH, excess enzymatic activities, and rich source of beneficial microorganisms such as nitrogen-fixing bacteria, plant growth stimulating, and enzyme-producing bacteria.

2 Components of Vermicompost

The components of vermicompost are generally divided into abiotic and biotic (Fig. 2). The abiotic components include all physical and chemical parameters present in vermicompost that can affect the success of vermicomposting. The physical component includes soil type, temperature, pH, moisture content, and aeration, while the chemical components comprise of carbon, nitrogen, phosphorus, potassium, calcium, sodium, etc. Biotic component of vermicompost includes all living organisms present and/or their active products present in the mixture. They include stocking density of diverse species of earthworms and microorganisms including bacteria, actinomycetes, and fungi. Products of these organisms are enzymes, plant growth promoters, and regulators as well as other important metabolites (Bolong & Saad, 2020).

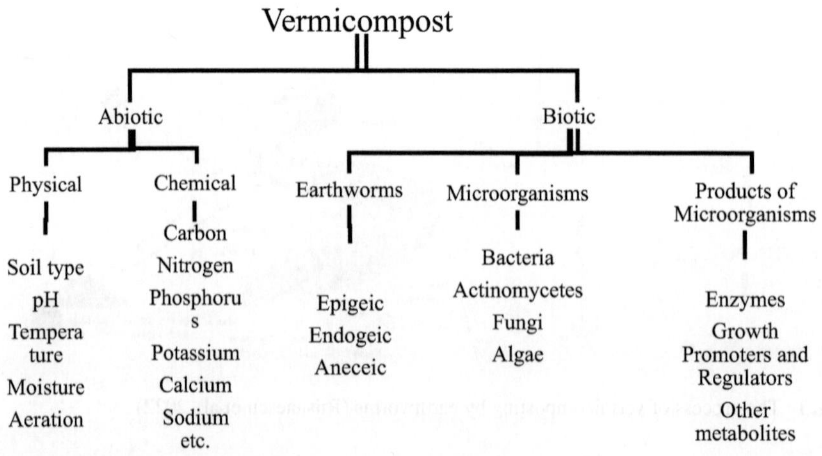

Fig. 2 Component of vermicompost

2.1 Abiotic Components of Vermicompost

2.1.1 Physical Components of Vermicompost

(a) Soil Type

The abundance and distribution of worms in addition to overall efficiency of vermicompost are affected by the nature of soil (Laossi et al., 2009). The texture of the soil used in vermicompost affect other characteristics such as nutrient availability, hydration, and ion exchange capacity. Loamy soil supports most species of earthworms in vermicompost than sandy and clay soils (Ahmed & Al-Mutairi, 2022). Although reports reveal clay soil sometimes support few species earthworms found in vermicompost such as *Aporrectodea caliginosa*, the salt content of the soil also affects the abundance of earthworm in the compost material (Hendrix et al., 2008).

(b) pH

The pH of vermicompost preparations varies from different sources from slightly acidic to neutral and slightly acidic in nature with pH value range of 6.5–8.2 (Bolong & Saad, 2020; Sinha et al., 2009). However, the pH of vermicompost can decrease or increase significantly until the process is complete. This may be due to the activities of earthworms present in the mixture, the presence of organic acid, high mineral content, application of fertilizer, nitrogen and phosphorus mineralization, and breakdown of organic matter by microorganisms to simultaneously produce organic, fulvic, and humic acids as well as carbon dioxide (Pathma & Sakthivel, 2012). However, pH lower than 7.2 may slow down decomposition and pH higher than 8.0 may result to the release of ammonia with an unpleasant odour (Vijayshankar et al., 2024).

(c) Temperature

The temperature of vermicompost greatly influences the metabolism, growth, reproduction, behaviour, and respiration of earthworms and other microorganisms that may be present in the mixture (Ahmed & Al-Mutairi, 2022; Hendrix et al., 2008). The survival of earthworms in vermicompost is at certain maximum range of temperatures; meanwhile, increasing temperature range above the limits kills off the earthworms. However, earthworms' tolerance to temperature changes in vermicompost varies from species to species (Ahmed & Al-Mutairi, 2022). Temperature range for optimum application of vermicompost has been reported to be 10–15 °C, but the most excellent condition for activities of earthworm is observed to be in the night where temperature of vermicompost is 10 °C and below (Duiker & Stehouwer, 2007).

(d) Moisture Content

The body weight of earthworms is made up of 75–90% water. Also, for adequate growth and development, the respiration of earthworms requires a moist skin and humid surface blood capillaries. Therefore, sufficient moisture is required in vermicompost for proper activity of earthworms and other microorganisms present in the compost material (Ahmed & Al-Mutairi, 2022; Bohlen et al., 2004). The need for a humid environment in vermicompost varies between species of earthworms found around the world. Generally, for optimum activity, about 60–70% humidity of the vermicompost material is necessary. Although some studies suggest that for optimum performance, the compost material should be retained at 45% moisture content (Bolong & Saad, 2020). In other words, the vermicompost activity using damp soils is higher than desiccated soil; hence, desiccation of the vermicompost matter is always prevented (Bohlen et al., 2004; Hendrix et al., 2008).

(e) Aeration

Aeration is a major physical component and factor that affects the nutritional quality and efficiency of vermicompost. Adequate aeration is required for all vermicompost materials to allow for oxygen supply and to make oxygen readily available to the earthworms and microorganisms present in the compost matter (Bolong & Saad, 2020).

2.1.2 Chemical Components of Vermicompost

Apart from environmental factors, the chemical composition of vermicompost is also dependent on other factors such as species and population of earthworm and microorganisms (amylolytic, lignolytic, cellulolytic, or nitrogen-fixers) as well as initial substrate found in the compost material (Vijayabharathi et al., 2015). The chemical constituents that make up most vermicompost materials (Fig. 3) include

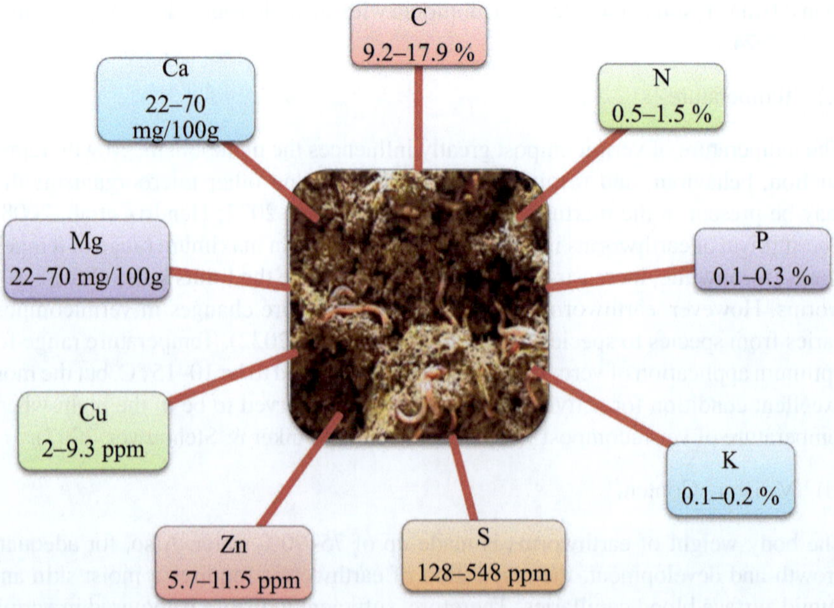

Fig. 3 Chemical components of vermicompost (Vijayabharathi et al., 2015)

organic carbon (C), phosphorus (P), nitrogen (N), potassium (K), magnesium (Mg), calcium (Ca), copper (Cu), zinc (Zn), and sulfur (S) (Kale, 1995; Vijayabharathi et al., 2015). The concentration of exchangeable K^{+}, Ca^{2+}, and Mg^{2+} in vermicompost is usually higher that the primary substrate which is an indication of the conversion of chemicals to readily available nutrient forms (Edwards, 1998; Vijayabharathi et al., 2015).

2.2 *Biotic Components of Vermicompost*

2.2.1 Earthworms

Earthworms are known to have great influence on volume, flora, and fauna of the soil. They have a crucial function in soil formation, turnover of carbon, degradation of cellulose, and accumulation of humus. Thus, the activities of earthworms greatly change the physical and chemical as well as the biological characteristics of the compost soil material. Although earthworms require little part of wastes for growth and multiplication, they voraciously feed on large portion of organic wastes and expel a great quantity of the consumed organic waste material in a semi-digested form (Jambhekar, 1992; Pathma & Sakthivel, 2012). The partly digested organic waste is rapidly transformed into vermicompost by microorganisms, hormones, and enzymes present in the intestine of the earthworm (Dominguez & Edwards, 2004; Nagavallemma et al., 2004; Pathma & Sakthivel, 2012).

Earthworms are classified under the phylum annelid as they are cylindrical, narrow, long, and bilaterally symmetrical. They are shimmering dark brown segmented invertebrates, enveloped in a fragile cuticle. Depending on ecological conditions and species type, they have a life span of 3–7 years. Earthworms are hermaphrodites, and they are categorized into epigeic, endogeic, and anecic (Fig. 4) on the basis of size, behaviour, nutrition, and trophic and ecological function (Bhatnagar & Palta, 1996; Eisenhauer & Eisenhauer, 2020) (Table 1).

Epigeic Earthworms

This class of earthworms is small in size with a short life cycle and high reproduction rate, and their body is evenly pigmented. They are found within litters on soil surface where they feed on and mineralize litters (Eisenhauer & Eisenhauer, 2020). Epigeic earthworms are phytophagous as they most often do not consume soil. Meanwhile, they rapidly convert organic matter into vermicompost with the aid of an active gizzard, making them an effective waste degraders and nutrient liberators. Epigeic earthworms used in vermicomposting include *E. fetida*, *Bimastus eiseni*, *Lumbricus castaneus*, *Lumbricus festivus*, *Lumbricus rubellus*, *Eiseniella tetraedra*,

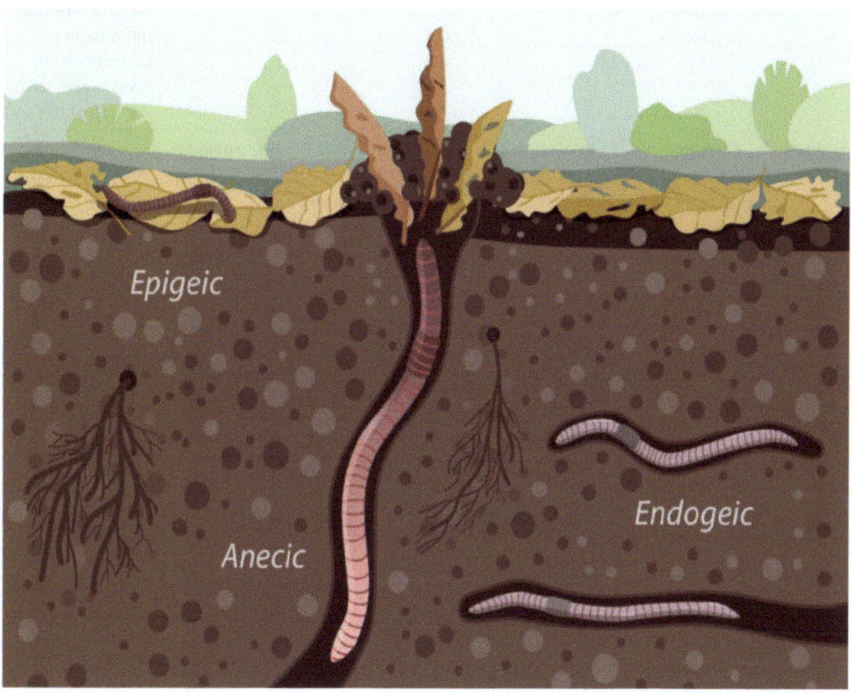

Fig. 4 The three main ecological classes of earthworm (Eisenhauer & Eisenhauer, 2020)

Table 1 Classification of earthworms, habitat, features, benefits, and examples (Pathma & Sakthivel, 2012)

Class of earthworm	Epigeics	Endogeics			Anecics
		Polyhumic	Mesohumic	Oligohumic	
Habitat	Soil surface layers, compost, and leaf litters	Uppermost soil (A1)	Horizons A and B	Horizons B and C	Deep permanent burrows in the soil
Features	Smaller size, uniformly pigmented body, active gizzard, short life cycle, high reproduction rate, and phytophagous	Small size, forms horizontal burrows, rich soil feeder	Medium size, forms extensive horizontal burrows, bulk (A1) soil feeder	Very large in size, forms extensive horizontal burrows, feeds on poor, deep soils	Large in size, dorsally pigmented, forms extensive, deep, vertical permanent burrows, low rate of reproductive, nocturnal, phytogeophagous
Benefits	Effective degraders and nutrient liberators, efficient compost producers	Facilitate prominent modifications in the physical makeup of the soil; efficient use of energy from poor soils for augmentation of the soil			Forms vertical burrows affecting water–air association and effective nutrient mixing through movement of nutrients from deep layers to the surface of the soil
Examples	*Eisenia fetida, Lumbricus rubellus, L. castaneus, L. festivus, Eiseniella tetraedra, Bimastus minusculus, B. eiseni, Dendrodrilus rubidus, Dendrobaena veneta, D. octaedra*	*Octolasion cyaneum, O. lacteum*	*Pontoscolex corethrurus, Allolobophora chlorotica*	*Amynthas* sp.	*L. terrestris, L. polyphemus, A. longa*

Bimastus minusculus, Dendrobaena octaedra, Dendrobaena veneta, Dendrodrilus rubidus (Pathma & Sakthivel, 2012).

Endogeic Earthworms

Endogeic earthworms cause elaborate changes in the soil's physical composition. The life cycle of this class of worm is medium, and they are characterized by faintly pigmented body (Eisenhauer & Eisenhauer, 2020). They vary in size from small to large with the formation of broad parallel burrows within the soil. Endogeic earthworms survive by feeding on particulate organic matter as well as the soil, hence making them geophagous in nature. This group has potentials for major soil enhancement since they are known to effectively exploit energy from poor soils. Examples of endogeic earthworm include *Aporrectodea trapezoides, A. caliginosa, Aporrectodea rosea, Octolasion lacteum, Octolasion cyaneum, Amynthas* sp., *Allolobophora chlorotica,* and *Pontoscolex corethrurus* (Pathma & Sakthivel, 2012). Endogeics are further categorized into three:

(i) Polyhumic: found on the top soil, small-sized, and feed on rich soil.
(ii) Mesohumic: established in the A and B horizon of the soil, medium-sized, and feed on bulk (A1) soil.
(iii) Oligohumic: dwell in B and C horizons of the soil, very large worms that feed on deep poor soil.

Anecic Earthworms

These are bigger, dorsally pigmented, nocturnal worms, having low reproduction rate and phytogeophagous mode of feeding. Anecics form midden, permanent, and extensively deep, vertical burrows in which they live in (Eisenhauer & Eisenhauer, 2020). The vertical burrows they form affect the air–water relationship by accelerating their movement from deep layers to the soil surface which facilitates the effective integration and mixing of nutrients. Examples of anecics include *Aporrectodea longa, Lumbricus terrestris,* and *Lumbricus polyphemus* (Kooch & Jalilvand, 2008). Epigeic and anecic earthworms have been greatly utilized in vermicomposting processes (Asha et al., 2008). The epigeic earthworms such as *E. fetida, E. eugeniae, Perionyx excavatus,* and *Eisenia andrei* have been exploited for the conversion of organic waste materials into vermicompost matter (Gebrehana et al., 2023; Hijam et al., 2022; Hussain et al., 2018; Munnoli et al., 2010; Suthar & Singh, 2008).

2.2.2 Microorganisms and Their Metabolites

Depending on the primary substrate of the vermicompost, the community of microorganisms including bacteria, fungi (yeast and moulds), and actinomycetes differs greatly with species of earthworm and subsequent vermicompost matter. Communities of mesophilic microorganisms have been isolated from vermicompost with earthworm species, for example *Allolobophora caliginosa, A. terrestris,* and *L. terrestris* (Vijayabharathi et al., 2015). The age of earthworm does not influence the microbial groups (Fernández-Gomez et al., 2012). However, the number of microorganisms may differ between species of earthworm due to environmental conditions, ecological group, food type, and their varying capabilities in digestion and assimilation of microbial biomass (Brown & Doube, 2004). Combination of these vermicompost features makes it a hotspot of microorganisms. *E. fetida* is known to accommodate indigenous microflora especially in the gut of the earthworm (Toyota & Kimura, 2000; Vijayabharathi et al., 2015).

During the movement of ingested materials through the gut of earthworms, microbial populations present in the gut of the worms can be rapidly multiplied up to 1000-fold (Edwards & Fletcher, 1988). However, the structural characteristics of the microflora of vermicompost are affected by the interaction of the earthworm with the biological as well as physicochemical components of the compost material (Monroy et al., 2009).

Bacteria

Bacteria populations from the phyla Bacteroidetes, Actinomycetes, Firmicutes, Chlorobi, Proteobacteria, and Planctomycetes have been isolated the earthworm *E. fetida* in vegetable waste compost where the anaerobic group Bacteroidetes was the most predominant (Huang et al., 2013). In compost from goat manure, *Bacillus, Pseudomonas,* and *Microbacterium* are the predominant groups of bacteria (Pathma & Sakthivel, 2013). Homemade and commercial composts (municipal solid waste, poultry litter, and sewage sludge) contain a diverse group of bacteria such as Actinobacteria: *Cellulosimicrobium cellulans, M. oxydans,* and *Microbacterium* spp.; Proteobacteria: *Pseudomonas* spp. and *Pseudomonas libanensis*; *Firmicutes: B. cereus, Bacillus benzoevorans, Bacillus megaterium, Bacillus macroides, B. subtilis, Bacillus pumilus,* and *Bacillus licheniformis;* and ungrouped genotypes: *Sphingomonas* spp. and *Kocuria palustris* (Vaz-Moreira et al., 2008). Meanwhile, Bacteroidetes, Alphaproteobacteria, Betaproteobacteria, Flavobacteria, *Pseudomonas* spp., Alphaproteobacteria, and *Alcaligenes* spp. (Betaproteobacteria) have been isolated from the earthworm vermicompost of *A. caliginosa* and *L. terrestris* (Nechitaylo et al., 2010).

Fungi

Fungi species are also present in vermicompost matter from which the phyla Saccharomycetes, Tremellomycetes, and Lecanoromycetes are found to be the most dominant within the primary materials of the vermicompost matter (Bonito et al., 2010). Other fungi groups found in vermicompost include Sordariomycetes, Pezizomycetes, Agaricomycetes, Saccharomycetes, Orbiliomycetes, and Eurotiomycetes (Bonito et al., 2010; Huang et al., 2013). *Dactylaria biseptata, Paecilomyces* spp., *Cephaliophora tropic, Williopsis californica, Trichoderma* spp., and *Geotrichum* spp. are some beneficial fungi established in vermicompost (Harman, 2006; Siddiqui & Mahmood, 1996; Vaz-Moreira et al., 2008).

Actinomycetes

Actinomycetes are another important group of microorganisms associated with vermicomposts, specifically the gut flora of earthworm (Jayasinghe & Parkinson, 2009). The antagonistic effect of actinomycetes makes them more predominant than fungi in vermicompost (Jayasinghe & Parkinson, 2009; Pathma & Sakthivel, 2013). Major actinomycetes group found in vermicompost includes *Streptomyces* spp., *Micromonospora* spp., and *Rhodococcus* (Huang et al., 2013; Yasir et al., 2009). *Streptomyces* spp. is a well-known antibiotic producer. Other microbial populations present in the earthworm's gut especially pathogenic fungi, Gram-positive bacteria, and litter-decomposing organisms are inhibited by actinomycetes, which make actinomycetes and other microorganisms resistant to antibiotics predominant in the gut of the earthworm; hence, they are applied as biocontrol agents against various pathogens of plants (Pathma & Sakthivel, 2013).

Enzymes in Vermicompost

The activities of amylase, cellulose, protease, invertase, protease, peroxidase, phosphatase, xylanases, dehydrogenase, urease, and various other enzymes are present in vermicompost matter (Devi et al., 2009). Availability of these enzymes in vermicompost is attributed to metabolic activities of microorganisms present in the earthworms' gut (Pathma & Sakthivel, 2013). Although there is a wide range of enzymes in vermicompost, the vermicomposting process gives rise to fluctuations in their activities which is attributed to the changing population of microorganisms at every stage of the process (Jouquet et al., 2006).

Plant Growth Promoters and Other Metabolites

Vermicompost matter contains plant growth promoting metabolites, for example auxins, cytokinins, gibberellins, and indole acetic acid (IAA) in addition to humic acids. These compounds are in significant concentrations and from microbial source (Atiyeh et al., 2002). However, the liquid extract of vermicompost known as the 'vermiwash' contains an assortment of excretory metabolites from earthworm as well as microorganisms (Pathma & Sakthivel, 2013). These metabolites have biocidal and plant growth promoting characteristics due to the occurrence of antibiotic compounds in addition to macro- and micronutrient (Fig. 5). Thus, vermicompost and its liquid portion (vermiwash) are ample source of plant growth-promoting substances (Edwards et al., 2004).

3 Phases of Vermicomposting

By combining the efforts of earthworms and microorganisms, vermicomposting is a special process that takes place in an earthworm's gut to turn organic wastes into organic fertilizer or vermicompost (El-Haddad et al., 2014). In a nutshell, when an earthworm consumes organic waste, the substrate travels through its stomach and

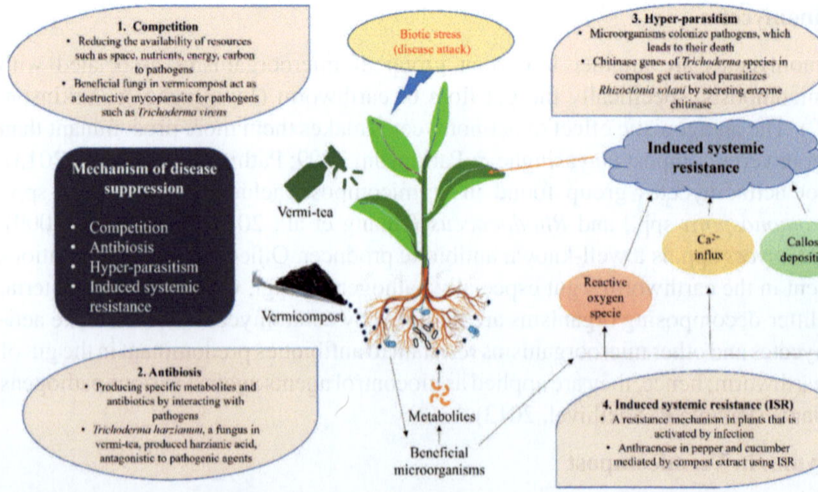

Fig. 5 Mechanisms of disease suppression in plants using vermicompost and its derivative (Rehman et al., 2023)

is broken down by helpful bacteria in the worm's intestine. Vermicompost, a finely split peat-like substance that is easily accessible to plants, is created in the digestive system by the breakdown of substrate by mucus or chemical secretions, enzymes, and antibiotics (Adhikary, 2012; Naidoo et al., 2017; Pathma & Sakthivel, 2012, 2014).

Earthworms consume fungus and other readily available microbes throughout this intricate breakdown process to create vermicompost, which is more enriched in comparison with the starting state of the digested wastes. Both physical and biological elements are involved in the vermicomposting process. Fragmentation, turnover, and aeration are examples of physical processes. Enzymatic digestion, nitrogen enrichment, and the transformation of both inorganic and organic materials are examples of biochemical processes (Edwards & Lofty, 1972; Pierre-Louis et al., 2021).

Vermicomposting is different from composting in that it uses earthworms and microorganisms to bio-oxidize and stabilize organic waste. There are two stages in the mesophilic phase of vermicomposting: active and maturation phases (Pathma & Sakthivel, 2012).

3.1 Active Phase of Vermicomposting

Active phase is the first phase of vermicomposting, and it is characterized by intense activity of earthworm (Gómez-Brandón et al., 2019). Red wigglers, the most common earthworm used for vermicomposting, may eat up to 75% of their body weight per day. Earthworms need oxygen and water, which they exchange through their skin. They weigh around 0.2 g. Earthworms decompose organic waste in this phase by consuming, breaking up, and decreasing the amount of the waste as well as by absorbing quickly biodegradable substances which are sugars, organic acids, proteins, and peptides (Domínguez et al., 2019; Fornes et al., 2012).

Similar to composting, the length of the active phase varies and depends on the kind and density of earthworms, which are the process's primary drivers, as well as the rates at which they ingest and digest waste (Domínguez et al., 2010).

The impact of earthworms on the breakdown of organic waste during the vermicomposting process is mostly caused by mechanisms related to gut-associated processes (GAPs). All the changes that the bacteria and decomposing organic waste go through throughout the intestinal transit are included in these processes. Addition of sugars and other substances such as enzymes (proteases, lipases, cellulases, chitinase), alteration of microbial diversity and activity, alteration of microfauna populations, homogenization, and intrinsic processes of digestion, assimilation, and production of mucus and excretory substances like urea and ammonia, which constitute a readily assimilable pool of nutrients for microorganisms, are some of these modifications (Domínguez et al., 2009; Madhushani et al., 2024).

Additionally, the endosymbiotic microorganisms found in earthworms' guts facilitate decomposition. These microorganisms create extracellular enzymes that break down cellulose and phenolic chemicals, speeding up the breakdown of material that has been consumed. Aeration and homogeneity of the substrate are two further physical changes made by earthworm burrowing that encourage microbial activity and accelerate decomposition (Domınguez, 2004).

3.2 *Maturation Phase of Vermicomposting*

The maturation phase is characterized by the earthworms moving away from older layers of digested waste, when bacteria take up the job of breaking down the already processed waste by the earthworm (Domınguez, 2004).

After GAPs are finished, the resulting earthworm castings proceed through cast-associated processes (CAPs), which are more intimately related to ageing processes, the activity of the microflora and microfauna present in the substrate, and the physical change of the ejected materials (Aira et al., 2005; Madhushani et al., 2024).

The earthworm's burrowing and ingestion of organic matter help to mix and aerate the soil, promoting the growth of microorganisms. As the earthworm digests organic matter, it also excretes castings that are rich in nutrients, which serve as a food source for the microorganisms. The earthworm's displacement to a new layer allows for the microorganisms to continue to decompose the previously processed biological material. This creates a dynamic and interconnected system where the earthworm and microorganisms work together to break down and recycle organic matter, leading to the improvement of soil health and fertility (Fig. 6).

Through vermicomposting, important plant nutrients like nitrogen, potassium, phosphorus, and calcium that are present in the organic waste are transformed into

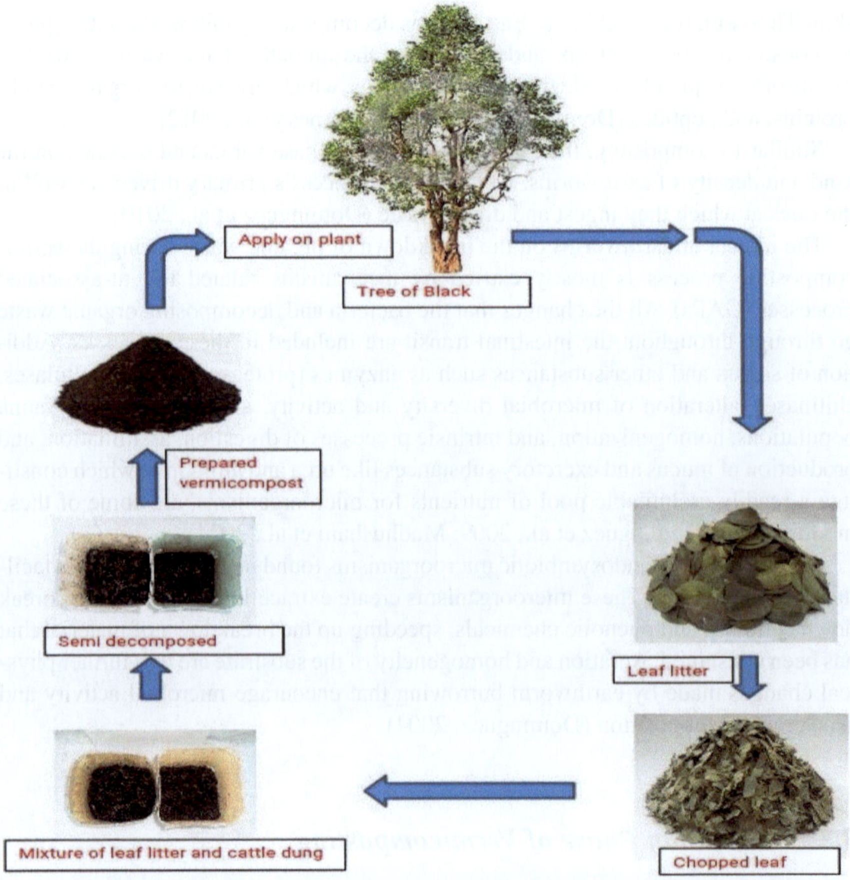

Fig. 6 Stages of vermicomposting with leaves and manure (Ritu et al., 2017)

inorganic forms through these additional microbial activities, making them considerably more soluble and accessible to the plants than those in the parent substrate (Jusselme et al., 2019; Ndegwa & Thompson, 2001).

Therefore, the use of vermicompost might both lessen the need for chemical fertilizers and the negative impacts they have on the soil and other natural resources, as well as lessen the quantity of organic waste that is disposed of in landfills (Chanda et al., 2011).

4 Symbiotic Relationship Between Bacteria and Earthworms in Vermicompost

4.1 Symbiotic Relationship

Bacteria establish symbiotic associations with earthworms in the drilosphere (portion of the soil impacted by the activity of earthworms and composed their secretions, cast and burrow) as well as with the earthworms' gut as endosymbionts (Gudeta et al., 2021, 2022). These associations result in the transformation of organic matter in vermicompost, where earthworms breakdown organic materials and soil particles, thereby increasing their availability for degradation and vermicompost formation by aerobic and anaerobic microorganisms (Pathma & Sakthivel, 2012).

Organic matter decomposition capacity of earthworms is linked to the communities of microorganisms colonizing their intestinal track (Medina-Sauza et al., 2019). Bacteria and other microorganisms are ingested by earthworms from decomposing organic matter as well as from the soil. Earthworms selectively stimulate microorganisms in the soil which in turn assist them in organic matter digestion, since earthworms lack well-developed organs for digestion and sufficient biodegradative enzymes (Hong et al., 2011; Medina-Sauza et al., 2019).

The activity and biomass of microorganisms are stimulated by the earthworm through organic matter aeration and fragmentation which increases it surface area and aid microbial degradation. Therefore, the structural and compositional characteristics of microbial communities in vermicompost influenced the activity of the earthworms (Domínguez et al., 2019).

Physiological conditions of the earthworm gut and diets regulate the microbial community of its gut (Gong et al., 2022). Organic compound and moisture contents of the earthworm gut make it favourable for bacterial proliferation, dormant bacteria activation, and endospore germination (Pathma & Sakthivel, 2012). Dormant soil bacteria are activated by energy-rich mucus produced by earthworms in the drilosphere; this activation results in accelerated microbial activities and processes. The glycoproteins containing mucus are found in the earthworm's intestine and on their cutaneous surface (Medina-Sauza et al., 2019).

The number and diversity of beneficial microbial community in vermicompost are enhanced by the activities of the earthworms. Furthermore, activities of earthworms at the same time suppress the pathogenic microorganisms, for example, reduction in number of *B. cereus* var. *mycoides* and complete elimination of pathogens (*E. coli* and *Serratia marcescens*) after earthworm gut passage (Pathma & Sakthivel, 2012).

Microorganisms associated with the gut of earthworms reciprocate by increasing the earthworm fitness (Ankrah & Douglas, 2018; Gong et al., 2022; Shapira, 2016) and aiding in the digestion of wide variety of compound by producing enzymes such as cellulase, amylase, and protease, which aid digestion (Munnoli et al., 2010). They also protect the earthworms from pathogenic microorganisms (Gudeta et al., 2021). The resistance earthworm to mercury (a toxic trace metal) and its bioaccumulation as methyl-Hg is reported to be linked to the activity of microbiome colonizing the earthworm intestine (Brantschen et al., 2020; Martín-Doimeadios et al., 2017). Thus, the gut microbiomes of earthworm play a vital role in the utilization of resources, metabolism, energy generation, and defence against pathogens (Dishaw et al., 2014). Hong et al. (2011) reported that growth and casts production rates of earthworms in vermicompost are increased by *Paenibacillus motobuensis* WN9.

4.2 Bacteria Species Associated with Vermicompost Earthworms

Composition of the bacterial communities in earthworm gut is dependent on the species of the earthworm, feeding habit, and features of the adjacent environment (Suna et al., 2020). Eight bacteria phyla, namely Actinobacteria, Acidobacteria, Bacteroidetes, Chloroflexi, Firmicutes, Proteobacteria, Planctomycetes, and Nitrospirae, and form the core network of bacterial community that associate with earthworms (Medina-Sauza et al., 2019). However, the dominant bacterial taxa commonly reported in the gut of earthworm are Proteobacteria, Acidobacteria, Actinobacteria, Firmicutes (Liu et al., 2018; Ma et al., 2017; Medina-Sauza et al., 2019; Sapkota et al., 2020), and Bacteroidetes (Suna et al., 2020).

Bacteria species such as *Photobacterium ganghwense, Aeromonas hydrophila,* and *P. motobuensis* are reported to colonize the intestine of *E. fetida* (an epigeic earthworm). These bacterial species produce a variety of enzymes which enhance organic matter biodegradation. These enzymes include amylase, cellulase, lipases, and proteases (Hong et al., 2011).

Pseudomonas oxalaticus with oxalate degradation ability was isolated from gut of *Pheretima* species, a vermicompost earthworm (Khambata & Bhat, 1953; Khyade, 2018). Nitrogen-fixing anaerobic bacteria species, *Clostridium beijerinckii, Clostridium butyricum,* and *Clostridium paraputrificum,* were reported to colonize the gut of vermicompost earthworm *E. fetida* (Citernesi et al., 1977; Khyade, 2018). Nitrogen fixing and phosphate solubilizing bacterial species are also reported among symbiotic beneficial bacteria associated with earthworms in vermicompost (Adhikary, 2012). Nitrogen-fixing and phosphorus-solubilizing species of *Serratia* and *Bacillus* as well as nitrogen-fixing *Kluyvera ascorbata* were isolated from the gut of earthworm (Hussain et al., 2016).

The nutrient-rich and limiting oxygen condition of earthworm intestinal tract is favourable for the proliferation of facultative anaerobic, anaerobic bacteria, and archaea (Koubová et al., 2015). So also, bacteria with the ability of aromatic compounds degradation under limiting oxygen supply such as member of the genera *Bacillus* and *Paenibacillus* are common colonizers of the earthworm gut (König, 2006). Bacteria species of the genus *Verminephrobacter* have been reported as symbiont of most lumbricid earthworms and are harboured in their nephridia (Lund et al., 2014).

Earthworm gut is reported to favour the proliferation of some specific groups of microorganisms, namely glucose fermenters such as members of *Aeromonadaceae, Clostridiaceae,* and *Enterobacteriaceae* (Meier et al., 2018; Wust et al., 2011); utilizers of cellobiose such as *Cellulosimicrobium funkei* (Kim et al., 2016); methanogens such as members of *Methanobacteriaceae, Methanomicrobiaceae,* and *Methanosarcinaceae* (Depkat-Jakob et al., 2013); and denitrifying bacteria such as *Bradyrhizobium, Dechloromonas, Pseudomonas,* and *Flavobacterium* (Drake & Horn, 2007).

4.3 Beneficial Role of the Symbiotic Association Between Microbes and Earthworms in Vermicompost

The rich and diverse microbial community and enzymes in the intestine of earthworm facilitates the rapid decomposition of organic materials making vermicomposting faster (approximately 4–8 weeks) than composting process (approximately 20 weeks) involving only microorganisms (Nagavallemma et al., 2004; Pathma & Sakthivel, 2012).

Symbiotic relationship between earthworms and microorganisms results in increased availability of rhizospheric bacteria in soil (Pathma & Sakthivel, 2012). The favourable gut microenvironment of earthworm activates and increases the population of rhizospheric bacteria that are ingested alongside the rhizosphere soil. These rhizospheric bacteria are the plant growth-promoting rhizobacteria, and they include *Azospirillium, Azotobacter Bacillus, Pseudomonas,* and *Rhizobium* (Sinha et al., 2010). Increase in rhizospheric bacteria population stimulates plant growth and health directly by enhancing nutrient availability, accelerating nutrient assimilation, and producing plant growth hormones and indirectly by suppressing plant pathogenic microorganisms (Elnahal et al., 2022).

The symbiotic relation between endosymbiotic microbes and earthworm during vermicomposting enhances organic matter decomposition and mineralization; this results in increased availability of nutrients for plant (Emperor & Kumar, 2015). Vermicompost increases the mineral content of the soil and favours the survival of beneficial soil microbes; hence, it application results in remarkable increase in the yield (Gudeta et al., 2022) (Fig. 7).

This relationship results in the production of peat like material that is highly porous, rich in nutrient with high water-holding capacity and low carbon: nitrogen ratio (Domínguez et al., 2019). The activities of the earthworms and their gut microbiome contribute to the structure, health, and sustainability of the soil. These activities also play a vital role in nutrient circulation within the ecosystem (Bi et al., 2021).

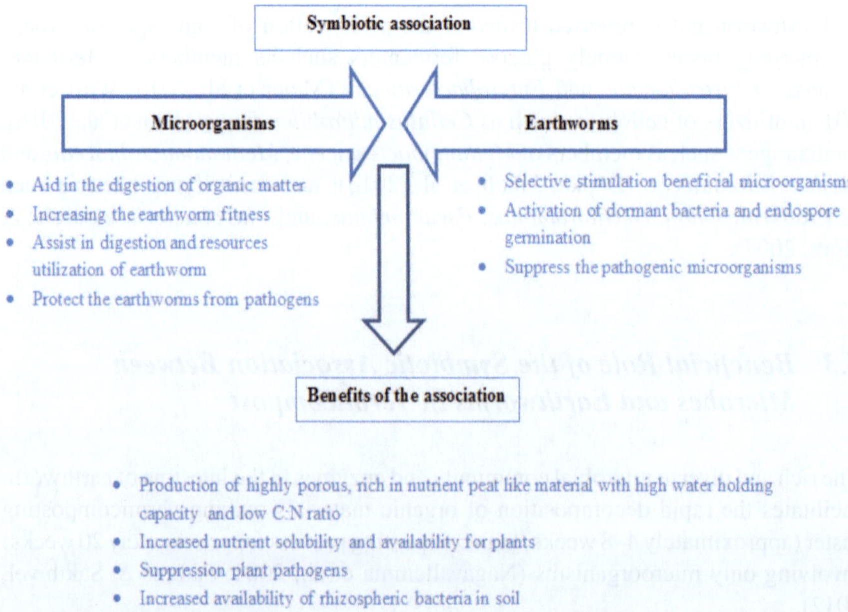

Fig. 7 Symbiotic association of bacteria and earthworm in vermicompost as well as benefits of the association

5 Symbiotic Relationships Between Fungi and Earthworm in Vermicompost

The biodegradation and conversion processes during composting are mostly a result of the activities of the resident microbial community, among which fungi play an important role. In vermicomposting, fungi species are an important component; they play a major role in the degradation and conversion processes of composting (Sahoo et al., 2015) including the decomposition of lingo-cellulosic materials such as cellulose, hemicellulose, and lignin, which brings about the maturation of compost matter. Fungi are known to utilize numerous source of carbon and can survive extreme conditions as such they mainly are responsible for compost maturation. Vermicomposting has been reported to modify the microbial community of organic matter in diverse ways (Pathma & Sakthivel, 2012).

Earthworms stabilize organic residues and mitigate pathogenic organisms; they also greatly affect the fungal communities. Earthworms select fungal species by influencing spore germination and creating microsites favourable or unfavourable for fungal development. Fungi are an important part of the natural diet of earthworms, although earthworm can digest fungi species, thereby increasing the number of fungi during gut transit of compost material in the earthworm (Vyas et al., 2022). Viable fungal count in casts produced by earthworms was reported to be greater than that of the initial waste substrates, suggesting that not all the fungi were digested; in fact, the rate of germination of fungal spores has been reported to be enhanced under the favourable condition of earthworm's gut (Pramanik et al., 2011).

Studies on the feeding preferences of species of earthworm disclosed selective feeding strategies in various species of earthworm to certain fungi (Neilson & Boag, 2003). Zirbes et al. (2012) reported a study carried out by Cooke and Luxton in 1980 on the effect of microbes on food selection by an earthworm species (*L. terrestris*) which showed that the earthworm species *L. terrestris* preferred apple leaves and paper discs inoculated with microorganisms, particularly those inoculated with the fungi species, *Mucor* sp. and *Penicillium* sp., indicating that the growth of fungi on organic substrates may enhance the availability of carbohydrates and nitrogenous compounds to earthworms. Similarly, Bonkowski et al. (2000) studied the preference of earthworm species to a variety of soil fungi. They concluded that earthworms use early successional fungal species as cues to detect fresh organic food sources in soil but the nature of this preference remains unknown. Aside bacteria, the biological, geological and chemical composition of vermicompost is also affected by fungi, but information on their relationship with earthworms and subsequent outcome of this interaction is scanty (Xiang & Li, 2014; Zhang et al., 2018). Majority of researches on earthworm–fungi relationship is usually focused on arbuscular mycorrhizae, and this has been found to transform the physical, chemical, biological composition of vermicompost soil (Milleret et al., 2009; Zhang et al., 2016), nutrient uptake by plants (Aghababaei et al., 2014; Li et al., 2012), and the type and distribution of fungi in the compost material (Cao et al., 2018) (Fig. 8). Although the symbiotic relationship of earthworm and mycorrhizae in vermicompost is not well understood, the available data suggest that there is a synergistic effect of this relationship, which results in increased performance on plants growth (Li et al., 2013a).

Earthworms significantly boost root mycorrhizal fungal colonization in soils contaminated with heavy metals like Cd and As (Cheng et al., 2007; Meng et al., 2021), and the beneficial effects were dependent on the quantity of earthworms and the timing of earthworm inoculation. A recent study found that only in the presence of no-till and straw removal methods can the inoculation of earthworms in soil boost the presence of mycorrhizal fungi in wheat roots, stimulating mycorrhizal absorption (Yang et al., 2020). According to Li et al. (2015), the interaction between earthworms and arbuscular mycorrhizal fungi (AMF) enhances the shoot and root biomasses in maize plants by encouraging phosphate acquisition. To boost the efficiency of N and P in the soil, earthworms in sweet potato plants controlled the activities of soil urease and alkaline phosphatase, and the AMF promoted the activities of soil phosphatase and increased root phosphorus absorption (Li et al., 2016). These findings imply that the relationship between earthworms and mycorrhizal fungus is advantageous for both parties involved.

Meanwhile, on the basis of the ecological classes of earthworms (anecic, endogeic, and epigeic), there is a considerable variation in the concentration and dispersal of infective units of mycorrhizal fungi depending on the species of earthworm utilized in the vermicompost material (Medina-Sauza et al., 2019). Although the ability of the different ecological classes of earthworm to effectively disperse infective units of mycorrhizal fungi in vermicompost is still under investigations, some earthworm

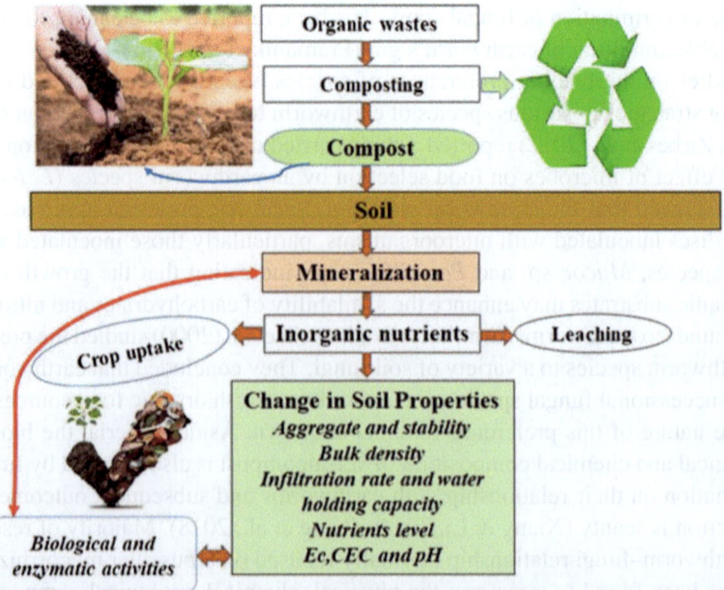

Fig. 8 Schematic diagram of compost mineralization after application to soil (Sayara et al., 2020)

species currently investigated include endogeic earthworm such as *Aporrectodea trapezoids*, *P. corethrurus*, *Octochaetona phillotti*, *Lampito mauritii*, and anecic earthworm *L. terrestris* (Fig. 9) (Li et al., 2013b; Medina-Sauza et al., 2019).

Epigeic and anecic species deposit vermicast on soil surfaces which consequently favours the dispersal of infective units of mycorrhizae and in turn favours root colonization by the fungi (Curry & Schmidt, 2007). However, there is a decrease in the dispersal of mycorrhizal infective units by endogeic species which in turn decreases root colonization by the fungi.

Reports suggest this group of worms use these fungi as their food source (Chen et al., 2023). Furthermore, the colonization of plant roots by mycorrhizal fungi can be promoted by earthworms in the vermicompost materials through favouring the growth and proliferation of some certain microbial groups present in the soil which may assist in root colonization by fungi (Zhang et al., 2016). Though there is scarce information with regard to this association, Gram positive have been reported to be involved in this interaction with mycorrhizae (Dempsey et al., 2013). A limitation of this earthworm–mycorrhizal interaction in vermicompost materials is that only a few species of fungi have been reported to be used successfully in several researches. This include *Funneliformis mosseae*, *Rhizophagus irregularis*, *Rhizophagus intraradices*, *Acaulospora* sp., *Claroideoglomus claroideum*, *Glomus caledonium*, *Glomus etunicatum*, and *Glomus geosporum* (Medina-Sauza et al., 2019). There is a partial comprehension of the dynamics of this relationship with these fungal species. Also, a wider variety of fungal species from different families need to be studied with potentially increased rates of colonization, abilities to transfer

Fig. 9 Earthworm species involved in concentration and dispersal of mycorrhizal spores

nutrients in the vermicompost soil to their host plants, and increased growth of mycelium (Cao et al., 2018). Meanwhile, information on the role of earthworms to potentially affect the composition and community structures of the mycorrhizal fungi in the vermicompost materials is not well documented (Medina-Sauza et al., 2019).

6 Conclusion and Future Prospects

Earthworms are the key players in the ecological process of waste biodegradation and soil improvement through their synergistic relationship with microorganism; however, microorganism carries out the decomposition of organic matter. The symbiotic relationship between the earthworms and the microbial communities has led to organic waste management resulting into availability of organic fertilizers that can serve as an substitute to or lessen the application of chemical fertilizers, reducing adverse pollution, climatic conditions, soil structure, pest and disease control and improved plant yields. This relationship also provides bacteria capable of nitrogen fixation, phosphate solubilization, and production of enzymes capable of biodegradation of waste, converting the organic substances and making them easily accessible to plants. Therefore, it is worthy to note that vermicompost has a huge potential in agriculture and its application should be largely extensive in the near future alone or in combination with inorganic fertilizers. This will be a step closer to achieving economic, environmental, social, and global sustainability. However, the nutrient factor of soil vermicompost does not provide enough support to explain the enhance crop production, indicating that other materials influencing plant growth within the vermicompost are present. These materials includes humic acids and plant growth hormones like auxins, gibberellins, and cytokins among the substances with the ability to enhance plant growth. Despite the plant growth-promoting potential of vermicompost, application of high concentration could impede the growth of plant negatively due to high concentration of salt soluble; therefore, the application of moderate concentration could provide maximum crop yield. Regardless of the mature being organic and safe for use, it is time-consuming and may take about six months to complete the process of converting organic waste to usable vermicompost. Setting up is also expensive as specialized containers are required to carry out vermicomposting. The process also requires lots of care and attention; invariably, a professional training is required to handle the process. Additionally, a lot of space is required to generate an ample amount of compost. Because organic fertilizer retail prices are higher than those of synthetic fertilizer, farmers now use it at a relatively low rate. Vermicomposting is largely dependent on and mediated by earthworms and microorganisms interacting with the soil biotic and abiotic components. The knowledge of vermicompost has been heavily informed by the mechanical, physical, bio-chemical, and ecological process; therefore, effective optimization and sustainability of the process due to partial understanding of the microbial communities and their structural diversity may not be achieved. Recent advancement in technologies such as omics approach can be employed through bioinformatics, metabolomics, genomics, and metagenomics to better understand the structure of the microbial communities and the earthworm for optimal vermicast production.

References

Abdelsattar, M., Kumar, V., & Abdelwahed, M. A. (2023). Green fertilizer technologies in agriculture. In *Green chemistry in agriculture and food production* (pp. 56–67). CRC Press.

Adhikary, S. (2012). Vermicompost, the story of organic gold: A review. *Agricultural Sciences, 3*(7), 905–917. https://doi.org/10.4236/as.2012.37110

Aghababaei, F., Raiesi, F., & Hosseinpur, A. (2014). The significant contribution of mycorrhizal fungi and earthworms to maize protection and phytoremediation in Cd-polluted soils. *Pedobiologia (Jena)., 57*, 223–233. https://doi.org/10.1016/j.pedobi.2014.09.004

Ahmed, N., & Al-Mutairi, K. A. (2022). Earthworms effect on microbial population and soil fertility as well as their interaction with agriculture practices. *Sustainability, 14*(13), 7803.

Aira, M., & Domínguez, J. (2008). Optimizing vermicomposting of animal wastes: Effects of rate of manure application on carbon loss and microbial stabilization. *Journal of Environmental Management, 88*(4), 1525–1529.

Aira, M., Monroy, F., & Domınguez, J. (2005). Ageing effects on nitrogen dynamics and enzyme activities in casts of *Aporrectodea caliginosa* (Lumbricidae). *Pedobiologia, 49*, 467–473.

Aira, M., Sampedro, L., Monroy, F., & Domınguez, J. (2008). Detritivorous earthworms directly modify the structure, thus altering the functioning of a micro-decomposer food web. *Soil Biology and Biochemistry, 40*, 2511–2516.

Ali, U., Sajid, N., Khalid, A., Riaz, L., Rabbani, M. M., Syed, J. H., & Malik, R. N. (2015). A review on vermicomposting of organic wastes. *Environmental Progress & Sustainable Energy, 34*(4), 1050–1062.

Ankrah, N. Y. D., & Douglas, A. E. (2018). Nutrient factories: Metabolic function of beneficial microorganisms associated with insects. *Environmental Microbiology, 20*, 2002–2011.

Aransiola, S. A, Ijah, U. J. J. Abioye, O. P., & Bala, J. D. (2022). Vermicompost-assisted phytoremediation of toxic trace element-contaminated soil in Madaka, Nigeria using *Melissa officinalis* L and *Sida acuta. International Journal of Environmental Science and Technology*. https://doi.org/10.1007/s13762-022-04105-y

Asha, A., Tripathi, A., & K, & Soni, P,. (2008). Vermicomposting: A better option for organic solid waste management. *Journal of Hum Ecology, 24*, 59–64.

Atiyeh, R. M., Lee, S., Edwards, C. A., Arancon, N. Q., & Metzger, J. D. (2002). The influence of humic acids derived from earthworm-processed organic wastes on plant growth. *Bioresource Technology, 84*, 7–14.

Bhat, S. A., Singh, J., & Vig, A. P. (2017). Earthworms as organic waste managers and biofertilizer producers. *Waste and Biomass Valorization, 9*(7), 1073–1086. https://doi.org/10.1007/s12649-017-9899-8

Bhatnagar, R. K., & Palta, R. K. (1996). *Earthworm-vermiculture and vermicomposting*. Kalyani Publishers.

Bi, Q.-F., Jin, B.-J., Zhu, D., Jiang, Y.-G., Zheng, B.-X., O'Connor, P., Yang, X.-R., Richter, A., Lin, X.-Y., & Zhu, Y.-G. (2021). How can fertilization regimes and durations shape earthworm gut microbiota in a long-term field experiment? *Ecotoxicology and Environmental Safety, 224*, Article 112643. https://doi.org/10.1016/j.ecoenv.2021.112643

Bianco, A., Fancello, F., Garau, M., Deroma, M., Atzori, A. S., Castaldi, P., Zara, G., & Budroni, M. (2022). Microbial and chemical dynamics of brewers' spent grain during a low-input pre-vermicomposting treatment. *Science of the Total Environment, 802*, Article 149792. https://doi.org/10.1016/j.scitotenv.2021.149792

Bohlen, P. J., Pelletier, D. M., Groffman, P. M., Fahey, T. J., & Fisk, M. C. (2004). Influence of earthworm invasion on redistribution and retention of soil carbon and nitrogen in northern temperate forests. *Ecosystems, 7*, 13–27.

Bolong, N., & Saad, I. (2020). Characterization of university residential and canteen solid waste for composting and vermicomposting development. *Green engineering for campus sustainability*. Springer Nature Singapore. 193–206. https://doi.org/10.1007/978-981-13-7260-5_14

Bonito, G., Isikhuemhen, O. S., & Vilgalys, R. (2010). Identification of fungi associated with municipal compost using DNA-based techniques. *Bioresource Technology, 101*, 1021–1027.

Bonkowski, M., Griffiths, B. S., & Ritz, K. (2000). Food preferences of earthworms for soil fungi. *Pedobiologia, 44*(6), 666–676.

Brantschen, J., Gygax, S., Mestrot, A., & Frossard, A. (2020). Soil Hg contamination impact on earthworms' gut microbiome. *Applied Sciences, 10*, 2565. https://doi.org/10.3390/app10072565

Brown, G. G., & Doube, M. (2004). Functional interactions between earthworms, microorganisms and organic matter and plants. In C. A. Edwards (Ed.), *Earthworm ecology* (pp. 213–240). Lucie Press, Boca Raton, FL.

Cao, J., Wang, C., Dou, Z., Liu, M., & Ji, D. (2018). Hyphospheric impacts of earthworms and arbuscular mycorrhizal fungus on soil bacterial community to promote oxytetracycline degradation. *Journal of Hazardous Materials, 341*, 346–354. https://doi.org/10.1016/j.jhazmat.2017.07.038

Chanda, G. K., Bhunia, G., & Chakraborty, S. K. (2011). The effect of vermicompost and other fertilizers on cultivation of tomato plants. *Journal of Horticulture and Forestry., 3*, 42–45.

Chen, Y., Zhang, Y., Shi, X., Xu, L., Zhang, L., & Zhang, L. (2022). The succession of GH6 cellulase-producing microbial communities and temporal profile of GH6 gene abundance during vermicomposting of maize stover and cow dung. *Bioresource Technology, 344*, Article 126242.

Chen, Y., Zhang, Y., Shi, X., Shi, E., Zhao, T., Zhang, Y., & Xu, L. (2023). The contribution of earthworms to carbon mineralization during vermicomposting of maize stover and cow dung. *Bioresource Technology, 368*, Article 128283.

Cheng, J. M., Yu, X. Z., & Huang, M. H. (2007). Effect of earthworm-mycorrhiza interaction on transformation of Cdfrom soil to plant. *Acta Scientiae Circum Stantiae, 27*(2), 228–234.

Chowdhury, S. D., Bandyopadhyay, R., & Bhunia, P. (2022). Techno-economic analysis and life-cycle assessment of vermi-technology for waste bioremediation. In *Biomass, biofuels, biochemicals* (pp. 315–349). Elsevier.

Citernesi, U., Neglia, R., Seritti, A., Lepidi, A. A., Filippi, C., Bagnoli, G., Nuti, M. P., & Galluzzi, R. (1977). Nitrogen fixation in the gastro-enteric cavity of soil animals. *Soil Biology and Biochemistry, 9*, 71–72.

Curry, J. P., & Schmidt, O. (2007). The feeding ecology of earthworms – A review. *Pedobiologia, 50*, 463–477. https://doi.org/10.1016/j.pedobi.2006.09.001

Das, P., & Paul, K. (2023). A review on integrated vermifiltration as a sustainable treatment method for wastewater. *Journal of Environmental Management, 328*, Article 116974.

Dempsey, M. A., Fisk, M. C., Yavitt, J. B., Fahey, T. J., & Balser, T. C. (2013). Exotic earthworms alter soil microbial community composition and function. *Soil Biology and Biochemistry., 67*, 263–270. https://doi.org/10.1016/j.soilbio.2013.09.009

Depkat-Jakob, P. S., Brown, G. G., Tsai, S. M., Horn, M. A., & Drake, H. L. (2013). Emission of nitrous oxide and dinitrogen by diverse earthworm families from Brazil and resolution of associated denitrifying and nitrate-dissimilating taxa. *FEMS Microbiology Ecology., 83*(2), 375–391.

Devi, S. H., Vijayalakshmi, K., Jyotsna, K. P., Shaheen, S. K., Jyothi, K., & Rani, M. S. (2009). Comparative assessment in enzyme activities and microbial populations during normal and vermicomposting. *Journal of Environmental Biology., 30*, 1013–1017.

Dishaw, L. J., Cannon, J. P., Litman, G. W., & Parker, W. (2014). Immune-directed support of rich microbial communities in the gut has ancient roots. *Developmental and Comparative Immunology, 47*(1), 36–51.

Dohaish, E. J. A. B. (2020). Vermicomposting of organic waste with *Eisenia fetida* increases the content of exchangeable nutrients in soil. *Pakistan Journal of Biological Sciences: PJBS, 23*(4), 501–509. https://doi.org/10.3923/pjbs.2020.501.509

Domınguez, J. (2004). State of the art and new perspectives on vermicomposting research. In C. A. Edwards (Ed.), *Earthworm ecology* (2nd ed., pp. 401–424). CRC.

Dominguez, J., & Edwards, C. A. (2004). Vermicomposting organic wastes: A review. In: S. H. Shakir Hanna & W. Z. A. Mikhail (Eds.), *Soil zoology for sustainable development in the 21st century*, Cairo (pp. 369–395).

Domínguez, J., Aira, M., & Gómez-Brandón, M. (2009). Vermicomposting: Earthworms enhance the work of microbes. *Microbes at Work*. https://doi.org/10.1007/978-3-642-04043-6_5

Domínguez, J., Aira, M., Kolbe, A. R., Gómez-Brandón, M., & Pérez-Losada, M. (2019). Changes in the composition and function of bacterial communities during vermicomposting may explain beneficial properties of vermicompost. *Scientific Reports, 9*(1), 9657. https://doi.org/10.1038/s41598-019-46018-w

Domínguez, J., Aira, M., & Gómez-Brandón, M. (2010). Vermicomposting: Earthworms enhance the work of microbes. In H. Insam, I. Franke-Whittle & M. Goberna, (Eds.), *Microbes at work: From wastes to resources* (pp. 93–114). Springer, ISBN 978-3-642-04042-9, Heidelberg, Germany.

Drake, H. L., & Horn, M. A. (2007). As the worm turns: The earthworm gut as a transient habitat for soil microbial biomes. *Annual Review of Microbiology, 61*, 169–189.

Ducasse, V., Capowiez, Y., & Peigné, J. (2022). Vermicomposting of municipal solid waste as a possible lever for the development of sustainable agriculture. A Review. *Agronomy for Sustainable Development, 42*(5), 89.

Duiker, S., & Stehouwer, R. (2007). *Earthworms.* Available online: http://pubs.cas.psu.edu/freepubs/pdfs/uc182.pdf

Edwards, C. A. (1998). *Earthworm ecology*. CRC Press LLC, Boca Raton, FL.

Edwards, C. A., & Arancon, N. Q. (2022). Interactions between earthworms, microorganisms, and other invertebrates. In *Biology and Ecology of Earthworms* (pp. 275–301). Springer US.

Edwards, C. A., Dominguez, J., & Arancon, N. Q. (2004). The influence of vermicomposts on pest and diseases. In S. H. S. Hanna & W. Z. A. Mikhail (Eds.), *Soil zoology for sustainable development in the 21st century*, Cairo, pp 397–418.

Edwards, C. A., & Fletcher, K. E. (1988). Interaction between earthworms and microorganisms in organic matter breakdown. *Agriculture, Ecosystems & Environment, 20*, 235–249.

Edwards, C. A., & Lofty, J. R. (1972). *Biology of earthworms* (pp. 1–288). Chapman and Hall.

Eisenhauer, N., & Eisenhauer, E. (2020). The intestines of the soil: The taxonomic and functional diversity of earthworms—A review for young ecologists. *Research Ideas and Outcomes, 5*, 1–6.

El-Haddad, M. E., Zayed, M. S., El-Sayed, G. A. M., Hassanein, M. K., & Abd El-Satar, A. M. (2014). Evaluation of compost, vermicompost and their teas produced from rice straw as affected by addition of different supplements. *Annals of Agricultural Sciences, 59*(2).

Elnahal, A. S., El-Saadony, M. T., Saad, A. M., Desoky, E. S. M., El-Tahan, A. M., Rady, M. M., et al. (2022). The use of microbial inoculants for biological control, plant growth promotion, and sustainable agriculture: A review. *European Journal of Plant Pathology, 162*(4), 759–792.

Emperor, G. N., & Kumar, K. (2015). Microbial population and activity on vermicompost of *Eudrilus eugeniae* and *Eisenia fetida* in different concentrations of tea waste with cow dung and kitchen waste mixture. *International Journal Current Microbiology and Applied Science, 4*(10), 496–507.

Fernández-Gomez, M. J., Nogales, R., Insam, H., Romero, E., & Goberna, M. (2012). Use of DGGE and COMPOCHIP for investigating bacterial communities of various vermicomposts produced from different wastes under dissimilar conditions. *Science of the Total Environment, 414*, 664–671.

Fornes, F., Mendoza-Hernández, D., García-de-la-Fuente, R., Abad, M., & Belda, R. M. (2012). Composting versus vermicomposting: A comparative study of organic matter evolution through straight and combined processes. *Bioresource Technology, 118*, 296–305. https://doi.org/10.1016/j.biortech.2012.05.028

Gebrehana, Z. G., Gebremikael, M. T., Beyene, S., Sleutel, S., Wesemael, W. M., & De Neve, S. (2023). Organic residue valorization for Ethiopian agriculture through vermicomposting with native (*Eudrilus eugeniae*) and exotic (*Eisenia fetida* and *Eisenia andrei*) earthworms. *European Journal of Soil Biology, 116*, Article 103488.

Gómez-Brandón, M., Fornasier, F., de Andrade, N., & Domínguez, J. (2022). Influence of earthworms on the microbial properties and extracellular enzyme activities during vermicomposting of raw and distilled grape marc. *Journal of Environmental Management, 319*, Article 115654.

Gomez-Brandon, M., Lazcano, C., Lores, M., & Domínguez, J. (2010). Detritivorous earthworms modify microbial community structure and accelerate plant residue decomposition. *Applied Soil Ecology, 44*(3), 237–244.

Gómez-Brandón, M., Lores, M., & Domínguez, J. (2012). Species-specific effects of epigeic earthworms on microbial community structure during first stages of decomposition of organic matter. *PLoS ONE, 7*(2), Article e31895.

Gómez-Brandón, M., Lores, M., Insam, H., & Domínguez, J. (2019). Strategies for recycling and valorization of grape marc. *Critical Reviews in Biotechnology, 39*, 437–450. https://doi.org/10.1080/07388551.2018.1555514

Gong, X., Chen, T.-W., Zhang, L., Pižl, V., Tajovský, K., & Devetter, M. (2022). Gut microbiome reflect adaptation of earthworms to cave and surface environments. *Animal Microbiome, 4*, 47. https://doi.org/10.1186/s42523-022-00200-0

Gopal, M., Gupta, A., Sunil, E., & Thomas, G. V. (2009). Amplification of plant beneficial microbial communities during conversion of coconut leaf substrate to vermicompost by *Eudrilus* sp. *Current Microbiology, 59*(1), 15–20.

Gudeta, K., Bhagat, A., Julka, J. M., Sinha, R., Verma, R., Kumar, A., Kumari, S., Ameen, F., Bhat, S. A., Amarowicz, R., & Sharma, M. (2022). Vermicompost and its derivatives against phytopathogenic fungi in the soil: A review. *Horticulturae, 8*, 311. https://doi.org/10.3390/horticulturae8040311

Gudeta, K., Julka, J. M., Kumar, A., Bhagat, A., & Kumari, A. (2021). Vermiwash: An agent of disease and pest control in soil, a review. *Heliyon, 7*, Article e06434.

Hendrix, P. F., Callaham, M. A., Drake, J. M., Huang, C. Y., James, S. W., & Snyder, B. A. (2008). Pandora's box contained bait: The global problem of introduced earthworms. *Annual Review of Ecology Evolution and Systematics, 39*, 593–613.

Hijam, S. D., Thounaojam, R. S., Sharma, C. K., & Talukdar, N. C. (2022). Comparative Study on Reproductive Biology of an Exotic (Eisenia andrei) and Indigenous Earthworm (Perionyx Sp.). *Annals of Multidisciplinary Research, Innovation and Technology* (Vol. 1(2), pp. 1–5).

Hong, S. W., Lee, J. S., & Chung, K. S. (2011). Effect of enzyme producing microorganisms on the biomass of epigeic earthworms (*Eisenia fetida*) in vermicompost. *Bioresource Technology, 102*(2011), 6344–6347. https://doi.org/10.1016/j.biortech.2011.02.096

Huang, K., Li, F., Wei, Y., Chen, X., & Fu, X. (2013). Changes of bacterial and fungal community compositions during vermicomposting of vegetable wastes by *Eisenia foetida. Bioresource Technology, 150*, 235–241.

Hussain, N., Das, S., Goswami, L., Das, P., Sahariah, B., & Bhattacharya, S. S. (2018). Intensification of vermitechnology for kitchen vegetable waste and paddy straw employing earthworm consortium: Assessment of maturity time, microbial community structure, and economic benefit. *Journal of Cleaner Production, 182*, 414–426.

Hussain, N., Singh, A., Saha, S., Kumar, V. S., & M., Bhattacharyya, P., Sundar Bhattacharya, S.,. (2016). Excellent N-fixing and P-solubilizing traits in earthworm gut-isolated bacteria: A vermicompost based assessment with vegetable market waste and rice straw feed mixtures. *Bioresource Technology, 222*, 165–174. https://doi.org/10.1016/j.biortech.2016.09.115

Iven, H., Walker, T. W. N., & Anthony, M. (2023). Biotic interactions in soil are underestimated drivers of microbial carbon use efficiency. *Current Microbiology, 80*, 13. https://doi.org/10.1007/s00284-022-02979-2

Jambhekar, H. (1992). Use of earthworm as a potential source of decompose organic wastes. *Proc Nat Sem Org Fmg, Coimbatore* (pp. 52–53).

Jayasinghe, B. A. T. D., & Parkinson, D. (2009). Earthworms as the vectors of actinomycetes antagonistic to litter decomposer fungi. *Applied Soil Ecology, 43*, 1–10.

Jouquet, P., Dauber, J., Lager, J., Lavelle, P., & Lepage, M. (2006). Soil invertebrates as ecosystem engineers: Intended and accidental effects on soil and feedback loops. *Applied Soil Ecology, 32*, 153–164.

Jusselme, M. D., Pruvost, C., Motard, E., Giusti-Miller, S., Frechault, S., Alphonse, V., Balland-Bolou-Bi, C., Dajoz, I., & Mora, P. (2019). Increasing the ability of a green roof to provide ecosystem services by adding organic matter and earthworms. *Agriculture Ecosystem, 143*, 61–69.

Kale, R. D. (1995). Vermicomposting has a bright scope. *Indian Silk, 34*, 6–9.

Karapantzou, I., Mitropoulou, G., Prapa, I., Papanikolaou, D., Charovas, V., & Kourkoutas, Y. (2023). Physicochemical changes and microbiome associations during vermicomposting of winery waste. *Sustainability, 15*(9), 7484.

Kaur, T. (2020). Vermicomposting: An effective option for recycling organic wastes. *Organic Agriculture*. https://doi.org/10.5772/intechopen.91892

Khambata, S. R., & Bhat, J. V. (1953). Studies on a new oxalate-decomposing bacterium, *Pseudomonas oxalaticus*. *Journal of Bacteriology, 66*, 505–507.

Khyade, V. B. (2018). Bacterial diversity in the alimentary canal of earthworms. *Journal of Bacteriology & Mycology: Open Access, 6*(3), 183–185.

Kim, D. Y., Lee, M. J., Cho, H. Y., Lee, J. S., Lee, M. H., Chung, C. W., Shin, D. H., Rhee, Y. H., Son, K. H., & Park, H. Y. (2016). Genetic and functional characterization of an extracellular modular GH6 endo-beta-1,4-glucanase from an earthworm symbiont, *Cellulosimicrobium funkei* HY-13. *Antonie Van Leeuwenhoek, 109*(1), 1–12.

König, H. (2006). *Bacillus* species in the intestine of termites and other soil invertebrates. *Journal of Applied Microbiology, 101*, 620–627. https://doi.org/10.1111/j.1365-2672.2006.02914.x

Kooch, Y., & Jalilvand, H. (2008). Earthworm as ecosystem engineers and the most important detritivors in forst soils. *Pakistan Journal of Biological Sciences, 11*, 819–825.

Koubová, A., Chronáková, A., PiŽl, V., Sánchez-Monedero, M. A., & Elhottová, D. (2015). The effects of earthworms *Eisenia* spp. on microbial community are habitat dependent. *European Journal of Soil Biology., 68*, 42–55. https://doi.org/10.1016/j.ejsobi.2015.03.004

Laossi, K. R., Noguera, D. C., Bartolomé-Lasa, A., Mathieu, J., Blouin, M., & Barot, S. (2009). Effects of endogeic and Anecic earthworms on the Earthworm services for cropping systems 565 competition between four annual plants and their relative reproduction potential. *Soil Biology & Biochemistry, 41*, 1668–1673.

Li, H., Du, Z. Y., Liu, Q., & Shi, Y. X. (2016). Effect of earthworm-mycorrhiza interaction on soil enzyme activities. *Journal of Plant Nutrition and Fertilizer, 22*(1), 209–215.

Li, H., Wang, C., Li, X., & Xiang, D. (2013a). Inoculating maize fields with earthworms (*Aporrectodea trapezoides*) and an arbuscular mycorrhizal fungus (*Rhizophagus intraradices*) improves mycorrhizal community structure and increases plant nutrient uptake. *Biol. Fert. Soils, 49*, 1167–1178. https://doi.org/10.1007/s00374-013-0815-5

Li, H., Wang, C., Li, X., Christie, P., Dou, Z., Zhang, J., et al. (2013b). Impact of the earthworm *Aporrectodea trapezoides* and the arbuscular mycorrhizal fungus Glomus intraradices on N-15 uptake by maize from wheat straw. *Biology and Fertility of Soils, 49*, 263–271. https://doi.org/10.1007/s00374-012-0716-z

Li, H., Wang, C., & Wang, S. Y. (2015). Interaction of earthworms and amfungi on maize growth, and nutrogen and phosphorus uptake. *Journal of Plant Nutrition and Fertilizer, 21*(4), 920–926.

Li, H., Xiang, D., Wang, C., Li, X., & Lou, Y. (2012). Effects of epigeic earthworm (Eisenia fetida) and arbuscular mycorrhizal fungus (*Glomus intraradices*) on enzyme activities of a sterilized soil-sand mixture and nutrient uptake by maize. *Biology and Fertility of Soils, 48*, 879–887. https://doi.org/10.1007/s00374-012-0679-0

Liu, D., Lian, B., Wu, C., & Guo, P. (2018). A comparative study of gut microbiota profiles of earthworms fed in three different substrates. *Symbiosis, 74*, 21–29. https://doi.org/10.1007/s13199-017-0491-6

Lund, M. B., Kjeldsen, K. U., & Schramm, A. (2014). The earthworm-*Verminephrobacter* symbiosis: An emerging experimental system to study extracellular symbiosis. *Frontiers in Microbiology, 5*, 1–6.

Ma, L., Xie, Y., Han, Z., Giesy, J. P., & Zhang, X. (2017). Responses of earthworms and microbial communities in their guts to Triclosan. *Chemosphere, 168*, 1194–1202. https://doi.org/10.1016/J.chemosphere.2016.10.079

Martín-Doimeadios, R. C. R., Mateo, R., & Jiménez-Moreno, M. (2017). Is gastrointestinal microbiota relevant for endogenous mercury methylation in terrestrial animals? *Environmental Research, 152*, 454–461.

Madhushani, G., Walahakularachchi, I. L., & Kumara, M. P. (2024). A review of the new technologies in vermicompost production. *Journal of Research Technology and Engineering., 5*(3), 143–152.

Medina-Sauza, R. M., Álvarez-Jiménez, M., Delhal, A., Reverchon, F., Blouin, M., Guerrero-Analco, J. A., Cerdán, C. R., Guevara, R., Villain, L., & Barois, I. (2019). Earthworms building up soil microbiota, a review. *Frontiers in Environmental Science, 7*, 81. https://doi.org/10.3389/fenvs.2019.00081

Meier, A. B., Hunger, S., & Drake, H. L. (2018). Differential engagement of fermentative taxa in gut contents of the earthworm *Lumbricus terrestris*. *Applied Environmental Microbiology, 84*(5), e01851-e1917.

Meng, L., Srivastava, A. K., Kuča, K., Giri, B., Rahman, M. M., & Wu, Q. (2021). Interaction between earthworms and arbuscular mycorrhizal fungi in plants: A review. *Phyton, 90*(3), 687.

Milleret, R., Le Bayon, R. C., Lamy, F., Gobat, J. M. A., & Boivin, P. (2009). Impact of roots, mycorrhizas and earthworms on soil physical properties as assessed by shrinkage analysis. *Journal of Hydrology, 373*, 499–507. https://doi.org/10.1016/j.jhydrol.2009.05.013

Monroy, F., Aira, M., & Domínguez, J. (2009). Reduction of total coliform numbers during vermicomposting is caused by short-term direct effects of earthworms on microorganisms and depends on the dose of application of pig slurry. *Science of the Total Environment, 407*, 5411–5416.

Munnoli, P. M. (2015). Role of microbes in vermicomposting: A review. *Bioprospects of Coastal Eubacteria*. https://doi.org/10.1007/978-3-319-12910-5_14

Munnoli, P. M., Da Silva, J. A. T., & Saroj, B. (2010). Dynamics of the soil-earthworm-plant relationship: A review. *Dynamic Soil, Dynamic Plant, 4*(1), 1–21.

Nagavallemma, K. P., Wani, S. P., Stephane, L., Padmaja, V. V., Vineela, C., Babu Rao, M., & Sahrawat, K. L. (2004). Vermicomposting: Recycling wastes into valuable organic fertilizer. Global Theme on Agrecosystems Report no.8. Patancheru 502324. International Crops Research Institute for the Semi-Arid Tropics, Andhra Pradesh, p. 20.

Naidoo, K., Swatson, H., Yobo, K. S., & Arthur, G. D. (2017). Boosting our soil with green technology: Conversion of organic waste into 'Black Gold'. In: *Food bioconversion* (pp. 491–510). Elsevier Inc.

Ndegwa, P. M., & Thompson, S. A. (2001). Integrating composting and vermicomposting in the treatment and bioconversion of biosolids. *Bioresource Technology, 76*, 107–112.

Nechitaylo, T. Y., Yakimov, M. M., Godinho, M., Timmis, K. N., Belogolova, E., Byzov, B. A., & Golyshin, P. N. (2010). Effect of the earthworms *Lumbricus terrestris* and *Aporrectodea caliginosa* on bacterial diversity in soil. *Microbial Ecology, 59*, 574–587.

Neilson, R., & Boag, B. (2003). Feeding preferences of some earthworm species common to upland pastures in Scotland. *Pedobiologia, 47*(1), 1–8.

Pathma, J., & Sakthivel, N. (2012). Microbial diversity of vermicompost bacteria that exhibit useful agricultural traits and waste management potential. *Springerplus, 1*, 1–19, 26. https://doi.org/10.1186/2193-1801-1-26

Pathma, J., & Sakthivel, N. (2013). Molecular and functional characterization of bacteria isolated from straw and goat manure based vermicompost. *Applied Soil Ecology, 70*, 33–47.

Pathma, J., & Sakthivel, N. (2014). Microbial and functional diversity of vermicompost bacteria. In: *Bacterial diversity in sustainable agriculture* (pp. 205–225). Springer.

Pierre-Louis, R. C., Kader, M. A., Desai, N. M., & John, E. H. (2021). Potentiality of vermicomposting in the South Pacific island countries: A review. *Agriculture, 11*(9), 876.

Pižl, V., & Nováková, A. (2003). Interactions between microfungi and *Eisenia andrei* (Oligochaeta) during cattle manure vermicomposting. *Pedobiologia, 47*(5–6), 895–899. https://doi.org/10.1078/0031-4056-00277

Pramanik, P., Ghosh, G. K., Ghosal, P. K., & Banik, P. (2007). Changes in organic–C, N, P and K and enzyme activities in vermicompost of biodegradable organic wastes under liming and microbial inoculants. *Bioresource Technology, 98*(13), 2485–2494.

Pramanik, P., Kim, S. Y., & Kim, P. J. (2012). Changes in fungal and bacterial diversity during vermicomposting of industrial sludge and poultry manure mixture: Detecting the mechanism of plant growth promotion by vermicompost. *Biomass–Detection, Production and Usage* (pp. 113–124).

Pramanik, P., Yoon, S., & Joo, P. (2011). Changes in fungal and bacterial diversity during vermicomposting of industrial sludge and poultry manure mixture: Detecting the mechanism of plant growth promotion by vermicompost. *INTECH*. https://doi.org/10.5772/16445

Raza, S. T., Wu, J., Rene, E. R., Ali, Z., & Chen, Z. (2022). Reuse of agricultural wastes, manure, and biochar as an organic amendment: A review on its implications for vermicomposting technology. *Journal of Cleaner Production, 360*, Article 132200.

Rehman, S. ur, De Castro, F., Aprile, A., Benedetti, M., & Fanizzi, F. P. (2023). Vermicompost: Enhancing plant growth and combating abiotic and biotic stress. *Agronomy, 13*(4), 1134. MDPI AG. Retrieved from https://doi.org/10.3390/agronomy13041134

Ritu, N., Anurag, T., & Praveesh, B. (2017). Prolific utilization of Earthworm species to convert green leaf of Jamun (Black plum) into soil nutrient. *Academy of Agriculture Journal, 3*, 240–245.

Rusănescu, C. O., Rusănescu, M., Voicu, G., Paraschiv, G., Biriş, S. Ştefan, & Popescu, I. N. (2022). The recovery of vermicompost sewage sludge in agriculture. *Agronomy, 12*(11), 2653. MDPI AG. Retrieved from https://doi.org/10.3390/agronomy12112653

Sahoo, H. R., Behera, S., Sahoo, M., Baboo, M., & Gupta, N. (2015). Analysis of fungal flora, physicochemical and antimicrobial properties of Vermicompost and Vermiwash developed through green waste digestion by *Eudrilus eugeniae*-a night crawler earthworm. *Agricultural Research and Technology: Open Access Journal, 1*(2), 555–556.

Sampedro, L., & Domınguez, J. (2008). Stable isotope natural abundances ($\delta^{13}C$ and $\delta^{15}N$) of the earthworm *Eisenia fetida* and other soil fauna living in two different vermicomposting environments. *Applied Soil Ecology, 38*, 91–99.

Sapkota, R., Santos, S., Farias, P., Krogh, P. H., & Winding, A. (2020). Insights into the earthworm gut multi-kingdom microbial communities. *Science of the Total Environment, 727*(2020), Article 138301. https://doi.org/10.1016/j.scitotenv.2020.138301

Satpathy, J., Saha, M. H., Mishra, A. S., & Mishra, S. K. (2020). Characterization of bacterial isolates in vermicompost produced from a mixture of cow dung, straw, Neem leaf and vegetable wastes. *bioRxiv*.

Sayara, T., Basheer-Salimia, R., Hawamde, F., & Sánchez, A. (2020). Recycling of organic wastes through composting: Process performance and compost application in agriculture. *Agronomy, 10*(11), 1838.

Shapira, M. (2016). Gut microbiotas and host evolution: Scaling up symbiosis. *Trends in Ecology and Evolution, 31*, 539–549.

Siddiqui, Z. A., & Mahmood, I. (1996). Biological control of plant parasitic nematodes by fungi: A review. *Bioresource Technology, 58*, 229–239.

Singh, A., Karmegam, N., Singh, G. S., Bhadauria, T., Chang, S. W., Awasthi, M. K., & Ravindran, B. (2020). Earthworms and vermicompost: An eco-friendly approach for repaying nature's debt. *Environmental Geochemistry and Health*. https://doi.org/10.1007/s10653-019-00510-4

Singleton, D. R., Hendrix, P. F., Coleman, D. C., & Whitman, W. B. (2003). Identification of uncultured bacteria tightly associated with the intestine of the earthworm *Lumbricus rubellus* (Lumbricidae; Oligochaeta). *Soil Biology and Biochemistry, 35*(12), 1547–1555.

Sinha, R. K., Agarwal, S., Chauhan, K., & Valani, D. (2010a). The wonders of earthworms & its vermicompost in farm production: Charles Darwin's 'friends of farmers', with potential to replace destructive chemical fertilizers. *Agricultural Sciences, 1*(02), 76. https://doi.org/10.4236/as.2010.12011

Sinha, R., Herat, S., Valani, D., & Chauhan, K. (2009). Earthworms vermicompost: A powerful crop nutrient over the conventional compost & protective soil conditioner against the destructive chemical fertilizers for food safety and security. *American-Eurasian Journal of Agricultural & Environmental Sciences, 5*(5), 01–55.

Srivastava, V., Squartini, A., Masi, A., Sarkar, A., & Singh, R. P., (2021). Metabarcoding analysis of the bacterial succession during vermicomposting of municipal solid waste employing the earthworm *Eisenia fetida*. *Science of the Total Environment, 766*, 144389. https://doi.org/10.1016/j.scitotenv.2020.144389

Suna, M., Chao, H., Zheng, X., Deng, S., Ye, M., & Hu, F. (2020). Ecological role of earthworm intestinal bacteria in terrestrial environments: A review. *Science of the Total Environment, 740*(2020), Article 140008. https://doi.org/10.1016/j.scitotenv.2020.140008

Suthar, S., & Singh, S. (2008). Vermicomposting of domestic waste by using two epigeic earthworms (*Perionyx excavatus* and *Perionyx sansibaricus*). *International Journal of Evnironmental Science and Technology, 5*, 99–106.

Thakur, S. S., Lone, A. R., Yellaboina, S., Tambat, S., Yadav, A. N., Jain, S. K., & Yadav, S. (2022). Metagenomic insights into the gut microbiota of *Eudrilus eugeniae* (Kinberg) and its potential roles in agroecosystem. *Current Microbiology, 79*(10), 295. https://doi.org/10.1007/s00284-022-02988-1

Toyota, K., & Kimura, M. (2000). Microbial community indigenous to the earthworm *Eisenia foetida. Biology and Fertility of Soils, 31*, 187–190.

Vaz-Moreira, I., Silva, M. E., Manaia, C. M., & Nunes, O. C. (2008). Diversity of bacterial isolates from commercial and homemade composts. *Microbial Ecology, 55*, 714–722.

Vijayabharathi, R., Sathya, A., & Gopalakrishnan, S. (2015). *Plant growth-promoting microbes from herbal vermicompost* (pp. 71–88). Springer International Publishing Switzerland.

Vijayshankar, G. R., Dhivya, K., Jyothi, K., Hassan, M. M., & Reddy, K. S. K. (2024). Collection and management of vermicomposting of market waste. In *E3S Web of Conferences* (Vol. 564, p. 11012). EDP Sciences. https://doi.org/10.1051/e3sconf/202456411012

Villar, I., Alves, D., & Mato, S. (2017). Product quality and microbial dynamics during vermicomposting and maturation of compost from pig manure. *Waste Management, 69*, 498–507. https://doi.org/10.1016/j.wasman.2017.08.031

Vyas, P., Sharma, S., & Gupta, J. (2022). Vermicomposting with microbial amendment: Implications for bioremediation of industrial and agricultural waste. *Biotechnologia, 103*(2), 203.

Wust, P. K., Horn, M. A., & Drake, H. L. (2011). *Clostridiaceae* and *Enterobacteriaceae* as active fermenters in earthworm gut content. *The ISME Journal, 5*(1), 92–106.

Xiang, D., & Li, H. (2014). Nutrient uptake in mycorrhizal plants - role of earthworms. *Acta Agriculture Scandinavica Section B, 64*, 434–441. https://doi.org/10.1080/09064710.2014.920412

Yadav, K. D., Dayanand, S., & Rajnikant, P. (2022). Challenges and opportunities for disposal of floral waste in developing countries by using composting method. *Advanced Organic Waste Management Sustainable Practices and Approaches*. https://doi.org/10.1016/B978-0-323-85792-5.00018-6

Yadav, P. K., & Singh, N. S. (2023). Role of earthworms in soil health and variables influencing their population dynamic: A review. *The Pharma Innovation Journal, 12*(4), 1961–1965.

Yang, H. S., Zhou, J. J., Weih, M., Li, Y. F., Zhai, S. L., et al. (2020). Mycorrhizal nitrogen uptake of wheat is increased by earthworm activity only under no-till and straw removal conditions. *Applied Soil Ecology, 155*, Article 103672. https://doi.org/10.1016/j.apsoil.2020.103672

Yasir, M., Aslam, Z., Kim, S. W., Lee, S. W., Jeon, C. O., & Chung, Y. R. (2009). Bacterial community composition and chitinase gene diversity of vermicompost with antifungal activity. *Bioresource Technology, 100*(19), 4396–4403.

Zhang, H., Xue, D., Huang, X., Wu, H., & Chen, H. (2023). Earthworms modify the soil bacterial community by regulating the soil carbon, enzyme activities, and pH. *Journal of Soil Science and Plant Nutrition*, 1–14.

Zhang, W., Cao, J., Zhang, S., & Wang, C. (2016). Effect of earthworms and arbuscular mycorrhizal fungi on the microbial community and maize growth under salt stress. *Applied Soil Ecology, 107*, 214–223. https://doi.org/10.1016/j.apsoil.2016.06.005

Zhang, W., Wang, C., Lu, T., & Zheng, Y. (2018). Cooperation between arbuscular mycorrhizal fungi and earthworms promotes the physiological adaptation of maize under a high salt stress. *Plant and Soil, 423*, 125–140. https://doi.org/10.1007/s11104-017-3481-9

Zirbes, L., Thonart, P., & Haubruge, E. (2012). Microsale interactions between earthworms and microorganisms, a review. *Biotechnologie, Agronomie, Société et Environnement, 16*(1).

Vermicast, Vermiwash: A Suitable Alternative to Chemical Fertilizers

Bellemkonda Ramesh, Srinivasan Kameswaran, Sudhakara Gujjala, Gopi Krishna Pitchika, Manjunatha Bangeppagari, B. Swapna, and M. Ramakrishna

Abstract An effective way to turn organic matter into useful compounds is through vermicomposting. This procedure uses earthworms' gastrointestinal tracts to stabilize organic wastes. The result is an odorless, clean, peat-like material that has a good form, the ability to retain moisture, and organic matter that contains generally appropriate amounts of nitrogen, phosphorus, potassium, and other minerals necessary for plant development. Vermicompost's outcome is abundant in important nutrients, macro and micro, as well as simple microbes. The use of vermicompost enhances both fertility and the condition of the soil while also promoting better plant development and total production. Plant growth regulator vermiwash is a liquid that is sprayed on leaves and has a significant amount of macro and micronutrients, vitamins, and hormones like auxins and gibberellins. By enhancing the "biological, chemical, and physical features" of the soil while also sustaining humus and organic matter in the soil, such inputs assist in preserving soil health and may be utilized to encourage the growth of useful soil microbes. It additionally aids in the uptake of humic acids or their breakdown products, which influence the overall growth and metabolism of plants. It also improves the hormonal and biochemical activities of humus substances. Water holding capacity, soil structure, drainage, seed germination, base exchange capacity, and soil erosion are all improved. For improving the

B. Ramesh
Department of Internal Medicine, University of Nebraska Medical Center, Omaha, NE, USA

S. Kameswaran (✉) · B. Swapna · M. Ramakrishna
Department of Botany, Vikrama Simhapuri University College, Kavali, Andhra Pradesh, India
e-mail: kambharath@gmail.com

S. Gujjala
Department of Biochemistry, Sri Krishnadevaraya University, Ananthapuramu, Andhra Pradesh, India

G. K. Pitchika
Department of Zoology, Vikrama Simhapuri University College, Kavali, Andhra Pradesh, India

M. Bangeppagari
Department of Cell Biology and Molecular Genetics, Sri Devaraj Urs Academy of Higher Education and Research, Kolar, Karnataka, India

S. A. Aransiola and N. R. Maddela (eds.), *Vermitechnology: Economic, Environmental and Agricultural Sustainability*, https://doi.org/10.1007/978-3-031-96212-7_14

health of soil in organic farming, they may thus be used extremely effectively as biotic indicators.

Keywords Vermicast · Vermiwash · Biological wastes · Vermi-biotechnology · Vermicomposting · Biofertilizer · Earthworm

1 Introduction

Excessive usage of chemical-based fertilizers affects both the environment and human health while also changing the physicochemical characteristics and texture of the soil. Chemical-based fertilizers have presented a severe threat to the ecosystem and destroyed beneficial insects, worms, and microbes in soil (Ansari & Ismail, 2001). Food ingredients lose their flavors and tastes with time. Food grains, fruits, and vegetables lose some of their ability to be stored and kept, making them more prone to disease. Increased concentrations of various chemicals and metals are caused by increased usage of chemical fertilizers, which ultimately have an impact on crops and the watershed. Protein, vitamins, amino acids, ascorbic acid, and other nutrients are depleted in the soil as a result of the overabundance of phosphorus, nitrogen-rich, and potassium fertilizers, compromising the fertility and conservation of the soil and possibly accelerating desertification. In addition, nitrogenous fertilizers damage water and food supplies, endangering human health. The human body can develop stomach cancer as a result of consuming meals and beverages with high nitrate $(NO)_3{}^-$ concentrations. While just for a brief time and for a single crop, these chemical fertilizers produce amazing results. Use of synthetic fertilizers in excess has a negative impact on soil quality and crop output (Gupta, 2005). Potash, phosphate, and nitrogen fertilizers are used excessively, which pollutes the water and food supplies and has a major negative impact on health (Bhattacharya, 2004).

In addition to altered soil textures and physicochemical properties, the uncontrolled use the pesticides and chemical fertilizers also results in the destruction of useless insects, microbes, and bugs within the soil itself. Nitrogenous fertilizers in the nitrite and nitrate form contaminate food and water, causing major issues. These problems led to a disruption in biological matter cycling, a decline in native soil health, and a plateau in production (Mondal et al., 2017; Singh & Chauhan, 2015).

The usage of biofertilizers can address these issues. The application of non-chemical-based fertilizers, as well as insecticides, represents a few of the common practices that are being adopted using innovative farming practices that utilize biofertilizers as one of their components. Biological wastes can be effectively managed to produce biofertilizers. The application of vermiwash or vermicompost with biopesticides in agricultural farms satisfies the plants' need for nutrients as well as their need for resistance (Bhattacharya, 2004; Eastman et al., 2001; Gupta, 2005; Leh-Togi Zobeashia et al., 2018).

According to Trivedi and Goel (1984), there was a great deal of pressure to increase grain yields through the utilization of contemporary agricultural methods. Field soil is significantly harmed by contemporary agricultural practices. Organic materials are not completely recycled, which has an impact on the soil's ability to replenish nutrients and, eventually, on crop yield. The use of heavily chemical fertilizers to boost crop productivity is among the primary causes of the destruction of soil's flora and fauna, which are responsible for their natural quality.

The primary focus of agricultural scientists has been the creation of modern agriculture that can be free from chemicals as well as secure for both people and animals. Organic farming refers to a variety of nontraditional agricultural techniques. Vermicomposting and organic farming are the steps that will enable us to coexist peacefully with nature. Sustainable agriculture is essential to minimizing contamination of the environment, keeping soils healthy, limiting the loss of soil, as well as decreasing the usage of fossil fuels and natural resources by following proper conservation principles (Garg et al., 2006; Lalandera et al., 2015; Cito Namulisa et al., 2022).

Utilization of organic farming practices enhances agricultural product quality, conserves resources, and preserves soil fertility. A biotechnological practice called organic farming may be able to address the urgent shortage of plant nutrients required for long-term productivity. Biofertilizer applications are claimed to be optimal for a sustainable future and safe for the environment. The majority of biofertilizers, like vermicomposts, are often made from biological wastes. Understanding biological wastes is preferable before talking about vermicomposting (Gupta, 2005).

2 Biological Waste

Millions of tons of agricultural and household trash, as well as cow manure, are produced each year in India. Several odor and environmental issues were brought on by these wastes in the neighborhood. Industrial waste from gurgum released hazardous gases, which led to major environmental issues. If they are not properly managed, the wastes from horses, goats, and sheep can cause unpleasant problems.

Another big waste problem in the field is the issue of postharvest waste from several native crops, such as vegetable garbage, golden gram (*Vigna radiata),* wheat (*Triticum aestivum*), sorghum (*Sorghum vulgare*), and pearl millet (*Pennisetum typhoides*). The issues with managing solid waste have gotten worse as a result of the high rate of industrialization. These problems span from rural areas, where agricultural wastes, including cattle and vegetable wastes, are generated, to urban and industrial areas, where municipal garbage, textile and sugar mill wastes, wine industry wastes, sludge from dairy wastes, and other waste materials are generated. By microbial breakdown, these wastes contribute to a number of diseases, odor, and pollution issues (Bhartiya & Singh, 2012a).

3 Natural Farming

Natural farming represents one of the greatest possibilities for waste management in ecologically friendly agriculture. It offers significant assistance in boosting food output, income sources, soil texture upkeep, and environmental preservation. In India, organic farming has been practiced for a long time. The importance of using animal feces as manure was emphasized in classical texts like the Rigveda. Before 1000 BC, green manures were a common practice, as the Atharava Veda noted. Animal excrement, oil cake, and other types of dung were documented in Kautilya's Arthashastra. According to the available literature, around 1900, "Sir Albert Howard, a British agronomist," started organic farming in a village in North India (Gupta, 2005).

Up to 50% of agriculture in the United States suffered a substantial loss of its precious, nutrients-rich topsoil, according to a survey by the USDA in the late 1970s. Each hectare loses over 15 tons of topsoil on average; the average person may find the numbers confusing. The United Nations, which monitors soil erosion with satellites, believes that losses per hectare in northern India exceed those in the United States. These global statistics are equally concerning. The farm will suffer terrible consequences from this topsoil erosion. Wind and water carry away the smaller particles, leaving behind a coarser subsoil that needs more fertilizer to support plant growth because there is lesser agricultural waste capable of breaking down the soil (Ranganathan, 2006; Suthar, 2007).

The supply of water for agriculture in dry and semiarid areas must be increased by irrigation because water is a necessary component for plant growth. The FAO estimates that 70% of the water used by humans worldwide is used to irrigate 270 million acres of land. Organic agriculture is an agricultural method that promotes the utilization of as much organic material as possible while discouraging the use of artificially produced agro-inputs in order to maintain soil fertility, and productivity, along with insect control for circumstances that ensure environmentally friendly resources and an ecologically sound environment (Becagli et al., 2022; Gupta, 2005; Ranganathan, 2006).

In natural agriculture, vermicomposting is a superior technique that helps improve soil quality and allows earthworms to manage waste. That constitutes some of the more interesting components because it helps to create a larger interaction between nutrition, the quality of the environment, as well as the safety for the health of people and animals. One better method of waste management than microorganisms is using earthworms. Compost is made from agricultural waste and is a valuable product that can be utilized in farming like compost for increasing food output. Various species of earthworms have transformed various natural and human-made garbage into beneficial compost. Earthworms consume a variety of materials and ingest them as vermicompost, which is a peat-like substance. These substances include waste from sewers, rice stubble, flue ash, water hyacinth, leaves of mango, waste from the vine fruit business, municipal solid garbage, slurry from paper manufacturing plants, waste from farming, plant litter, and sewage from cattle and dairy manufacturing operations. Earthworm vermicomposting helped solve the waste problem by waste recycling (Garg et al., 2006).

4 Earthworm

Earthworms are members of the class Oligochaeta, phylum Annelida. The cylindrical body has ring annuli that are distributed more or less evenly along its length. The dorsal side of the body can be identified by its deeper color and black middorsal line due to the presence of a strong lateral blood artery that is placed directly below its semi-translucent skin. Genital apertures and papillae, which are found in the body's anterior area, are characteristics of the ventral surface. The eyes, hearing, and lungs are gone. They are driven through the soil's surface as a result of the rain's saturation and breathe the air which has been diffused through their thin skins into the spaces between the particles of soil. The earthworms have been hermaphrodites, meaning they have both sex organs on a single person but need a partner to mate because the sexes are located in distinct segments. The clitellum, which surrounds young earthworms that are reproducing, secretes sacs after which other worms' sperm is kept. The sperm and eggs are enclosed inside the worm as the mucus passes over it. The worm seals at both ends when it is released, forming a 1/8-inches-long lemon-shaped cocoon. Although they have over three thousand different types of earthworms, only a few epigeic varieties are used in vermicomposting. Several species efficiently work the soil and recycle organic substances to help plants develop (Sinha, 2009).

The normal earthworm is endozoic, epigeic, and anecic. Anecic worms are nocturnal organisms that emerge from their vast tunnels deep under the mineral layer of the earth to collect food. The endozoic worms likewise burrow, though considerably less frequently, rarely emerge from their tunnels above ground, and they eat soil organic materials. The majority of endozoic creatures have extremely extended life cycles and weak regeneration abilities. The epigeic species eat decomposing organic waste and dwell in the surface litter. They have a rapid potential for regeneration and are quite active (Aransiola et al., 2022; Kumar, 2014).

In order to stabilize the inorganic materials for plants into organic molecules and increase the fertility of the soil, earthworms are essential. Earthworms multiply their inorganic substances, in addition to a few hormones and vitamins that promote plant development, by mixing the worms' cast into compost. They also assist in the management of solid organic waste that is decomposing and dumped in landfills, which causes odor issues and pollutes the water, soil, air, and human population's health. The benefits of earthworms are well understood thanks to extensive scientific studies. According to Charles Darwin, "the worms seem to be the main promoters of vegetation," perforation and loosening soil, and introducing straw, stalks of leaves, and twigs into it to make it more permeable to rain and plant fiber. Primarily through producing an endless supply of "worm-cast," which is fine fertilizer for grains and grass made of their waste (Gupta, 2005; Mondal et al., 2017; Ranganathan, 2006).

Worms are primarily detritivores and omnivores; however, they usually have particular food preferences. They consumed a variety of living and dead organic materials, including fungi, bacteria, algae, nematodes, diatoms, and protozoa. The trash was totally broken down, and it was stabilized in organic form. The worms prefer the soil when it is sufficiently moist, warm, and full of organic debris that is decaying at night because they are nocturnal. Earthworms were known as the "intestine of the earth" by Aristotle (Ranganathan, 2006).

Earthworms act like biological indicators of soil quality and are sometimes referred to be farmers' buddies. Earthworms sustain an adequate amount of microbes, fungi, protozoa, actinomycetes, spiders, insects, and millipedes needed to maintain a healthy soil. The earthworms are well recognized as having a significant role in the development of rich soil. Maintaining soil quality results in soil fertility. The decomposition process is moving forward at a fast enough rate to discharge nutrients that plants need for healthy development. Since most life processes were dependent on extreme temperatures, earthworms, for example, are thermostatic fermentation systems and have sophisticated systems for regulating their body temperatures (Ranganathan, 2006; Suthar, 2007).

A few of the factors influencing earthworm distribution in the soil are soil textures, temperature, aeration, moisture, organic minerals, pH, decaying matter, feces, as well as reproduction potentials (Garg et al., 2006; Suthar, 2006). The condition of the earthworm's diapause is greatly impacted by the pH (Aalok et al., 2008). These are the primary tunneling crustaceans as well as their activity has increased the soil's ability to retain water. They also provide the best conditions for bacterial and plant root growth in an aerobic environment (Wurst et al., 2003). The earthworm may serve as helpful bio-indicator for soil metal contamination (Khyade & Pawar, 2016).

Earthworms are an integral part of the food chain and may extract harmful compounds through soils (Handriks et al., 1995; Spurgeon & Hopkin, 1996). Their earthworm population is mostly controlled by the soil's nutrient resources and the external biotic parameter (Albanell et al., 1988; Edwards & Bohlen, 1996). The decomposition of degradable wastes poses environmental problems, and the earthworms are known to collect insecticides as well as heavy metals like cadmium, nickel, lead, and mercury in their tissues (Bhartiya, 2013; Bhartiya & Singh, 2012a, 2012b; Kaplan et al., 2011; Ojuolape et al., 2015; Samadhiya et al., 2013).

According to Ismail (1993) and Aalok et al. (2008), the earthworm's gizzards allow them to masticate organic materials, which increases the surface area available for microbial attack on feces. They contribute to keeping the pH of the soil stable by using the biocatalysts in their digestive tract, which include amylase, lipases, lichenases, proteases, cellulose, and chitinases. All enzymes are extremely active within a specific pH range (6.3 to 7.3). Earthworms have the capacity to separate oxygen from air, supplying the gut microflora in the process. This encourages several oxygenated waste preservation processes and eradicates pathogens from the soil. They boost the activity of nitrogen fixation in the soil by providing the ideal circumstances of food, air, and moisture to the nitrogen-fixing bacteria.

From ancient times, people have known that earthworms may convert organic wastes from the soil into products with value-added that can fertilize the land. As the first contemporary scientific investigation into this subject, Charles Darwin's research on the beneficial role of earthworms in soil ecology was published in 1881. Earlier in the modern era, vermicomposting was said to have been invented in the United States of America. Based on the scientific usage of vermicomposting, earthworm growing was promoted in the late 1940s as a productive way for farmers to increase the fertility of their soil. In the United States, Japan, Britain, and France, research on earthworms' potential to improve garbage management and land reclamation began in 1970. According to Science 1889, Cuba heavily relies on vermicomposting being its main form of fertilization for the elimination of both farm and urban solid waste. As a result, the extensive research shows that vermicomposting is a highly promising waste management technique. With more than a calendar year of practical knowledge in the management of natural resources, "Grace Mckellar Centre, Geelang, Victoria, Australia," offered consultancy services. Each week, this facility vermicomposted around 13 cubic yards of garden trimmings, paper waste, and kitchen scraps.

Earthworms, as well as their ash, have also been used as a tooth powder and stimulating agent for head hair growth in addition to being utilized as a natural antipyretic. This is used to cure stones in the bladder, a high fever, jaundice, smallpox, piles, and other conditions. Earthworms and their extract have antioxidant effects, and they can also open the bronchi and treat rheumatism, impotence, and other conditions. The gel and its *Eisenia fetida* extract prevent damage from oxidation because the earthworm tissue contains significant amounts of antioxidants like glutathione (GSH) and GSH peroxide. On albino rats, glutathione peroxidase (GPx), reduced GSH, and enzymes are shown to be increased by the paste of *Lampito mauritii*, whereas lipid peroxidation is decreased. The oral use of the earthworm glue of *L. mauritii* has restored gastrointestinal injury by lowering stomach acid secretion, reducing acidity, and elevating pH. Also, it had boosted the activity of antioxidant enzymes like GSH, GPx, and CAT to stop albino rat stomach mucous membrane damage. In both the acute and chronic phases, earthworm paste and its extract exhibit anti-inflammatory activities. Earthworm *L. mauritii*'s ethanolic and petroleum extracts have anti-inflammatory effects in albino rats. The dermal extraction phosphagen of *Lumbricus terrestris* inhibits the growth of mammary cancers in SHN mice. Strong antibacterial capabilities may be found in the tissue extract and coelomic fluid of earthworms. *Pseudomonas aeruginosa, Salmonella enteritidis, Streptococcus pyogenes,* and *Escherichia coli* few disease-causing and nonpathogenic microorganisms that are resistant to several earthworm extracts (Ansari et al., 2015; Balamurugan, 2006; Garg et al., 2006).

Due to their high proteins (65%), lipids (14%), carbohydrates (41%), and ashes (3%), earthworms have been utilized as a feeding component for aquatic animals. They are also used as animal feed in the poultry sector. Earthworms have long been a staple food for several native tribes in New Guinea and Africa. Due to their high protein level, earthworms are utilized as a supplement in the Philippines for recipes like adobo and dinuguan (Ghosh, 2004).

According to Albanell et al. (1988), earthworms sped up the process of mineralization and transformed manure into castings that had a greater level of humification and nutritional content. In order to make fishing baits, chicken feed, as well as aquarium fish food that complied with industry requirements, Cynthia and Rajeshkumar (2012) examined whether the research team *L. mauritii* had a significant amount of protein. The bio-potential of the earthworm organisms *Perionyx excavates,* over controlling waste as well as earthworm cocoons, and biomass generation is best maximized by a 3:1:1 mixture of cow dung, sawdust, and guar-gum waste from industries (a ligno-cellulosic waste that is of guar *Cyamposis tetragonaloba*) (Suthar, 2006). According to Ismail (1993), the best species in both vermicomposts and soil preservation in the southern part of India are *Perionyx excavatus* and *L. mauritii.*

Earthworms serve as essential in protecting soil fertility, managing soil processes, and fostering the cycling of nutrients together with different creatures (Chauhan, 2013; Ismail, 1993; Kumar, 2014). These epigeic species of earthworms are frequently used in vermicomposting to control trash. The earthworm species *Lumbricus rubellus, E. fetida, Peryonix excavates, Eudrilus eugeniae, L. mauritii,* and *Perionyx sansibaricus* are among those recommended for the decomposition of organic materials (Suthar, 2007; Talashilkar & Dosani, 2005). Two species, *E. eugeniae* (night crawler), and *E. fetida* (redworm) as well as foreign species like *P. excavatus, Denderobaena veneta,* and *L. rubellus* are commonly used for vermiculture in India (Edwards et al., 1995; Lalandera et al., 2015).

4.1 Earthworms that Are Suitable for Vermicomposting

Together, earthworms as well as the bacteria were connected with preserved nutrition in the vermicast and managed the biodegradable wastes created through metabolism. The utilization of five different species of earthworms—*D. veneta, E. fetida, E. eugeniae, Peryonix excavatus,* and *L. rubellus Hoffmeister*—has been proposed for the degradation of organic materials. India makes extensive use of exotic species like *P. excavatus, L. rubellus,* and *D. veneta,* as well as domestic species like *Eudrilus eugineae* and *E. fetida*, for vermiculture. Annual applications of animal manure and poultry litter increase the concentration of nutrients in soil (Aransiola et al., 2022; Suthar, 2007).

Parthasarathi and Ranganathan (2000) claim that the production by vermicomposts with various blends of fly ash and dung from cattle significantly increased the level of plant nutrients following the inoculation of *E. eugeniae.* In the stomach and cast of *P. excavatus, L. mauritii,* and *E. eugeniae, Aspergillus* spp., *Micrococci* spp., *Pseudomonas* spp., and *Bacillus* spp. were identified as phosphate-dissolving microbes (Kinberg). In Eastern UP, India, the usage of *E. fetida* (Savigny) has been advised for vermicomposting. The species *P. excavatus* and *L. mauritii* are suggested as being ideal for the effective composting process and management of soil in South India. The *Perionyx sensibaricus, Denderobaena vaneta, L. rubellus,* and *P. excavatus* are suitable for the management of solid wastes. The earthworm's best function was also seen in the bioaccumulation of various heavy metals from various animal excrement. With every combination of cattle dung and agricultural waste, the levels of toxic metals dramatically decreased after vermicomposting (Ismail, 1993).

4.2 India Uses Certain Species of Earthworms

The use of epigeic species that live on the surface for vermicomposting was advised by certain vermitechnologists, whereas endogeic and anecic species that burrow were advised by others. The following epigeic species are widely employed: *E. fetida, E. eugeniae, P. sansibaricus,* and the endemic *P. sansibaricus* Michaelsen. Most likely, the best one for vermicomposting is *E. fetida.* Additionally, the "Institute of National Organic Agriculture (INORA)" in Pune, India, encourages the use of surface-living worms since they quickly multiply and consume a variety of trash (Garg et al., 2006; Julka et al., 2009).

According to Lalandera et al. (2015), *P. excavatus* and *L. mauritii* are suitable species for the composting process in the southern region of India. For the treatment of solid waste, *D. veneta, P. sansibaricus, L. rubellus,* and *P. excavatus* are ideal. According to Bhartiya and Singh (2012a, 2012b), earthworms play a key role in the bioaccumulation of many heavy metals from animal excrement, including chromium, nickel, cobalt, lead, cadmium, and arsenic. After vermicomposting, the levels of toxic metals significantly decreased in all mixtures of cattle dung and agriculture waste.

The quantity of earthworms in the soil is influenced by their temperature, texture, moisture, aeration, pH, inorganic salts, organic matter, dung, and litter, as well as by their ability for reproduction and dispersal. The proportion of earthworms in soil is larger in temperate areas, whereas in tropical areas, humus feeder worms outnumber organic feeder worms. The main factors limiting the numbers of earthworms are the external biotic parameters and inadequate nutritional resources. Because a nitrogen-rich diet promotes growth and reproduction, earthworms prefer it. The majority of worms were found in soil that was between 12 and 45% moist. The key regulating elements for their population growth are the soil's nutrient resources and the external biotic parameters. The majority of pesticides and heavy metals are very difficult to harm earthworms. Worms could counteract the effect by producing more mucus, limiting their motility, and boosting their reproductive potential up to a specific concentration. Pesticides and heavy metals are also known to accumulate in the tissues of earthworms. *E. fetida* was capable of concentrating sodium but not lead. According to soil characteristics (Bhartiya & Singh, 2014), Ca and pH properties, earthworm tissues bioaccumulate toxins differently (Nath et al., 2009).

According to Edwards (1998), the symbiotic relationship between earthworms and bacteria serves as a method for regulating enzymes throughout processes and maintains the nutritional content of vermin-cast. The usage of the following five earthworm species has been recommended: *D. veneta, P. excavatus, E. fetida, E. eugeniae,* and *L. rubellus*. In India, vermiculture is commonly practiced using indigenous species, including *P. excavatus, D. veneta,* and *Lumbricus,* as well as domestic species like *Eudrilus eugineae* (night crawler) and *E. fetida* (redworm). In loamless manures for horticulture, worm-digested cattle dung acts as a complement to peat. An increased nutritional concentration of metals in soil is the result of repeated annual treatments of animal waste and poultry litter. Modern biotechnology, known as vermiculture, involves the development of the earthworm *E. fetida* as well as the use of their casting as a vital tool for trash management and the composting process. For efficient and environmentally responsible waste management, earthworms serve as natural bioreactors.

Regarding the treatment of the organic matter in the south of India, *L. mauritii, L. rubellus, P. sansibaricus,* and *P. excavatus* are appropriate species, as are *P. excavatus, L. rubellus,* and *D. veneta.* Vermicomposts frequently use the epigeic species *E. eugeniae* to break down organic waste. When contrasting the various ratios of the cattle dung and fly ash using inoculating of *E. eugeniae,* the level of nutrient accessibility was considerably higher in the 1:3 (fly ash to cow dung) treatments. *Pseudomonas* sp., *Bacillus* sp., *Micrococci,* and *Aspergillus* sp. found among the phosphate-solubilizing bacteria discovered in the casts and intestines of *P. excavatus, E. eugeniae,* and *L. mauritii.* The vermin-cast of *E. eugeniae* was abundant in cellulolytic bacteria that fix nitrogen and were capable of saturating rocks with phosphate (Venkatesh & Eevera, 2008).

According to *L. mauritii,* the species prefers areas with high C/N ratios for the treatment of activated sludge in order to prevent environmental pollution. *Eudrilus eugeniae*'s population and growth are directly impacted by time and space. The earthworms are a crucial biotic component of soil that, by breaking down organic matter that is decaying, maintains soil health and complies with environmental regulations. Earthworms serve as a source of food for the catfish *Mystus Vittatus* since they are so rich in nutrients. Vermicomposting of paper mill sludge using *L. mauritii.* The only meal that *E. eugeniae* did not favor was water hyacinth, whereas the addition of cow dung at a rate of about 14% had a good effect on biomass growth and hatchling output. Under experimental conditions, the *L. terrestris* favors solid sludge from paper mills. Even though earthworm growth was slow, solid paper mill sludge had no harmful effects on them in a laboratory setting. Several exotic epigeic species are employed for vermicomposting in India, including *E. eugeniae* and *P. excavates* (Parthasarathi and Ranganathan 2000; Samadhiya et al., 2013; Suthar, 2007).

5 What is Vermicast

Vermicomposting or vermicasting is the practice of employing earthworms to break down organic waste. Unlike conventional composting, it is a natural, odorless, aerobic process. Earthworms consume garbage and subsequently excrete castings, which are granules of nutrient- and organically rich soil that are dark, odorless, and useful for improving the soil. Earthworm castings serve as a fertilizer that may be applied more quickly than compost since the nutrients are released more slowly for developing plants (Awadhpersad et al., 2021; Hassan et al., 2022; Ortega-Torres et al., 2023) (Fig. 1).

However, feeding garbage to the earthworms causes nitrogen to mineralize, which is then followed by the phosphorus, along with sulfur decomposition after digestion. Vermicast nutrient content varies with the type of earthworm diet. Casts typically contain between 75 and 80% moisture and have a C:N ratio of 12–15:1, 1.5% to

Fig. 1 **a** A model vermicompost bed with sheds; **b** large-scale vermicompost production

2.5% N, 1.25% to 2.25% P_2O_5, and 1% to 2% K_2O. Nutrients can be given to plants in line with their needs because of the controlled-release granule form of earthworm castings.

6 Vermicomposting and Vermiculture

According to Ghosh (2004), vermiculture is a cutting-edge form of biotechnology in which earthworm *E. fetida* breeding and multiplication, as well as the use of its castings, have become crucial methods for recycling trash and creating vermicompost. Earthworms are necessary for the breakdown of biodegradable wastes such as cattle, pig, and dairy solids, poultry, sludge, game, rabbit, and horse dung; agricultural wastes; municipal waste; and urban waste (Chauhan & Singh, 2012; Singh & Chauhan, 2015). Earthworms can be used as inexpensive and environmentally friendly bioreactors for waste management, according to research by Aalok et al. (2008) (Table 1).

Vermiculture is the large-scale industrial farming of a particular species of earthworm for the benefit of both the soil and man in a selected area or containers that mimic semi-natural circumstances. When it comes to their capacity for composting, different earthworm species have varying potential. Several species might be utilized in the production of high-protein aquaculture feed. The protein content of dead earthworm (*E. fetida*) powder ranges from 62 to 64%, along with amino acids, 4.3% lysine, 2.3% cystine, and 2.2% methionine. A mixture of soil microbes and an efficient species of earthworm is used in verticulture (Edwards & Arancon, 2006; Ghosh, 2004; Singh & Chauhan, 2015).

An eco-biotechnical method called vermicomposting uses earthworms to convert complex organic materials with high energy content into stabilized humus-like products. Since they modify the raw materials as well as alter the process of biological

Table 1 Physical and chemical properties of vermicompost

S. No.	Nutrient*	Vermicompost
1	C:N	8.37
2	Total nitrogen (%)	4.2
3	Total phosphorus (%)	0.83
4	Total potassium (mg kg^{-1})	920.0
5	Total calcium (%)	3.9
6	Total magnesium (%)	3.2
7	Total manganese (mg kg^{-1})	320.95
8	Total zinc (mg kg^{-1})	200.12
9	Total Fe (mg kg^{-1})	5986
10	Total copper (mg kg^{-1})	42.62
11	Organic carbon (%)	20.5
12	EC (dSm^{-1})	7.41
13	pH	7.64

*These values are subject to variation depending on the type of organic waste

movement, earthworms are important participants in the biochemical breakdown of organic materials by bacteria (Ansari et al., 2015; Garg et al., 2006).

There are a number of epigeic earthworms that feed on detritus and can be raised in organic waste. The top contenders for the organic waste recycling market have been identified as the earthworm species: *P. excavatus, E. eugeniae*, and *E. fetida.* Through aeration, bioconversion, their excretions, and a qualitative or quantitative impact on the telluric microflora, earthworms speed up the conversion of organic waste into stabilized forms (Kaviraj & Sharma, 2003).

According to Chauhan and Singh's (2014) research, vermicomposting might be a good strategy for turning organic materials into worthwhile goods using earthworms. Crucial nutrients for plants incorporated into feed substances, such as N, P, PO_4^{3-}, and Ca, are converted during the process into forms with greater solubility as well as absorbability for plant microbes compared to those found in parent material. By ingesting various solid organic wastes, earthworms produce vermicompost, a peat-like substance that is considerably more fractured, porous, and conducive to microbial activity than the original waste was. This is due to the thorough humidification as well as the degradation that vermicompost has undergone. The process of composting is a rapid yet efficient way to reuse rubbish from farms, cities, and kitchens. It also bioconverts organic waste into compost that is rich in nutrients, thanks to the activity of earthworms, which also have a large population of actinomycetes and helpful microorganisms. It boosts drainage capacity, porosity, and aeration, which lowers the amount of water needed to irrigate crops. It boosts the availability of nutrients and might be used to make complicated fertilizer pellets. Vermicomposts used for green gram, chickpea, and field peas increased percentages of germination by 92, 90, and 93, respectively, along with raised crop yields by 9, 14, and 12 q/ha, according to Baghel et al. (2018).

Vermicomposts significantly reduce the number of dangerous pathogenic bacteria. It is commonly accepted that during the thermally sensitive stage of the process of composting, harmful organisms are eliminated. Nevertheless, composting with earthworms boosts and speeds up the rate of nitrogen mineralization. The humification process is more pronounced and rapid during the maturation stage of vermicomposting (Fig. 1). Moreover, compared to composting, this technique may result in a larger reduction of heavy metals that are bioavailable. In terms of trash management, vermicomposting is more useful than composting. It is one of the most important measures in resolving this ecological problem and stabilizing organic waste (Mishra et al., 2014).

Vermicomposting uses fine-separated peat mass materials that have high permeability, excellent air circulation, water flow, and water-retaining capability. Vermicomposts have been shown to have higher levels of organic matter and total organic carbon, readily available NPK alongside additional micronutrients, and microbiological activity of enzymes, as well as plant growth hormones. Vermicomposts have a balanced pH, a higher ability to exchange cations, a lower ability to dissolve ions, and a higher concentration of humic acids. The plants rapidly absorb the nutrients present. Vermicomposts are more durable than the original material and have improved the physicochemical characteristics of the soil. They also have increased nutrient availability (Mondal et al., 2017; Sinha, 2009).

Vermicomposting, according to Sharma and Garg (2018), is a biochemical breakdown method used for organic waste that requires the relationship between bacteria and earthworms. Vermicomposting products contain higher levels of plant-available nutrients and a considerably wider variety of agricultural and aquacultural probiotics than traditional compost. Vermicompost is therefore regarded as a microbial fertilizer used primarily on agricultural grounds. Animal manure can be used as a substrate to create vermicomposts. Vermicomposts encourage the development of soil qualities such as granulation, fine dirt, effective aeration, simple root penetration, and increased water holding capacity. Organic manure may enhance the soil's physical properties and its organic content, water transfer and retention, mass density, as well as accumulation. Field crops, decorative and flowering plants, cereals, legumes, sugarcane, and vegetables are just a few of the crops where vermicompost has been shown to affect plant growth and production. They also showed that the increased response by plants was only apparent while vermicompost had been applied around 10% to 40% of the quantity in the growth medium; larger dosages of vermicompost, however, do not promote plant development due to the excessive salt quantity (Bansal & Kapoor, 2000; Hand et al., 1988).

7 Vermiwash

Vermiwash, a liquid biofertilizer, has a pale yellow appearance. This is a powerful fertilizer made from an extract of earthworms' mucus and excretory secretions, which help plants grow and produce more (Fig. 3). It has micronutrients, hormones, vitamins, and the ability to resist disease. A honey-brown substance, including heterotrophic microbes, fungal organisms as well as nitrogen-fixing organisms and phosphate solubilizers, macro- and micronutrients, hormones, enzymes, and vitamins, is vermiwash (Table 2). It is a liquid organic biofertilizer pesticide. Vermiwash, liquid manure, is especially beneficial as a foliar spray for fostering the development and productivity of plants and also for halting the spread of disease. Using the vermiwash complex, lawns, nurseries, and orchids can all be successfully grown. Vermiwash, a coelomic fluid extraction, is a source of many enzymes, auxines, cytokinins, and gibberellins, among other plant growth regulators. Additionally, it contains vitamins, especially B12. To encourage crop development and production, the resulting manure is obtained in the form of liquid and sprayed onto plants as foliar fertilizer. Moreover, it makes crops more resistant to dangerous diseases. Vermiwash boosts crop output by providing foliar manure for the roots of the plants. Administration of vermicompost alongside vermiwash, which slowly releases nutrients for absorption together with other nutrients like auxins, gibberellins, and cytokinin, has been linked to improved plant development and higher yield (Manyuchi & Nyamunokora, 2014; Nath & Singh, 2016). Vermiwash supports plant growth by providing a wealth of hormones, enzymes, vitamins, macronutrients, and micronutrients (Ortega-Torres et al., 2023; Verma et al., 2018) (Figs. 2 and 3).

The bacteriostatic substances in the earthworm-produced vermiwash protect the vegetation against pathogenic bacteria. Vermiwash is well-known to be used in the morning, just before sunrise, as well as evening, just after sunset, for spray-on foliage applications. Additionally, this is asserted that mixing liquid manure with biopesticides in a 1:1 ratio and diluting them ten times in water will affect cow urine and vermiwash (Ismail, 1993).

Table 2 General decomposing bacteria and fungi forming a symbiotic relationship with earthworms found in vermiwash and vermicasting

Genera	Bacteria/fungus	References
Absidia glauca	Fungus	Tiwari and Mishra (1993)
Actinomycetes	Bacteria	Singh et al. (2015), Balachandar et al. (2018a, 2018b)
Actinobacteria	Bacteria	Singh et al. (2015)
Agrobacterium spp.	Bacteria	Tripathi et al. (2005)
Alternaria alternate	Fungus	Tiwari and Mishra (1993)
Azotobacter spp.	Bacteria	Tripathi et al. (2005)
Bacillus circulans	Bacteria	Idowu et al. (2008)
Clostridium absonum	Bacteria	Idowu et al. (2008)
Firmicutes	Bacteria	Singh et al. (2015)
Mycorrhizae	Fungus	Singh et al. (2015)
Penicillium absonum	Fungus	Tiwari and Mishra (1993)
Proteobacteria	Bacteria	Singh et al. (2015)
Rhizobium spp.	Bacteria	Tripathi et al. (2005)
Staphylococcus aureus	Bacteria	Idowu et al. (2008)

Fig. 2 Vermicomposting techniques are used in the production of organic tea

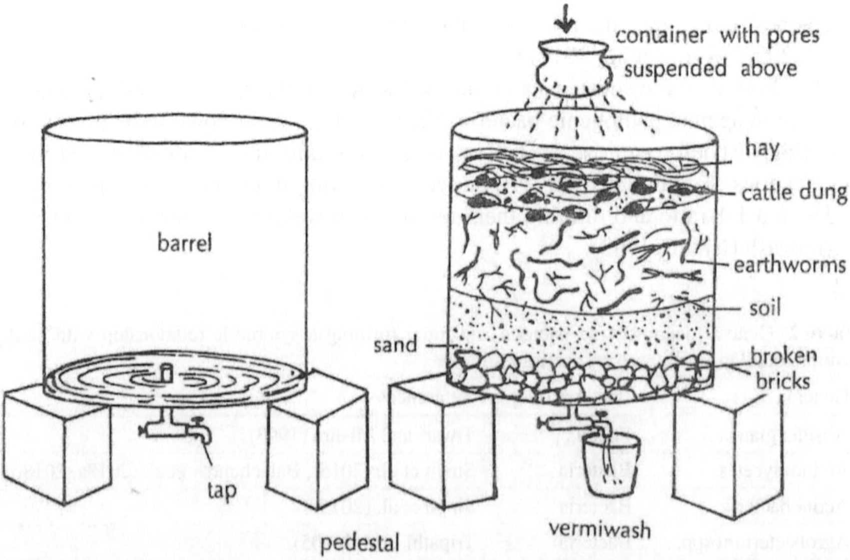

Fig. 3 Vermiwash preparation steps

According to Zaller (2006), vermicompost extract foliar spraying has been demonstrated to provide a number of advantages for fruit flavor, including the reduction of "late blight" in farm-grown tomatoes. Moreover, it stops the microorganisms that cause illness in tomato plants. It has been demonstrated that the use of aqueous compost extract can lessen both necrotrophs' and biotrophs' elocote illness. The soil-borne diseases and pests have been found to be suppressed by aqueous extracts of vermicomposts. Given their high nutritional content, vermicompost extract and composts make sense as potential spray fertilizers. Generally speaking, spraying onto foliage provides an easier way of supplying fertilizers to more vigorous vegetation than root treatment.

In dry conditions, foliar fertilization is significantly more successful than soil treatment, according to Mishra et al. (2014). Vermiwash has reportedly been sprayed on different varieties of tomatoes, significantly increasing the plant's growth and fruit production. Foliar sprays containing nutrients can also be used to counteract the decline of the absorption of nutrients through roots that occurs as the reproductive phase begins, as an outcome of increased sinks' competition for sugars. The quality and conditioning of the soil have increased with the judicious and selective use of organic additions, including vermicomposts, vermiwash, mulch (farm wastes), and green manures.

It has been demonstrated that adding fluids from pots containing earthworms has an impact on rye-grass generation of dry matter. The fruit quality of plants can be impacted by the microbial population in conventional thermophilic composts, humic compounds in vermiwash, or both. Vermiwash is applied to plant leaf regions to control pest arthropods and plant parasitic nematodes while enhancing plant growth, productivity, and seed germination. The increased microbial activity in vermiwash results in the significant production of plant-promoting hormones like auxin, cytokinins, and gibberellins. Humic acids are created in large quantities during vermicomposting and leach out of vermicomposts during vermiwash extraction. Humic acid promotes the growth of plants. Thermophilic compost extract has been shown to be beneficial against a number of fungal infections that affect fruits and leaves (Awadhpersad et al., 2021; Sudha et al., 2003).

According to Shukla and Singh (2010), nitrogen, which is present in vermiwash as a type of fluid, nitrogen-containing elimination compounds, hormones that regulate growth, and enzymes, has a substantial impact on the germination of seed and the growth of bean seedlings. Vermiwash's black gramme spraying showed a remarkable growth. Okra plant growth and productivity are dramatically impacted by vermicomposed weeds and their aqueous extract. Vermiwash is a natural plant growth stimulant for crops, including coconuts, horticulture, and tea. A natural fertilizer with inorganic N and K is called vermiwash.

Vermiwash made from vegetable waste and cow dung has been shown to be highly effective against the cow pea powdery mildew illness. The vermiwash also contains nitrogen-fixing bacteria such as *Azobacter, Agrobacterium*, and *Rizobium* species, as well as a cocktail of enzymes including proteases, amylases, ureases, and phosphatases. The vermiwash's microbe population had a big impact on how phosphorus, which is found in organic substances, moved through its biological cycle. The biogeochemical cycle of phosphorus is dissected, then mineralization using enzyme complexes made by microorganisms, such as phosphatase (Balam, 2000).

Vermiwash is an excellent foliar spray and insecticide when mixed with diluted cow's pee (8 L of water, 1 L of vermiwash, and 1 L of cow's urine). By using vermiwash topically, soya bean (*Glycine max*) growth and yield were improved (Rana, 2000). Marigold plants' growth and output significantly increased as a result of using vermiwash. Vermiwash was used weekly, which resulted in a 7.3% increase in radish output. Additionally, it may be utilized as a liquid foliar fertilizer at a 20–30% concentration and stops dangerous fungi from developing their mycelium. This is clear that vermiwash can be a useful tool for encouraging plant development while preventing a variety of bacterial illnesses.

The effects of vermiwash created from various waste materials alone and in binary mixtures of both plant developments along with crop production have not yet been studied, in spite of the topic's significant investigation. When various wastes from cattle are mixed into agricultural also food waste separately, along with bipolar configurations, there is a significant effect on certain vegetable crops (Chauhan & Singh, 2014; Rao, 2005; Thangavel et al., 2003).

8 Conclusion

The employment of epigeic earthworms will significantly reduce the pollution risk brought on by the decomposition of organic wastes, as is evident from the aforementioned accounts. The appropriate method for converting trash into rich organic manure is vermicomposting. Utilizing vermicompost as well as vermiwash in the fields on the farm can help the plants grow and be more productive. If earthworms are introduced into farming regions, various wastes can be recycled more effectively. Therefore, we may infer that vermi-biotechnology, along with the use of earthworms in fields of crops and vermicompost/vermiwash, is an effective method for promoting environmentally friendly development.

References

Aalok, A., Tripathi, A. K., & Soni, P. (2008). Vermicomposting: A better option for organic solid waste management. *Journal of Human Ecology, 24*(1), 54–64.

Albanell, E., Plaixats, J., & Cabrero, T. (1988). Chemical changes during vermicomposting (*Eisenia fetida*) of sheep manure mixed with cotton industrial wastes. *Biology and Fertility of Soils, 6*(3), 266–269.

Ansari, A. A., & Ismail, S. A. (2001). A case study on organic farming in Uttar Pradesh. *Journal of Soil Biology & Ecology, 27*, 25–27.

Ansari, A. A., Pereira, M., & Jaikishun, S. (2015). Effect of vermiwash obtained from different sources (neem, rice straw and bagasse) and standardised hydroponics solution on the growth of *Colocasia esculenta* (Australian poi) in Guyana. *American Journal of Experimental Agriculture, 7*(5), 275–283.

Aransiola, S. A., Ijah, U. J. J., Abioye, O. P., & Bala, J. D. (2022). Vermicompost-assisted phytoremediation of toxic trace element-contaminated soil in Madaka, Nigeria using Melissa officinalis L and Sida acuta. *International Journal of Environmental Science and Technology*. https://doi.org/10.1007/s13762-022-04105-y

Awadhpersad, V. R., Ori, L., & Adil Ansari, A. (2021). Production and effect of vermiwash and vermicompost on plant growth parameters of tomato (*Lycopersicon esculentum* Mill.) in Suriname. *International Journal of Recycling Organic Waste in Agriculture, 10*(4), 397–413.

Baghel, B., Sahu, R., & Pandey, D. (2018). Vermicomposting an economical enterprise for nutrient and waste management for rural agriculture. *International Journal of Current Microbiology and Applied Sciences, 7*(2), 3754–3758.

Balachandar, R., Karmegam, N., & Subbaiya, R. (2018a). Extraction, separation and characterization of bioactive compounds produced by Streptomyces isolated from vermicast soil. *Research Journal of Pharmacy and Technology, 11*(10), 4569–4574.

Balachandar, R., Karmegam, N., Saravanan, M., Subbaiya, R., & Gurumoorthy, P. (2018b). Synthesis of bioactive compounds from vermicast isolated actinomycetes species and its antimicrobial activity against human pathogenic bacteria. *Microbial Pathogenesis, 121*, 155–165.

Balam. (2000). *Studies on biopesticidal activity of vermiwash in control of some foliar pathogens* [MSc (Agri.) thesis, Dr. B.S.K.K.V., Dapoli].

Balamurugan, M. (2006). *Effect of earthworm paste Lampito mauritii, (Kinberg) on the anti-inflammatory, antioxidative, haematological and serum biochemical indices of rat (Rattus norvegicus)* [M.Phil. thesis, Annamalai University].

Bansal, S., & Kapoor, K. K. (2000). Vermicomposting of crop residues and cattle dung with *Eisenia fetida. Bioresource Technology, 73*(2), 95–98.

Becagli, M., Arduini, I., & Cardelli, R. (2022). Using biochar and vermiwash to improve biological activities of soil. *Agriculture, 12*(2), 178.

Bhartiya, D. K. (2013). *Studies on bioaccumulation of heavy metals by earthworm Eisenia fetida in different wastes and field soils* [Thesis awarded, Department of Zoology, Deen Dayal Upadhyaya Gorakhpur University, Gorakhpur, India].

Bhartiya, D. K., & Singh, K. (2012a). Heavy metals accumulation from municipal solid wastes with different animal dung through vermicomposting by earthworm *Eisenia fetida. World Applied Sciences Journal, 17*(1), 133–139.

Bhartiya, D. K., & Singh, K. (2012b). Heavy metals remediation from maize (*Zea mays*) crop by the use of vermicomposts through vermicomposting by *Eisenia fetida. American-Eurasian Journal of Agricultural & Environmental Sciences, 12*(9), 1215–1222.

Bhartiya, D. K., & Singh, K. (2014). Co, Cr and Pb accumulate by *Eisenia fetida* (Savigny) from animal dung, soil and pea (*Pisum sativum* L.) grain through vermic-activity. *Scholarly Journal of Agricultural Science, 5*(5), 154–164.

Bhattacharya, P. (2004). *Organic food production in India.* Agrobios.

Chauhan, H. K. (2013). *Effect of different combinations of animal dung and agro wastes on the reproduction and development of earthworm Eisenia fetida* [Thesis awarded, Department of Zoology, DeenDayal Upadhyaya Gorakhpur University, Gorakhpur, India].

Chauhan, H. K., & Singh, K. (2012). Effect of binary combinations of buffalo, cow and goat dung with different agro wastes on reproduction and development of earthworm *Eisenia fetida. World Journal of Zoology, 7*(1), 23–29.

Chauhan, H. K., & Singh, K. (2014). Potency of vermiwash with *Azadirachta indica* A. Juss on yield of gram (*Cicer arietinum*) and infestation of *Helicoverpa armigera* (Hübner). *American-Eurasian Journal of Toxicological Sciences, 6*(4), 87–93.

Cito Namulisa, P., Heiling, M., Grand, A., Resch, C., Dercon, G., & Hood-Nowotny, R. (2022). Greenhouse gas emissions from worm-compost-biochar combinations from farm to production to fork. In *EGU general assembly conference abstracts* (pp. EGU22-7741).

Cynthia, J. M., & Rajeshkumar, K. T. (2012). A study on sustainable utility of sugar mill effluent to vermicompost. *Advances in Applied Science Research, 3*(2), 1092–1097.

Eastman, B. R., Kane, P. N., Edwards, C. A., Trytek, L., Gunadi, B., Stermer, A. L., & Mobley, J. R. (2001). The effectiveness of vermiculture in human pathogen reduction for USEPA biosolids stabilization. *Compost Science and Utilization, 9*(1), 38–49.

Edwards, C. A. (1998). The use of earthworms in the breakdown and management of organic wastes. In: C. A. Edwards (Ed.), *Earthworm ecology* (pp. 327–354). CRC Press.

Edwards, C. A., & Arancon, N. Q. (2006). The science of vermiculture: The use of earthworms in organic waste management. In R. D. Guerrero III, & M. R. A. Guerrero-del Castillo (Eds.), *Proceedings of the international symposium-workshop on vermi technologies for developing countries, Philippine Fisheries Association, Inc., Los Baños, Laguna, Philippines, 16–18 November 2005, pp. 1–30.*

Edwards, C. A., & Bohlen, P. J. (1996). *Biology and ecology of earthworms.* Chapman and Hall.

Edwards, C. A., Bohlen, P. J., Linden, D. R., & Subler, S. (1995). Earthworms in agro ecosystems. In P. F. Hendrix (Ed.), *Earthworm ecology and biogeography in North America* (pp. 185–213). Lewis Publisher.

Garg, P., Gupta, A., & Satya, S. (2006). Vermicomposting of different types of waste using *Eisenia fetida*: A comparative study. *Bioresource Technology, 97*(3), 391–395.

Ghosh, C. (2004). Integrated vermi-pisciculture: An alternative option for recycling of solid municipal waste in rural area. *Bioresource Technology, 93*(1), 71–75.

Gupta, P. K. (2005). *Vermicomposting for sustainable agriculture* (pp. 11–163). Bharat Printing Press.

Hand, P., Hayes, W. A., Satchell, J. E., & Frankland, J. C. (1988). The vermicomposting of cow slurry. *Pedobiologia, 31*(3), 199–209.

Handriks, A. J., Ma, W. C., Brounds, J. J., Ruiter-Dijkman, E. M. D., & Gart, R. (1995). Modeling and monitoring organochloride and heavy metal accumulation in soils, earthworms, and shrews in Rhine-Delta floodplains. *Archives of Environmental Contamination and Toxicology, 29*(1), 115–127.

Hassan, S. A., Taha, R. A., Zaied, N. S., & Essa, E. M. (2022). Effect of vermicompost on vegetative growth and nutrient status of acclimatized Grand Naine banana plants. *Heliyon, 8*(10).

Idowu, A. B., Edema, M. O., & Adeyi, A. O. (2008). Gut microflora and microfauna of earthworm species in the soils of the research farms of the University of Agriculture, Abeokuta, Nigeria. *Biological Agriculture & Horticulture, 25*(3), 185–200.

Ismail, S. A. (1993). Keynote papers and extended abstracts. *Congress on Traditional Sciences and Technologies of India, pp. 1027–1030, I.I.T., Bombay*.

Julka, J. M., Paliwal, R., & Kathireswari, P. (2009). Biodiversity of Indian earthworms—An overview. In C. A. Edwards, R. Jeyaraaj, & I. A. Jayraaj (Eds.), *Proceedings Indo-US workshop on vermitechogy in human welfare, Kongunadu Arts and Science College (Autonomous), Coimbatore, India, pp. 36–56*.

Kaplan, O., Yildrim, N. C., Yildirim, N., & Cimen, M. (2011). Toxic elements in animal products and environmental health. *Asian Journal of Animal and Veterinary Advances, 6*(3), 228–232.

Kaviraj, S. S. (2003). Municipal solid waste management through vermicomposting employing exotic and local species of earthworms. *Bioresource Technology, 90*(2), 169–173.

Khyade, V. B., & Pawar, S. R. (2016). Physical, nutritional and biochemical status of vermiwash produced by two earthworm species *Lampito mauritii* (L) and *Eudrillus eugeniae* (L). *World Scientific News, 42*, 228–255.

Kumar, Y. (2014). *Studies on diversity and ecology of earthworms of Eastern Uttar Pradesh* [Thesis awarded, Department of Zoology, Deen Dayal Upadhyaya Gorakhpur University, Gorakhpur, India].

Lalandera, C. H., Komakecha, A. J., & Vinnerasa, B. (2015). Vermicomposting as manure management strategy for urban small-holder animal farms—Kampala case study. *Waste Management, 39*, 96–103.

Leh-Togi Zobeashia, S. S., Aransiola, S. A., Ijah, U. J. J., & Abioye, O. P. (2018). Anaerobic digestion and agricultural application of organic wastes. *Advances in Environmental Research, 7*(2), 73–85. http://www.techno-press.org/content/?page=article&journal=aer&volume=7&num=2&ordernum=1

Manyuchi, M. M., & Nyamunokora, M. (2014). Granulation of vermicompost using vermiwash as a binding media. *Global Journal of Engineering Science and Researches, 1*(1), 4–6.

Mishra, K., Singh, K., & Tripathi, C. P. M. (2014). Management of infestation of pod borer (*Leucinodes Orbonalis* Guenee) and productivity enhancement of brinjal (*Solanum melongena*) through vermiwash with biopesticide. *International Journal of Advanced Research, 2*(1), 780–789.

Mondal, T., Datta, J. K., & Mondal, N. K. (2017). Chemical fertilizer in conjunction with biofertilizer and vermicompost induced changes in morpho-physiological and biochemical traits of mustard crop. *Journal of the Saudi Society of Agricultural Sciences, 16*(2), 135–144.

Nath, G., Singh, K., & Singh, D. K. (2009). Effect of different combinations of animal dung, and agro/kitchen wastes on growth and development of earthworm *Eisenia fetida. Australian Journal of Basic and Applied Sciences, 3*(4), 3672–3676.

Nath, S., & Singh, K. (2016). Analysis of different nutrients status of liquid bio-fertilizer of different combinations of buffalo dung with gram bran and water hyacinth through vermicomposting by *Eisenia fetida. Environment Development and Sustainability, 18*(3), 645–656.

Ojuolape, O. T., Aransiola, S. A., & Obideyi, O. A. (2015) Vermitechnology: A solution to environmental problems; A review. *Journal of Global Ecology and Environment, 3*(3), 127–135. http://www.ikpress.org/abstract/4516

Ortega-Torres, A. E., Herrera-Matallana, T. S., & Rico-Garcia, E. (2023). Vermiliquer as a biostimulant and antioxidant in hydroponic lettuce (*Lactuca sativa*) production. *Horticultural Science, 50*(1).

Parthasarathi, K., & Ranganathan, L. S. (2000). Aging effect on enzyme activities in pressmudvermicast of *Lampito mauritii* (Kinberg) and *Eudrilus eugeniae* (Kinberg). *Biology and Fertility of Soils, 30*(4), 347–350.

Rana, A. S. (2000). *Effect of organic manures and foliar spray of nutrients and growth regulators on the growth and yield of soybean (Glycine max. L.), Merril* [MSc (Ag) thesis, Tamil Nadu Agriculture University, Coimbatore].

Ranganathan, L. S. (2006). Vermicomposting enhances humification, mineralization and chelation. *Journal of Annamalai University Sciences, 42*, 1–14.

Rao, B. R. C. (2005) *Vermicomposting*. IEC CELL-KUDCEMP.

Samadhiya, H., Dandotiya, P., Chaturvedi, J., & Agrawal, O. P. (2013). Effect of vermiwash on growth and development of leaves and stem of tomato plants. *International Journal of Current Research, 5*(10), 3020–3023.

Sharma, K., & Garg, V. K. (2018). Comparative analysis of vermicompost quality produced from rice straw and paper waste employing earthworm *Eisenia fetida* (Sav.). *Bioresource Technology, 250*, 708–715.

Shukla, R. C., & Singh, K. (2010). Effect of vermicompost of different organic wastes germination, growth and productivity of certain crops. *Journal of Solid Wastes Technology and Management, 36*(3), 170–181.

Singh, K., & Chauhan, H. K. (2015). Potency of vermiwash with neem plant parts on the infestation of *Earias vittella* (Fabricius) and productivity of okra (*Abelmoschus esculentus*) (L.) Moench. *Asian Journal of Research in Pharmaceutical Sciences, 5*(1), 36–40.

Singh, A., Singh, D. P., Tiwari, R., Kumar, K., Singh, R. V., Singh, S., Prasanna, R., Saxena, A. K., & Nain, L. (2015). Taxonomic and functional annotation of gut bacterial communities of Eisenia foetida and Perionyx excavatus. *Microbiology Research, 175*, 48–56.

Sinha, R. K. (2009). Earthworms: The miracle of nature (Charles Darwin's 'unheralded soldiers of mankind and farmer's friends'). *The Environmentalist, 29*(4), 339–340.

Spurgeon, D. J., & Hopkin, S. P. (1996). The effects of metal contamination on earthworm populations around a smelting works: Quantifying species effects. *Applied Soil Ecology, 4*(2), 147–160.

Sudha, R., Ganesh, P., Mohan, M., Saleem, S. S., & Vijaylaxmi, G. S. (2003). Effect of vermiwash on the growth of black gram (*Vigna mungo*). *Agrobios News Letter, 30*(1), 77–79.

Suthar, S. (2006). Potential utilization of Guar gum industrial wastes in vermicomposts productions. *Bioresource Technology, 97*(18), 2474–2477.

Suthar, S. (2007). Nutrient changes and biodynamics of epigeic earthworm *Perionyx excavatus* (Perrier) during recycling of some agriculture wastes. *Bioresource Technology, 98*(8), 1608–1614.

Talashilkar, S. C., & Dosani, A. A. K. (2005). *Earthworms in agriculture*. Agrobios, Vedams eBooks (P) Ltd.

Thangavel, P., Balagurunathan, R., Divakaran, J., & Prabhakaran, J. (2003). Effect of vermiwash and vermicast extraction soil nutrient status, growth and yield of paddy. *Advances of Plant Sciences, 16*, 187–190.

Tiwari, S. C., & Mishra, R. R. (1993). Fungal abundance and diversity in earthworm casts and in uningested soil. *Biology and Fertility of Soils, 16*(2), 131–134.

Tripathi, Y. C., Hazarika, P., & Pandey, B. K. (2005). Vermicomposting: An Ecofriendly Approach to Sustainable Agriculture. In: *Verms and Vermitechnology, SB Nangia* (pp. 22–39). APH Publishing Corp., New Delhi.

Trivedi, R. C., & Goel, D. (1984). *Water pollution and physiochemical properties* (2nd ed., pp. 304–346). Pragati Prakashan.

Venkatesh, R. M., & Eevera, T. (2008). Mass reduction and recovery of nutrients through vermicomposting of fly ash. *Applied Ecology and Environmental Research, 6*(1), 77–84.

Verma, S., Babu, A., Patel, A., Singh, S. K., Pradhan, S. S., Verma, S. K., Singh, J. P., & Singh, R. K. (2018). Significance of vermiwash on crop production: A review. *Journal of Pharmacognosy and Phytochemistry, 7*(2), 297–301.

Wurst, S., Langel, R., Reineking, A., Bonkowski, M., & Scheu, S. (2003). Effects of earthworms and organic litter distribution on plant performance and aphid reproduction. *Oecologia, 137*(1), 90–96.

Zaller, J. G. (2006). Foliar spraying of vermicompost extracts: Effects on fruit quality and indications of late-blight suppression of field grown tomatoes. *Biological Agriculture and Horticulture, 24*(2), 165–180.

The Role of Vermitechnology in Plant Growth and Nutrient Enhancement

Gabriel Gbenga Babaniyi, Ademola Bisi-Omotosho, and Ulelu Jessica Akor

Abstract Rapid urbanization, economic expansion, and population growth have intensified the problem of managing organic waste from domestic agriculture and firms, posing serious environmental and financial dangers. The improper disposal of waste endangers the environment and increases the risk of disease spread. In order to address this global issue, this study reviews the role of vermitechnology in plant growth and nutrient enhancement. The chapter explores the process of vermicomposting for both organic and inorganic materials, including the production of vermiwash as a biocide and fertilizer. It covers the management of organic solid waste (OSW) and earthworms, factors affecting vermicomposting efficiency, and the reduction in material weight during the process. Additionally, it discusses moisture requirements, the interaction between aeration and oxygen availability, the influence of raw material particle size on composting, and plant growth in relation to nutrient availability. The chapter also examines direct nutrient uptake by plants from the soil and the role of symbiotic relationships with soil microorganisms in nutrient acquisition. Notably, the chapter describes vermicomposting as an environmentally friendly technological approach for managing OSW. More so, the chapter emphasizes the critical necessity of vermitechnology to be a method that emerges as a sustainable solution for agricultural and sanitary applications by utilizing the potential of earthworms to digest organic waste. This chapter concludes that the problems created by rising organic waste can be efficiently addressed through the incorporation of vermitechnology, paving the way for a more ecologically conscious and resource-efficient future.

G. G. Babaniyi (✉)
Department of Agricultural Development and Management, Agricultural and Rural Management Training Institute, Ilorin, Nigeria
e-mail: gabrielbabs25@gmail.com

A. Bisi-Omotosho
Department of Sustainability Management and Innovation, University of Westminster, London, UK

U. J. Akor
Department of Crop Science, Faculty of Agriculture, University of Abuja, Abuja, Nigeria

S. A. Aransiola and N. R. Maddela (eds.), *Vermitechnology: Economic, Environmental and Agricultural Sustainability*, https://doi.org/10.1007/978-3-031-96212-7_15

Keywords Vermitechnology · Vermicompost · Vermicast · Vermiwash · Vermiculture

1 Introduction to Vermitechnology

Utilizing local species of earthworms on the surface and below for soil management and composting is known as vermitechnology (Aransiola et al., 2022). Vermicompost improves soil nutrient content, boosts soil microbial activity, increases oxygen availability, sustains natural soil temperature, promotes soil porosity and water infiltration, and enhances plant growth, yield, and quality. Vermitechnology is the study and practical use of processes that use earthworms to break down organic waste for sanitary and agricultural purposes. Earthworms can decompose and stabilize solid organic waste as well as organic waste that is suspended or dissolved in water (Ansari & Ismail, 2012). The role of an earthworm in improving soil health may be unmatched by that of any other living thing in the soil. Earthworms enhance plant growth by improving soil aeration, water infiltration, soil structure, nutrient cycling, and water movement. They are also one of the primary organic matter decomposers (Labenz, 2022; Rekha et al., 2018). More specifically, earthworms serve as key detritus feeders and are important for soil metabolism and the decomposition of organic materials (OMs), making them excellent markers of the health of the soil. According to Ansari and Ismail (2012), soil fertility is improved by a complicated process that involves the partial breakdown of OMs and combining them with earthworm cast, which is a mixture of mucus and gut microbial flora. They have an impact on the soil by creating the drilosphere, which increases soil porosity. For plant growth and productivity, earthworms play a very vital and useful role. Because of the importance of earthworms, vermitechnology, which uses native kinds of surface and underground earthworms for composting and soil management, has been developed (Ojuolape et al., 2015).

But according to Labenz (2022), a certain kind of soil's capacity to work within the restrictions of an uncontrolled or carefully controlled ecosystem in a way that promotes animal and plant productivity, upholds or enhances the air and water quality, as well as promotes health and habitation of humans is known as soil health. An earthworm may be the most important living thing in the soil, contributing significantly to the improvement of soil health. By enhancing soil aeration, infiltration, structure, nutrient cycling, and water movement, earthworms promote plant growth. With large quantities of humus, nitrogen (2–3%), phosphorus (1.55–2.25%), potassium (1.85–2.25%), and other micronutrients, along with more favorable soil microbes such as "nitrogen-fixing bacteria" and mycorrhiza fungus, vermicompost is also a nourishing organic fertilizer (Charan et al., 2024). Vermicompost is considered a remarkable plant growth enhancer, as supported by scientific studies (Chaoui et al., 2003; Guerrero, 2010). According to Ansari and Ismail (2012), vermicast produced by worms contains 19.58% phosphorus as P_2O_5 and 7.37% nitrogen. Earthworms are among the primary decomposers of organic matter, and they rely on microorganisms living in OM and soil for their food. As earthworms burrow and consume soil, they create tubular pathways that can persist in the ground for a long time (Charan et al., 2024). These burrows enhance soil porosity, allowing greater air and water infiltration, which in turn reduces bulk density and supports root growth. The casts produced by earthworms enrich the soil with essential nutrients such as nitrogen, phosphorus, potassium, and magnesium, thereby improving its fertility. Aristotle aptly referred to earthworms as "the intestines of the earth and the replenishing agents of soil fertility" (Shipley, 1970). They act as biological soil quality indicators since a healthy earthworm population suggests that the soil is healthy, given that it is home to numerous bacteria, viruses, fungi, insects, spiders, and other species (Lachnicht & Hendrix, 2001). However, when OM is digested in their intestines and becomes more abundant, earthworm casts also contain microbes. Therefore, it is widely understood how earthworms contribute to soil structure, soil productivity, agriculture, and environmental management of organic waste. Microbe development and the recirculation of nutrients from organic matter promote plant growth. The soil structure as well as aggregate stability are also improved by earthworm casts and the binding chemicals they release (Ismail, 2005).

Ansari and Hanief (2013) also state that there are three distinct kinds of earthworms, each of which has a unique lifestyle in the soil and a unique way of feeding and burrowing. Epigeic species are those that inhabit the soil's surface, are often small, eat decomposing plant matter, and have evolved to withstand the variations in temperature and moisture that occur there. Endogeic species inhabit the top layer of the soil and eat soil components and OM. As they crawl through the soil, they create short-lived burrows that are lined with worm casts. Deep-burrowing anecic species create long, enduring tunnels that can reach several feet into the ground. They mostly pull surface material back into their burrows to eat. They block the burrow's entrance with OM or worm casts. The optimal conditions are created by no-till farming and other conservation techniques that increase plant residue and soil structure. An abundance of earthworms is a reliable sign of healthy soil. It is unquestionably true that Darwin (1892) studied their actions in detail and came to the conclusion that "it may be questioned if there are any other animals that have played such a vital part in the history of the planet as these lowly organized creatures." For a while, it has been understood how crucial earthworms' job is for agriculture. Along with other organisms, earthworms have played a critical role in regulating soil processes, maintaining soil fertility, and fostering nutrient cycling (Ismail, 1997). Earthworms are essential to the structure of the soil because they produce aggregates and enhance the physiological conditions that enable plant growth and nutrient uptake. By expediting the decomposition of organic matter and plant litter and, as a result, releasing nutrients in a form that plants can absorb, they help increase soil fertility.

2 Vermicomposting of Organic and Inorganic Materials Processes

Vermicomposting is a technique wherein earthworms and other microorganisms break down and stabilize organic waste through biological means to produce vermicompost. These are a crucial component of modern organic farming. Its outstanding qualities make preparation simple, and it is safe for plants. The microbiological activity is substantially stimulated, the organic waste substrates are broken up, and mineralization rates are increased by the earthworms. By lowering pollution, they also have a great chance to support organic farming and reclaim land. Additionally, a lack of soil fertility has always been a barrier to increasing agricultural output (Aransiola et al., 2024; Pariyar et al., 2022). Vermicompost, according to Pariyar et al. (2022), quickly transforms trash with more and more diverse microbial activity, forming humus-like substances that have a finer structure than thermophilic composts. As the earth's intestine, earthworms are potential bioreactors that not only assist in the removal of solid organic waste produced by households, municipalities, or the agricultural sector but also contribute to the enhancement of the soil's physical and chemical composition, texture, and microbial population (Aransiola et al., 2022). Therefore, when added to clay soil, vermicompost, an organic substance that is stable and in the form of fine granules, helps to loosen the soil and improve the air entry path (Fig. 1).

In order to avoid waterlogging and improve water-holding capacity, the cast's hydroscopic mucus absorbs water. Vermincompost contains organic carbon, which releases nutrients into the system more gradually, allowing for easier plant absorption. Vermincompost enriches the soil with additional elements not present in synthetic fertilizers (Ahangar & Keshtehgar, 2015). Additionally, vermicomposting provides a way to recycle and utilize the massive organic agricultural waste that farmers traditionally burn so as to advance our agricultural development in a more effective, cost-effective, and ecologically responsible way. Technology has advanced since Darwin (1892) first highlighted the significance of earthworms in the handling of organic solid waste (OSW) and the production of the useful bio-product vermicompost (Ismail, 2005; Yatoo et al., 2020).

Similar to this, vermicomposting utilizes epigeic earthworms (Fig. 2) such as *Eudrilus eugeniae, Lumbricus rubellus, Eisenia fetida,* and *Perionyx excavatus*; however, in tropical or subtropical environments, *P. excavatus* has been shown to be an effective composting earthworm (Udayakumar & Parthasarathi, 2021). Vermitech is a technique for vermicomposting that combines regional earthworms of the epigeic and anecic species: *P. excavatus* and *Lampito mauritii* (Ismail, 1993, 2005). Vermicompost is a type of compost created by adding earthworms, and Vermitech is a technique for composting that involves employing a local species of earthworm (Ismail, 1993). Vermicompost normally consists of a finely divided ground peat-like substance with excellent structure, porosity, aeration, drainage, and moisture-retentive capacity (Lal & MS, 2024). Additionally, vermicompost boosts soil microbial activity, raises oxygen availability, keeps soil temperature normal, increases soil

Fig. 1 Organic waste management through vermincomposting

porosity and water infiltration, as well promotes plant development, yield, and quality (Arora et al., 2011). So, the input material has a significant impact on the nutrient content of vermicompost. Most mineral elements, in accessible forms, are typically present in greater amounts than in the source material (Edwards & Bohlen, 1996).

Biological, chemical, and physical qualities of soil are improved by vermicompost (Kale, 1998). Research has shown that vermicompost positively impacts all yield characteristics of various crops, including sugarcane, rice, and wheat. There is strong evidence that it encourages plant growth. Earthworm farming is known as vermiculture, and the waste these worms create is known as vermicast (Lalitha et al., 2000). Compost, cow dung, and other animal excretions are widely used in agriculture to cultivate plants. The problem of eliminating waste from our industries, homes, etc., is one we face in today's society. We can use vermicomposting technology to efficiently manage our trash. By using this method, we may compost the biodegradable waste while also using the compost's byproducts to improve crop

Fig. 2 Sample of vermicomposting (Planet Natural, 2001)

growth without using synthetic fertilizers. According to Ansari and Ismail (2012), using synthetic fertilizers throughout time has led to degraded soil health, decreased yields, an increase in insect and disease occurrences, and environmental contamination. The vermitechnology has emerged as the most effective corrective tool to address these grave issues (Edwards & Bohlen, 1996; Kumar, 2005).

2.1 *Vermiwash as Biocide and Fertilizer*

It is highly recommended to utilize vermiwash, a liquid created when water passes through a worm-action column, as a foliar spray. It contains micronutrients produced from organic molecules in the soil, together with excretory mucus and mucus secreted by earthworms. In the natural ecosystem, they are passed on to the plant's leaves, young shoots, and other parts. When properly collected, vermiwash is a transparent, translucent fluid with a little yellow hue (Ismail, 1997; Zambare et al., 2008). A liquid

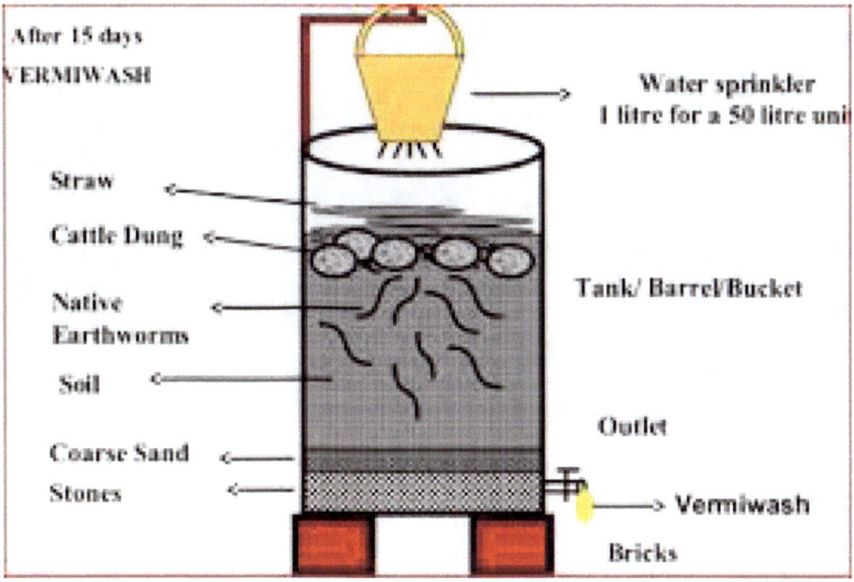

Fig. 3 Verminwash production (Bendalam & Kaviti, 2020)

fertilizer called vermiwash (Figs. 3 and 4) is obtained after water has flowed through a column of active worms, often known as a foliar spray or worm column. Agricultural waste, kitchen waste, and nitrogen-rich material like cattle dung, goat droppings, and pig manure are possible frequent sources of earthworm nourishment (Ismail, 2005). It is made up of the main micronutrients in the soil, organic compounds in the soil that are advantageous to plants, earthworm excretory and secretory products, as well as other components. Vermiwash appears to have the capacity to act as a modest biocide and a fertilizer simultaneously (Ismail, 1997).

In addition, vermicast (Fig. 5) is the feces that earthworms excrete, and vermiculture is the practice of raising them (Ansari & Kumar, 2010; Ismail, 1997). Compost, cow dung, and other animal wastes are used extensively in agriculture to cultivate plants. Getting rid of waste from our homes, businesses, and other sources is a challenge in today's world. We can use vermicomposting technology to handle our garbage in an efficient manner. With the help of this procedure, we can compost the biodegradable waste while also using the compost's byproducts to increase crop productivity and do away with the need for chemical fertilizers. According to Ansari and Ismail (2001), over time, the use of synthetic fertilizers has led to soil health deterioration, decreased yields, an increase in insect and disease outbreaks, and environmental pollution. The most efficient corrective method to deal with these serious problems has been identified as the vermintechnology (Das et al., 2014; Kumar, 2005). Therefore, organic farming contributes to a number of benefits, including the elimination of chemical fertilizers and pesticides, the recycling and regeneration of waste into wealth, the improvement of soil, plant, animal, and human health, and

Fig. 4 Agriculture liquid organic vermiwash (Agriculture Farming, 2020)

the development of environmentally friendly, economically viable biosystem models (Ansari & Ismail, 2001; Gudeta et al., 2021).

Similar to this, the bioactive macromolecules from an earthworm's skin secretion, coelomic fluid, and mucus were directly able to protect pathogenic soil bacteria against the worm, eradicating the disease from the environment. Earthworms create symbiotic relationships with bacteria, provide a critical component for plant growth, and control plant root disease (Gudeta et al., 2021). More specifically, Vermiwash application has a synergistic effect in reducing insect infestations such as thrips

Fig. 5 Vermicasting

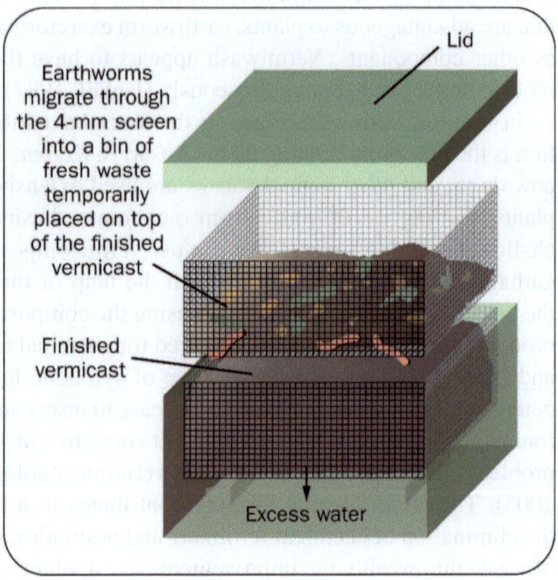

and mites and producing a lot of healthy pods to boost production (Kanchan et al., 2013). The production of crops is significantly harmed by parasitic arachnids called acarids. For instance, Aghamohammadi et al. (2016) showed that under laboratory conditions, vermiwash-treated bean leaf (*Phaseolus vulgaris L.*) displayed a strong repellent outcome against Acarina (*Tetranychus urticae*) as compared to leaves that are not treated. Vermiwash can also be used in agricultural fields being an insecticide to manage infection with red spider mites, according to Thakur and Sood's (2019) research that it effectively suppressed the growth of red spider mite (*T. urticae*) eggs. Vermiwash's repulsive qualities are related to the mucus that is present in it (Nadana et al., 2020). When an earthworm is irritated, coelomic fluid is discharged as mucus through dorsal pores as a coping technique. This bioactive fluid has been found to have antifungal, insecticidal, and pesticidal properties (Nadana et al., 2020).

2.2 Management of Organic Solid Waste and Earthworms

Organic waste dumping from home, agricultural, and industrial sources has recently resulted in significant environmental risks and economic issues (see Fig. 1). The control of organic waste has become a major problem worldwide as a result of rapid urbanization, economic expansion, and population rise. The improper organic wastes can harm the ecosystem, perhaps spread disease. Landfills, open dumping, and incineration are not environmentally sustainable disposal methods because of the production of some leaching and hazardous gases, which can contaminate the environment. For maintaining a pollution-free environment, managing organic waste is a critical challenge (Bhat et al., 2018). Burning organic wastes significantly increases environmental pollution, which results in contaminated air, water, and land. Along with dust particles, this process also releases a substantial measure of atmospheric carbon dioxide, which is a significant factor in global warming. Burning also eliminates the soil's OM, wipes off microbial life, and changes the soil's physical characteristics (Livan & Thompson, 1997; Siddiqui et al., 2022). Earthworms can be used to process the waste from the food, wood, and paper sectors as well as domestic waste, municipal trash, sewage sludge, and other types of waste (Ansari & Ismail, 2012). For managing OSW in the tropical and subtropical environments, the best earthworms for vermicomposting are *E. eugeniae* and *P. excavatus*. (Kale, 1998). Earthworms are used in the composting process to speed up vermicompost production and shorten the time it takes for waste to stabilize. Because it places a strong emphasis on soil health and food quality, more and more people are turning to organic farming. Vegetables with other crops have been successfully grown using Vermiculture and vermicompost alongside other biological inputs. It was discovered that these methods are productive and affordable (Ansari & Ismail, 2012). In this respect, it is possible to recycle organic waste to create beneficial organic manure for use in agriculture. Compost plays an increasing role in the drive to boost food production in an environmentally responsible way.

Compost is evolving significantly in an effort to increase food yield in an environmentally responsible manner. Tons of organic agricultural waste are burned by farmers every year, and vermicomposting provides a way for the reuse and recycling of waste in order to further agricultural development in a more effective, cost-effective, and ecologically friendly way (Rini et al., 2020). The management of increasing amounts of solid waste in an environmentally friendly manner is a serious issue that must be addressed at all costs. Both the rice and the sugar sectors burn their waste, which greatly contributes to environmental pollution and contaminates the air, water, and land. Along with dust particles, this process also releases a considerable amount of atmospheric carbon dioxide, which is a significant factor in global warming. Therefore, organic farming contributes a number of benefits, including the exclusion of chemical fertilizers and pesticides, the recycling and regeneration of waste into wealth, the improvement of soil, plant, animal, and human health, and the development of environmentally friendly, economically viable biosystem models (Ansari & Ismail, 2012).

As a result, a lot of attention is being paid to adopting environmentally sound, commercially viable, and socially acceptable tools that will strengthen the potential recovery of nutrients with the least amount of pollution. The most effective way to manage waste is by recycling or composting it and then using it in agriculture. Converting industrial waste into vermicompost offers dual benefits, as it reduces pollution while transforming waste into a valuable fertilizer (Bhat et al., 2015). Nevertheless, one of the most effective methods for recovering priceless nutrient elements from OSW is composting. Nutrients are biologically stabilized during composting, which is a regulated biological disintegration process in which bacteria transform OM-based products into products that have been stabilized and sterilized. Earthworm technology and vermiculture are terms used to describe the breeding of earthworms in organic substances and earthworms' biological conversion of OMs. One of the greatest and richest organic fertilizers in nutrients available is vermicompost. Vermicomposting is an excellent method for waste management, as it involves recycling or composting waste and using it in agriculture. Transforming industrial waste into vermicompost provides dual benefits by reducing pollution and converting waste into a nutrient-rich fertilizer (Bhat et al., 2015; Soobhany, 2019).

One of the finest qualities and richest in nutrients organic fertilizers available is vermicompost. Earthworms are also referred to as "nature's ploughman" or "the farmer's friend. "Earthworms were referred to as the "unheralded soldiers of mankind" by Sir Charles Darwin, the "protector and producer" by Sir Anatoly Igonin, and the "intestine of earth" by Aristotle because they preserved the soil's fertility and could digest a variety of organic substrates (Dada & Balogun, 2023; Martin, 1976). Earthworms are segmented, bilaterally symmetrical, hermaphrodites with a body color of dark brown and a clitellum for creating cocoons that are classified under the phylum Annelida and class Oligochaeta. Earthworms' gizzards break down OMs into smaller pieces. Microbial symbionts (bacteria, protozoa, and fungi) live inside the earthworm's gut and are in charge of degrading organic matter (Munnoli et al., 2010).

3 Factors Influencing Vermicomposting

Earthworms consume various organic waste products and turn them into "Castings." Because it includes nutrients that are stable but still accessible to plants, the finished product is greatly valued as a soil amendment. Farmers may be offered castings. Worms are another commodity that can be directly sold for the purpose of fish baits or as a supplement to livestock feed (Ndegwa et al., 2000). Thus, vermin composting is an ecologically friendly method for managing OSW, avoiding the bad odor and other ecological problems that come with dumping or landfilling garbage, while creating an equivalent or greater amount of income from the sale of its products. Commercial vermin composting is a relatively new idea that has been used for at least a decade. Earthworms can be used as a treatment method for a variety of waste streams, according to several scientific studies (Amaravathi & Reddy Mallikarjuna, 2015). Earthworms engage in both physical/mechanical and biological activity during this process. Aeration, mixing, and actual grinding of the substrate are among the physical/mechanical processes. The substrate is broken down by microorganisms in the earthworms' intestines to carry out the biochemical reaction. The most expensive part of a conventional microbial composting process is typically these physical or mechanical operations. To guarantee healthy growing worm populations, environmental elements like moisture, temperature, and aerobic conditions in the nutrient medium must be maintained. The selection of the best species for the available feed material and the production and growth rates of the worms are all necessary for a profitable production.

3.1 Reduction of Weight During Vermicomposting

Earthworms may obtain enough nourishment from the soil to thrive even in difficult conditions and ingest a variety of OMs. Because of bio-oxidation and the stability of organic matter brought about by microorganisms and earthworm interactions, the weight of the substrate used for vermicomposting decreases as the population of earthworms increases. Even though microorganisms are primarily in charge of the biochemical breakdown of organic matter, earthworms are crucial to the process because they condition and fragment the substrate, which increases the microbial surface area and changes the biological activity of the organism (Compose-Turner.net, 2022). High earthworm population densities in vermincomposting systems lead to a quick conversion of new OM into earthworm castings and a greater weight reduction of the vermincompost. The dry matter or weight of the compost utilized will decrease for vermincomposting due to the development of earthworms and biodegradation by microorganisms (Pattnaik & Reddy, 2010).

3.2 *Moisture Requirements*

All types of earthworm species depend on moisture for survival. Moisture within the worms' bodies provides them with structure, allows them to move, and helps them collect oxygen. It is one of the prerequisites for the composting process since nutritious sources are only capable of being absorbed by microbes when combined with water. The majority of worm species have a moisture range of 60–85%, which enables the worm to retain all moisture that may otherwise be lost by evaporation (Amaravathi & Reddy, 2014; Kováčik et al., 2018). The ideal water content for composting raw materials is often between 50 and 60%. Insufficient moisture (less than 30%) would hinder microbe activity and make it difficult for OMs to break down, whereas excessive moisture would slow composting, cause anaerobic decomposition, odor creation, and nutrient bleeding (Amaravathi & Reddy, 2013). Therefore, the moisture content of the composting material has a direct impact on both the ventilation capacity of the compost turner machine and the structural integrity of the compost material. Air will not be allowed to pass through if there is more than 60% water content, causing the materials to become compact and the compost to develop a direction toward anaerobes. In this instance, ventilation needs to be improved. Meanwhile, microbes will be rejected and have an impact on their growth if the moisture content is less than 20%. As a result, moisture is needed in compost piles for microbial activity necessary for proper decomposition to take place (Nagavallemma et al., 2006). To ensure an effective decomposition rate, residents of locations with limited rainfall must periodically water the compost pile. The pile should be moistened with just enough water, but it is important to prevent overwatering because that might cause the air in the pile to be replaced with water, which would suffocate the microorganisms. Due to the anaerobic conditions created, the decomposition process will be slowed down and unpleasant odors will be produced (Amaravathi & Reddy, 2013). To enhance air, piles that are overly damp should be turned often. If a pile is excessively dry, it should be turned and watered.

3.2.1 Interaction Between Aeration and Oxygen Supply

The amount of aeration and oxygen supply, one of the crucial factors in effective composting, is correlated with the organic content of the composting materials, i.e., the higher the content of organic carbon, the higher the rate of oxygen consumption. When the oxygen content falls below 18%, the microorganism's life activity in the composting process will be reduced, and is likely to result in a stink (Xia et al., 2019). The issue with the oxygen supply argument is not about the availability of oxygen in the air, but rather ensuring a consistent supply without relying on excessive ventilation at the same time. On the other hand, compost workers should prevent overly prolonged waste disposal and poor odors brought on by insufficient oxygen, as well as temperature drops in waste dumps and excessive energy and operating costs brought on by unnecessary aeration (Hénault-Ethier et al., 2016).

As previously stated, oxygen is therefore necessary for the microorganisms engaged in aerobic composting to break down organic waste effectively. Anaerobic composting can achieve decomposition without oxygen, but such decomposition processes are exceedingly sluggish, emit unpleasant smells, and frequently produce chemical compounds that are poisonous to plants. The best method for residential settings is aerobic composting due to the unpleasant smells that anaerobic composting emits (Jjagwe et al., 2019). However, there are various ways to provide oxygen to the pile. It is normal to be able to speed up decomposition by simply mixing or turning the pile a few times each month to give the decomposers the required oxygen. The pile's decomposition will proceed more quickly if it is raised off the ground, allowing ventilation from both the pile's bottom and top. Using PVC pipes with holes already drilled in them and putting them through the compost container is an additional option. According to Gómez Brandón et al., (2019), fresh air can enter the center of the pile through the PVC pipes' perforations and ends.

3.3 Temperature

The optimal temperature range for the majority of worm species is between 15 and 27 °C, although each species has different tolerances and requirements for specific temperatures. So, the amount of moisture in the system has a significant impact on the worm's tolerance of temperatures outside its optimal range. For vermin composting, heat is more problematic than cold. Worms generally prefer chilly temperatures. In the spring, as the temperature warms and cools, they are at their peak in terms of activity and reproduction (Kováčik et al., 2018). However, a critical element in the success of composting is the temperature's impact on microorganism development. Because so many physical and chemical interactions are occurring inside the pile, heat is produced as the pile decomposes. The biological activity depends on the temperature of the compost pile. There is widespread agreement that thermophilic bacteria are more effective in degrading OMs than mesophilic bacteria. When mesophilic bacteria have had a chance to function for one or two days, the temperature of the composting process rises to 50 to 65 °C, where it can achieve the necessary safe conditions and eliminate the majority of dangerous bacteria in about five to six days (Hasan et al., 2021). In a nutshell, too low a temperature would significantly lengthen the thrashing process, while too high (above 70 °C) would have a negative effect on the microorganisms that break down organic waste during composting. Slow decomposition may be caused by inadequate heat. Microbes and invertebrates essential to the process can be killed by excessive heat (Vuković et al., 2021). Microbes that contribute to decomposition primarily fall into one of two categories:

- Mesophilic creatures that thrive between 70 and 100 °F.
- creatures that are thermophilic and thrive between 113 and 155 °F.

Table 1 Process parameters for aerobic composting, acceptable, and ideal conditions

Condition	Acceptable	Ideal
Nutrient balance (carbon to nitrogen ratio)	20:1–40:1	25:1–35:1
Moisture content by weight (%)	40–65	45–60
Oxygen content (%)	> 5%	> 10
Temperature (°C)	43–66	55–60
pH	5.5–9.0	6.5–8.0
Porosity (%)	45–65	45–65
Bulk density (kg/m^3)	600	600

Moderate temperatures encourage the growth and activity of mesophilic bacteria, which are the most effective decomposers. Higher temperatures kill disease organisms and weed seeds. For accelerated decomposition, pile temperatures between 90 and 140 °F are desirable. However, many of the microbes and invertebrates will perish or become less active if temperatures rise above 140 °F. The pile should be turned at this point to improve airflow and reduce the temperature. The injection of nitrogen and/or water may be necessary if a pile never reaches 120 °F, since cold weather can also hinder the pile from heating (Jafari et al., 2021).

3.4 Conditions for pH

In general, the substrate's pH has a significant role in regulating the development of earthworms and the microorganisms found in their guts. The substrate's pH must be neutral or close to 7.0 for the best results to improve microbial and earthworm activity (Amaravathi & Reddy, 2015). Therefore, a neutral or weak alkali has the best pH value for microorganisms, and a pH value that is excessively high or too low may make disposing of compost challenging. Additionally, the pH level has an impact on nitrogen loss; when the pH is above 7.0, nitrogen volatilizes as ammonia (Vuković et al., 2021). However, pH values between 6.0 and 7.5 are ideal for bacterial development, whereas those between 5.5 and 8.0 are ideal for fungal growth, and these values have a key function in the process and the size's quality of the composting source material's particles (Table 1).

3.5 Electrical Conductivity (EC)

The quantity at any temperature of dissolved salts present inside the compost or soil will be revealed by the electrical conductivity (EC) measurement. Minerals in the form of soluble salts make it easier for plants to take in the necessary soil-derived salts. Hence, substrate or vermin compost's enhanced EC is a benefit for enhancing soil fertility. According to Kováčik et al., (2018), vermicompost with a higher EC is thought to be of superior quality.

3.6 *The Carbon to Nitrogen Ratio*

The carbon and nitrogen ratio is the most crucial among the several elements required by microorganism degradation. The temperature of compost is correlated with the carbon–nitrogen ratio. The growth of bacteria and other microorganisms will be constrained, the rate at which OMs decompose will be slow, and fermentation will be the result if there is a high material ratio, i.e., if nitrogen is often insufficient while carbon is more frequent. Additionally, the high carbon-to-nitrogen ratio in the source materials will quickly result in a similarly high carbon–nitrogen ratio in the finished compost. This imbalance can deplete the soil's nitrogen content, leading to nitrogen deficiency, which ultimately hinders plant growth (Nurhidayati et al., 2016). But if the ratio of carbon to nitrogen is too low, particularly less than a ratio of 20:1, carbon elements that can be consumed are scarce while nitrogen elements are comparatively abundant. As a result, the nitrogen in materials turns into ammoniacal nitrogen and volatilizes, which lowers fertilizer efficiency because there is a significant amount of nitrogen loss. Therefore, the composting should achieve the ideal carbon–nitrogen ratio (25–35:1) needed for microorganisms in order to ensure that the microorganism nutrients of organic breakdown are balanced (Alavi et al., 2017).

3.7 *Composting Raw Materials' Particle Size*

Reduced pellet size will enhance surface area, encourage microorganism activity, and speed up the composting process because microorganisms carry out their activities on the surface of organic pellets. However, overly thin materials also restrict airflow and lower the oxygen level during composting, which slows the rate of microorganism activity. As a result, it is necessary to reduce the size of raw materials on the basis of air ventilation (Hosseinzadeh et al., 2020). However, completed compost needs to be stable and mature in order to be packaged and transported without risk and not have negative consequences when used. Smaller particles, however, have a larger surface area than larger ones. The rate of decomposition increases with the amount of surface area that microorganisms have access to. Because of this, reducing the size of composting material will hasten decomposition (Vuković et al., 2021). Larger materials can be reduced in size using a shredder or a woodchipper. Simply mulching leaves with a lawnmower before raking them might minimize their size.

4 Plant Growth in Relation to Nutrients

The composition and quantity of mineral nutrients present in the soil play a significant role in determining plant growth and development. Because of their relative immobility, plants frequently encounter substantial difficulties in receiving an appropriate supply of essential nutrients to meet the requirements of fundamental cellular activities. Reduced plant fertility or productivity may occur from a deficiency in any one of them. A lack of nutrients can cause stunted growth, the death of plant cells, or chlorosis caused by a decrease in the chlorophyll pigment production needed for photosynthesis (Gupta et al., 2023; Morgan & Connolly, 2013). Deficiency of certain nutrients can have a big influence on agriculture, lowering plant quality or crop yield. Due to the fact that plants are the primary producers in the majority of food webs, nutrient deficiency can also result in a decrease in overall biodiversity. However, McDonald (1994) asserts that there are two universal characteristics of plant nutrition. First, because all physiological processes in the plant ultimately depend on the integration of one or more mineral nutrients in a form appropriate to the underlying biochemistries, an increase in the size of plant organs and their proper functioning ultimately depend on an appropriate availability of essential nutrients. Second, the quantities and availability of nutrients that are stored and the extent of recycling within the plant will determine the degree to which current nutrient intake that is available externally influences growth processes (Hungria et al., 2021).

The availability of some nutrients can differ as a result of changes in the climate and atmosphere, which can have a negative impact on plants. It is crucial to comprehend how plants have evolved to overcome some of these challenges in a world where the climate is constantly changing. Therefore, phenomena like storage and recycling can be crucial for an individual's survival and fitness in nutrient-poor environments or under conditions of changing nutrient supply (Abbott & Robson, 2018). Seeds, as an example, can contain enough nutrients to allow for a significant increase in plant size without further nutrient uptake, in contrast to plantlets developed from cell culture. Additionally, when crop output rises, soil nutrients are removed from the soil at a faster rate, underscoring the significance of replenishing soil fertility through effective and efficient fertilizer management. The availability of native soil nutrients depends on the soil's capacity to absorb nutrients lost during crop removal. N, S, and micronutrient supplies from mineralization of soil organic fractions are finite, whereas P, K, Ca, Mg, and micronutrient supplies are replenished by mineral dissolution and surface exchange reactions (Ahanger et al., 2016; Rao et al., 2024). The degree to which nutrients are mobile in soil affects ion transport to plant roots, assessments of the nutrients' availability to plants, and ultimately decisions about nutrient management. Soil testing is necessary to determine the crop's nutrient needs and the soil's ability to supply nutrients in order to manage nutrients effectively (Havlin, 2020). The nutrient stored in micro-propagated plantlets will be limited, and current growth will predictably depend heavily on the current availability and uptake of mineral nutrients.

Mineral nutrients are typically taken up by plant roots from the soil, although a variety of conditions might alter how well nutrients are taken up. First, some soils' chemistry and composition may make it more difficult for plants to absorb nutrients. In some soils, the nutrients might not be present, or they might be in a form that the plants can't use. These issues may be made worse by the soil's characteristics like water content, pH, and compaction (Yadav et al., 2021). Second, certain plants have processes or structural characteristics that help them thrive in specific kinds of nutrient-poor soils. In fact, in an effort to overcome nutrient scarcity, the majority of plants have developed nutrient uptake mechanisms that are tailored to their local soils. A change in root structure, which may increase the root's overall surface area to boost nutrient uptake or may lengthen the root system to reach additional nutrient sources, is one of the most common responses to nutrient-limited soils. These adjustments may result in a larger allocation of resources to total root growth, which would raise the ratio of roots to shoots in nutrient-limited plants (López-Bucio et al., 2003). Additionally, different nutrient shortages are known to cause distinct responses in plants, and these responses can differ between species. Preventing the growth of the main roots (typically linked to Phosphorus shortage), a development and density of lateral roots increasing (frequently associated with N, P, Fe, and S deficient), an increased density and development of root hair are the most frequent changes, according to Gebremikael et al. (2016) (often associated with P and Fe deficiency).

4.1 Direct Plant Uptake of Nutrients from the Soil

Potassium (K): Potassium is the most prevalent cation in plant cells and is regarded as a macronutrient for plants. In plants, potassium has a lot of crucial roles, including regulating the charges of cellular anions, activating enzymes (Aransiola et al., 2023; Babaniyi et al., 2023), regulating stomatal opening and closing, and acting as an osmoticum for cellular development. As a result, plants cultivated in sandy soil usually experience potassium deficiency, which manifests as a variety of symptoms such as restricted growth and fertility, leaves browning, leaf tips curling, and leaves yellowing (i.e., showing chlorosis) (Rana et al., 2020). Additionally, for many years, potassium uptake mechanisms have been the focus of extensive research. Early research suggested that plants directly absorb potassium from the soil using both high and low-affinity transport mechanisms. When potassium levels in the soil are enough for plant growth and development, low-affinity transport systems typically work. This process allows K^+ to passively transfer from areas of higher external concentration into plant cells, where the K^+ concentration is lower. This process is mediated by ion channels in the plasma membrane of root cells. The availability of potassium does not seem to have a substantial impact on the expression of these low-affinity transporters (Abdelgawad et al., 2019).

Conversely, when potassium is scarce, plants typically generate high-affinity K^+ transport channels. High-affinity potassium transport likely involves a large number of proteins. Two proteins have been identified in Arabidopsis as important transporters in this process (Khan et al., 2019). The AKT1 is a protein channel that likely mediates a transport mechanism passively with an enhanced K^+ affinity under conditions of potassium shortage. Whereas, AtHAK5, a carrier protein, is assumed to mediate active transport of potassium into plant roots (Aransiola et al., 2023; Pyo et al., 2010). Meanwhile, in order to transfer potassium throughout the plants, research has shown that a number of transport systems are present in plants (Ge & Zhang, 2019). Even though there is still much to study about the uptake of potassium, as well study plant translocation. It is evident that mechanisms are intricate and carefully regulated to aid plants absorb enough potassium from the soil in a variety of situations (Ahanger et al., 2017; Hmaeid et al., 2019).

Iron (Fe): A cofactor for proteins needed for several essential metabolic functions, such as photosynthesis and respiration, iron (Fe) is essential for the development and growth of plants. Iron is the fourth most common element in the crust of the Earth, but it is frequently a limiting element for plants because it frequently forms insoluble complexes in aerobic soils with neutral to basic pH (Guerinot & Yi, 1994). It is believed that up to 30% of soils worldwide pose a problem of iron limitation for plants. Therefore, many plants face challenges due to insufficient iron availability. Plants with low levels of iron frequently exhibit interveinal chlorosis, a condition in which the leaf's veins continue to be green and the spaces between them turn yellow. As a result of iron's low solubility in various soils, prior to transporting iron throughout the plant, plants frequently must deploy iron in the rhizosphere, also referred to as the region of soil that surrounds and affects the roots (Walker & Connolly, 2008).

4.2 Plant Nutrient Acquisition Through Symbiotic Relationships with Soil-Based Microorganisms

Nitrogen and phosphorus are considered among the most limiting elements for plant growth and productivity because they are often available in limited quantities or in forms that plants cannot readily utilize. As a result, symbiotic partnerships with soil-borne microbes have evolved during the course of many plant species' evolutionary processes. Both the host plant and the microbe symbiont benefit from these partnerships because they allow them to share vital resources that are necessary for their own productivity and survival (Khan et al., 2016).

Nitrogen fixation: Despite being the most abundant gaseous element in the atmosphere, nitrogen in its N_2 form cannot be directly used by plants. In soils with low nitrogen content, plants may suffer from nitrogen deficiency. Nitrogen deficiency severely restricts plant productivity since both proteins and nucleic acids must have it to function (Khanna et al., 2019b). In order to boost the availability of nutrients, crop productivity, and overcome nitrogen deficit in an agricultural environment, nitrogen-rich fertilizers can be added to agricultural soil. However, as surplus nutrients frequently wind up in groundwater, eutrophication and the resulting depletion of oxygen in related water-based ecosystems, this can be a risky practice (Khanna et al., 2019a). So, nitrate and ammonium can indeed be absorbed by plants directly from the soil. However, some species of Fabaceae (legume) plants begin symbiotic partnerships with a genus of microorganisms that fix nitrogen, known as Rhizobia, when nitrogen sources have been absent. These interactions are rather specialized and necessitate chemical signal recognition between the bacterium and the host plant. When a plant releases substances called flavonoids into the soil, the bacteria are drawn to the root and the relationship begins (Kumar et al., 2017).

The bacteria react by releasing substances known as Nod Factors (NF), which locally alter the root's and its root hairs' structure. To enclose germs in a small corner, the root hair coils sharply. An infection thread, formed by the invasion of the plant cell membrane and the breakdown of the plant cell wall, extends to the root cortex's cells. As bacteria develop into what are known as bacteroids, they are encased in a membrane generated from plants (Sethi et al., 2023). These structures are permitted access to cortical cells' cytoplasm, where they transform nitrate from the air into ammonia, which plants can utilize. The bacteroids get carbohydrates produced by photosynthetic processes in exchange, which they might employ to produce energy (Ferguson et al., 2010; Limpens & Bisseling, 2003).

5 Conclusion

Environmental hazards are exacerbated by the buildup of organic waste from various sources, such as household, agricultural, and industrial wastes, which can be recycled using crude and unrefined technology. Vermicomposting is a step toward environmentally sustainable organic farming and is a key component of vermitechnology that can be used to grow various fruits, vegetables, and other crops. Innovations in organic waste management can lead to zero-waste farms, eliminating the need for burning or discarding organic waste. Instead, they focus on recycling and repurposing valuable OM, helping to restore and preserve natural resources through biotechnology. Additionally, since plants are immobile and often face environmental nutrient deficiencies, they have developed advanced systems to acquire essential macro- and micronutrients for healthy growth, development, and reproduction. These adaptations include changes in their growth patterns and root structures, which enhance their ability to absorb critical nutrients from the soil. They also introduce systems that help plants absorb nutrients more efficiently and form partnerships with other organisms to help them get the nutrients they need. All of these processes help plants get more nutrients from the soil. By combining these systems, plants can increase their ability to absorb nutrients while preventing the development of extra nutrients that could be damaging to the plant. There is no doubt that crop yields, ecosystem health, the composition of plant communities, soil ecology, and biodiversity are all indisputable indicators of agricultural success and are significantly influenced by a plant's capacity to use these mechanisms. Finally, in order to utilize waste in agriculture, the most efficient approach is to first compost or recycle it. Vermicompost is among the best and richest in nutrients available organic fertilizers in the market. Hence, vermicompost is one of the best organic fertilizers available in terms of quality and nutrient content. Also, vermiwash appears to have the capacity to serve as both a modest biocide and a fertilizer.

References

Abbott, L. K., & Robson, A. D. (2018). The effect of VA mycorrhizae on plant growth. In *VA mycorrhiza* (pp. 113–130). CRC Press.

Abdelgawad, F. K., El-Mogy, M. M., Mohamed, M. I. A., Garchery, C., & Stevens, R. G. (2019). Increasing ascorbic acid content and salinity tolerance of cherry tomato plants by suppressed expression of the ascorbate oxidase gene. *Agronomy, 9*(2), 51.

Aghamohammadi, Z., Etesami, H., & Alikhani, H. A. (2016). Vermiwash allows reduced application rates of acaricide azocyclotin for the control of two spotted spider mite, Tetranychusurticae Koch, on bean plant (Phaseolus vulgaris L.). *Ecological Engineering, 93*, 234–241.

Agriculture Farming. (2020, January 11). Vermiwash preparation process, benefits, cost. *Agri Farming*. www.agrifarming.in. https://www.agrifarming.in/vermiwash-preparation-process-benefits-cost

Ahangar, A. G., & Keshtehgar, A. (2015). Influence of heavy metals on earthworms, soil microbial community and mammals. *Journal of Novel Applied Sciences, 4*, 125–134.

Ahanger, M. A., Morad-Talab, N., Abd-Allah, E. F., Ahmad, P., & Hajiboland, R. (2016). Plant growth under drought stress: Significance of mineral nutrients. *Water Stress and Crop Plants: A Sustainable Approach, 2*, 649–668.

Ahanger, M. A., Tomar, N. S., Tittal, M., Argal, S., & Agarwal, R. (2017). Plant growth under water/salt stress: ROS production; antioxidants and significance of added potassium under such conditions. *Physiology and Molecular Biology of Plants, 23*(4), 731–744.

Alavi, N., Daneshpajou, M., Shirmardi, M., Goudarzi, G., Neisi, A., & Babaei, A. A. (2017). Investigating the efficiency of co-composting and vermicomposting of vinasse with the mixture of cow manure wastes, bagasse, and natural zeolite. *Waste Management, 69*, 117–126.

Amaravathi, G., & Reddy, R. M. (2013). Reproductive characters of earthworms useful for vermincomposting of municipal solid waste. *National Journal of Life Science, 10*(1), 47–51.

Amaravathi, G., & Reddy, R. M. (2014). Effect of substrate composition on the nutrients of vermicompost prepared by different types of earthworms. *American International Journal of Contemporary Scientific Research, 1*(3), 85–95.

Amaravathi, G., & Reddy Mallikarjuna, R. (2015). Environmental factors affecting vermicomposting of municipal solid waste. *International Journal of Pharmacy and Biological Sciences, 5*(3), 81–93.

Ansari, A. A., & Ismail, S. A. (2001). A case study on organic farming in Uttar Pradesh. *Journal of Soil Biology and Ecology, 27*, 25–27.

Ansari, A. A., & Ismail, S. A. (2012). Role of earthworms in vermitechnology. *Journal of Agricultural Technology, 8*(2), 403–415.

Ansari, A. A., & Kumar, S. (2010). Effect of vermiwash and vermicompost on soil parameters and productivity of okra (Abelmoschus esculentus) in Guyana. *Current Advances in Agricultural Sciences (An International Journal), 2*(1), 1–4.

Ansari, A., &Hanief, A. (2013). *Microbial succession during vermicomposting* (No. 537-2016-38589).

Aransiola, S. A., Afolabi, F., Joseph, F., & Maddela, N. R. (2023). Soil enzymes: Distribution, interactions and influencing factors. In P. Jeschke, & E. B. Starikov (Eds.), *Agricultural biocatalysis: Enzymes in agriculture and industry* (1st ed., pp. 303–333). ISBN 9789814968478, Jenny Stanford Publishing. https://www.routledge.com/Agricultural-Biocatalysis-Enzymes-in-Agriculture-and-Industry/Jeschke-Starikov/p/book/9789814968478

Aransiola, S. A., Ijah, U. J. J., Abioye, O. P., & Bala, J. D. (2022). Vermicompost-assisted phytoremediation of toxic trace element-contaminated soil in Madaka, Nigeria using Melissa officinalis L and Sida acuta. *International Journal of Environmental Science and Technology*. https://doi.org/10.1007/s13762-022-04105-y

Aransiola, S. A., Ijah, U. J. J., Abioye, O. P., & Bala, J. D. (2023). Vermicompost-assisted phytoremediation of toxic trace element-contaminated soil in Madaka, Nigeria, using Melissa officinalis L and Sida acuta. *International Journal of Environmental Science and Technology, 20*(2), 1823–1836.

Aransiola, S. A., Ijah, U. J. J., Abioye, O. P., Bala, J. D., Rivadeneira-Mendoza, R. F., Luque, R., Rodríguez-Díaz, J. M., & Maddela, N. R. (2024). Micro and vermicompost assisted remediation of heavy metal contaminated soils using phytoextractors. *Case Studies in Chemical and Environmental Engineering., 9*(2024), Article 100755. https://doi.org/10.1016/j.cscee.2024.100755

Arora, V. K., Singh, C. B., Sidhu, A. S., & Thind, S. S. (2011). Irrigation, tillage and mulching effects on soybean yield and water productivity in relation to soil texture. *Agricultural Water Management, 98*(4), 563–568.

Babaniyi, G. G., Olagoke, O. E., & Aransiola, S. A. (2023). Extracellular enzymatic activity of bacteria in aquatic ecosystems. In N. R. Maddela, L. K. W. Eller, & R. Prasad (Eds.), *Microbiology for cleaner production and environmental sustainability*. CRC Press. ISBN: 9781032496061. https://www.routledge.com/Microbiology-for-Cleaner-Production-and-Environmental-Sustainability/Maddela-Eller-Prasad/p/book/9781032496061

Bendalam, P., & Kaviti, V. L. (2020). *Vermiwash*. https://justagriculture.in/files/magazine/dec/012%20VERMIWASH.pdf

Bhat, S. A., Singh, J., & Vig, A. P. (2015). Potential utilization of bagasse as feed material for earthworm Eisenia fetida and production of vermicompost. *Springerplus, 4*(1), 1–9.

Bhat, S. A., Singh, J., & Vig, A. P. (2018). Earthworms as organic waste managers and biofertilizer producers. *Waste and Biomass Valorization, 9*(7), 1073–1086.

Chaoui, H. I., Zibilske, L. M., & Ohno, T. (2003). Effects of earthworm casts and compost on soil microbial activity and plant nutrient availability. *Soil Biology and Biochemistry, 35*(2), 295–302.

Charan, K., Bhattacharyya, P., & Bhattacharya, S. S. (2024). Vermitechnology transforms hazardous red mud into benign organic input for agriculture: Insights on earthworm-microbe interaction, metal removal, and soil-crop improvement. *Journal of Environmental Management, 354*, Article 120320.

Compose-Turner.net. (2022). *Factors affecting composting process, compost quality and maturity*. RichenTek. https://compost-turner.net/composting-technologies/factors-affect-composting-process-and-compost-quality.html

Dada, E. O., & Balogun, Y. O. (2023). Vermitechnology: An underutilised agro-tool in Africa. In *Vermicomposting for Sustainable Food Systems in Africa* (pp. 127–143). Springer Nature Singapore.

Darwin, C. (1892). *The formation of vegetable mould, through the action of worms: with observations on their habits*. J. Murray.

Das, S. K., Avasthe, R. K., & Gopi, R. (2014). Vermiwash: Use in organic agriculture for improved crop production. *Popular Kheti, 2*(4), 45–46.

Edwards, C. A., & Bohlen, P. J. (1996). *Biology and ecology of earthworms* (Vol. 3). Springer Science & Business Media.

Ferguson, B. J., Indrasumunar, A., Hayashi, S., Lin, M. H., Lin, Y. H., Reid, D. E., & Gresshoff, P. M. (2010). Molecular analysis of legume nodule development and autoregulation. *Journal of Integrative Plant Biology, 52*(1), 61–76.

Ge, H., & Zhang, F. (2019). Growth-promoting ability of Rhodopseudomonas palustris G5 and its effect on induced resistance in cucumber against salt stress. *Journal of Plant Growth Regulation, 38*(1), 180–188.

Gebremikael, M. T., Steel, H., Buchan, D., Bert, W., & De Neve, S. (2016). Nematodes enhance plant growth and nutrient uptake under C and N-rich conditions. *Scientific Reports, 6*(1), 1–10.

Gómez Brandón, M., Aira, M., Kolbe, A. R., De Andrade, N., Pérez-Losada, M., & Domínguez, J. (2019). Rapid bacterial community changes during vermicomposting of grape marc derived from red winemaking. *Microorganisms, 7*(10), 473.

Gudeta, K., Julka, J. M., Kumar, A., Bhagat, A., & Kumari, A. (2021). Vermiwash: An agent of disease and pest control in soil, a review. *Heliyon, 7*(3), Article e06434.

Guerinot, M. L., & Yi, Y. (1994). Iron: Nutritious, noxious, and not readily available. *Plant Physiology, 104*(3), 815.

Guerrero, R. D., III. (2010). Vermicompost production and its use for crop production in the Philippines. *International Journal of Global Environmental Issues, 10*(3–4), 378–383.

Gupta, R., Mago, M., & Garg, V. K. (2023). Sustainable utilization and treatment of barnyard grass (Echinochloa crus-galli) weed biomass using vermitechnology. *Tropical Ecology*, 1–10.

Hasan, S. S., Roy, S., Saha, S., & Hoque, M. Z. (2021). Assessment of the farmers' perception on vermicompost as waste management practice and economic return in some areas of Bangladesh. *European Journal of Agriculture and Food Sciences, 3*(3), 14–20.

Havlin, J. L. (2020). Soil: Fertility and nutrient management. In *Landscape and land capacity* (pp. 251–265). CRC Press.

Hénault-Ethier, L., Martin, V. J., & Gélinas, Y. (2016). Persistence of Escherichia coli in batch and continuous vermicomposting systems. *Waste Management, 56*, 88–99.

Hmaeid, N., Wali, M., Mahmoud, O. M. B., Pueyo, J. J., Ghnaya, T., & Abdelly, C. (2019). Efficient rhizobacteria promote growth and alleviate NaCl-induced stress in the plant species Sulla carnosa. *Applied Soil Ecology, 133*, 104–113.

Hosseinzadeh, A., Baziar, M., Alidadi, H., Zhou, J. L., Altaee, A., Najafpoor, A. A., & Jafarpour, S. (2020). Application of artificial neural network and multiple linear regression in modeling nutrient recovery in vermicompost under different conditions. *Bioresource Technology, 303*, Article 122926.

Hungria, M., Rondina, A. B. L., Nunes, A. L. P., Araujo, R. S., & Nogueira, M. A. (2021). Seed and leaf-spray inoculation of PGPR in brachiarias (Urochloa spp.) as an economic and environmental opportunity to improve plant growth, forage yield and nutrient status. *Plant and Soil, 463*(1), 171–186.

Ismail, A. (1997). *Vermicology: The biology of earthworms*. Orient Longman.

Ismail, S. A. (1993). *Composting through earthworms*. Shri AMM Murugappa Chettiar Research Centre, Photosynthesis and Energy Division.

Ismail, S. A. (2005). *The earthworm book* (p. 101). Other India Press.

Jafari, F., Khademi, H., Shahrokh, V., Angel, F. A. Z., Acosta, J. A., & Khormali, F. (2021). Biological weathering of phlogopite during enriched vermicomposting. *Pedosphere, 31*(3), 440–451.

Jjagwe, J., Komakech, A. J., Karungi, J., Amann, A., Wanyama, J., & Lederer, J. (2019). Assessment of a cattle manure vermicomposting system using material flow analysis: A case study from Uganda. *Sustainability, 11*(19), 5173.

Kale, R. D. (1998). *Earthworm; Cinderella of organic farming*.

Kanchan, M., Keshav, S., & Tripathi, C. P. M. (2013). Management of pod borer (Helicoverpa armigera) infestation and productivity enhancement of gram crop (Cicer aritenium) through vermiwash with biopesticides. *World Journal of Agricultural Sciences, 9*(5), 401–408.

Khan, F., Ahmed, K. B. M., Shariq, M., & Siddiqui, M. A. (2019). Potentiality of plant growth-promoting rhizobacteria in easing of soil salinity and environmental sustainability. In *Salt stress, microbes, and plant interactions: Causes and solution* (pp. 21–58). Springer.

Khan, M. A., Gemenet, D. C., & Villordon, A. (2016). Root system architecture and abiotic stress tolerance: Current knowledge in root and tuber crops. *Frontiers in Plant Science, 7*, 1584.

Khanna, K., Jamwal, V. L., Kohli, S. K., Gandhi, S. G., Ohri, P., Bhardwaj, R., Wijaya, L., Alyemeni, M. N., & Ahmad, P. (2019a). Role of plant growth promoting Bacteria (PGPRs) as biocontrol agents of Meloidogyne incognita through improved plant defense of Lycopersiconesculentum. *Plant and Soil, 436*(1), 325–345.

Khanna, K., Jamwal, V. L., Sharma, A., Gandhi, S. G., Ohri, P., Bhardwaj, R., Al-Huqail, A. A., Siddiqui, M. H., Ali, H. M., & Ahmad, P. (2019b). Supplementation with plant growth promoting rhizobacteria (PGPR) alleviates cadmium toxicity in Solanum lycopersicum by modulating the expression of secondary metabolites. *Chemosphere, 230*, 628–639.

Kováčik, P., Šalamún, P., & Wierzbowska, J. (2018). Vermicompost and Eisenia foetida as factors influencing the formation of radish phytomass. *Agriculture, 64*(2), 49.

Kumar, A. (2005). *Verms and vermitechnology* (pp. 129–131).

Kumar, D., Al Hassan, M., Naranjo, M. A., Agrawal, V., Boscaiu, M., & Vicente, O. (2017). Effects of salinity and drought on growth, ionic relations, compatible solutes and activation of antioxidant systems in oleander (Nerium oleander L.). *Plos One, 12*(9), e0185017.

Labenz, T. A. (2022). *Earthworm activity increases soil health | NRCS Kansas*. Natural Resources Conservation Service (NRCS) Kansas. www.nrcs.usda.gov. https://www.nrcs.usda.gov/wps/portal/nrcs/detail/ks/newsroom/features/?cid=stelprdb1242736

Lachnicht, S. L., & Hendrix, P. F. (2001). Interaction of the earthworm Diplocardia mississippiensis (Megascolecidae) with microbial and nutrient dynamics in a subtropical Spodosol. *Soil Biology and Biochemistry, 33*(10), 1411–1417.

Lal, A., & MS, S. (2024). Advanced vermicomposting modality: Microbial acceleration to improve micronutrient profile of agricultural residue derived compost.

Lalitha, R., Fathima, K., & Ismail, S. A. (2000). Impact of biopesticides and microbial fertilizers on productivity and growth of Abelmoschus esculentus. *Vasundhara the Earth, 1*(2), 4–9.

Limpens, E., & Bisseling, T. (2003). Signaling in symbiosis. *Current Opinion in Plant Biology, 6*(4), 343–350.

Livan, M. A., & Thompson, W. (1997). NARI Annual Report: Evaluation of vermicompost. In *Proc. Nat. Semi. Org. Fmg. MPKV, Pune.*

López-Bucio, J., Cruz-Ramırez, A., & Herrera-Estrella, L. (2003). The role of nutrient availability in regulating root architecture. *Current Opinion in Plant Biology, 6*(3), 280–287.

Martin, J. P. (1976). *Darwin on earthworms: The formation of vegetable moulds*. Bookworm Publishing.

McDonald, A. J. S. (1994). Nutrient supply and plant growth. In *Physiology, growth and development of plants in culture* (pp. 47–57). Springer.

Morgan, J. Á., & Connolly, E. Á. (2013). Plant-soil interactions: Nutrient uptake. *Nature Education Knowledge, 4*(8), 2.

Munnoli, P. M., Da Silva, J. A. T., & Saroj, B. (2010). Dynamics of the soil-earthworm-plant relationship: A review. *Dynamic Soil, Dynamic Plant, 4*(1), 1–21.

Nadana, G. R. V., Rajesh, C., Kavitha, A., Sivakumar, P., Sridevi, G., & Palanichelvam, K. (2020). Induction of growth and defense mechanism in rice plants towards fungal pathogen by eco-friendly coelomic fluid of earthworm. *Environmental Technology & Innovation, 19*, Article 101011.

Nagavallemma, K. P., Wani, S. P., & Stephane, L. (2006). Vermicomposting: Recycling wastes into valuable organic fertilizer. *Journal of SAT Agricultural Research, 2*(1), 1–17.

Ndegwa, P. M., Thompson, S. A., & Das, K. C. (2000). Effects of stocking density and feeding rate on vermicomposting of biosolids. *Bioresource Technology, 71*(1), 5–12.

Nurhidayati, N., Ali, U., & Murwani, I. (2016). Yield and quality of cabbage (Brassica oleracea L. var. Capitata) under organic growing media using vermicompost and earthworm Pontoscolex corethrurus inoculation. *Agriculture and Agricultural Science Procedia, 11*, 5–13.

Ojuolape, O. T., Aransiola, S. A., & Obideyi, O. A. (2015) Vermitechnology: A solution to environmental problems; A review. *Journal of Global Ecology and Environment, 3*(3), 127–135. http://www.ikpress.org/abstract/4516

Pariyar, P., Dutta, A., & Bhutia, D. (2022). A report on vermicomposting efficiency of earthworm species from Darjeeling hills and Eisenia fetida. *Acta Scientific Veterinary Sciences (ISSN: 2582–3183)*, *4*(1).

Pattnaik, S., & Reddy, M. V. (2010). Nutrient status of vermicompost of urban green waste processed by three earthworm species—Eisenia fetida, Eudriluseugeniae, and Perionyx excavatus. *Applied and Environmental soil science, 2010.*

Planet Natural. (2001, October 7). *Vermicomposting: All you need to know*. Retrieved January 29, 2024, from https://www.planetnatural.com/composting-101/indoor-composting/vermicomposting/

Pyo, Y. J., Gierth, M., Schroeder, J. I., & Cho, M. H. (2010). High-affinity K+ transport in Arabidopsis: AtHAK5 and AKT1 are vital for seedling establishment and postgermination growth under low-potassium conditions. *Plant Physiology, 153*(2), 863–875.

Rana, K. L., Kour, D., Kaur, T., Sheikh, I., Yadav, A. N., Kumar, V., Suman, A., & Dhaliwal, H. S. (2020). Endophytic microbes from diverse wheat genotypes and their potential biotechnological applications in plant growth promotion and nutrient uptake. *Proceedings of the National Academy of Sciences, India Section B: Biological Sciences, 90*(5), 969–979.

Rao, A. G., Singh, D. P., & Yadav, A. K. (2024). Optimizing vermicomposting methods involves improving efficiency, maximizing nutrient retention and minimizing environmental impacts: A review. *Uttar Pradesh Journal of Zoology, 45*(10), 50–56.

Rekha, G. S., Kaleena, P. K., Elumalai, D., Srikumaran, M. P., & Maheswari, V. N. (2018). Effects of vermicompost and plant growth enhancers on the exo-morphological features of Capsicum annum (Linn.) Hepper. *International Journal of Recycling of Organic Waste in Agriculture, 7*(1), 83–88.

Rini, J., Deepthi, M. P., Saminathan, K., Narendhirakannan, R. T., Karmegam, N., & Kathireswari, P. (2020). Nutrient recovery and vermicompost production from livestock solid wastes with epigeic earthworms. *Bioresource Technology, 313*, Article 123690.

Sethi, D., Kusumavathi, K., Ravindran, B., Panda, N., Padhan, K., Dash, S., Sahoo, T. R., Mangaraj, S., Dhal, A., Swain, S. K., Sarkar, S., Febrisiantosa, A. (2023). Bioconversion of organic wastes into wealth by vermitechnology: a review. In *Recent trends in solid waste management* (pp. 27–53).

Shipley, A. E. (1970). In S. F. Harmer, & A. E. Shipley (Eds.), *The Cambridge natural history*.

Siddiqui, M. A., Neeraj, A., & Hiranmai, R. Y. (2022). Vermitechnology: An eco-friendly approach for organic solid waste management and soil fertility improvement—A review. In *Strategies and tools for pollutant mitigation: Research trends in developing nations* (pp. 91–112).

Soobhany, N. (2019). Insight into the recovery of nutrients from organic solid waste through biochemical conversion processes for fertilizer production: A review. *Journal of Cleaner Production, 241*, Article 118413.

Thakur, S., & Sood, A. K. (2019). Lethal and inhibitory activities of natural products and biopesticide formulations against Tetranychusurticae Koch (Acarina: Tetranychidae). *International Journal of Acarology, 45*(6–7), 381–390.

Udayakumar, S., &Parthasarathi, K. (2021). Influence of pongamia leaf litter on growth and reproduction of earthworms, Eisenia fetida (SAVIGNY) AND Lampitomauritii (KINBERG). *Uttar Pradesh Journal of Zoology*, 67–74.

Vuković, A., Velki, M., Ečimović, S., Vuković, R., Štolfa Čamagajevac, I., & Lončarić, Z. (2021). Vermicomposting—Facts, benefits and knowledge gaps. *Agronomy, 11*(10), 1952.

Walker, E. L., & Connolly, E. L. (2008). Time to pump iron: Iron-deficiency-signaling mechanisms of higher plants. *Current Opinion in Plant Biology, 11*(5), 530–535.

Xia, H., Chen, J., Chen, X., Huang, K., & Wu, Y. (2019). Effects of tetracycline residuals on humification, microbial profile and antibiotic resistance genes during vermicomposting of dewatered sludge. *Environmental Pollution, 252*, 1068–1077.

Yadav, A. N., Kour, D., Kaur, T., Devi, R., Yadav, A., Dikilitas, M., Abdel-Azeem, A. M., Ahluwalia, A. S., & Saxena, A. K. (2021). Biodiversity, and biotechnological contribution of beneficial soil microbiomes for nutrient cycling, plant growth improvement and nutrient uptake. *Biocatalysis and Agricultural Biotechnology, 33*, 102009.

Yatoo, A. M., Rasool, S., Ali, S., Majid, S., Rehman, M. U., Ali, M., Eachkoti, R., Rasool, S., Rashid, S. M., & Farooq, S. (2020). Vermicomposting: An eco-friendly approach for recycling/management of organic wastes. In *Bioremediation and biotechnology* (pp. 167–187). Springer.

Zambare, V. P., Padul, M. V., Yadav, A. A., & Shete, T. B. (2008). Vermiwash: Biochemical and microbiological approach as ecofriendly soil conditioner. *ARPN Journal of Agricultural and Biological Science, 3*(4), 1–5.

Organic Farming—The Role of Vermitechnology

Anjorin Ezekiel Adeyemi, Raufu Olusola Sanusi, and Daji Morumda

Abstract The impact of climate change is having severe consequences all over the world, but even more significantly in sub-Saharan Africa. The region's heavy reliance on rain-fed agriculture, coupled with the poor soil nutrient status of many farmlands, has compounded the effects of this crisis. The disruptions caused by recent events have affected many farming households and communities, leading to problems with agricultural food chains, food security, and socioeconomic well-being. Some researchers have suggested vermitechnology as a sustainable solution to address these issues. Vermitechnology, which is the commercial application of technologies that utilize earthworms for degrading waste organic materials to produce vermicast for agricultural reuse, may offer great potential, but several factors may impede these potentials, especially within the African context. In this chapter, we discuss the potentials, the limitations, the role of earthworms, and the advantages of vermicomposted foods.

Keywords Organic farming · Vermitechnology · Food · Health · Consumer

Iyanu Adebola—Deceased.

A. E. Adeyemi (✉)
Department of Agriculture, Fort Hays State University, Hays, USA
e-mail: aeadeyemi@fhsu.edu

R. O. Sanusi
Department of Agriculture, Phoenix University, Agwada, Nigeria
e-mail: raufu.sanusi@phoenixuniversity.edu.ng

D. Morumda
Department of Microbiology, Federal University Wukari, Wukari, Nigeria
e-mail: morumda@fuwukari.edu.ng

S. A. Aransiola and N. R. Maddela (eds.), *Vermitechnology: Economic, Environmental and Agricultural Sustainability*, https://doi.org/10.1007/978-3-031-96212-7_16

1 Vermicompost Organic Farming in Sub-Saharan Africa: A Myth, an Illusion, or an Imperative?

The climate change crisis impacts many parts of the world, especially sub-Saharan Africa (Bichisa et al., 2023; Blanc, 2012; Calzadilla et al., 2013; Serdeczny et al., 2016; Watson et al., 1998). The region is already experiencing higher temperatures, more frequent and intense heatwaves, and more prolonged droughts, which are all affecting crop yields and the overall productivity of agricultural systems which can have severe consequences for agriculture and food security (Falloon & Betts, 2009; FAO, 2011; Reilly et al., 2003; World Bank, 2007; Zhou, 2023).

In addition, rising sea levels and more frequent extreme weather events, such as floods and hurricanes, are also posing a threat to the region. These challenges are making it more difficult for farmers to grow food and provide for their families, and they are also contributing to the already high levels of poverty and food insecurity in the region (Betts et al., 2007; IPCC, 2007; Long et al., 2006).

Several coping, adaptive, and mitigating strategies against climate change have been suggested such as (Paavola & Adger, 2006; Ringler, 2008):

1. Adopting sustainable agriculture practices: This can include practices such as agroforestry, conservation agriculture, and drip irrigation, which can help to improve the productivity and sustainability of farms.
2. Implementing water conservation measures can include improving irrigation efficiency, harvesting rainwater, and reusing greywater.
3. Promoting renewable energy sources: This can include things like solar panels, wind turbines, and small hydroelectric projects, which can help to reduce reliance on fossil fuels.
4. Enhancing disaster preparedness and response can include investing in early warning systems, strengthening infrastructure, and developing evacuation plans.
5. Planting trees and other vegetation can help absorb carbon dioxide from the atmosphere and provide other ecosystem services such as erosion control and wildlife habitat.
6. Reducing greenhouse gas emissions: This can include improving energy efficiency, transitioning to low-carbon transportation options, and transitioning to low-carbon energy sources.

While implementing the above can be very beneficial to the agricultural food chain, the environment, and farming households, the complexity and dynamics of the implementation can sometimes be daunting, too technical, and not affordable to farmers. This is particularly true in most parts of sub-Saharan Africa, where there is now a growing focus on the potential of vermicomposting in helping to address some of the challenges facing agriculture in the region (Aranda et al., 1999; Maboeta & Rensburg, 2003; Mainoo, 2007; Ndegwa & Thompson, 2001).

Vermicomposting, or the use of worms to decompose organic matter and produce compost (Fig. 1), can be a valuable tool for improving soil health and increasing crop yields in sub-Saharan Africa. It is a sustainable and environmentally friendly

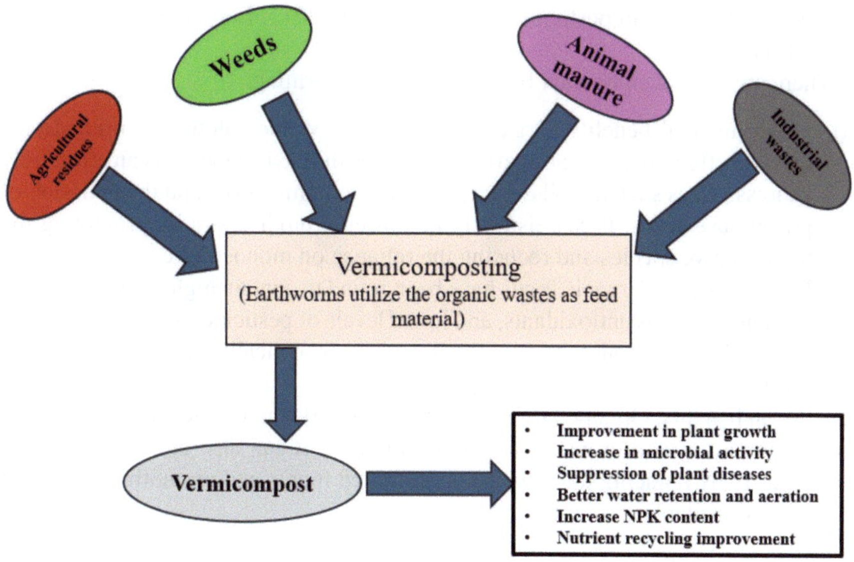

Fig. 1 The vermicomposting process for different types of organic waste. Reproduced from Singh et al. (2020)

method of waste management that can also provide an additional source of income for smallholder farmers (Akinnuoye-Adelabu et al., 2019; Aransiola et al., 2022; Lim et al., 2015; Maturi et al., 2022; Singh et al., 2020).

Some of the benefits of vermicomposting in sub-Saharan Africa include:

(a) Improving soil structure and water-holding capacity: Vermicompost contains a higher concentration of nutrients and organic matter than traditional compost, which can help improve the design and fertility of the soil. It also helps to increase the soil's ability to hold water, which is essential in areas with irregular rainfall patterns.
(b) Reducing the need for synthetic fertilizers: Vermicompost can be used as a natural alternative to synthetic fertilizers, which can be expensive and harmful to the environment. It can also help to reduce the reliance on chemical inputs, which can be a significant cost for smallholder farmers in sub-Saharan Africa.
(c) Reducing food waste: Vermicomposting can help reduce the amount of food waste generated by households and restaurants, a significant problem in many parts of sub-Saharan Africa. By turning this waste into valuable compost, vermicomposting can help close the food waste loop and reduce the pressure on landfills.

Organic farming is an agricultural method that prioritizes sustainability, environmental friendliness, and humane practices (Luttikholt, 2007; Reganold & Wachter, 2016; Seufert et al., 2017). This approach utilizes natural processes such as vermicomposting to improve land health and productivity instead of synthetic inputs like chemical fertilizers and pesticides (Gattinger et al., 2012; Lori et al., 2017; Muller et al., 2017).

There are several significant benefits of organic farming:

(a) Environmental benefits: Organic farming is a viable solution to mitigate the harmful effects of conventional agriculture on the environment. It can effectively address issues such as soil depletion, water contamination, and the emission of greenhouse gases. It can also help to conserve biodiversity by promoting the use of native species and reducing the reliance on monoculture.
(b) Health benefits: Organic foods have been found to contain higher levels of some nutrients, such as antioxidants, and lower levels of pesticides and other contaminants. They may also be less likely to have genetically modified organisms (GMOs).
(c) Social benefits: Organic farming can help support small-scale farmers and rural communities and promote fair labor practices. It can also contribute to food security by reducing the reliance on imported inputs and promoting the use of locally available resources.

Thus, vermicompost organic farming combines the strength of vermicomposting with that of organic farming to deliver healthy foods produced in the most environmentally sustainable way, which has economic potential for all the stakeholders in the value chain.

While vermicomposting organic farming is not a silver bullet solution to all the challenges facing agriculture in sub-Saharan Africa, it can be an essential part of an integrated approach to improving soil fertility and productivity in the region. It is just one tool that can be used as part of an integrated approach to improving soil health and increasing crop yields. Other important factors for the success of agriculture in the developing world include access to credit, training and extension services, and infrastructure, such as roads and storage facilities. It is also essential to recognize that different regions and farming systems will have different needs and challenges and that a one-size-fits-all approach is unlikely to be effective. Instead, it will be necessary to carefully assess a particular region's needs and constraints and develop tailored methods considering these factors.

Overall, vermicompost organic farming has the potential to make a significant contribution to the sustainable development of agriculture in sub-Saharan Africa. While there may be challenges to implementing vermicomposting on a larger scale in the region, it is an important technology that should be considered as part of an integrated approach to improving soil health and increasing crop yields.

2 Limitations of Vermitechnology and Organic Farming in Nigeria

Organic agriculture was the practice of the people of Africa in the years gone by. Human interrelationship with their immediate environment initially had little or no adverse effects. However, the population explosion and technological advancement have upset the natural ecosystem. Applying machines and chemical farm inputs disturbs soil health and destroys beneficial bio-organisms. This continuous disruption and depletion of arable land, with the resultant health implications on human/animal health, gave rise to calls for organic farming, of which vermitechnology is a branch.

Vermicomposting is a great way to create high-quality organic fertilizer from plant waste (Ojuolape et al., 2015). Other methods of utilizing plant waste, such as composting and anaerobic digestion, are more complex and expensive and may not be able to handle all types of plant waste (Abbasi et al., 2012; Yaser & Lamaming, 2022). Landfills, which currently operate the most biodegradable waste, are already overwhelmed and unsuitable for the disposal of plants (Annepu, 2012). Vermireactors with high rates have a lot of advantages, such as high earthworm densities, pulse-fed operation, and high surface area-to-volume ratios. These reactors optimize the use of space and ensure better substrate agitation, uniform moisture distribution, and no leachate generation. Moreover, they prevent the formation of anaerobic pockets commonly found in traditional vermireactors.

Vermicomposting is a unique method because it is aerobic. When phytomass is used for vermicomposting, only 40% of the carbon is converted into CO_2, while the rest becomes vermicompost, which can enrich the soil. This process also helps with carbon sequestration because the CO_2 released comes from already sequestered carbon. Although vermicomposting has potential benefits, it has only been studied in laboratory feasibility studies and has yet to be implemented on a larger scale for phytomass utilization. In Nigeria, the low awareness of vermitechnology among farm families significantly contributes to the current situation. While farmers are familiar with earthworms as valuable allies, the knowledge of vermitechnology is almost nonexistent. Farmers generally know that earthworms enrich the soil with nutrients due to their burrowing and conversion of organic matter into vermicast. However, the awareness that it could be cultured on a large scale to produce soil-friendly organic matter for commercial farming is missing. Most of those who (primarily researchers) know this technology are still practicing it at experimental stations. Most of the earthworms cultured in Nigeria are used as bait by anglers and as part of raw materials for livestock feed.

Also, there is neither known public support for organic farming research nor effective and holistic support for smallholders in Nigeria. This has severely hindered the awareness and subsequent widespread adoption of innovations. With no definite government policy to encourage organic farming and vermitechnology, much is therefore not expected from the citizen regarding adoption.

Also, the requisite skills and know-how about the vermicomposting technique are missing. From Fig. 2, there is no doubt that technical know-how is necessary for

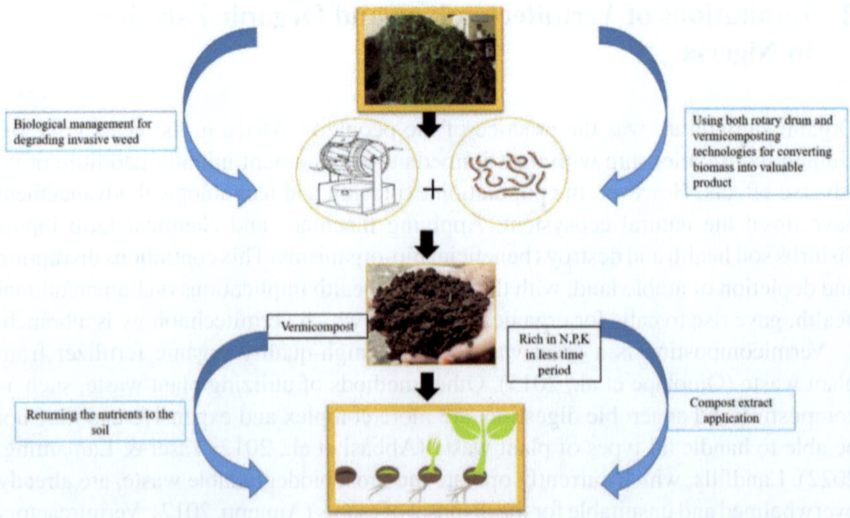

Fig. 2 The vermicomposting process. Reproduced from Kauser and Khwairakpam (2022)

success in any endeavor. The procedural process for achieving vermicast in a large quantity such that a farmer could satisfactorily adopt organic farming is yet to be mastered. The concept of high-rate vermicomposting and its associated knowledge has been developed by researchers and authors, as evidenced by the works of Abbasi et al. (2011). Furthermore, there is limited availability of organic matter for continuous enhancement of the soil nutrients because biomass and organic waste materials are finding other uses. Simply put, organic farming methods are more expensive than traditional farming systems. This could discourage many would-be organic farmers.

Also, it takes time before the complete restoration of depleted soil nutrients in organic farms. Two or more years could pass before an organic farm can produce crops in full. The time it takes for nutrient recovery causes many to shun the idea of organic farming. People must feed daily to survive. The fallow period could pose severe adverse consequences in the farming-feeding continuum.

Moreover, vermitechnology depends on natural and climatic conditions such as soil fertility, moisture content, temperature, and the right combination of substrates. For instance, the substrate should be void of oil, salt, and harmful chemicals, and the required temperature to produce vermicast must also be observed. Keeping these conditions could be a challenge for an average would-be organic farmer.

Research conducted by Abbasi et al. (2012) has shown that adding enormous amounts of cow dung to plant-based biomass (phytomass) is not a practical solution (Onwosi et al., 2022). This is because processing the phytomass requires an equivalent amount of animal manure, which is difficult to obtain due to its high demand for various purposes. Animal manure is not free like waste phytomass, but it can still be cost-effective. However, collecting and transporting it can release harmful greenhouse gases like methane and nitrous oxide, as well as pollutants such as ammonia. Therefore, using manure in large-scale phytomass vermicomposting can have negative consequences such as pollution and global warming (Abbasi et al., 2015). Vermicomposting is very slow and takes time! Since the efficiency of any innovation is a function of the rate of its process, the rationale for its continuous usage is diminished. It typically takes at least 65 days to convert phytomass to vermicast unless pre-composting is done beforehand. However, pre-composting can increase the process's time and cost, making it less economically viable.

Furthermore, there is a lack of a clear standard to determine the output of a vermireactor and how to confirm that the initial materials have been completely transformed into the desired outcome. This lack of criteria makes optimizing the efficiency and cost-effectiveness of the vermireactor's operation difficult. Previous efforts have mainly focused on batch reactors with lengthy and unsupported solid retention times (SRTs) (Abbasi et al., 2015).

3 The Role of Earthworms in a Successful Vermicompost Organic Farming

Earthworms' importance and functional role in the soil have received much attention recently. The earthworm is lauded by Aristotle as "the intestine of the earth," thus making them a very special invertebrate with unique roles on the earth's surface, especially in the soil (Fig. 3). They are referred to as farmers' friends that work day and night without any cost. They plow the field without cost during their burrowing activities, which is one of their significant roles in organic farming.

Earthworms are a crucial component of soil in various regions across the globe, both temperate and tropical. Their presence is essential to agroecosystems' overall sustainability, as Pelosi et al. (2014) pointed out. In many land ecosystems, these species are crucial because they affect the soil in various ways. They impact soil structure, including porosity and water infiltration, and promote root penetration through compacted layers. Additionally, they play a role in soil fertility, biochemical cycles, and other organisms, according to Brussaard et al.'s (2007) research. Table 1 shows a list of different species of earthworms identified across Africa and their suitability or otherwise for vermicomposting (Table 1).

4 Classes of Earthworms

In organic farming, earthworms play various roles depending on their ecology. Bouche's (1977) and Curry's (2004) classifications have identified three main ecological groups: epigeic, endogeic, and anecic. These groups collaborate and share functions to fulfill their respective roles, as explained by Briones (2014). It is crucial to understand this for effective farming practices. The epigeic (Fig. 4a) are the topsoil

Fig. 3 African nightcrawler. *Source* Original picture by authors

dwellers that mainly feed on the litter and the microflora that are found with and joined in close association with them (Blouin et al., 2013). Their primary function is to significantly impact the breakdown of organic material and the circulation of nutrients, increasing available space and interacting with other organisms in the soil. Hence, they help in the soil detoxification process, waste disposal, pest control, affect the soil surface roughness, and increase water infiltration rate, and all these bring about soil fertility to cultivate crops, which is very important (Rüdisser et al., 2021; Smith et al., 2008).

Endogeic earthworms, or mineral dwellers, reside and nourish themselves within the soil (see Fig. 4b). These creatures create horizontal burrows and consume soil with high amounts of organic matter. Some consume minerals and organic particles, while a few oligohumic endogeic species feed on low-quality organic matter found in deep horizons of the tropical soil. They are supported by mutual relationships with microflora, as stated by Bernard et al. (2012) and Römbke et al. (2005). Their activities stimulate the growth of plant roots, microbial community, and induction of organic matter decomposition, hence, helping in soil structure maintenance by stabilizing and controlling erosion rates, decomposition, and fixing process, biological population control, and aiding plant production (Brown et al., 2004; Rüdisser et al., 2021).

Figure 4c shows Anecic species, which are vertical burrowers that construct permanent vertical burrows in the soil. Their presence significantly enhances water infiltration. At the entrance of their burrows, they build middens on the soil surface

Table 1 Earthworm species distribution across Africa

Species	Country/Region	Suitable for vermicomposting
Agastrodrilus opisthogynus	Ivory Coast	No
Amynthas minimus	South Africa, Asia	No
Amynthas rodericensis	South Africa, Asia	Yes
Chuniodrilus palustris	Ivory Coast	No
Chuniodrilus zielae	Ivory Coast	No
Dichogaster affinnis	West Africa	No
Dichogaster agilis	Ivory Coast	No
Dichogaster annae	West Africa	Yes
Dichogaster bolaui	West Africa	Yes
Dichogaster gracilis	West Africa	Unknown
Dichogaster grafii	Congo	No
Dichogaster itolienses	Rwanda	No
Dichogaster modigliani	West Africa	No
Dichogaster saliens	West Africa	No
Eminoscolex lavellei	Rwanda	No
Eudrilus eugeniae	Ghana, Ivory Coast, Nigeria, West Africa	Yes
Gordiodrilus peguanus	Central Africa	No
Hyperiodrilus africanus	Ghana, Ivory Coast, Nigeria, West Africa	Yes
Millsonia anomala	Ivory Coast	No
Millsonia ghanensis	Ivory Coast	Unknown
Millsonia inermis	Burkina Faso	No
Millsonia lamtoiana	Ivory Coast	No

Source Fragoso et al. (1999), Hauser (1993), Blanchart et al. (1997), Rossi (2003)

Fig. 4 Types of earthworms. **a** Epigeic earthworms. **b** Endogeic earthworms. **c** Anecic earthworms. *Source* www.trees.com

that contain visible pieces of organic matter and casts, as reported by Medina-Sauza et al. (2019). Their presence brings about large macrospores, which in turn contribute to the water flow regulations in the soil (Fischer et al., 2014; Van Schaik et al., 2014). They help curb rain's profound effect on soils and plants (Andriuzzi et al., 2015). According to Johnston et al. (2015), they can break down and create soil structures. Unlike other species, endogeic species can impact the entire soil profile. They work alongside other species to maintain water flow, stabilize mass, control erosion rates, and aid in decomposition and fixing processes.

4.1 Earthworm and Nutrient Cycling

Using organic amendments in farming can improve nutrient availability and promote the breakdown of organic matter, especially when earthworms are involved. Bertrand et al. (2015) have demonstrated that cast can enhance nutrient use efficiency and decrease nitrate leaching in soil. The rich nutrient content in the cast can effectively enrich the soil. Calcium, potassium, and magnesium, some of the water-soluble nutrients, are also enhanced as soil organic matter and litter pass through the gut of earthworms. The grinding of organic minerals in the earthworm gut dissolves these nutrients, as explained by Carpenter et al. (2007).

Studies have shown that certain types of earthworms, specifically the endogeic (geophagous) groups, can enhance the soil's carbon and nitrogen mineralization process. This has been documented in various studies conducted by Lavelle et al. (1998), Araujo et al. (2004), Coq et al. (2007), and Gopal et al. (2017). Meanwhile, epigeic earthworms have been found to increase phosphorus levels in soil or substrate, as noted in a study by Medina-Sauza et al. (2019). Research suggests that the cast (excrement) produced by earthworms contains high levels of essential nutrients, particularly mineralized carbon and readily available phosphorus, compared to the soil around it. This is due to the priming effect caused by the earthworms' ingestion and digestion and their gut microbiome's influence on the decomposition rate of organic matter in the soil (Athmann et al., 2017; Ros et al., 2017). Organic matter decomposition in earthworms is credited to the microorganisms in their digestive tract and the structures they create, forming the drilosphere. This unique location supports microbial activities, as Barois et al. and Bernard et al. (2012) noted. According to Taheri et al. (2018), the positive priming effect helps recycle nutrients, particularly organic Nitrogen and Phosphorus, in the soil's organic matter. Hence, by enhancing nutrient availability, earthworm benefits plant growth in organic farming.

4.2 Earthworm and Plant Growth

Earthworms enhance plant growth and yield by actively improving soil conditions, promoting root penetration, nutrient absorption, and gas exchange. They play a significant role in nitrogen release in organic matter, facilitated by excretion, mucus secretion, and organism death. Furthermore, species that feed on litter effectively transfer nitrogen from surface detritus to plants, thus making them essential for successful plant growth (Amador & Görres, 2005; Costello & Lamberti, 2008; Van Groenigen et al., 2014).

Other mechanisms can change gene expression, such as releasing signal molecules when earthworms are present. These mechanisms are responsible for earthworms' positive effects on plant growth, as explained by Puga-Freitas et al. (2012). Signal molecules are generated by plants and soil organisms, including microorganisms and soil fauna (Puga-Freitas & Blouin, 2015). The molecules in plants, such as sugar, organic acids, and vitamins, play a crucial role in signaling pathways that trigger the production of phytohormones like auxins, gibberellins, cytokinins, ethylene, and abscisic acid (ABA). They also activate the plant's immune system and regulate its growth and development through secondary metabolites. Puga-Freitas et al. (2012) discovered that Humic acids, Indole acetic acids (IAAs), and aminocyclopropane-L-carboxylate (ACC) are among the signal molecules produced by earthworms.

In 2019, Hernández reported the presence of jasmonic acid (JA), salicylic acid (SA), and ABA in the vermicompost of *Eisenia fetida*. The impact of signal molecules on plants in the presence of earthworms has been extensively observed. These observations, particularly about vermicompost, have consistently demonstrated increased growth and yield and the development of flowers and fruits. Furthermore, these effects have been linked to significantly improved tolerance to both biotic and abiotic stresses.

Certain types of earthworm species, such as *E. fetida*, *Aporrectodea caliginosa*, and *Aporrectodea rosea*, have been found to positively impact plant growth by producing IAA, ACC, and humic acids with the help of associated bacteria, according to Medina-Sauza et al. (2019). Furthermore, studies have demonstrated that earthworms play a role in seed burial and the growth of seedlings, and they can potentially impact the makeup of plant communities by aiding certain species in out-competing others, whether through their function or taxonomy (Milcu et al., 2006; Schmidt & Curry, 1999).

5 Earthworm and Microorganisms' Biodiversity

Earthworms' interaction with soil microbes is one of their prominent role in biodiversity. They modify the environment of soil microorganisms (Lavelle et al., 1997) and regulate the availability of resources to other organisms by causing physical, chemical, and biological changes in the biotic and abiotic materials (Jones et al., 1994).

Thereby impacting the structure of the soil microbial community (Egert et al., 2004). Earthworm guts inhabit many microorganisms with varied taxonomical affiliations and functions (Ahmed & Al-Mutairi, 2022). Earthworms can digest microbes and even encourage the growth of certain soil microbes that aid in the digestion of organic matter. This enriches the soil with bacteria that can break down the materials that earthworms feed on, as well as bacteria that can survive in the low-oxygen environment of the earthworm gut and help reduce nitrate levels (Chapuis-Lardy et al., 2010; Fujii et al., 2012; Hong et al., 2011; Nechitaylo et al., 2010; Shan et al., 2013).

Earthworms can affect the abundance or serve as a distinct microhabitat of some Protists, Nematodes, and other invertebrates (Cameron et al., 2013; Stromberger et al., 2012; Tiunov et al., 2001). Medina-Sauza et al. (2019) have found that the effect of earthworms on microbial communities depends on the earthworm species and microhabitat. The impact can be beneficial, detrimental, or have no effect at all.

Due to the presence of microorganisms in earthworms' intestines and excrement, the soil benefits significantly from high levels of nitrogen fixation. This leads to a significant increase in the percentage of nitrogenase casts and promotes more excellent nitrification in the soil, resulting in a higher amount of nitrogen being present in the soil (Atiyeh et al., 2000; Chan et al., 2004). Soil microorganisms play a crucial role in soil organic matter decomposition and mineralization. The structure of soil microbial communities is changed by earthworms to accelerate soil organic matter decomposition and mineralization (Scheu et al., 2002).

6 Earthworm and Soil Structure

Different authors have proposed earthworms to be an excellent indicator of soil health (Fusaro et al., 2018; Kibblewhite et al., 2008) and a sustainability check in agricultural landscapes (Bispo et al., 2009; Paoletti, 1999; Turbe et al., 2010). The abundance and diversity of earthworms were rated high as an indicative value of fifty (50) biological soil parameters (Rutgers et al., 2012). Their activities can improve ecosystem productivity, such as increased soil fertility, flood and erosion control, and restoration, increasing agricultural output (Van Groenigen et al., 2014). Soil structure is crucial for soil fertility and other essential functions. Earthworms are vital contributors in this regard as they significantly shape soil structure and nutrient cycling. These ecosystem engineers are incredibly beneficial for themselves and other organisms that rely on them for creating habitats (Blouin et al., 2013; Jones et al., 1994).

Consuming soil minerals and organic components, earthworms have the power to boost the stability of soil structure significantly. This is achieved through the creation of casts and increased carbon mineral associations, both of which work together to improve the overall stability of soil aggregates (Deeb et al., 2017; Oades, 1993). Assessing the aggregate strength of soil is a critical factor in determining its quality. This is because it influences the distribution of pore sizes in the soil, ultimately affecting air and water movement through it. The presence of well-aerated soil is crucial for the growth and survival of plants and microbes.

Research suggests that earthworms play a crucial role in plant growth and decomposition and contribute to the nutrient cycle (Johnsen et al., 2005; Fujii et al., 2012). Earthworms are estimated to produce up to 100 tons of casts, significantly contributing to stable soil aggregates forming (Brown & Doube, 2004). As earthworms move through the soil, feed on it, and release material, they significantly impact the soil structure's reorganization (Chapuis-Lardy et al., 2010).

7 Earthworms, Pests, and Diseases

It is worth noting that earthworms are crucial in managing pests and plant diseases. They enhance nutrient availability and even boost the plant's resistance through their actions. According to Ahmad and Al-Mutairi's research in (2022), earthworms play a vital role in maintaining healthy plant growth. Earthworm activity has been found to increase plant growth and suppress disease incidence. Stephens and Davoren (1997) reported a decrease in disease incidence in some plants like Clover, grains, and grapes caused by *Rhizoctonia spp* and *Gaeumannomyces spp* (Clapperton et al., 2001) when exposed to earthworms. Also, diseases were suppressed in three vegetables infested with *Fusarium oxysporum* spp., *Asparagi, F. proliferatum* (Asparagus), F. oxysparum f.sp. Lycopersicum (Tomato), and *Verticillium dahlia* (Eggplant) when augmented with *Lumbricus terrestris*. However, it was also observed that the disease suppression may have been mediated through microbial activity (Elmer, 2009). Furthermore, earthworms reduced the *Fusarium* wilt of strawberries, and this was done by regulating microorganisms and degrading phenolics using *Metaphire guillemi* and *E. fetida*; in this case, *M. guillemi* was found to be more effective in reducing the *Fusarium* wilt of strawberries (Bi et al., 2018).

8 Earthworm Burrows

Research has consistently demonstrated that earthworms burrow deeper into cultivated soils than forest and grassland soils. This was reported by Brown et al. (2000) and Kuzyakov and Blagodatskaya (2015). Earthworms significantly reduce soil erosion in temperate and tropical soils, according to studies conducted by Le Bayon et al. (2002). Their burrowing and casting activities help to increase the soil's structural stability and porosity and influence soil mechanical and hydraulic properties. This creates a pathway for water flow and generates macropores, crucial for water infiltration and supplying crops with water. Earthworms also aid soil aeration and control surface runoff and erosion (Bertrand et al., 2015; Laossi et al., 2010; Ritsema & Dekker, 2000; Spurgeon et al., 2013).

The walls of their burrow contain higher levels of Carbon and Nitrogen than the surrounding soil. These walls are also crucial for the mineralization and nutrient turnover (Binet & Trehen, 1992; Don et al., 2008), creating a suitable habitat for microbial and plant communities and other processes driven by microbes, which is essential (Monard et al., 2011). Furlong et al. (2002) have observed that fresh casts and middens are characterized by elevated microbial activity. Moreover, the gut of earthworms houses thriving microbial communities that flourish in such conditions. (Thakuria et al., 2010).

9 Earthworm and Livestock Farming

Earthworms are found to be helpful in livestock farming. They are a good source of protein with nutritional value like fish nutrient profile (Sogbesan et al., 2007; Jabir et al., 2012). These seeds can be used either fresh or milled, and they are a great source of nutrients (Chiripasi et al., 2013). Various countries have used them to add nutrition to poultry, fish, and swine feed (Vieira et al., 2004; Rawling et al., 2012).

10 Factors Affecting Earthworm Population

Cultural practices, including crop rotation, irrigation, drainage, organic matter inputs, and tillage, can impact the earthworm population. As earthworms reside in the soil, these practices can influence their numbers. The amount of earthworm biomass depends on crop management (Pelosi et al., 2009; Bertrand et al., 2015). The impact of earthworms on plant growth and soil pathogens is likely to be influenced by soil properties like water content and soil organic matter, which are also affected by cultural practices.

Organic manure is highly beneficial for earthworms, as it is their primary food source. However, mechanical weeding can negatively impact their physical structure and reproductive functions. The diversity and distribution of earthworms are also influenced by various factors such as climate, soil quality, moisture content, temperature, and pH. Farmers must choose the right tillage system and utilize organic fertilizers to maximize earthworm abundance in their fields. Numerous studies have highlighted the advantages of these practices, including research conducted by Bertrand et al. (2015), and Ahmed & Al-Mutairi (2022).

11 The Nutritional or Health Benefits of Organic Foods Produced Through Vermitechnology

Several studies have been done to compare the nutritional or health benefits of organic foods and foods produced through vermitechnology, and below are some of the benefits.

11.1 Higher Content of Antioxidants

Antioxidants prevent damage to the cell due to the activities of reactive oxygen or nitrogen species, such as peroxides, dioxygen, and free radicals, by neutralizing them. It has been reported in several studies that organically produced foods or foods produced by vermitechnology contain a higher content of antioxidants compared to those produced by conventional agricultural practices (Lester et al., 2007; Reganold et al., 2010). Antioxidants such as SA and phenols were demonstrated to be more available in organic foods. Antioxidants are essential in promoting the health of both humans and plants due to their protective activities in preventing cell damage from the destructive activities of these reactive species or radicals. Lycopene is the primary carotenoid in tomatoes and is the main component responsible for the red coloration of this vegetable. Its concentration was discovered to be higher in tomatoes produced via vermitechnology, meaning that the anticancer, antidiabetic, anti-inflammatory, antiatherogenic, anti-allergenic, antithrombotic, antimicrobial, vasodilator, and cardioprotective effects (Ali et al., 2020). Shankar et al. (2008) observed a significantly higher lycopene content in tomatoes produced by vermitechnology compared to other conventional practices.

12 Higher Concentration of Bioactive Compounds

Foods produced by vermitechnology have been discovered to contain higher amounts of bioactive compounds that promote health, such as ascorbic acid, minerals, sugars, lycopene, phenols, nitrates, and pectin. These nutrients perform many vital functions in the body, such as preventing constipation, stimulating blood circulation, reducing high blood pressure, maintaining the structure of bones, removing toxins from the body, and regulating lipid profiles (Ali et al., 2020). According to a study by Ahirwar and Hussain (2015), organically produced tomatoes contained more ascorbic acid than conventionally produced ones. This is because ascorbic acid synthesis requires glucose, which is more readily available in organic farming. Furthermore, using chemical fertilizers in conventional farming practices can decrease the amount of ascorbic acid and other bioactive secondary metabolites in crops.

13 Prevention of Many Chronic Degenerative Diseases

The concentration of many critical secondary metabolites of plants that play an essential role in preventing chronic degenerative diseases such as cancer, cardiovascular diseases, and neurodegenerative diseases is higher in foods produced by vermitechnology than those produced by traditional conventional farming. Secondary metabolites such as carotenoids like lycopene and β-carotenoids, tocopherol, ascorbic acid, and other bioactive phenolic compounds can reduce and prevent many diseases, especially chronic diseases (Navarro-González et al., 2018), as they are involved in preventing reactive oxygen species (ROS) by neutralizing free radicals, prevention of cellular proliferation and damage, expression of cytokine and signal transduction pathways, modulation of enzymatic activities, preventing apoptosis as well as metal chelation (Ali et al., 2020).

14 Higher Nutritional Value

In conventional agriculture, more attention is given to taste and crop yield rather than to nutritional content, which decreases the nutrient quality of such products. Organically produced foods contain high nutrient levels because they are supplied with nutrients and conditions in their natural forms and are given sufficient time for their development. Using vermitechnology offers optimal conditions for plants to absorb essential nutrients required for their growth, potentially clarifying the reason behind this phenomenon (Prasad, 2021).

15 They are Chemical Free with Excellent Storage Value

Food products produced by vermitechnology are free from chemicals and thus considered safer than those produced by conventional agriculture. It has also been discovered that the storage quality or keeping quality of vermitechnology produce, such as flowers, fruits, vegetables, and food grains, is higher and better than in conventional agriculture (Chakrabarty et al., 2009). Using chemicals in traditional farming can result in unpredictable outcomes and leave harmful residue, which may lead to various health risks. Such risks include damage to the nervous system, cancer, infertility, immune disorders, blue baby syndrome, brain cancer, leukemia, and congenital disabilities in children, as stated by Sharma and Agarwal (2014).

An example is the discovery that foods produced by conventional agriculture contained higher amounts of nitrates, which can easily be transformed into nitrites with the capability of binding with oxygen molecules in the blood due to their reactive nature. This makes oxygen unavailable for binding by the hemoglobin, leading to methemoglobinemia and anoxia. Studies have shown that combining nitrites with secondary amines can produce nitrosamines, which are known to be highly carcinogenic. Meanwhile, Huuml et al. (2011) research found that organic farming methods have better quality tomatoes that maintain their firmness than nonorganic farming techniques.

16 Higher and Better Sensory Benefits

Vermitechnology-produced fruits and vegetables boast flavonoids, phenolics, and anthocyanin compounds that create natural pigmentation, aroma, and flavors. This delivers an enhanced sensory experience for consumers. Tomatoes cultivated via vermitechnology showcase a juicier and more robust tomato aroma; they are less mealy than conventionally grown tomatoes (Sharma & Agarwal, 2014). It is vital to avoid excessive nitrogen as it can hurt taste and flavor by decreasing carbohydrate synthesis and glucose content, which is common in foods produced through chemical fertilizers.

17 Conclusion

Vermicomposting, an evolving environmentally friendly technology and it is a cheap but rich source of organic nutrients to the soil. Although global adoption of the technology is low at the moment due to inadequate awareness that vermicast could be cultured in commercial quantities, there is nonetheless great future for it as awareness intensifies. The nutritional and health benefits of vermitechnology-produced food brighten its foreseeable future acceptance in the world of agriculture, given the support of government and other stakeholders in the field of agriculture.

References

Abbasi, T., & Abbasi, S. A. (2012). Is the use of renewable energy sources an answer to the problems of global warming and pollution? *Critical Reviews in Environment Science and Technology, 42*, 99–154.

Abbasi, T., Tauseef, S. M., & Abbasi, S. A. (2011). The inclined parallel stack continuously operable vermireactor. *Official Journal of the Patent Office, 22*, 9571.

Abbasi, T., Tauseef, S. M., & Abbasi, S. A. (2012). *Biogas energy* (pp. 1–10). Springer. https://doi.org/10.1007/978-1-4614-1040-9_1

Abbasi, S. A., Nayeem-Shah, M., & Abbasi, T. (2015). Vermicomposting of phytomass: Limitations of the past approaches and the emerging directions. *Journal of Cleaner Production, 93*, 103–114.

Abd Rahman Jabir, M. D., Razak, S. A., & Vikineswary, S. (2012). Nutritive potential and utilization of super worm (*Zophobas morio*) meal in the diet of Nile Tilapia (*Oreochromis niloticus*) juvenile. *African Journal of Biotechnology, 11*(24), 6592–6593.

Ahirwar, C. S., & Hussain, A. (2015). Effect of vermicompost on growth, yield, and quality of vegetable crops. *International Journal of Applied and Pure Science and Agriculture, 1*(8), 49–56.

Ahmed, N., & Al-Mutairi, K. A. (2022). Earthworms' effect on microbial population and soil fertility as well as their interaction with agriculture practices. *Sustainability, 14*, 7803.

Akinnuoye-Adelabu, D. B., Steenhuisen, S., & Bredenhand, E. (2019). Improving pea quality with vermicompost tea and aqueous biochar: Prospects for sustainable farming in Southern Africa. *South African Journal of Botany, 123*(2019), 278–285. https://doi.org/10.1016/j.sajb.2019.03.009

Ali, M. Y., Sina, A. A. I., Khandker, S. S., Neesa, L., Tanvir, E. M., Kabir, A., Khalil, M. I., & Gan, S. H. (2020). Nutritional composition and bioactive compounds in tomatoes and their impact on human health and disease: A review. *Foods, 10*(1), 45.

Amador, J. A., & Görres, J. H. (2005). Role of the anecic earthworm *Lumbricus terrestris* L. in the distribution of plant residue nitrogen in a corn (*Zea mays*)–soil system. *Applied Soil Ecology, 30*, 203–214.

Andriuzzi, W., Pulleman, M., Schmidt, O., Faber, J., & Brussaard, L. (2015). Anecic earthworms (*Lumbricus terrestris*) alleviate negative effects of extreme rainfall events on soil and plants in field mesocosms. *Plant Soil*, 1–11.

Annepu, R. K. (2012). *Sustainable solid waste management in India* (pp. 1–189). MSc Thesis, Department of Earth and Environmental Engineering, Columbia University.

Aranda, E., Barois, I., Arellano, P., Irisson, S., Salazar, T., Rodriguez, J., & Patron, J. C. (1999). Vermicomposting in the tropics. In P. Lavelle, L. Brussard, & P. Hendrix (Eds.), *Earthworm management in tropical agro-ecosystems* (pp. 253–288). CABI Publishing.

Aransiola, S. A., Ijah, U. J. J., Abioye, O. P., & Bala, J. D. (2022). Vermicompost-assisted phytoremediation of toxic trace element-contaminated soil in Madaka, Nigeria using *Melissa officinalis* L. and *Sida acuta*. *International Journal of Environmental Science and Technology*. https://doi.org/10.1007/s13762-022-04105-y

Araujo, Y., Luizão, F. J., & Barros, E. (2004). Effect of earthworm addition on soil nitrogen availability, microbial biomass and litter decomposition in mesocosms. *Biology and Fertility of Soil, 39*, 146–152.

Athmann, M., Kautz, T., Banfield, C., Bauke, S., Hoang, D. T., & Lüsebrink, M., et al. (2017).

Atiyeh, R. M., Domínguez, J., Subler, S., & Edwards, C. A. (2000). Changes in biochemical properties of cow manure during processing by earthworms (*Eisenia andrei*, Bouché) and the effects on seedling growth. *Pedobiologia, 44*, 709–724.

Bouché, M. B. (1977). Strategies lombriciennes. *Ecological Bulletins*, 122–132.

Bernard, L., Chapuis-Lardy, L., Razafimbelo, T., Razafindrakoto, M., Pablo, A.-L., Legname, E., Poulain, J., Bruls, T., O'Donohue, M., Brauman, A., et al. (2012). Endogeic earthworms shape bacterial functional communities and affect organic matter mineralization in a tropical soil. *ISME Journal, 6*, 213–222.

Bertrand, M., Barot, S., Blouin, M., Whalen, J., de Oliveira, T., & Estrade, J. R. (2015). Earthworm services for cropping systems: A review. *Agronomy for Sustainable Development, 35*(2), 553–567.

Betts, R. A., Boucher, O., Collins, M., Cox, P. M., Falloon, P. D., Gedney, N., Hemming, D. L., Huntingford, C., Jones, C. D., Sexton, D. M., & Webb, M. J. (2007). Projected increase in continental runoff due to plant responses to increasing carbon dioxide. *Nature, 448*(7157), 1037–1041.

Bi, Y., Tian, G., Wang, C., Zhang, Y., Wang, D., Zhang, F., Zhang, L., & Sun, Z. (2018). Differential effects of two earthworm species on fusarium wilt of strawberry. *Applied Soil Ecology, 126*, 174–181.

Bichisa, E., Lachore, S. T., & Uncha, A. (2023). Vulnerability and responses of rural households to climate variability: Agro-climatic-based evidence from southern Ethiopia. *Sustainability and Climate Change, 16*(4), 286–301.

Binet, F., & Trehen, P. (1992). Experimental microcosm study of the role of *Lumbricus terrestris* (Oligochaeta: Lumbricidae) on nitrogen dynamics in cultivated soils. *Soil Biology and Biochemistry, 24*, 1501–1506.

Bispo, A., Cluzeau, D., Creamer, R., Dombos, M, Graefe, U., Krogh, P. H., Sousa, J. P., Peres, G., Rutgers, M., Winding, A., et al. (2009). Indicators for monitoring soil biodiversity. *Integrated Environmental Assessment and Management, 5*, 717–719.

Blanc, E. (2012). The impact of climate change on crop yields in sub-Saharan Africa. *American Journal of Climate Change, 1*, 1–13. https://doi.org/10.4236/ajcc.2012.11001

Blanchart, E., Lavelle, P., Braudeau, E., Le Bissonnais, Y., & Valentin, C. (1997). Regulation of soil structure by geophagous earthworm activities in humid savannas of Côte d'Ivoire. *Soil Biology and Biochemistry, 29*(3-4), 431–439.

Blouin, M., Sery, N., Cluzeau, D., Brun, J.-J., & Bédécarrats, A. (2013). Balkanized research in ecological engineering revealed by a bibliometric analysis of earthworms and ecosystem services. *Environmental Management, 52*, 309–320.

Briones, M. J. (2014). Soil fauna and soil functions: A jigsaw puzzle. *Frontiers of Environmental Science, 2*, 7.

Brown, G. G., Barois, I., & Lavelle, P. (2000). Regulation of soil organic matter dynamics and microbial activity in the drilosphere and the role of interactions with other edaphic functional domains. *European Journal of Soil Biology, 36*, 1–23.

Brown, G. G., & Doube, B. (2004). Functional interactions between earthworms, microorganisms, organic matter, and plants. In *Earthworm ecology*, (2nd ed., pp. 213–240). CRC Press.

Brown, G. G., Edwards, C. A., & Brussaard, L. (2004). How earthworms affect plant growth: Burrowing into the mechanisms. In C. A. Edwards (Ed.), *Earthworm ecology* (pp. 13–49). CRC Press.

Brussaard, L., de Ruiter, P. C., & Brown, G. G. (2007). Soil biodiversity for agricultural sustainability. *Agriculture, Ecosystems and Environment, 121*, 233–244.

Calzadilla, A., Zhu, T., Rehdanz, K., Tol, R. S. J., & Ringler, C. (2013). Economywide impacts of climate change on agriculture in Sub-Saharan Africa. *Ecological Economics, 93*, 150–165. https://doi.org/10.1016/j.ecolecon.2013.05.006

Cameron, E. K., Knysh, K. M., Proctor, H. C., & Bayne, E. M. (2013). Influence of two exotic earthworm species with different foraging strategies on abundance and composition of boreal microarthropods. *Soil Biology and Biochemistry, 57*, 334–340.

Carpenter, D., Hodson, M. E., Eggleton, P., & Kirk, C. (2007). Earthworm induced mineral weathering: Preliminary results. *European Journal of Soil Biology, 43*, S176–S183.

Chakrabarty, D., Das, S. K., Das, K. M., Biswas, P., & Karmegam, N. (2009). Application of vermitechnology in aquaculture. *Dynamic Soil, Dynamic Plant, 3*(2), 41–44

Chan, K. Y., Baker, G. H., Conyers, M. K., Scott, B., & Munro, K. (2004). Complementary ability of three European earthworms (Lumbricidae) to bury lime and increase pasture production in acidic soils of Southeastern Australia. *Applied Soil Ecology, 26*, 257–271.

Chapuis-Lardy, L., Brauman, A., Bernard, L., Pablo, A. L., Toucet, J., Mano, M. J., Weber, L., Brunet, D., Razafimbelo, T., Chotte, J. L., et al. (2010). Effect of the endogeic earthworm Pontoscolex corethrurus on the microbial structure and activity related to CO_2 and N_2O fluxes from a tropical soil (Madagascar). *Applied Soil Ecology, 45*, 201–208.

Chiripasi, S. C., Moreki, J. C., Nsoso, S. J., & Letso, M. (2013). Effect of feeding mopane worm meal on mineral intake, retention and utilization in guinea fowl under intensive system. *International Journal of Poultry Science, 12*(1), 19–28.

Clapperton, M. J., Lee, N. O., Biset, F., & Conner, R. L. (2001). Earthworm indirectly reduce the effect of take all (*Gaunnomyces graminis* var) on soft white spring wheat (*Triticum aestivum* cv Felder). *Soil, Biology and Biochemistry, 33*, 1531–1538.

Coq, S., Barthès, B. G., Oliver, R., Rabary, B., & Blanchart, E. (2007). Earthworm activity affects soil aggregation and organic matter dynamics according to the quality and localization of crop residues—An experimental study (Madagascar). *Soil Biology and Biochemistry, 39*, 2119–2128.

Costello, D. M., & Lamberti, G. A. (2008). Non-native earthworms in riparian soils increase nitrogen flux into adjacent aquatic ecosystems. *Oecologia, 158*, 499–510.

Curry, J. P. (2004). Functional interactions between earthworms, microorganisms, organic matter, and plants. In C. A. Edwards (Ed.), *Earthworm ecology* (2nd ed.). CRC Press.

Deeb, M., Desjardins, T., Podwojewski, P., Pando, A., Blouin, M., & Lerch, T. Z. (2017). Interactive effects of compost, plants, and earthworms on the aggregations of constructed technosols. *Geoderma, 305*, 305–313.

Don, A., Steinberg, B., Schoning, I., Pritsch, K., Joschko, M., & Gleixner, G. (2008). Organic carbon sequestration in earthworm burrows. *Soil Biology and Biochemistry, 40*, 1803–1812.

Doran, J. W., Jones, A. J., & Parkin, T. B. (1996). Quantitative indicators of soil quality: A minimum data set methods for assessing soil quality. *Soil Science Society of America*, 25–37.

Egert, M., Marhan, S., Wagner, B., Scheu, S., & Friedrich, M. W. (2004). Molecular profiling of 16S rRNA genes reveals diet-related differences of microbial communities in soil, gut, and casts of *Lumbricus terrestris* L. (Oligochaeta: Lumbricidae). *FEMS Microbiology Ecology, 48*, 187–197.

Elmer, W. H. (2009). Influence of earthworm activity on soil microbes and soilborne diseases of vegetable. *Plant Diseases, 93*, 175–179.

Falloon, P. D., & Betts, R. A. (2009). Climate impacts on European agriculture and water management in the context of adaptation and mitigation—The importance of an integrated approach. *Science of the Total Environment*. https://doi.org/10.1016/j.scitotenv.2009.05.002

Fischer, C., Roscher, C., Jensen, B., Eisenhauer, N., Baade, J., Attinger, S., Scheu, S., Weisser, W. W., Schumacher, J., & Hildebrandt, A. (2014). How do earthworms, soil texture, and plant composition affect infiltration along an experimental plant diversity gradient in grassland? *PLoS ONE, 9*(6), Article e98987.

Food and Agricultural Organization of the United Nations. (2011). *Strengthening capacity for climate change adaptation in agriculture: Experience and lessons from Lesotho*. Food and Agricultural Organization of the United Nations.

Fragoso, C., Lavelle, P., Blanchart, E., Senapati, B. K., Jimenez, J. J., Martinez, M. D. L. A., & Tondoh, J. (1999). Earthworm communities of tropical agroecosystems: Origin, structure and influence of management practices. *Earthworm management in Tropical Agroecosystems,* (pp. 27–55).

Fujii, K., Ikeda, K., & Yoshida, S. (2012). Isolation and characterization of aerobic microorganisms with cellulolytic activity in the gut of endogeic earthworms. *International Microbiology, 15*, 121–130.

Furlong, M. A., Singleton, D. R., Coleman, D. C., & Whitman, W. B. (2002). Molecular and culture-based analyses of prokaryotic communities from an agricultural soil and the burrows and casts of the earthworm *Lumbricus rubellus. Applied Environment and Microbiology, 68*, 1265–1279.

Fusaro, S., Gavinelli, F., Lazzarini, F., & Paoletti, M. G. (2018). Soil biological quality index based on earthworms (QBS-e). A new way to use earthworms as bioindicators in agro-ecosystems. *Ecological Indicators, 93*, 1276–1292.

Ganeshkumar, T., Premalatha, M., Gajalakshmi, S., & Abbasi, S. A. (2014). A new process for the rapid and direct vermicomposting of the aquatic weed salvinia (*Salvinia molesta*). *Bioresources and Bioprocessing, 1*, 1–5.

Gattinger, A., Muller, A., Haeni, M., Skinner, C., Fliessbach, A., Buchmann, N., Mader, P., Stolze, M., Smith, P., Scialabba, N.-E.-H., & Niggli, U. (2012). Enhanced topsoil carbon stocks under organic farming. *Proceedings of the National Academy of Sciences, 109*(44), 18226–18231.

Gopal, M., Bhute, S. S., Gupta, A., Prabhu, S. R., Thomas, G. V., Whitman, W. B., et al. (2017). Changes in structure and function of bacterial communities during coconut leaf vermicomposting. *Antonie Leeuwenhoek, 110*, 1339–1355.

Hauser, S. (1993). Distribution and activity of earthworms and contribution to nutrient recycling in alley cropping. *Biology and Fertility of Soils, 15*(1), 16–20.

Hendrix, P. F., Callaham, M. A., Drake, J. M., Huang, C. Y., James, S. W., Snyder, B., & Zhang, W. (2008). Pandora's box contained bait: The global problem of introduced earthworms. *Annual Review of Ecology, Evolution and Systematics, 39*, 593–613.

Hong, S. W., Lee, J. S., & Chung, K. S. (2011). Effect of enzyme producing microorganisms on the biomass of epigeic earthworms (*Eisenia fetida*) in Vermicompost. *Bioresources and Technology, 102*, 6344–6347.

Huuml, U. N., Halime, O. Z. U. N., Yaar, K., & Huuml, P. (2011). Influence of organic and conventional production systems on the quality of tomatoes during storage. *African Journal of Agricultural Research, 6*(3), 538–544.

Intergovernmental Panel on Climate Change. (2007). *Climate change 2007: The physical science basis. Summary for Policy Makers*. Intergovernmental Panel on Climate Change Secretariat.

Johnsen, A., Wick, L. Y., & Harms, H. (2005). Principles of microbial PAH-degradation in soil. *Environmental Pollution, 133*, 71–84.

Johnsonmaynard, J., Umiker, K., & Guy, S. (2007). Earthworm dynamics and soil physical properties in the first three years of no-till management. *Soil Tillage Research, 94*, 338–345.

Johnston, A. S., Sibly, R. M., Hodson, M. E., Alvarez, T., & Thorbek, P. (2015). Effects of agricultural management practices on earthworm populations and crop yield: Validation and application of a mechanistic modelling approach. *Journal of Applied Ecology, 52*, 1334–1342.

Jones, C. G., Lawton, J. H., & Shachak, M. (1994). Organisms as ecosystem engineers. *Oikos, 69*, 373–386.

Kauser, H., & Khwairakpam, M. (2022). Organic waste management by two-stage composting process to decrease the time required for vermicomposting. *Environmental Technology and Innovation, 25*(2022), Article 102193. https://doi.org/10.1016/j.eti.2021.102193

Kibblewhite, M. G., Ritz, K., & Swift, M. J. (2008). Soil health in agricultural systems. *Philosophical Transaction of the Royal Society B, 363*, 685–701.

Kuzyakov, Y., & Blagodatskaya, E. (2015). Microbial hotspots and hot moments in soil: Concept & review. *Soil Biology and Biochemistry, 83*, 184–199.

Laossi, K. R., Decaëns, T., Jouquet, P., & Barot, S. (2010). Can we predict how earthworm effects on plant growth vary with soil properties? *Applied Environmental Soil Science, 6*, 1–6.

Lavelle, P., Bignell, D., Lepage, M., Wolters, V., Roger, P., Ineson, P., et al. (1997). Soil function in a changing world: The role of invertebrate ecosystem engineers. *European Journal of Soil Biology, 33*, 159–193.

Lavelle, P., Pashanasi, B., Charpentier, F. C. G., Rossi, J. P., Derouard, L., André, J., et al. (1998). Large-scale effect of earthworms on soil organic matter and nutrient dynamics. In C. A. Edwards (Ed.), *Earthworm ecology* (pp. 103–122). St. Lucie Press.

Le Bayon, R. C., Moreau, S., Gascuel-Odoux, C., & Binet, F. (2002). Annual variations in earthworm surface-casting activity and soil transport by water runoff under a temperate maize agro-ecosystem. *Geoderma, 106*, 121–135.

Lester, G. E., Manthey, J. A., & Buslig, B. S. (2007). Organic vs. conventionally grown Rio Red whole grapefruit and juice: comparison of production inputs, market quality, consumer acceptance, and human health-bioactive compounds. *Journal of Agricultural and Food Chemistry, 55*(11), 4474–4480.

Lim, S. L., Wu, T. Y., Lim, P. N., & Shak, K. P. Y. (2015). The use of vermicompost in organic farming: Overview, effects on soil and economics. *Journal of the Science of Food and Agriculture, 95*, 1143–1156.

Lombriculture Au Sud Vietnam. *Biotechnology, Agronomy and Society and Environment, 7*(3–4), 171–175.

Long, S. P., Ainsworth, E. A., Leakey, A. D. B., Nösberger, J., & Ort, D. R. (2006). Food for thought: Lower-than-expected crop yield stimulation with rising CO_2 concentrations. *Science, 312*(5782), 1918–1921.

Lori, M., Symnaczik, S., M¨ader, P., De Deyn, G., Gattinger, A., & Lehman, R. M. (2017). Organic farming enhances soil microbial abundance and activity—A meta-analysis and meta-regression. *PLoS ONE 12*(7), e0180442.

Luttikholt, L. W. M. (2007). The international federation of organic agriculture movements formulated principles of organic agriculture. *NJAS-Wageningen Journal of Life Sciences, 54*(4), 347–360.

Maboeta, M. S., & Van Rensburg, L. (2003). Vermicomposting of industrially produced woodchips and sewage sludge utilizing *Eisenia fetida. Ecotoxicology and Environmental Safety, 56*(2), 265–270.

Mainoo, N. K. (2007). *Feasibility of low-cost vermicompost production in Accra, Ghana.* Thesis submitted to McGill University.

Maturi, K. C., Haq, I., & Kalamdhad, A. S. (2022). Composting techniques: Utilization of organic wastes in urban areas of Indian cities. In *Sciencedirect* (43–55p).

Medina-Sauza, R. M., Álvarez-Jiménez, M., Delhal, A., Reverchon, F., Blouin, M., Guerrero-Analco, J. A., Cerdán, C. R., Luc Villain, G., & Barois, I. (2019). Earthworms building up soil microbiota, A review. *Frontiers in Environmental Science, 7*(81), 1–20.

Milcu, A., Schumacher, J., & Scheu, S. (2006). Earthworms (*Lumbricus terrestris*) affect plant seedling recruitment and microhabitat heterogeneity. *Functional Ecology, 20*, 261–268.

Monard, C., Vandenkoornhuyse, P., Le Bot, B., & Binet, F. (2011). Relationship between bacterial diversity and function under biotic control: The soil pesticide degraders as a case study. *The ISME Journal, 5*, 1048–1056.

Muller, A., Schader, C., Scialabba, N.-E.-H., Brüggemann, J., Isensee, A., Erb, K.-H., Smith, P., Klocke, P., Leiber, F., Stolze, M., & Niggli, U. (2017). Strategies for feeding the world more sustainably with organic agriculture. *Nature Communications, 8*, 1290.

Navarro-González, I., García-Alonso, J., & Periago, M. J. (2018). Bioactive compounds of tomato: Cancer chemopreventive effects and influence on the transcriptome in hepatocytes. *Journal of Functional Foods, 42*, 271–280.

Ndegwa, P. M., & Thompson, S. A. (2001). Integrating composting and vermicomposting in the treatment and bioconversion of biosolids. *Bioresource Technology, 76*, 107–112.

Nechitaylo, T. Y., Yakimov, M. M., Godinho, M., Timmis, K. N., Belogolova, E., & Byzov, B. A. (2010). Effect of the earthworms *Lumbricus terrestris* and *Aporrectodea caliginosa* on bacterial diversity in soil. *Microbial Ecology, 59*, 574–587.

Oades, J. M. (1993). The role of biology in the formation, stabilization, and degradation of soil structure. *Geoderma, 56*(1–4), 377–400.

Ojuolape, O. T., Aransiola, S. A., & Obideyi, O. A. (2015) Vermitechnology: A solution to environmental problems; A review. *Journal of Global Ecology and Environment, 2454–2644, 3*(3), 127–135. http://www.ikpress.org/abstract/4516

Onwosi, C. O., Ozoegwu, C. G., Nwogu, T. N., Nwobodo, T. N., Eke, I. E., Igbokwe, V. C., Ugwuolu, E. T., & Ugwuodo, C. J. (2022). Cattle manure as a sustainable bioenergy source: Prospects and environmental impacts of its utilization as a major feedstock in Nigeria. *Bioresource Technology Reports, 19*(101131), 270–277.

Paavola, J., & Adger, W. N. (2006). Fair adaptation to climate change. *Ecological Economics, 56*, 594–609.

Paoletti, M. G. (1999). The role of earthworms for assessment of sustainability and as bioindicators. *Agriculture Ecosystem and Environment, 74*, 137–155.

Peigné, J., Cannavaciuolo, M., Gautronneau, Y., Aveline, A., Giteau, J. L., & Cluzeau, D. (2009). Earthworm populations under different tillage systems in organic farming. *Soil and Tillage Research, 104*, 207–214.

Pelosi, C., Barot, S., Capowiez, Y., et al. (2014). Pesticides and earthworms. A review. *Agronomy for Sustainable Development, 34*, 199–228. https://doi.org/10.1007/s13593-013-0151-z

Pelosi, C., Bertrand, M., Capowiez, Y., Boizard, H., & Roger-Estrade, J. (2009). Earthworm collection from agricultural fields: Comparisons of selected expellants in presence/absence of hand-sorting. *European Journal of Soil Biology, 45*, 176–183.

Prasad, K. (2021). Advantages and nutritional importance of organic agriculture produces food on human, soil, and environmental health in modern lifestyle for sustainable development. *Aditum Journal of Clinical and Biomedical Research, 5*(2), 1–7.

Puga-Freitas, R., Barot, S., Taconnat, L., Renou, J.-P., & Blouin, M. (2012). Signal molecules mediate the impact of the earthworm *Aporrectodea caliginosa* on growth, development and defence of the plant *Arabidopsis thaliana*. *PLoS ONE, 7*, Article e49504.

Puga-Freitas, R., & Blouin, M. (2015). A review of the effects of soil organisms on plant hormone signalling pathways. *Environmental and Experimental Botany, 114*, 104–116.

Rawling, M. D., Merrifield, D. L., Snellgrove, D. L., Kühlwein, H., Adams, A., & Davies, S. J. (2012). Haemato-immunological and growth response of mirror carp (*Cyprinus carpio*) fed a tropical earthworm meal in experimental diets. *Fish & Shellfish Immunology, 32*(6), 1002–1007.

Reganold, J. P., Andrews, P. K., Reeve, J. R., Carpenter-Boggs, L., Schadt, C. W., Alldredge, J. R., & Zhou, J. (2010). Fruit and soil quality of organic and conventional strawberry agro-ecosystems. *PLoS ONE, 5*(9), e12346.

Reganold, J. P., & Wachter, J. M. (2016). Organic agriculture in the twenty-first century. *Nature Plant, 2*, 1–8.

Reilly, J., Tubiello, F., McCarl, B., Abler, D., Darwin, R., Fuglie, K., Hollinger, S., Izaurralde, C., Jagtap, S., Jones, J., Learns, L., Ojima, D., Paul, E., Paustian, K., Riha, S., Rosenberg, N., & Rosenzweig, C. (2003). U.S. agriculture and climate change: New results. *Climatic Change, 57*, 43–69.

Ringler, C. (2008). *The impact of climate variability and climate change on water and food outcomes: A framework for the analysis*. IFPRI Research Briefs 15-I. International Food Policy Research Institute.

Ritsema, C., & Dekker, L. (2000). Preferential flow in water repellent sandy soils: Principles and modeling implications. *Journal of Hydrology, 231–232*, 308–319.

Römbke, J., Jansch, S., & Didden, W. (2005). The use of earthworms in ecological soil classification and assessment concepts. *Ecotoxicology and Environmental Safety, 62*, 249–265.

Ros, M. B., Hiemstra, T., Van Groenigen, J. W., Chareesri, A., & Koopmans, G. F. (2017). Exploring the pathways of earthworm-induced phosphorus availability. *Geoderma, 303*, 99–109.

Rossi, J. P. (2003). Clusters in earthworm spatial distribution: The 7th international symposium on earthworm ecology. Cardiff· W ales· 2002. *Pedobiologia, 47*(5–6), 490–496.

Rüdisser, J., Tasser, E., Peham, T., Meyer, E., & Tappeiner, U. (2021). Hidden engineers and service providers: Earthworms in agricultural land-use types of south Tyrol, Italy. *Sustainability, 13*, 312.

Rutgers, M., Van Wijnen, H. J., Schouten, A. J., Mulder, C., Kuiten, A. M. P., Brussaard, L., & Breure, A. M. (2012). A method to assess ecosystem services developed from soil attributes with stakeholders and data of four arable farms. *Science of the Total Environment, 415*, 39–48.

Scheu, S., Schlitt, N., Tiunov, A. V., Newington, J. E., & Jones, H. T. (2002). Effects of the presence and community composition of earthworms on microbial community functioning. *Oecologia, 133*, 254–260.

Schmidt, O., & Curry, J. P. (1999). Effects of earthworms on biomass production, nitrogen allocation and nitrogen transfer in wheat-clover intercropping systems. *Plant and Soil, 214*(2), 187–198.

Serdeczny, O., Adams, S., Baarsch, F., Coumou, D., Robinson, A., Hare, W., Schaeffer, M., Perrette, M., & Reinhardt, J. (2016). Climate change impacts in sub-Saharan Africa: From physical changes to their social repercussions. *Regional Environmental Change, 17*, 1585–1600.

Seufert, V., Ramankutty, N., & Mayerhofer, T. (2017). What is this thing called organic? How organic farming is codified in regulations? *Food Policy, 68*, 10–20.

Shan, J., Liu, J., Wang, Y., Yan, X., Guo, H., Li, X., & Ji, R. (2013). Digestion and residue stabilization of bacterial and fungal cells, protein, peptidoglycan, and chitin by the geophagous earthworm metaphire Guillermo. *Soil Biology and Biochemistry, 64*, 9–17.

Shankar, K. S., Sumathi, S., & Shankar, M. (2008). Minerals and microbiological quality of organically and conventionally grown vegetables. *Indian Journal of Dryland Agricultural Research and Development, 23*(1), 87–95.

Sharma, J., & Agarwal, S. (2014). Vermihorticulture: A horticulturally viable and environmentally sustainable technology to chemical farming. *Cibtech Journal of Bio-Protocols, 3*(2), 7–20.

Sheahan, M., & Barrett, C. B. (2017). Ten striking facts about agricultural input use in Sub-Saharan Africa. *Food Policy, 67*, 12–25.

Singh, S., Singh, J., Kandoria, A., Quadar, J., Bhat, S. A., Chowdhary, A. B., & Vig, A. P. (2020). Bioconversion of different organic waste into fortified vermicompost with the help of earthworm: A comprehensive review. *International Journal of Recycling of Organic Waste in Agriculture, 9*, 423–439.
Sinha, R. K., Agarwal, S., Chauhan, K., Chandran, V., & Soni, B. K. (2010). Vermiculture technology: Reviving the dreams of Sir Charles Darwin for scientific use of earthworms in sustainable development programs. *Technology and Investment, 1*(3), 155–172.
Smith, R. G., McSwiney, C. P., Grandy, A. S., Suwanwaree, P., Snider, R. M., & Robertson, G. P. (2008). Diversity and abundance of earthworms across an agricultural land-use intensity gradient. *Soil Tillage Research, 100*, 83–88.
Sogbesan, A. O., & Ugwumba, A. A. A. (2008). Nutritional evaluation of termite (*Macrotermes subhyalinus*) meal as animal protein supplements in the diets of *Heterobranchus longifilis* (Valenciennes, 1840) fingerlings. *Turkish Journal of Fisheries and Aquatic Sciences, 8*(1), 149–158.
Sogbesan, A. O., Ugwumba, A. A. A., & Madu, C. T. (2007). Productivity potentials and nutritional values of semi-arid zone earthworm (Hyperiodrilus euryaulos; Clausen, 1967) cultured in organic wastes as fish meal supplement. *Pakistan Journal of Biological Sciences, 10*, 2992–2997.
Spurgeon, D., Keith, A., Schmidt, O., Lammertsma, D., & Faber, J. (2013). Land-use and land-management change: Relationships with earthworm and fungi communities and soil structural properties. *BMC Ecology, 13*, 46.
Staff, T. (2020). *Different types of earthworms with pictures & facts*. Trees.Com. https://www.trees.com/gardening-and-landscaping/types-of-earthworms.
Stephens, P. M., & Davoren, C. W. (1997). Influence of earthworm *Aporrectodea trapezoids, A. rosea* on disease severity of rhizoctonia Solani on Subterranean clover and rye grass. *Soil Biology and Biochemistry, 29*, 511–516.
Stromberger, M. E., Keith, A. M., & Schmidt, O. (2012). Distinc Microbial and faunal communities and translocated carbon in *Lumbricus terrestris* Drilospheres. *Soil Biology and Biochemistry, 46*, 153–162.
Taheri, S., Pelosi, C., & Dupont, L. (2018). Harmful or useful? A case study of the exotic peregrine earthworm morphospecies *Pontoscolex corethrurus*. *Soil Biology & Biochemistry, 116*, 277–289. https://doi.org/10.1016/j.soilbio.2017.10.030
Taherzadeh, M., Bolton, K., Wong, J., & Pandey, A. (Eds.). (2019). *Sustainable resource recovery and zero waste approaches*. Elsevier.
Thakuria, D., Schmidt, O., Finan, D., Egan, D., & Doohan, F. M. (2010). Gut wall bacteria of Earthworms: A natural selection process. *ISME Journal, 4*, 357–366.
Tiunov, A., Bonkowski, M., Alphei, J., & Scheu, S. (2001). Microflora, protozoa and nematoda in *Lumbricus terrestris* burrow walls: A laboratory experiment. *Pedobiologia, 45*, 46–60.
Turbé, A., De Toni, A., Benito, P., Lavelle, P., Ruiz, N., Van der Putten, W. H., Labouze, E., & Mudgal, S. (2010). *Soil biodiversity: Functions, threats, and tools for policy makers*. Report for European Commission; DG Environment.
Van Groenigen, J. W., Lubbers, I. M., Vos, H. M. J., Brown, G. G., de Deyn, G. B., & Van Groenigen, K. J. (2014). Earthworms increase plant production: A meta-analysis. *Scientific Reports, 4*, 6365.
Van Schaik, L., Palm, J., Klaus, J., Zehe, E., & Schröder, B. (2014). Linking spatial earthworm distribution to macropore numbers and hydrological effectiveness. *Ecohydrology, 7*, 401–408.
Vieira, M. L., Ferreira, A. S., & Donzelle, J. L. (2004). Digestibilidade da farinha de minhoca para suínos. *Boletim de Indústria Animal, 61*(1), 83–89.
Watson, R. T., Zinyowera, M. C., & Moss, R. H. (Eds.). (1998). *The regional impacts of climate change: An assessment of vulnerability*. Cambridge University Press.
World Bank. (2007). *World development report 2008: Agriculture for development*. World Bank.
Wurst, S., Dugassa-Gobena, D., Langel, R., Bonkowski, M., & Scheu, S. (2004). Combined effects of earthworms and vesicular-arbuscular mycorrhizas on plant and aphid performance. *New Phytology, 163*, 169–176.

Yaser, A. Z., & Lamaming, J. (2022). Composting and anaerobic digestion of food waste and sewage sludge for campus sustainability: A review. *International Journal of Chemical Engineering.* https://doi.org/10.1155/2022/6455889

Zhou, Z. (2023). Climate change impact and risks: Insights for tourism development in Victorial Falls, Zimbabwe. *Social Sciences, 3*(9), 23–49.

Vermiwash: A Vermicompost By-Product for Sustainable Agriculture

Pawan Kumar Rose, Sivaraman Balaji, Surojit Bera, Sandip Mondal, Manjeet Bansal, Mithilesh Kumar Jha, and Sagnik Chakraborty

Abstract The extensive dependence on agrochemicals in conventional agricultural practices in today's scenario needs a comprehensive evaluation of organic approaches to reduce the adverse impacts of their application. The implementation of vermiwash is such an organic approach with various potential. The earthworm-rich media used to create vermicompost also yields the brown-colored, odorless liquid extract known as "vermiwash." It has a high concentration of bacteria, mucus, vitamins, various bioavailable minerals, hormones, enzymes, and antimicrobial peptides. The present chapter elucidates several techniques for manufacturing vermiwash and explores its

---

The original version of the chapter has been revised. A correction to this chapter can be found at https://doi.org/10.1007/978-3-031-96212-7_19

P. K. Rose
Department of Energy and Environmental Sciences, Chaudhary Devi Lal University, Sirsa, Haryana, India

S. Balaji
Division of Zoonotic Diseases Programme, National Centre for Disease Control, New Delhi, India

S. Bera
Department of Microbiology, School of Bioengineering and Biosciences, Lovely Professional University, Phagwara, Punjab, India

S. Mondal
School of Pharmaceutical Technology, School of Medical Sciences, ADAMAS University, Kolkata, West Bengal, India

M. Bansal
Department of Civil Engineering, Maharaja Ranjit Singh Punjab Technical University, Bathinda, Punjab, India

M. K. Jha
Department of Chemical Engineering and Centre for Energy and Environment, Dr. B. R. Ambedkar National Institute of Technology, Jalandhar, Punjab, India

S. Chakraborty (✉)
Indian Council of Medical Research Division of Epidemiology and Communicable Diseases, New Delhi, India
e-mail: sagnikbio@gmail.com

S. A. Aransiola and N. R. Maddela (eds.), *Vermitechnology: Economic, Environmental and Agricultural Sustainability*, https://doi.org/10.1007/978-3-031-96212-7_17

potential as an organic fertilizer for sustainable agriculture. The impact of vermiwash treatment on agricultural and vegetable crops is also emphasized. The chapter concluded with the potential of vermiwash in enhancing soil fertility.

Keywords Vermiwash · Vermicompost · Organic fertilizers · Sustainable agriculture · Soil fertility

1 Introduction

In 1881, renowned scientific visionary Sir Charles Darwin brought widespread attention to earthworms' vital role in decomposing organic matter in his book "The formation of vegetable mould through the action of worms, with observations on their habits." Professor Otto von Graff in Germany performed the first fundamental study on earthworms' potential for recycling organic wastes into organic fertilizer. In the United States, Hartenstein and Mitchell of the State University of New York initially employed Earthworms and vermiculture technologies to dispose of municipal sewage sludge (Singh & Sinha, 2022). The application of earthworms includes a food source, biocontrol agents against pests and disease, and degradation products as biofertilizers. Vermiculture refers to the agricultural practice of cultivating earthworms, while vermicomposting involves the process of organic material decomposition facilitated by earthworms. The vermicomposting yields two products based on the decomposition technique: vermicompost and vermiwash. Vermicompost is a solid product of vermicomposting, whereas vermiwash is a liquid filtered from the water wash of earthworms (Aransiola et al., 2022; Rose et al., 2024). Vermiwash, a product of vermiculture, is essentially a liquid wash passed through the bodies of earthworms. Earthworm mucus and other excretory materials, including micronutrients and organic compounds from the soil, are commonly found in vermiwash (Akazawa et al., 2023; Domínguez et al., 2017; Sulaiman & Mohamad, 2020). In accordance with the findings of Clause et al. (2014), it has been determined that the quality of vermicast is influenced to a greater extent by the type of soil (62%) as compared to the species of earthworms (10%). Nath and Singh (2016) observed a similar finding regarding the nutritional quality of vermiwash in response to various combination of feed material. Vermiwash is rich in vitamins, amino acids, minerals (such as potassium, calcium, zinc, copper, nitrogen, iron, and magnesium), beneficial microorganisms, and growth hormones (such as cytokinin and auxins) (Aghamohammadi et al., 2016; Gudeta et al., 2021). Vermiwash has been extensively studied as a liquid fertilizer and a spray due to its high nutritional quality. Vermiwash application in the soil increases its carbon content, cation exchange capacity, nutrient content, bulk density, and water-holding capacity (Akazawa et al., 2023; Mishra et al., 2014). Therefore, vermiwash is an organic product that supports environmental conservation. One advantage of this substance is its homogeneity, which allows it to be applied uniformly to both soil and plants through spraying. Moreover, vermiwash provides various vital nutrients to plants without causing leaching and works well as a fertilizer, whether used alone or in conjunction with other types of fertilizers, both organic and inorganic. The use of vermiwash has been associated with enhanced seed germination, plant growth, yield, and nutritional composition of agricultural products (Joshi et al., 2023; Patnaik et al., 2022). Vermiwash, being an organic fertilizer, might potentially serve as a viable solution to the problems associated with chemical fertilizers. The prospective substitution of chemical inputs in crop production can reduce economic expenditures and possibly facilitate the development of organic goods, which may possess higher market value. Organic products, or crops raised with natural fertilizers, are in high demand because of rising concerns about food safety and environmental impact (Rose et al., 2022a; Ramnarain et al., 2019). Hence, the current chapter offers an in-depth examination of the techniques used in producing vermiwash and its potential as a liquid organic fertilizer.

2 Vermiwash Preparation Methods

Vermicompost is the solid product of earthworms, attributed to the biodegradation of organic raw materials such as leaf litter, cow dung, or other organic materials. However, vermiwash is a liquid product that results from the vermicomposting process. Vermiwash can be produced in both batch and continuous modes at various scales, including large and small. The continuous method involves the consistent provision of vermiwash after introducing worms to a continuous supply of raw materials, whereas the batch mode requires periodic inoculation of worms (Gudeta et al., 2021; Thirunavukkarasu et al., 2023).

2.1 Method-I

The conventional method of vermiwash production is carried out in a large plastic or brick container with a hole in the bottom (Fig. 1). This was developed by the Ecoscience Research Foundation. The container is often referred to as vermipit or vermireactor, comprised of diverse configurations of various materials organized in layers, extending from the bottom to the top, as shown in Fig. 1 (setup for 250 L container). The bottom layer consists of barrels, gravels, or broken small pieces of brick up to a height of about 25 cm, followed by a second layer of coarse sand of the exact measurement. Some studies recommend the use of a loam soil layer above the coarse sand (Chattopadhyay, 2015). The first layer acts as a filter and absorbs excess water from the vermipit added from the top. The second layer is filled with sandy soil to prevent the accumulation of extra water in the medium. The third layer consists of vermicompost material, which comprises moistened pre-decomposed organic waste or 10-day-old cow dung and a dense number of adult or young earthworms, along with organic soil (a loamy soil layer of 30–45 cm thickness). An equal number of epigeic (surface) and anecic (sub-surface) earthworm species are commonly used

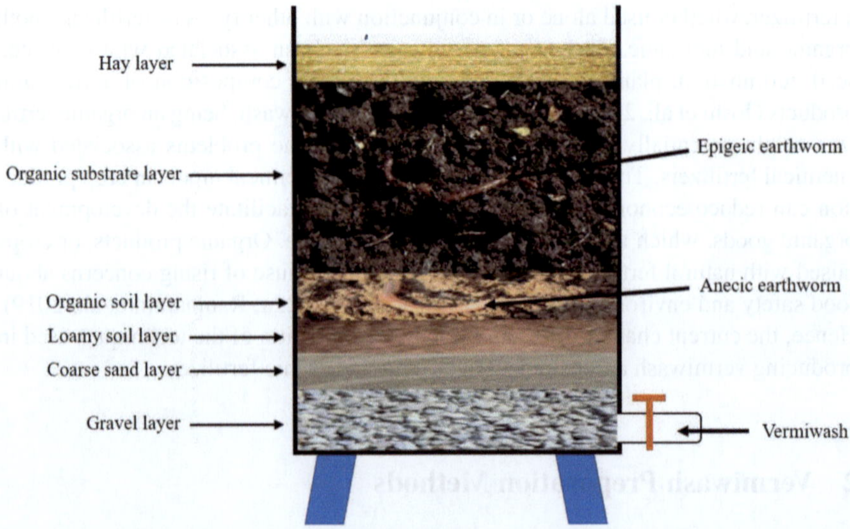

Fig. 1 Conventional procedure for the preparation of vermiwash

in the conventional method. *Lampito mauritii* (anecic indigenous), *Eisenia fetida* (epigeic, red wigglers), and *Eudrilus eugeniae* (exotic, African nightcrawler) are commonly employed earthworm species to produce vermiwash (Tharmaraj et al., 2011; Thirunavukkarasu et al., 2023). A covering layer consisting of cow dung pats and hay shields the vermipit from direct sunlight and ensures adequate moisture retention within the medium. Some studies recommend the application of coconut fronds and jute cloth as the top layer of vermipit (Tharmaraj et al., 2011). Daily additions of fresh water ensure that the vermicompost stays damp.

Moreover, the pit contents must be turned once every two or three days to improve aeration during vermicomposting. After 60 days of composting with earthworms, the resulting vermiwash can be harvested. Nevertheless, the duration of this interval may differ among various studies. After 30 days of incubation, Gopal et al. (2010) harvested vermiwash produced by *Eudrilus* sp. on a substrate of partially decomposed coconut leaf litter and cow dung (10:1 w/w basis). The vermiwash filtering and accumulation compartment consists of a 10 cm bottom layer of smooth pebbles, 10 cm middle layer of clean, coarse gravel, and a final 10 cm top layer of clean beach sand. A substrate weighing 100 kg was hydrated by adding water until it reached a consistent moisture content of 40%. Nayak et al. (2019) extracted the vermiwash in a 15–20 L plastic container with an initial foundation layer of medium-sized bricks or stones 10–15 cm high, followed by a coarse (15 cm) and fine sand (12.5 cm) layer. Subsequently, indigenous earthworms of the species *Eisenia foetida* are introduced into the experimental setup along with a mixture of fertile soil, partially digested cow dung (20–25 cm), and organic waste (40–45 cm). A daily supply of 2 L of fresh water is provided. The vermiwash preparation process starts in the unit after 16–20 days, with a daily output rate of around 1–2 L.

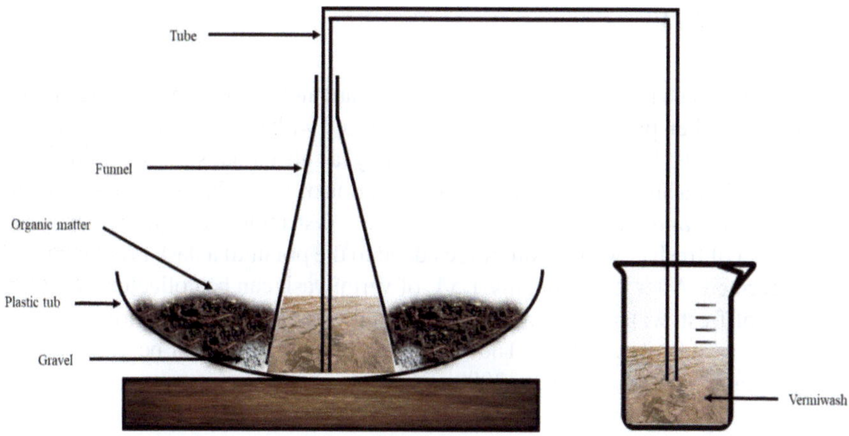

Fig. 2 A household system to produce vermiwash

2.2 Method-II

The Sri Paramakalyani Centre for Environmental Studies in Alwarkurichi, India, developed a household system to produce vermiwash. The apparatus mainly comprises a 10–15 L capacity plastic tub with a wide base, a plastic perforated funnel, and a hand pump (siphon) (Fig. 2). The funnel is placed in the middle of the plastic tub with its wide-open end covered with a nylon net to only allow passage of liquid extract. The percolation of extra water is regulated by placing a layer of stones and sand around the perforated plastic container. About 3 kg of vermicompost is distributed outside the plastic funnel in the tub and moistened throughly with three liters of water. After 24 h, the vermiwash in the funnel is extracted using a tube and syphon mechanism, which yields a honey-brown liquid containing heterotrophic bacteria, fungi, and actinomycetes, including nitrogen fixers and phosphate solubilizers (Lourduraj & Yadav, 2005).

2.3 Method-III

This method utilizes a 1000 L plastic barrel and is suitable for producing a large volume of vermiwash, i.e. 20 L/day. Gravel serves as the foundation, and on top of that, a series of layers are placed, including biogas slurry, sand, red soil, compost derived from partially decomposed biomass, and, ultimately, fresh green leaves as the uppermost layer. Adding more compost, worms, and trash helps to regulate the vermipit's temperature. After three months, 20 L of fresh water is added to the vermireactor to produce 20 L of vermiwash, which can be diluted in a 1:3 or 1:5 ratio before application.

2.4 Method-IV

Making vermiwash in this manner is straightforward and inexpensive. The technique includes an earthen pot with a 10 kg capacity packed with stone fragments to a depth of 10 cm, followed by a plastic net. Subsequently, a substantial stratum of coir fiber, accompanied by a humus encompassing a population of 1500–2000 worms belonging to *Eudrillus euginae* or *Eisenia foetida*, is evenly distributed. Daily, kitchen waste and a splash of fresh cow dung slurry are added to the pot until a dark brown or black mass is formed. After 24 h, roughly 1.5 L of vermiwash can be collected by adding two liters of fresh water to the pot. This procedure can be repeated until the brown hue of the wash begins to fade. The solid compost in the pot can be gathered and used as manure (Alexander et al., 2009).

2.5 Method-V

In this method, earthworms undergo heat stress to produce vermiwash. The methodology used in this approach is based on the research conducted by Karuna et al. (1999). Briefly, about 30 g of adult earthworms (*Eisenia foetida*) are introduced into a glass beaker containing 500 ml of warm (40 °C) distilled water and stirred for 5–6 min. After that, they are promptly taken out and transferred to another plastic beaker containing sterilized water at room temperature. Any leftover excretory and secretory materials that have clung to the worms' bodies are collected after thorough rinsing. The pale straw-yellow contents of the glass and plastic beakers are combined, labelled as vermiwash, and kept in a sterile dark-color container at 4 °C. Kale (1998) prepared vermiwash from *Eisenia euginiae* without mixing the casts, and this study included 1 kg of adult earthworms (about 1000 worms), which were released into 500 ml of lukewarm distilled water (37–40 °C) and stirred for 2 min. The worms are extracted, subjected to a second rinsing process using 500 ml of water at room temperature (30 °C), and reintroduced into the container. The earthworm's bodies produce sufficient mucus and body fluids when subjected to agitation in lukewarm water, referred to as true vermiwash. Pattnaik et al. (2015) utilized *Eisenia foetida* weighing 7–9 g to create vermiwash by soaking them for 5 min in 100 ml of lukewarm distilled water. After removing the earthworms, the filtrate can be used as 100% vermiwash.

2.6 Method-VI

This method produces vermiwash by subjecting the earthworm to cold stress. According to Parmanik (2010), 25 fully mature worms of *Eisenia foetida,* roughly equal in length are subjected to cold stress by introducing them into a beaker filled with ice cubes that maintain a temperature of about −5 °C for 3–4 min. Subsequently, the worms are transferred to a glass beaker with a capacity of 500 ml, containing cold distilled water for 7–8 min, and subjected to stirring occasionally. Afterwards, the worms are moved to a sterile beaker containing distilled water at room temperature. The light yellow exudates from both containers are combined, designated as vermiwash, and stored in a dark-colored, sterilized glass container at 4 °C (Chattopadhyay, 2015).

2.7 Method-VII

The vermiwash is prepared by collecting water that passes through a column of worm action. This method includes two bags concealed in a vertical column. The outside cover is a storage bag for food items, while the inside cover is a sheet of 100 × 50 cm black polythene. The black polythene is stitched into a funnel shape and fitted with the plastic funnel to facilitate drainage through a muslin cloth. The column contains 15 cm of subsequent layers of coarse sand, garden soil, and rotten powdered cow dung. The culture material (plant: cow dung (1:1)) is piled on top of the layer until it reaches a height of 25 cm. The excess water is removed after a short soak. Every day, the unit is moistened (80% moisture). After 15 days, drainage is collected from the top of the column by spraying half a liter of water (Mayooran & Mikunthan, 2012).

3 Characteristic of Vermiwash

In general, vermiwash appears like a honey-brown colored worm coelomic liquid extract comprising various enzymes, plant growth regulators such as indole-3-acetic acid (IAA), cytokinin, gibberellic acid A3 (GA3), and vitamins. Earthworm excretory materials and other organic waste components are essential for plant growth and development (Jayabhaye & Bhalerao, 2015). In addition to the nonliving compounds, it has a unique microbiota essential for litter decomposing and mineralizing several nitrogenous and phosphorous components available in the soil by producing enzymes. It has been reported that vermiwash is abundant in genera such as *Azotobactor*, *Agrobacterium*, *Rhizobium*, phosphate solubilizing (PBS) (*Pseudomonas*, *Bacillus*, *Micrococcus*, *Aspergillus*), and urease-producing microbes (Gammaproteobacteria except for *Cupriaviduss* species) (Chandukishore et al., 2023; Gudeta et al., 2021; Zambare et al., 2008). The various characteristics of the components of vermicompost are given in the following paragraph.

3.1 Facilitate Plant Growth and Productivity

Several studies have shown that vermiwash is a wonderful plant tonic, improving growth and development by supplementing the required macro- and micro-nutrients. The abundance of these nutrients hinders the survival of pathogenic bacteria. In contrast, decomposer bacteria can thrive by directly ingesting them through a saprophytic feeding mode (Das et al., 2014). Research has shown a higher concentration of nitrogen, phosphate, potassium, zinc, iron, manganese, copper, etc., in vermiwash compared to vermicompost made from cow manure (Gudeta et al., 2021). It also contains hormones like IAA and GA3, as well as enzymes like phosphatase, amylase, and cellulase, all of which are essential for plant growth and development. Various studies have shown that vermiwash improves the properties of plants, including root length, shoot length, biomass, leaf number, and fruit seed, and flower production (Nayak et al., 2019). Further, a minimal amount of vermiwash is adequate for seed germination and growth. In addition to directly applying to the soil, foliar spray is also effective in improving plant growth and development (Fathima & Sekar, 2014). Researchers also found that combining vermiwash with other agents, such as vermicompost, enhances plant growth compared to the individual applications of each.

3.2 Antimicrobial Properties

Various metabolites are present in vermiwash, which can enhance plant growth and development by reducing the load of pathogenic microorganisms and improving the nutritional value (Gudeta et al., 2021). One of the important secretory components is an antimicrobial peptide from earthworm's mucus and skin, including lysozyme, fetidins, eseniapore, bacteriostatins, lysenin, and coelomic cytolytic factor. These active components act specifically against pathogens through phagocytosis, encapsulation, agglutination, opsonisation, clotting, and lysis properties (Homa, 2018). For instance, the antimicrobial property of the earthworms' secretion enhanced plant growth and development by inhibiting pathogenic fungi such as *Fusarium graminearum*. In addition, the coelomic fluid in earthworms effectively controls protozoans. Interestingly, plants treated with showed better seed germination, shoot, and root length than those treated with GA3 (Kobayashi et al., 2004). These reports indicate that earthworm secretory material is a significant part of the vermiwash quality.

3.3 Suppressing Pathogenic Microorganisms

Research has shown that vermiwash contains a high concentration of decomposer bacteria, which can effectively diminish harmful organisms such as bacteria and fungi. A particular type of metabolite produced by the *Pseudomonas* species found in vermiwash can suppress the growth of fungal disease that affects the productivity of certain commercially essential plants (Gudeta et al., 2021; Kalantari et al., 2018). The vermiwash derived from cow dung and vegetable wastes using *Eisenia foetida* can effectively control mildew disease (Balam, 2000; Das et al., 2014). The beneficial microorganisms in vermiwash commonly serve as antagonistic agents competing with pathogenic organisms in the soil for space and nutrients. Hence, vermiwash significantly enhances soil health by eliminating harmful bacteria and creating a supportive environment for better plant growth and development (Ojuolape et al., 2015).

3.4 Biopesticides Properties

The use of vermiwash as a biopesticide has shown its efficacy as a significant proportion of the treated plants exhibited a notable absence of foliar damage caused by leaf-eating organisms. Vermiwash enhanced the yield of lablab beans and showed strong growth and more resistance to plant insects. Researchers reported that vermiwash is an ideal agent for productivity and management of *Lucinodes orbanalis* infection on *Solanum melongena* (brinjal crop) and *Leptocorisa varicornis* on *Oryza sativa* (Mishra et al., 2015). A moderate concentration of vermiwash effectively controls the *Leptocorisa* species that infects *Solanum lycopersicum,* i.e. tomato (Sayyad, 2017). In addition, the vermiwash produced from animal waste blended with gram bran and neem oil was an effective insecticide for managing the pod borer (*Helicoverpa armigera*) (Nath & Singh, 2015).

4 Vermiwash as an Organic Fertilizer

The first green revolution in India greatly increased crop productivity; however, the extensive use of chemically synthesized fertilizers over time has compromised soil robustness, decreased agricultural output, increased insect pest and disease rates, and led to environmental pollution. The persistent use of a wide range of agriculturally significant chemicals, such as fertilizers, plant growth enhancers, and pesticides, has deleterious consequences on ecosystems due to the contamination of soil, water, the food chain, and the genetic diversity of plants (Mukhi et al., 2022; Rose et al., 2022b; Ram et al., 2022). The second green revolution started with organic farming, now attracting a booming eco-friendly fertilizer market (Nayak et al., 2019). The application of vermiwash as an organic fertilizer is discussed in the following subsections.

4.1 In Agricultural Crop

According to Mishra et al. (2013), the pest *Helicoverpa armigera*, which threatens gram crops (*Cicer aritenium*), may be efficiently managed using vermiwash. This approach has been shown to significantly enhance the production of gram crops. Another study found that using vermiwash increased the grain by 11.21% and stover yield by 10.28% compared to the control for corn (*Zea mays* L.) (More et al., 2013). Chattopadhyay (2015) investigated the effect of vermiwash collected from the vermicompost unit infested with *Eisenia foetida* on seed germination of green mung (*Vigna radiata*). The seed germination increased up to 100% using vermiwash prepared with cold stress technology in a ratio of 1:5. A study on *Sorghum bicolor* under salt stress showed that the application of vermicompost and vermiwash together increased the crop yield by maintaining nutritive equilibrium in soil, delaying salt-mediated injury, and improving the growth (Sharif et al., 2016). Compared to the recommended chemical fertilizer, the foliar application of vermiwash amalgamated with vermicompost, significantly improved germination yield in *Linum usitatissimum* L. (Makkar et al., 2017, 2019). Suganya et al. (2018) reported that Zinc oxide (ZnO) nanoparticles modified with vermiwash of *Eudrilus eugeniae* improved seed germination in the green gram (*Vigna radiata*). Rathika et al. (2020) investigated the effect of combining vermiwash and citric acid on the biomass of *Sorghum bicolor* cultivated in lead and nickel contaminated soil. The biomass of *S. bicolor* was increased by applying vermiwash (24% and 26%) and citric acid (11% and 9%) on soil polluted with lead and nickel, respectively. The vermiwash treatment exhibited significant improvements in shoot and root lengths and chlorophyll concentrations compared to citric acid due to the potential of vermiwash as a chelator. A recent study compared the effect of different organic manures on the growth, yield, and quality of betelvine (*Piper betle* L.), including vermiwash, on a positive note (Ekka et al., 2023) (Fig. 3).

4.2 In Vegetable Crop

In 2013, Elumalai et al. investigated the use of vermiwash by applying liquid fertilizer directly to the leaves to determine plant growth, internode diameter, phyllosphere region, and leaf count, wet and dry weight of the shoot and root of *Abelmoschus esculentus,* and reported maximum efficiency at 15% serving concentration followed by 10% vermiwash, gibberellic acid (100 g/ml) and naphthalene acetic acid (100 g/ml). In another study, significant improvements in the growth, yield, and antimicrobial activity of *Andrographis paniculata* were observed with combinations of vermicompost, vermicompost extract, and vermiwash compared to the recommended dose of chemical fertilizer and the control sample (Vijayakumar & Muthuselvam, 2013). In another study on Bhut Jolokia (*Capsicum assamicum*), foliar spray with vermiwash modifies the arbuscular mycorrhizal dependency and nutrient stoichiometry of the plant, subsequently leading to improvement in crop growth (Khan et al., 2014). A

Fig. 3 The images showing the production of the vermiwash from an open vermireactor (**a**) and a closed container (**b**), and its application in a nursery (**c**) and in an open agriculture field (**d**) (Khadwala, 2024; Indiamart 2024)

study conducted on French dwarf beans (*Phaseolus vulgaris* L.) showed that combinations of vermicompost leachate and vermiwash can be used as fertilizer for sustainable bean cultivation by controlling electrical conductivity (Ayyobi et al., 2014). The combination of vermiwash obtained from different sources (neem, rice straw, and bagasse) and a standardized hydroponics solution can be the most effective treatment in hydroponics for promoting better plant growth and yield of *Colocasia esculenta* (Australian Poi) (Ansari et al., 2015). The concentration of vermiwash with humic acid (3:1.5%) appeared to be the most fruitful formulation, which increased overall growth, including branching efficiency, total sugar content, and total protein content of *Allium cepa* (Prasad et al., 2016). The vermicompost and vermiwash can mitigate the adverse effects of high salt concentrations on plant growth and bulb formation in potato plants (Perez-Gomez et al., 2017). A study on *Abelmoschus esculentus* (L.) with different concentrations of vermiwash reported improvements in chlorophyll and protein content, along with 100% seed germination (Senthilmurugan et al., 2018). The formulation of panchagavya and vermiwash improved the yield and quality of bitter gourd (*Momordica charantia* L.) by foliar application (Gajjela & Chatterjee, 2019). Yassen et al. (2020) reported that applying vermiwash via foliar spray effectively improved the maximum vegetative growth, yield, and nutrition status of lettuce (*Lactuca sativa* L.) without side effects. Rajput et al. (2021) examined the impacts of seed biopriming with *Trichoderma pseudokoningii* and vermiwash treatment on the nutrient content of tomatoes and the defense response against *Sclerotium rolfsii* under heat-stress conditions. The treatment reduced oxidative damage and pathogen infection, ultimately improving plant growth. Furthermore, vermiwash and drip irrigation techniques can positively influence the correlation coefficient between the root length and stem length of chili plants (*Capsicum annuum*) (Rao et al., 2022). Seeds cultivated by traditional farming methods often exhibit high responsiveness to fertilizers but may lack the necessary quality features needed to maintain agricultural yields in the face of shifting climate trends (Wani et al., 2023).

5 Vermiwash and Soil Properties

The biochemical characteristics of soil significantly affect soil fertility. The recycling of organic nutrients and bioenergy from waste materials is emerging towards achieving the goal of sustainable agriculture (Kobayashi et al., 2004). Recent research has shown that the combination of vermiwash with vermicompost enhances the biochemical characteristics of soil by enriching micronutrients and improving its physical and chemical properties (Tharmaraj et al., 2011). Compared to the untreated control, vermiwash-treated soil exhibits considerably higher values for pH, electrical conductivity, porosity, moisture content, water holding capacity, and macronutrients, including nitrogen, phosphorus, potassium, calcium, iron, and magnesium. Therefore, crop varieties treated with vermiwash showed rapid growth and considerable productivity (Nayak et al., 2019). As discussed earlier, vermiwash is rich in macro- and micronutrients, hormones, vitamins, beneficial microflora, antimicrobial agents, and bioinsecticidal compounds, thereby helping to improve the physiochemical characteristics of the soil. Tripathi et al. (2005) reported that the diversity of microflora increased in vermiwash-treated soil, which decomposes organic material in the soil and makes nutrients available to plants. In addition, these microflora enhance plant defence against prokaryotic and eukaryotic organisms (Tharmaraj et al., 2011). Moreover, the vermiwash-treated soil exhibited reduced pH compared to untreated soil, which enhances the plant's nutrient absorption ratio. The water-holding capacity, moisture content, and porosity of soil also improved with the vermiwash (Ansari & Kumar, 2010). These enhanced properties would certainly help the plants grow and develop better. Ansari and Kumar (2010) studied the effect of vermiwash on okra (*Abelmoschus esculentus*) productivity attributed to the enhanced soil properties of the vermiwash. Similarly, Tharmaraj et al. (2011) conducted a study on Oryza sativa and observed an improvement in soil properties, which led to a significant increase in the number of leaves, leaf length, plant height, and root length.

6 Conclusion

In conclusion, vermiwash possesses the inherent properties essential for enhancing plant growth, development, and productivity by providing defense against pathogenic microorganisms, annoying insects, and essential nutrients. It should be acknowledged that the selection of vermiwash compositions should be contingent upon the intended application, since it exhibits significant variation between plant species. The use of vermiwash has been shown to improve the physicochemical characteristics of soil, leading to an enhancement in soil fertility. The enhanced characteristics of the soil depend on the specific composition and concentration of the vermiwash used for soil treatment. The optimization of planting soil and crop variety is crucial in achieving optimal growth and development.

References

Aghamohammadi, Z., Etesami, H., & Alikhani, H. A. (2016). Vermiwash allows reduced application rates of acaricide azocyclotin for the control of two spotted spider mite, Tetranychus urticae Koch, on bean plant (Phaseolus vulgaris L.). *Ecological Engineering, 93*, 234–241. https://doi.org/10.1016/j.ecoleng.2016.05.041

Akazawa, S. I., Badamkhatan, T., Omiya, K., Shimizu, Y., Hasegawa, N., Sakai, K., Kamimura, K., Takeuchi, A., & Murakami, Y. (2023). The growth-promoting effect of earthworm vermiwash on house tomato plants. *Sustainability, 15*(13), 10327. https://doi.org/10.3390/su151310327

Alexander, D., Rajan, S., Rajamony, L., Ushakumari, K., & Kurien, S. (2009). The ad hoc Package of Practices recommendations for organic farming. In *Organic farming*. Kerala Agricultural University https://keralaagriculture.gov.in/wp-content/uploads/2018/12/package_2015.pdf

Ansari, A. A., & Kumar, S. (2010). Effect of vermiwash and vermicompost on soil parameters and productivity of okra (Abelmoschus esculentus) in Guyana. *Current Advances in Agricultural Sciences (An International Journal), 2*(1), 1–4.

Ansari, A. A., Pereira, M., & Jaikishun, S. (2015). Effect of vermiwash obtained from different sources (neem, rice straw and bagasse) and standardized hydroponics solution on the growth of *Colocasia esculenta* (Australian Poi) in Guyana. *American Journal of Experimental Agriculture, 7*(5), 275–283.

Aransiola, S. A, Ijah, U. J. J., Abioye, O. P, & Bala, J. D. (2022). Vermicompost-assisted phytoremediation of toxic trace element-contaminated soil in Madaka, Nigeria using Melissa officinalis L and Sida acuta. *International Journal of Environmental Science and Technology*. https://doi.org/10.1007/s13762-022-04105-y

Ayyobi, H., Hassanpour, E., Alaqemand, S., Fathi, S., Olfati, J. A., & Peyvast, G. (2014). Vermicompost leachate and vermiwash enhance French dwarf bean yield. *International Journal of Vegetable Science, 20*(1), 21–27.

Balam, S. B. (2000). Studies on bio pesticidal effect of vermiwash in control of some foliar pathogens. M.Sc. (Ag) thesis submitted to Dr. Balasaheb Sewant Konkan Krishi Vidyapeeth, Dapoli, Dist. Ratnagiri, Maharashtra.

Chandukishore, T., Samskrathi, D., Srujana, T. L., Rangaswamy, B. E., & Prabhu, A. A. (2023). Influence of plant extract-based vermiwash on plant growth parameters and biocontrol of Thrips (Scirtothrips dorsalis) in Capsicum annum. *Journal of Natural Pesticide Research, 5*, Article 100042. https://doi.org/10.1016/j.napere.2023.100042

Chattopadhyay, A. (2015). Effect of vermiwash of Eisenia foetida produced by different methods on seed germination of green mung, Vigna radiate. *International Journal of Recycling of Organic Waste in Agriculture, 4*, 233–237. https://doi.org/10.1007/s40093-015-0103-5

Clause, J., Barot, S., Richard, B., Decaëns, T., & Forey, E. (2014). The interactions between soil type and earthworm species determine the properties of earthworm casts. *Applied Soil Ecology, 83*, 149–158. https://doi.org/10.1016/j.apsoil.2013.12.006

Das, S. K., Avasthe, R. K., & Gopi, R. (2014). Vermiwash: Use in organic agriculture for improved crop production. *Popular Kheti, 2*(4), 45–46.

Domínguez, J., Sanchez-Hernandez, J. C., & Lores, M. (2017). Vermicomposting of winemaking by-products. In *Handbook of grape processing by-products* (pp. 55–78). Academic Press. https://doi.org/10.1016/B978-0-12-809870-7.00003-X

Ekka, N., Asati, B. S., Singh, J., & Kumar, M. (2023). Effect of different organic manures on growth, yield and quality of betelvine (Piper betle L.). *The Pharma Innovation Journal, 12*(3), 2617–2620.

Elumalai, D., Kaleena, P. K., Fathima, M., & Hemavathi, M. (2013). Influence of vermiwash and plant growth regulators on the exomorphological characters of *Abelmoschus esculentus* (Linn.) Moench. *African Journal of Basic and Applied Sciences, 5*(2), 82–90.

Fathima, M., & Sekar, M. (2014). Studies on growth promoting effects of vermiwash on the germination of vegetable crops. *International Journal of Current Microbiology and Applied Sciences., 3*(6), 564–570.

Gajjela, S., & Chatterjee, R. (2019). Effect of foliar application of Panchagavya and Vermiwash on yield and quality of bitter gourd (*Momordica charantia* L.). *International Journal of Chemical Studies, 7*(3), 218–224.

Gopal, M., Gupta, A., Palaniswami, C., Dhanapal, R., & Thomas, G. V. (2010). Coconut leaf vermiwash: a bio-liquid from coconut leaf vermicompost for improving the crop production capacities of soil. *Current Science*, 1202–1210.

Gudeta, K., Julka, J. M., Kumar, A., Bhagat, A., & Kumari, A. (2021). Vermiwash: An agent of disease and pest control in soil, a review. *Heliyon, 7*(3), Article e06434. https://doi.org/10.1016/j.heliyon.2021.e06434

Homa, J. (2018). Earthworm coelomocyte extracellular traps: Structural and functional similarities with neutrophil NETs. *Cell and Tissue Research, 371*(3), 407–414.

Indiamart. (2024). https://www.indiamart.com/proddetail/worm-vermiwash-24205681891.html. Accessed 12-January 2024

Jayabhaye, M. M. & Bhalerao, S. A. (20215). Influence of vermiwash on germination and growth parameters of seedlings of green gram (Vigna radiata L.) and black gram (Vigna mungo L.). *International Journal of Current Microbiology and Applied Sciences, 4*(9):635–643.

Joshi, D., Yadav, L. R., Ratore, B. S., Srivastava, H., Verma, R. S., Gurjar, B. S., Yadav, M., Sharma, C., & Karol, A. (2023). Enhancing nutrient uptake and economics of black gram through vermicompost and vermiwash application. *Internaional Journal of Environment Climate Change, 13*(9), 3186–3193. https://doi.org/10.9734/IJECC/2023/v13i92563

Kalantari, S., Marefat, A., Naseri, B., & Hemmati, R. (2018). Improvement of bean yield and Fusarium root rot biocontrol using mixtures of Bacillus, Pseudomonas and Rhizobium. *Tropical Plant Pathology, 43*, 499–505.

Kale, K. E. (1998). *Earthworm: Cinderella of organic farming* (p. 88). Prism books Pvt. ltd.

Karuna, K., Patil, C. R., Narayanswamy, P., & Kale, R. D. (1999). Stimulatory effect of earthworm body fluid (vermiwash) on crinkle red variety of Anthurium andreanum lind. *Crop Research, 17*(2), 253–257.

Khadwala. (2024). https://khadwala.com/products/greenedge-organic-vermiwash-5-liters-liquid-concentratefertilizer-nutrition-5-l. Accessed 12-January 2024.

Khan, M. H., Meghvansi, M. K., Gupta, R., Veer, V., Singh, L., & Kalita, M. C. (2014). Foliar spray with vermiwash modifies the arbuscular mycorrhizal dependency and nutrient stoichiometry of bhut jolokia (*Capsicum assamicum*). *PLoS ONE, 9*(3), Article e92318.

Kobayashi, H., Ohta, N., & Umeda, M. (2004). Biology of lysenin, a protein in the coelomic fluid of the earthworm Eisenia foetida. *International Review of Cytology, 236*, 45–99.
Lourduraj, A. C., & Yadav, B. K. (2005). Vermiwash production techniques. In *Verms & Vermitechnology* (p. 101). APH Publishing.
Makkar, C., Singh, J., & Parkash, C. (2017). Vermicompost and vermiwash as supplement to improve seedling, plant growth and yield in *Linum usitassimum* L. for organic agriculture. *International Journal of Recycling of Organic Waste in Agriculture, 6*, 203–218.
Makkar, C., Singh, J., & Parkash, C. (2019). Modulatory role of vermicompost and vermiwash on growth, yield and nutritional profiling of *Linum usitatissimum* L. (Linseed): A field study. *Environmental Science and Pollution Research, 26*, 3006–3018.
Mayooran, S., & Mikunthan, G. (2012). Influence of vermiwash on growth and development of the seedlings of vigna radiata.
Mishra, K., Singh, K., & Tripathi, C. P. M. (2013). Management of pod borer (Helicoverpa armigera) infestation and productivity enhancement of gram crop (*Cicer aritenium*) through vermiwash with biopesticides. *World Journal of Agricultural Sciences, 9*(5), 401–408.
Mishra, K., Singh, K., & Tripathi, C. M. (2014). Management of municipal solid wastes and production of liquid biofertiliser through vermic activity of epigeic earthworm Eisenia fetida. *International Journal of Recycling of Organic Waste in Agriculture, 3*, 1–7. https://doi.org/10.1007/s40093-014-0056-0
Mishra, K., Singh, K., & Tripathi, C. P. M. (2015). Organic farming of rice crop and management of infestation of Leptocorisa varicornis through combined effect of vermiwash with biopesticides. *Research Journal of Science and Technology., 7*(4), 205–211.
More, S., Deshmukh, S., Shinde, P., & Deshmukh, V. (2013). Effect of integrated nitrogen management with vermiwash in corn (*Zea mays* L.) on growth and yield. *African Journal of Agricultural Research, 8*(38), 4761–4765.
Mukhi, S. K., Nayak, M. P., Sardar, S. S., & Bar, N. (2022). Vermiwash: A potential tool for crop production in organic agriculture. *International Journal of Plant & Soil Science, 34*(23), 1650–1656.
Nath, G., & Singh, K. (2015). Combined effect of vermiwash with biopesticides against infestation of pod borer (Helicoverpa armigera Hub.). *International Journal of Zoological Investigations, 1*(1), 40–51.
Nath, S., & Singh, K. (2016). Analysis of different nutrient status of liquid bio-fertiliser of different combinations of buffalo dung with gram bran and water hyacinth through vermicomposting by Eisenia fetida. *Environment, Development and Sustainability, 18*, 645–656. https://doi.org/10.1007/s10668-015-9666-6
Nayak, H., Rai, S., Mahto, R., Rani, P., Yadav, S., Prasad, S. K., & Singh, R. K. (2019). Vermiwash: A potential tool for sustainable agriculture. *Journal of Pharmacognosy and Phytochemistry, 8*(5S), 308–312.
Ojuolape, O. T., Aransiola, S. A., & Obideyi, O. A. (2015). Vermitechnology: A solution to environmental problems; A review. *Journal of Global Ecology and Environment*, *3*(3), 127–135. http://www.ikpress.org/abstract/4516.
Parmanik, P. (2010). Quantification of hydraulic and proteolytic enzymes in the excreta of three epigeic earthworms and detection of thiocarbamic acid by GC-MS-MS. *The Environmentalist, 30*, 212–215.
Patnaik, P., Chyne, E., Abbasi, T., & Abbasi, S. A. (2022). Vermiwash: An organic fertilizer of great potential. In *Advances in Sustainable Development: Proceedings of HSFEA 2020* (pp. 15–27). Springer Singapore. https://doi.org/10.1007/978-981-16-4400-9_2
Pattnaik, S., Parida, S., Mishra, S. P., Dash, J., & Samantray, S. M. (2015). Control of phytopathogens with application of vermiwash. *Journal of Pure and Applied Microbiology, 9*(2), 1697–1701.
Pérez-Gómez, J. D. J., Abud-Archila, M., Villalobos-Maldonado, J. J., Enciso-Saenz, S., Hernández de León, H., Ruiz-Valdiviezo, V. M., & Gutiérrez-Miceli, F. A. (2017). Vermicompost and vermiwash minimised the influence of salinity stress on growth parameters in potato plants. *Compost Science & Utilisation, 25*(4), 282–287.

Prasad, U., Sunkar, S., NMD, S. K., Gala, A. A., & Kumar, A. (2016). Formulation of vermiwash and humic acid and its application on *Allium cepa. Biosciences Biotechnology Research Asia*, 13(1), 523–529.

Rajput, R. S., Singh, J., Singh, P., Vaishnav, A., & Singh, H. B. (2021). Influence of seed biopriming and vermiwash treatment on tomato plant's immunity and nutritional quality upon Sclerotium rolfsii challenge inoculation. *Journal of Plant Growth Regulation, 40*, 1493–1509.

Ram, R. A., Israr, A., Kumar, A., & Maurya, S. (2022). Microbial validation of organic preparations used in natural farming. *Journal of Eco-Friendly Agriculture., 17*(1), 11–16. https://doi.org/10.5958/2582-2683.2022.00003.X

Ramnarain, Y. I., Ansari, A. A., & Ori, L. (2019). Vermicomposting of different organic materials using the epigeic earthworm Eisenia foetida. *International Journal of Recycling of Organic Waste in Agriculture, 8*, 23–36. https://doi.org/10.1007/s40093-018-0225-7

Rao, K. R., Silveira, J., Anjali, N., Sneha, N., Sutisha, R., & Savita, R (2022). Effect of Vermiwash on the growth of Capsicum annuum. *Ecology, Environment and Conservation, 28*, (S420–S428).

Rathika, R., Khalifa, A. Y., Srinivasan, P., Praburaman, L., Kamala-Kannan, S., Selvankumar, T., Govarthanan, M., et al. (2020). Effect of citric acid and vermi-wash on growth and metal accumulation of Sorghum bicolor cultivated in lead and nickel contaminated soil. *Chemosphere, 243*, Article 125327.

Rose, P. K., Dhull, S. B., & Kidwai, M. K. (2022a). Cultivation of wild mushrooms using lignocellulosic biomass-based residue as a substrate. In *Wild mushrooms* (pp. 493–520). CRC Press.

Rose, P. K., Kidwai, M. K., & Dhull, S. B. (2022b). Food industry waste: Potential pollutants and their bioremediation strategies. In *Food processing waste and utilization* (pp. 343–359). CRC Press.

Rose, P. K., Kidwai, M. K., & Kantiwal, P. (2024). Green approaches for the valorization of olive mill wastewater. In *Green chemistry approaches to environmental sustainability* (pp. 313–336). Elsevier.

Sayyad, N. R. (2017). Utilisation of vermiwash potential against insect pests of tomato. *International Research Journal of Biological Sciences., 6*(1), 44–46.

Senthilmurugan, S., Sattanathan, G., Vijayan, P. P. K., & Tamizhazhagan, V. (2018). Evaluation of different concentration of vermiwash on seed germination and biochemical response in *Abelmoschus esculentus* (L.). *International Journal of Biology Research, Evaluation, 3*(1), 228–231.

Sharif, F., Danish, M. U., Ali, A. S., Khan, A. U., Shahzad, L., Ali, H., & Ghafoor, A. (2016). Salinity tolerance of earthworms and effects of salinity and vermi amendments on growth of *Sorghum bicolor*. *Archives of Agronomy and Soil Science, 62*(8), 1169–1181.

Singh, S., & Sinha, R. K. (2022). Vermicomposting of organic wastes by earthworms: Making wealth from waste by converting 'garbage into gold' for farmers. In *Advanced Organic Waste Management* (pp. 93–120). Elsevier. https://doi.org/10.1016/B978-0-323-85792-5.00004-6

Suganya, P., Rajamohan, C., & Mahalingam, P. U. (2018). Synthesis and surface modification of zinc nano rods using vermiwash of *Eudrilus eugeniae* and functionalisation to seed germination of green gram *Vigna radiata. Materials Research Express, 6*(2), Article 025409.

Sulaiman, I. S. C., & Mohamad, A. (2020). The use of vermiwash and vermicompost extract in plant disease and pest control. In *Natural remedies for pest, disease and weed control* (pp. 187–201). Academic Press. https://doi.org/10.1016/B978-0-12-819304-4.00016-6

Tharmaraj, K., Ganesh, P., Kolanjinathan, K., Suresh Kumar, R., & Anandan, A. (2011). Influence of vermicompost and vermiwash on physico chemical properties of rice cultivated soil. *Current Botany, 2*(3), 18–21.

Thirunavukkarasu, A., Raja, S., Nithya, R., Sathya, A. B., Priyadharshini, V., Premkumar, B., Muthuveni, M., & Sakthishobana, K. (2023). Sustainable organic waste management using vermicomposting: A critical review on the prevailing research gaps and opportunities. *Environmental Science: Processes & Impacts.* https://doi.org/10.1039/D2EM00324D

Tripathi, Y. C., Hazarika, P., & Pandey, B. K. (2005). Vermicomposting: An ecofriendly approach to sustainable agriculture. In A. Kumar (Ed.), *Verms and vermitechnology* (pp. 23–39). APH Publishing Corporation.

Vijayakumar, A., & Muthuselvam, K. (2013). Effect of vermicompost, vermicompost extract and vermiwash on growth, yield and antimicrobial activity of *Andrographis paniculata. Journal of Applicable Chemistry, 2*(5), 1255–1261.

Wani, I. M., Narayan, S., Magray, M. M., Khan, F. A., Mir, S. A., Yaseen, I., & Khurshid, A. (2023). Effect of organic manures and biofertilizers on seed quality of radish (Raphanus sativus L.) under temperate conditions of Kashmir. *Biological Forum—An International Journal, 15*(1), 315–319.

Yassen, A. A., Essa, E. M., Marzouk, N. M., & Zaghloul, S. M. (2020). Impact of vermicompost and foliar spray of vermiwash on growth, yield and nutritional status of lettuce plants. *Plant Archives, 20*(1), 449–455.

Zambare, V. P., Padul, M. V., Yadav, A. A., & Shete, T. B. (2008). Vermiwash: Biochemical and Biological approach as eco-friendly soil conditioner. *ARPN Journal of Agricultural and Biological Sciences., 3*(4), 28–37.

Future Direction of Research in Vermitechnology

S. A. Aransiola, I. O. Musa, A. E. Oyewumi, O. J. Oyedele, and Naga Raju Maddela

Abstract Globally, anthropogenic pollution and contamination from farming, industrial, and other activities are constantly being applied to the soil. The focus is increasingly turning to biological in situ alternatives since traditional physico-chemical cleanup technologies have substantial costs and environmental deterioration associated with them. The most recent vermicomposting model is eco-friendly, affordable, low maintenance, requires little space, protects against predators, uses little water, has a self-roof, and is simple to operate. Out of the approximately 3000 species of earthworms found globally, over 500 species have been identified in India. Earthworms use physical and biological mechanisms to carry out their vermicomposting activity. When compared to biochemical processes, which entail the break-down of waste by a variety of enzymes present in earthworms' guts and are regulated by bacteria present in their intestines, physical processes involve the aeration, mixing, and grinding of substrate. Vermiremediation is a viable long-term biological cleanup approach; however, it is only effective on soil that has only a minimal amount of contamination. It is necessary to therefore review the future direction of research on vermitechnology.

Keywords Vermitechnology · Global warming · Agriculture · Remediation

S. A. Aransiola (✉)
Department of Microbiology, University of Abuja, Abuja, Nigeria
e-mail: blessedabiodun@gmail.com

S. A. Aransiola · I. O. Musa · O. J. Oyedele
Bioresources Development Centre, National Biotechnology Development Agency, Ogbomoso, Nigeria

A. E. Oyewumi
Department of Microbiology, Alex Ekwueme Federal University, Ndufu-Alike, Ebonyi State, Nigeria

N. R. Maddela
Department of Biological Sciences, Faculty of Health Sciences, Universidad Técnica de Manabí, Portoviejo, Ecuador

I. O. Musa
Department of Microbiology, School of Sciences and Information Technology, Skyline University Nigeria, Kano, Nigeria

S. A. Aransiola and N. R. Maddela (eds.), *Vermitechnology: Economic, Environmental and Agricultural Sustainability*, https://doi.org/10.1007/978-3-031-96212-7_18

1 Introduction

Today's worldwide scientific community is looking for a technology that is socially acceptable, ecologically sustainable, and commercially feasible. Vermiculture technique includes all the positive traits and attributes (Sinha et al., 2010). Vermiculture research is undergoing a revolution for several environmental applications and development (Aransiola et al., 2024). For a very long time, vermiculture researchers from all over the world were aware of the importance of earthworms in managing soil fertility, managing waste, and managing trash. However, some relatively recent findings regarding their function in the treatment of contaminated soil and waste water and possible usage in present-day medication to safeguard human health, including dissolving blood clots for stroke, thinning blood, lowering of blood pressure, curing cancer, acting as an inflammatory agent and helping patients with heart diseases, curing arthritis and rheumatisms, providing antibiotics, and being a rich source of high-quality protein, have emerged.

Through intricate mechanical and biochemical interactions with soil's abiotic and biotic elements, earthworms increase plant growth and production (Sinha et al., 2008). The mechanical disintegration of soil particles caused by earthworms' ingestion of the soil increases the surface area available for biotic activity. Water, particles, nutrients, and aeration travel through, through, and through earthworm burrows. Earthworms' intestines are home to millions of enzymes and microorganisms that speed up the biochemical conversion and mineralization of soil organic materials, enriching the soil (Sinha et al., 2008). Increased plant growth and crop output are made possible by all of these mechanisms working in conjunction with other elements. Earthworms are known to have the ability to lessen a number of environmental problems. (Dada et al., 2016; Sinha et al., 2008).

Around the world, soil is contaminated and exposed to anthropogenic pollution from farming, industrial, and other activities. Chemicals, organic wastes, and inorganic substances or elements, notably metals, are common soil pollutants (Dada et al., 2015). The focus is increasingly turning to biological in situ alternatives since traditional physicochemical cleanup technologies have high costs and ecological and environmental instability.

It is well recognized that earthworms working as soil ecological engineers have a significant impact on nearby biological, physical, and chemical. Earthworms enhance agricultural output by increasing the soil's nutritional content, reducing the toxicity of wastes, and maybe even helping to detoxify polluted soil. Since the 1800s, earthworms have been extensively investigated for a variety of applications as a result of their role in maintaining the terrestrial ecosystems. Studies on its application in soil remediation may date back to the 1980s (Sinha et al., 2010). However, this subject has just lately experienced a rapid expansion, and the word "vermiremediation" has only recently been used (Sinha et al., 2008). Vermiremediation, a word used to describe the method by which earthworms remove toxins from soil. In order to alter, degrade, or eliminate pollutants from the soil environment, vermiremediation makes use of the abiotic and biotic interactions, life cycle, burrowing, and feeding activity of earthworms (Aransiola et al. 2022a).

2 Concept of Vermiremediation

Vermis (Latin for "worm") and remedium (Latin for "remedy") are combined to form the phrase "vermiremediation" (means to remove something bad). Edwards and Arancon coined the phrase "vermiremediation" initially (2006). However, Rodriguez-Campos et al. (2014) may be the first authors to define the phrase, which is "the use of earthworms for eliminating toxins from the soil or when earthworms assist in the degradation of nonrecyclable substances." A thorough definition was presented here with care because it seems that this definition is rather ambiguous. Vermiremediation is an earthworm-based bioremediation method that collects and extracts, transforms, or degrades toxins in the soil environment by the use of earthworms using the ability of burrowing, eating, secretion, and metabolism. Figure 1 illustrates these procedures (Aransiola et al., 2024).

Fig. 1 Strengths and weaknesses of vermiremediation

3 Organic Pollutants: Vermi-Transformation and Vermi-Degradation

When enzymes produced by earthworms, like CYP450, are used to transform organic pollutants, this is referred to as vermi-degradation/vermi-transformation. By this definition, one of the steps to "vermicomposting" and "vermi-conversion" is "vermi-transformation/vermi-degradation." Vermi-conversion is the name for the quick conversion of slowly biodegradable solid wastes into useful fertilizer elements using earthworms and microbes together (Sahariah et al., 2014). Vermicomposting is a method of biologically converting organic waste into stabilized organic fertilizer (Lim et al., 2016). It appears that the two words refer to the same bioconversion procedure. Therefore, using the contaminant removal or detoxification abilities of earthworms or "vermin-endophyte" in vermi-conversion and vermicomposting automatically involves the remediation of contaminants. Vermi-conversion and vermicomposting, on the other hand, focus on biodegradable solid wastes, whereas vermi-transformation and vermi-degradation directly address the treatment of chemicals (e.g., organic pollutants). There is research on vermin metamorphosis that has mostly examined the ecotoxicological consequences of organic pollutants on earthworms (Zhao et al., 2016).

4 Organic Xenobiotics Detoxification by Earthworms

Through their metabolic processes, earthworms can collect organic xenobiotics from contaminated environments. Various earthworm species, primarily *E. fetida*, have been reported to be able to metabolize some organic contaminants, including polycyclic aromatic hydrocarbons (PAHs), pesticides, herbicides, and 2, 4, 6-trinitrotoluene (Renoux et al., 2000). Scientists have recently been interested in the enantioselective breakdown of chiral insecticides in earthworms (Wang et al., 2014). There have also been a few organic pollutants' metabolites found in earthworms. In earthworms, 10:2 fluorotelomer alcohol may be converted to perfluorocarboxylic acids, perfluorodecanoate, perfluorononanate (Zhao & Zhu, 2017). In the soil, n-ethylperfluorooctane sulfonamide ethanol may be converted to perfluorooctane sulfonamide, perfluorooctane sulfonate, n-ethylperfluorooctane sulfonamide acetate, and this metabolite was the main one produced by earthworms (Zhao et al., 2016). However, given the intricacy of vermitransformation and vermidegradation processes, numerous metabolites must be identified. According to one description, the biochemical metabolic digestion of organic pollutants in earthworms involves a number of enzyme-catalyzed steps (Li et al., 2018; Saint-Denis et al., 1999).

The CYP450 genes of *L. terrestris* and *E. fetida* have been discovered. Direct conjugation of organic pollutants or conjugation of metabolites from Phase I processes with glutathione, amino acids, or sugars are two examples of Phase II conversion. By forming covalent bonds with any of these endogenous molecules, conjugation results in hydrophilic conjugates. For instance, the pyrene metabolites produced by *E. andrei* are all conjugates of 1-hydroxypyrene. Due to its great sensitivity and ecological importance, glutathione S transferase (GST) was one of the most researched enzymes in Phase II. GST aids in the elimination of reactive electrophiles by catalyzing the conjugation process between GSH and electrophilic xenobiotics. While this is happening, GST works as a crucial Phase II detoxifying enzyme to excrete and get rid of the byproducts of Phase I metabolism (Aransiola et al., 2019; Ojuolape et al., 2015).

5 Advantages of Vermi-Remediation

Vermiremediation provides a number of benefits over traditional physicochemical approaches, and subsequent research has shown that it is technologically possible. Vermiremediation, like phytoremediation, is primarily an aesthetically beautiful, ecologically acceptable method (Sinha et al., 2008). Earthworms typically clean up polluted ground without removing the topsoil, preserving or even increasing soil fertility and utility. The environment is just somewhat disturbed. It can also be used to remediate a variety of environmental pollutants. According to Rodriguez-Campos et al. (2014), the presence of earthworms in the soil can help eliminate pesticides, herbicides, PAHs, PCBs, and crude oil. Because vermiaccumulation for organic pollutants may contribute significantly, as opposed to phytoaccumulation, it is a comparatively efficient approach when compared to phytoremediation strategies. By increasing organic matter, nutrient concentrations, and biological activity, vermiremediation can also enhance soil quality (Sinha et al., 2008). Previous studies also demonstrated that even in contaminated soil, earthworms continue to serve as ecosystem engineers.

5.1 Limitations of Vermi-Remediation

Like other bioremediation techniques, the vermiremediation method has several limitations. Only weakly or moderately polluted soils that enable earthworm survival are suitable for vermiremediation. The survival of earthworms may be impacted by the toxicity of pollutants in severely polluted soil (Rodriguez-Campos et al., 2014). Additionally, depending on the ecological groups of earthworm species used, vermiremediation is only applicable to depths with earthworm activity. Earthworms are divided into three groups based on their preferred habitats: epigeic, anecic, and endogeic. Epigeic earthworms, like *E. fetida,* are classified as detritivores based on their feeding habits and are found on the upper surface of soils where they primarily consume plant litter and other organic debris.

Similar to *Anecics, endogeic* is known as a geophagous and a phyto-geophagous species, respectively. Since vermi-remediation mostly depends on the eating habits of the worm species, food availability may potentially limit its potential. However, if handled carelessly or disposed of improperly, toxins that have built up in earthworms might enter the food chain. Vermiremediation could also be used under tight guidelines. Due to their sensitivity, earthworms may not survive in certain climates, seasons, or other environmental circumstances, which might impede vermiremediation procedures.

6 Colonization and Inoculation of Earthworms in Polluted Site

Vermiremediation may need a lot of earthworms (Sinha et al., 2008). Therefore, the technique for introducing earthworms into contaminated soil to activate vermiremediation is the first issue that has to be resolved. The earthworm inoculation methods discussed by Butt and Grigoropoulou (2010) included grass cutting and relaying, chemical/physical extraction with broadcasting, and the earthworm inoculation unit technique. Additionally, both their benefits and drawbacks were clearly stated. However, a study of earthworm survival in polluted sites is required to confirm colonization and ensure the viability of vermiremediation.

7 The Fate of Earthworms Used for Vermi-Remediation

What happens to earthworms once they are utilized to remediate soil contaminated by organic matter is another significant subject. Earthworms used in vermiremediation should be managed carefully since they can acquire a lot of organic pollutants. Burning earthworms as hazardous trash in specialized landfills is a potentially easy approach for securely disposing of them, similar to the post-treatment of plants employed in phytoremediation. To progress the application of vermi-remediation, these difficulties must be further researched because they are rarely taken into account. Finally, vermicomposting has steadily gained recognition as a growing solution for the treatment of organically contaminated soil. Vermiremediation has been the subject of several investigations during the past ten years. In order to promote the general application of mirage remediation, further basic studies will be required in the future.

8 Future Direction/Research in Vermitechnology

India's success in agricultural output following the first Green Revolution in the late 1960s, led by Swami Nathan, Karwar, and Bourlaug, produced a huge increase in agricultural productivity. This is brought about by a growth in the use of contemporary agricultural inputs, including farm machinery, chemical fertilizers, pesticides, herbicides, hybrid varieties of seeds, and other inputs. Despite hardly any growth in net planted area, India's food grain output more than quadrupled during the first green revolution. From 95 million tons in 1967–1968 to 209 million tons in 1999–2000, it climbs. India became self-sufficient and self-independent in terms of agriculture and food supplies as a consequence of increased productivity, but this had a negative impact on the ecosystem and gene pool as a whole. The effects of its unsustainable use turned the emphasis back to organic farming. A paradigm change in agriculture is seen in this situation. Farmers are converting to environmentally friendly methods. For the generation of vermicompost and vermiwash, new vermitechnology models employ the surface and subsurface varieties of the earthworm *Eudrillus eugenie*. Agriculture has a huge need for the work being done on vermitechnology. Together with soil bacteria, earthworms play a crucial role in soil fertility and processing. In clay soil, the usage of vermicompost enhances air movement. The hygroscopic nature of mucus increases its capacity to store water by absorbing it. Vermicompost enriches the soil with extra nutrients that chemical fertilizers do not have (Kale, 1998). When water is passed through a column of worm activity, coelomic fluid and vermicomposting filtrate are recovered. It is extremely beneficial as a foliar spray and contains a variety of micronutrients, enzymes (Aransiola et al., 2022b), hormones, and microorganisms (Babaniyi et al., 2023). When properly collected, vermiwash is a translucent liquid with a light yellow hue. Agro waste, kitchen trash, and nitrogen-rich materials like cow, sheep, and pig manure can all be used as the earthworms' food sources.

In the second half of the twentieth century, the world's population more than quadrupled, from 2.5 billion in 1950 to 6 billion in 2000, and it will reach 12 billion people by the end of the century. The emerging and third-world nations of Asia, Africa, and South America have seen and will continue to see the majority of this population growth. In India, there were just 361 million people, and by the year 2000, that number had increased to 1004.5 million, virtually tripling in the second half of the twentieth century. India is already one of the largest producers and exporters of various agricultural commodities, but the negative effects of these advancements are quickly becoming apparent. Now it is evident that these developments have adversely affected on ecosystem at large. Effects are noticed not only in soil, water, produced food, and air, but also went way beyond and affected on gene pool of wild seeds and, in turn, affect biodiversity.

Due to increased environmental awareness, the negative impacts of agrochemicals, rising input costs, fluctuating commodity prices, and the evolution of pesticide resistance in pests. In India, agriculture has undergone a paradigm shift as a result of this altered situation. Organic agricultural techniques are now being used by more farmers. It appears that the second evergreen revolution is about to begin. In light of this, we have begun developing high-quality, affordable agricultural supplies to support the farming community.

9 Prospects of Vermitechnology in Solving Environmental and Agricultural Problems

Vermicompost is a soil conditioner and plant feeding material created by earthworm species like the red worm (*Eisenia foetida*) by altering the physical and chemical properties of cow feces and organic plant components. There are severe worries about the security of natural resources due to the widespread use of agro-chemicals, which endangers human health and the environment, degrades soil quality, and raises disease resistance (Kumar & Gupta, 2018). The need for effective organic goods as biological fertilizers and insecticides has prompted scientists to create sustainable agricultural production systems. In this area, aerobic compost and vermicompost products, which boost soil quality in all aspects, have acquired enormous significance. Chemical fertilizer usage has a major impact on environmental degradation since it uses up fossil fuels, produces carbon dioxide, and pollutes the atmosphere, water, and soil. The loss of soil fertility and the resulting negative effect on agricultural output are direct consequences of the improper use of chemical fertilizers.

Ecological and sustainable agricultural techniques are the only way to undo the damage done to the environment by the widespread use of chemical fertilizers, it has become clear in recent years. Soil, under normal conditions, is home to a wide variety of micro and macroorganisms that break down organic matter into humus and plant nutrients. The presence of earthworms in the soil is crucial because they aid in nitrogen cycling and boost soil fertility. Because of this, earthworms have been appropriately dubbed "the buddy of farmers" by Charles Darwin. Producing "Vermicest," a product considered to have a biological suppressive impact on plant development, plant nutrition, and rot factors, is made possible by vermicomposting techniques, which are a reliable, inexpensive, and sustainable approach for evaluating diverse organic wastes. Vermicompost (worm manure) allows for a low-input production method, which is crucial for small and medium-sized agricultural producers, and helps offset the initial drop in output seen when switching from conventional to organic farming. Food safety for people and animals is ensured by vermicomposting procedures, which also promote a sustainable agricultural production model that is both environmentally and economically beneficial (Demir et al., 2010).

As a means of boosting soil productivity, vermicompost helps produce aggregates in the soil. It enhances the soil's ability to store water and to absorb air, and it strengthens the soil's overall structure. The plant's root growth is aided by these as well. As a result, the plant is able to take up more of the soil's beneficial nutrients, leading to enhanced growth and a higher harvest. Also, the organic structure of the vermicompost put into the soil provides an increase in soil nutrients

(Mısırlıoğlu, 2011). Soil worms have a major impact on the health of a garden's soil, its fertility, and its yields. Their feeding and gallery opening activities have a number of beneficial effects on the soil, including a shift toward a more favorable nutrient-to-organic-matter ratio, deeper water penetration, and faster incorporation of topsoil-applied organic matter, lime, and fertilizers. Studies have also shown that they improve the quality of grain, enhance yields in fields, and decrease the prevalence of root infections (Tomati & Gali, 1995). A wide variety of plants from all over the globe, including those from Turkey, have been subjected to several vermicompost experiments. In most cases, vermicompost was applied alongside conventional fertilizers in the lab, and the outcomes were compared based on the status of the control groups. Vermicompost research mostly focuses on how it affects crop productivity and disease. Insufficient studies have been done in the economic field. The physical, chemical, biological, and microbiological changes it brings to the soils it is applied to make vermicompost a trustworthy organic fertilizer that increases agricultural productivity while decreasing nutrient loss. Among its many well-known advantages are that it acts as a soil conditioner, contains an adequate quantity of vital plant nutrients, helps prevent pesticide and disease buildup, boosts crop yields by improving soil quality, and costs less to use over the long run. Biological origin deficiency may be remedied by incorporating vegetable leftovers, farm manure, chicken manure, rubbish compost, and organic industrial wastes into the soil. There is a favorable correlation between the incorporation of these minerals into the soil and an increase in crop yield and quality (Entry et al., 1997; Pascual et al., 1997; Sönmez et al., 2002).

Research into vermiculture contributes to the fields of waste management, soil purification and revitalization, and eco-friendly agriculture. There are basically just two places where commercial vermiculture is done. For starters, there's worm biomass generation and processing vermicompost. Creating worm biomass is done so that worms may be used as a protein source in the poultry and fish farming industries. In contrast, vermicomposting is a method used to stabilize soils that have been contaminated by sewage, sludge, or other similar wastes in the process of stabilization. Grain yields, in particular, benefit from earthworms' presence, increasing by 35% (Baker, 1994). It's true that vermicompost applications are very new to the United States, in contrast to compost applications made from a broad variety of materials, which are quickly becoming ubiquitous here. Previous research using vermicompost has highlighted its beneficial impacts on plant production and quality, as well as its enhancement of the physical and biological structure of the soil. Further research and investigations are needed to disclose the full extent of vermicompost's acknowledgment in our nation and its impact on plant production. It has been shown via research those organic fertilizers are better for plants, soil, the environment, and the economy. The primary goals of these research are to improve plant production and to demonstrate that these improvements are sustainable in comparison to the use of chemical fertilizers. When applied to the rhizosphere, where microbial activity is greater than in other sections of the soil, bio fertilizer and vermicompost enhance the soil's physicochemical qualities and biological productivity. Soil biodiversity and biomass improve as a result, and less chemical fertilizers are required. An increase in maize plant dry matter was seen with a drop in soil pH after using worm compost made from garbage (Ferreira & Merchant, 1992).

The impact of vermicompost and chemical fertilizer on radish plant production and quality was investigated by Kumar and Gupta (2018). Vermicompost treatment resulted in the greatest gains in plant height, dry matter, tuber weight, and root length. Overall, it was shown that vermicompost treatment led to a greater increase in yield than the control group. The organic matter content of the soil, the aeration of the soil, the water holding capacity, the ease of nutrient absorption by plants, and the yield are all favorably impacted by vermicompost. Vermicompost improves soil porosity, facilitates root spread, and stimulates greater root growth, all of which contribute to a higher harvest (Jackson, 1967). Many people and groups are interested in sustainable and organic agriculture since vermicompost is considered an organic fertilizer, is produced in a natural way, and does not degrade the soil.

Some features and benefits of worms and vermicompost can be listed as follows.

- Boosts the resilience of plants and hastens their growth, allowing for an early harvest. With its granular form, it controls soil structure, boosts water retention capacity, and vastly improves soil aeration, providing an early harvest by roughly 15–20 days.
- Nitrogen-fixing microorganisms have benefited plants greatly.
- Since it makes the soil more pliable, it boosts soil productivity, and it allows plants to absorb more of the soil's nutrients, leading to higher-quality end products with a higher solids-to-water ratio.
- Profitability is achieved by a significant increase in output thanks to the incorporation of worm feces, which includes components like enzymes, amino acids, growth hormones, and vitamins in the fertilizer.
- It saves money since fewer herbicides and fungicides are needed because there are no weed seeds present. When chemical fertilizers are used less often, fertilizer expenditures are lowered.
- It alleviates pesticide-induced plant stress.
- Soil pH regulation improves plant health, which in turn raises yields and profits. It also reduces seed loss and expedites the sprouting process.
- Body fluids (coelom liquid) that worms transmit into manure promote resistance against plant illnesses and protect plants from cold.

Vermicompost has a positive impact on the bottom line because of the reasons listed above. Vermicompost helps farmers save money and make more money on their crops, according to the findings.

10 Applications of Vermitechnology in Global Warming Mitigation

Increases in both population and trash production are serious problems, especially in emerging nations. While composting and vermicomposting are low-impact ways to manage organic waste, they do have certain drawbacks, including the emission of greenhouse gases (GHGs). This overview provides a comprehensive analysis of the key factors driving GHG production, including aeration, C/N ratio, temperature, pH, bulking agent, and moisture content (Cao et al., 2020). In many circumstances, vermicomposting reduces GHG emissions compared to composting, especially methane (CH_4) emissions. Nevertheless, earthworms are also substantial contributors to nitrous oxide (N_2O) during vermicomposting; therefore, comparing the two processes in terms of GHG emission (GHGsE) is necessary. Carbon dioxide and other GHGs and Hydrogen Carbon Dioxide (CO_2).

Carbon dioxide (CO_2) is a major contributor to global warming because it acts as a GHG by trapping heat in the atmosphere. The current global CO_2 concentration is believed to be approximately 416 ppm, and NASA predicts that this number will continue to climb in the near future (NASA, 2021). There are two types of carbon dioxide (CO_2) created during the composting process: biogenic CO_2 and non-biogenic CO_2. In contrast to the CO_2 that was created by burning fossil fuels like coal and oil, the CO_2 that was produced by biological processes, including litterfall decomposition and soil respiration. Since CO_2 released during decomposition is already taken up as a food source by plants, it is not included for the GWP assessment of carbon dioxide gas for organic waste (Swati & Hait, 2018). Rapid CO_2 emissions during composting indicate a high level of microbial activity and overall organic matter decomposition (Aransiola et al., 2023). Modifying emission-causing factors may lead to a decrease in GHGsE. Increases or decreases in GHGsE might be drastically affected by the inclusion of changes. With a biochar dosage of 10%, Wang et al. (2014) showed that CO_2 emissions from composting pig dung could be reduced by 26.06%. There are two distinct phases in composting, the first thermophilic phase and the final maturation, or mesophilic, phase, during which the waste is stabilized. Both aerobic and anaerobic thermophiles ingest and metabolize readily available OM during the thermophilic phase. High temperatures (50–70 °C) and increased microbial activity characterize this phase, leading to the volatilization of OM and the release of CO_2 and NH_3. At the last stage of maturation, temperatures decrease to the 40 to 50s Celsius, suggesting that all biodegradable OM in the trash has been decomposed and that very little or no GHGsE remain. According to Awasthi et al. (2020), CO_2 emissions were quite high at first but then steadily declined. Several studies have revealed that CO_2 emissions from substrates such as municipal solid waste, green waste, and sewage sludge are greatest at the beginning of the composting or vermicomposting process and gradually decrease as the process progresses. Composting food waste produces a large amount of carbon dioxide (CO_2) during the first two weeks, but this emission gradually decreases as the process progresses (Wang et al., 2014).

Numerous other scientists have also seen a rise in composting temperatures. The first phases, as shown by Lleó et al., (2013), may reach temperatures as high as 55 °C before gradually cooling down. Thus, it is probable that the high temperatures during the first stages are linked to the increase in CO_2 emissions during the initial stages. Liu et al. (2021) also noted this, citing temperature spikes and the breakdown of organic matter as the primary causes of the rise in emissions. Although carbon dioxide emissions are more widespread, other gases, including nitrous oxide (N_2O) and chlorofluorocarbons (CH_4), contribute to ozone depletion and have GWPs that are 298 and 25 times higher than CO_2, respectively. As the CO_2 emitted during composting comes from the natural breakdown of plant matter, it has a zero global warming potential.

Methane (CH_4)

The production of CH_4 is second only to that of CO_2 among GHGs (Leh-Togi Zobeashia et al., 2018). In the previous 10 decades, CH_4's GWP has been seen to be 28 times higher than CO_2's. The pace of CH_4 emissions, the amount of material utilized, the moisture content, the temperature, and the activity of methanotrophic bacteria all have a role in the total amount of CH_4 emitted by composting operations. GHGsE is reduced by increasing the aeration rate, whereas high CH_4 emissions are encouraged by decreasing the aeration rate. It has also been shown that sporadic oxygen delivery helps cut GHG emissions. In addition, the compost pile's heterogeneity might lead to CH4 emissions in places with little access to air.

Nitrous Oxide (N_2O)

Composting generates a significant amount of N_2O via a process called denitrification. Turning the pile redistributes and transfers nitrates, and the solubility in the pile affects the emission rate. Nitrous oxide (N_2O) generation is stimulated by the availability of carbon during OM decomposition, which in turn impacts the availability of oxygen (O_2) at microsites. Researchers observed that both lignite and non-lignite windrows emitted negligible amounts of N_2O. The majority (76.7%) of lignite windrows' GHGsE were found to be N_2O. We measured 1.16 (0.3) g kg 1 DM of cumulative N_2O emission from the lignite windrow and – 0.52 (0.1) g kg 1 DM from the non-lignite windrow. The nitrification process is sluggish, as seen above. Composting manure results in less nitrogen (N) loss as N_2O when lignite is added to the mix. Another research by Swati and Hati (2018) demonstrated that N_2O emissions were significantly reduced when bamboo biochar was used as an additive during composting of poul 5t9 try manure. Biochar increases oxygenation of the compost's surface and the temperature, both of which reduce the activity of nitrifying microorganisms and enzymes and so reduce the production of nitrogen oxide (N_2O). In comparison to the overall loss of N, the loss of N_2O was just 3.15–0.58%. Excessive acidification (pH 6 and pH 5) was observed by Cao et al. (2020) to hasten N_2O emission by 18.6% and 17.6%, respectively. As the pH drops from 7 to 5.5, the denitrification process emits more nitrogen oxides. Vermicomposting using agricultural waste was used to demonstrate complete N_2O emission, and it was shown that vermicompost emits less N_2O than thermophilic compost. The configuration of the pile is what caused the increased release of N_2O in thermophilic compost. Vermicompost, on the other hand, has a 53% lower emission of N_2O because the earthworms eat away at the arrangement or stratification and homogenize the material. It was also established that the earthworm's gut underwent denitrification, leading to reduced N_2O emissions.

11 Conclusion

The future guidelines in vermitechnology are the formulation of policies by stakeholders and environmentalists in order to improve and adopt this environmentally friendly technology because earthworms have a variety of enzymes present in their guts and are regulated by bacteria present in their intestines. These enzymes could be extracted for other industrial and biomedical uses. Also, since earthworms use aeration, mixing, and grinding of substrate produce useful products, they possess a valuable nutrient in their by-product which could be a perfect answer to improve plant proliferation and curb global food insecurity.

References

Aransiola, S. A., Ijah, U. J. J., Abioye, O. P., & Bala J. D. (2022a). Vermicompost-assisted phytoremediation of toxic trace element-contaminated soil in Madaka, Nigeria using *Melissa officinalis* L and *Sida acuta. International Journal of Environmental Science and Technology* (2022). https://doi.org/10.1007/s13762-022-04105-y

Aransiola, S. A., Ijah, U. J. J., Abioye, O. P., & Bala, J. D. (2019). Microbial-aided phytoremediation of heavy metals contaminated soil: A review. *European Journal of Biological Research, 9*(2), 104–125. http://www.journals.tmkarpinski.com/index.php/ejbr/article/view/157

Aransiola, S. A., Ijah, U. J. J., Abioye, O. P., Bala, J. D., Rivadeneira-Mendoza, R. F., Luque, R., Rodríguez-Díaz, J. M., & Maddela, N. R. (2024). Micro and vermicompost assisted remediation of heavy metal contaminated soils using phytoextractors. *Case Studies in Chemical and Environmental Engineering., 9*(2024), Article 100755. https://doi.org/10.1016/j.cscee.2024.100755

Aransiola, S. A., Victor-Ekwebelem, M. O., Ajiboye, A. E., Leh-Togi Zobeashia, S. S., Ijah, U. J. J., & Oyedele, O. J. (2023). Ecological impacts and toxicity of micro and nanoplastics in agroecosystem. In N. R. Maddela, K. V. Reddy, & P. Ranjit (Eds.), *Micro and nanoplastics in soil*. Springer. https://doi.org/10.10007/978-3-031-21195-9_10

Aransiola, S. A., Victor-Ekwebelem, M. O., & Maddela, N. R. (2022b). Microbial enzymes for sustainable development—Future guidelines. In N. R. Maddela, S. A. Aransiola, & R. Prasad (Eds.), *Ecological interplays in microbial enzymology* (1st ed.) Springer Nature Singapore Pte Ltd. ISSN: 2662-1681. https://www.springer.com/series/16324

Awasthi, M. K., et al. (2020). Influence of bamboo biochar on mitigating greenhouse gas emissions and nitrogen loss during poultry manure composting. *Bioresources Technology, 303*, Article 122952.

Baker, G. H. (1994). Earthworm, new discoveries, rural research. *A CSIRO Quarterly, 163*, 19–23.

Babaniyi, B. R., Ogundele, O. D., Bisi-omotosho, A., Babaniyi, E. E., & Aransiola, S. A. (2023). Remediation approaches in environmental sustainability. In Maddela, N. R., Eller, L. K. W.,

Prasad, R. (Eds.), *Microbiology for cleaner production and environmental sustainability*. CRC Press. ISBN: 9781032496061. https://www.routledge.com/Microbiology-for-Cleaner-Production-and-Environmental-Sustainability/Maddela-Eller-Prasad/p/book/9781032496061

Butt, K. R., & Grigoropoulou, N. (2010). Basic research tools for earthworm ecology. *Applied and Environmental Soil Science*, 1–12.

Cao, R., et al. (2020). The profile of antibiotic resistance genes in pig manure composting shaped by composting stage: Mesophilic-thermophilic and cooling-maturation stages. *Chemosphere, 250*(2020), 126181, https://doi.org/10.1016/j.chemosphere.2020.126181

Dada, E. O., Njoku, K. L., Osuntoki, A. A., & Akinola, M. O. (2015). A review of current techniques of in situ physicochemical and biological remediation of heavy metals polluted soil. *The Ethiopian Journal of Environmental Studies and Management* [Internet]. 2015 [Cited 2017 October 2] *8*(5), 606615. https://doi.org/10.4314/ejesm.v8i5.13. https://www.ajol.info//index.php/ejesm/article/view/120489

Dada, E. O., Njoku, K. L., Osuntoki, A. A., & Akinola, M. O. (2016). Heavy metal remediation potential of a tropical wetland earthworm, Libyodrilus violaceus (Beddard). *Iranica Journal of Energy and Environment, [Internet], 2016* [Cited 2017 November 2], *7*(3), 247–254. https://doi.org/10.5829/idosi.ijee.2016.07.03.06. http://www.ijee.net/article_64635

Demir, H., Polat, E., & Sönmez, İ. (2010). Ülkemiz İçin Yeni Bir Organik Gübre: Solucan Gübresi. *Tarım Aktüel, 14*, 54–60.

Domínguez, J. (2018). *Earthworms and vermicomposting* (pp. 63–77). IntechOpen.

Edwards, C. A., & Arancon, N. Q. (2004). The use of earthworms in the breakdown of organic wastes to produce vermicomposts and animal feed protein. In C. A. Edwards (Ed.), *Earthworm ecology* (2nd Ed.). CRC Press.

Entry, J. A., Wood, B. H., Edwards, J. H., & Wood, C. W. (1997). Influence of organic byproducts and nitrogen source on chemical and microbiological status of an agricultural soil. *Biology Fertilizer Soil, 24*, 196–204.

Ferreira L. D. & Merchant K. A. (1992). Field research in management accounting and control: A review and evaluation. *Accounting, Auditing & Accountability Journal, 5*(4).

Jackson M. L. (1967). *Soil chemical analysis*. Prentice Hall of India Pvt. Ltd. 8.30, 2021 tarihinde. https://books.google.com.tr/books?hl=tr&lr=&id=falseadresindenalındı

Kale, R. D. (1998). *Earthworm: Cinderella of organic farming*. Prism.

Kumar, A. A., & Gupta, R. K. (2018). The effects of vermicompost on growth and yield parameters of vegetable crop radish (Raphanus sativus). *Journal of Pharmacognosy and Phytochemistry, 7*(2), 589–592.

Leh-Togi Zobeashia, S. S, Aransiola, S. A., Ijah, U. J. J., & Abioye, O. P. (2018). Anaerobic digestion and agricultural application of organic wastes. *Advances in Environmental Research, 7*(2), 73–85. http://www.techno-press.org/content/?page=article&journal=aer&volume=7&num=2&ordernum=1

Li, Y., Zhao, C., Lu, X., Ai, X., & Qiu, J. (2018). Identification of a cytochrome P450 gene in the earthworm Eisenia fetida and its mRNA expression under enrofloxacin stress. *Ecotoxicology and Environmental Safety, 150*, 70–75.

Lim, S. L., Lee, L. H., & Wu, T. Y. (2016). Sustainability of using composting and vermicomposting technologies for organic solid waste biotransformation: recent overview, greenhouse gases emissions and economic analysis. *Journal of Cleaner Production, 111*, 262–278.

Liu, T., et al. (2021). Evaluation of cornstalk as bulking agent on greenhouse gases emission and bacterial community during further composting. *Bioresource Technology, 340*, Article 125713. https://doi.org/10.1016/j.biortech.2021.125713

Lleó, T., et al. (2013). Home and vermicomposting as sustainable options for biowaste management. *Journal of Cleaner Production, 47*, 70–76. https://doi.org/10.1016/j.jclepro.2012.08.011

Mısırlıoğlu, M. (2011). Toprak Solucanları Biyolojileri, Ekolojileri ve Türkiye Türleri. Nobel Akademik Yayıncılık.

NASA. (2021). Carbon dioxide concentration | NASA global climate change. Climate Change Vital Signs Planet Nov 16. http://climate.nasa.gov/vital-signs/carbon-dioxide/

Ojuolape, O. T., Aransiola, S. A., & Obideyi, O. A. (2015) Vermitechnology: A solution to environmental problems; a review. *Journal of Global Ecology and Environment, 3*(3), 127–135. http://www.ikpress.org/abstract/4516

Pascual, J. A., Ayuso, M., Hernández, T., & García, C. (1997). Phytotoxicity and fertilizer value of different organic materials. *Agrochemical, 41*, 50–62.

Renoux, A. Y., Sarrazin, M., Hawari, J., & Sunahara, G. I. (2000). Transformation of 2, 4, 6-trinitrotoluene in soil in the presence of the earthworm Eisenia andrei. *Environmental Toxicology and Chemistry: An International Journal, 19*(6), 1473–1480.

Rodriguez-Campos, J., Dendooven, L., Alvarez-Bernal, D., & Contreras-Ramos, S. M. (2014). Potential of earthworms to accelerate removal of organic contaminants from soil: A review. *Applied Soil Ecology, 79*, 10–25.

Sahariah, B., Sinha, I., Sharma, P., Goswami, L., Bhattacharyya, P., Gogoi, N., & Bhattacharya, S. S. (2014). Efficacy of bioconversion of paper mill bamboo sludge and lime waste by composting and vermiconversion technologies. *Chemosphere, 109*, 77–83.

Saint-Denis, M., Narbonne, J., Arnaud, C., Thybaud, E., & Ribera, D. (1999). Biochemical responses of the earthworm Eisenia fetida andrei exposed to contaminated artificial soil: Effects of benzo (a) pyrene. *Soil Biology & Biochemistry, 31*, 1837–1846.

Shi, Z., Liu, J., Tang, Z., Zhao, Y., & Wang, C. (2020a). Vermiremediation of organically contaminated soils: Concepts, current status, and future perspectives. *Applied Soil Ecology*. [Internet]. 2020 [Cited 2020 July 29] *147*, 103377. https://doi.org/10.1016/j.apsoil.2019.103377

Shi, Z., Liu, J., Tang, Z., Zhao, Y., & Wang, C. (2020b). Vermiremediation of organically contaminated soils: Concepts, current status, and future perspectives. *Applied Soil Ecology, 147*, Article 103377.

Sinha, R. K., Bharambe, G., & Ryan, D. (2008). Converting wasteland into wonderland—A low-cost nature's technology for soil remediation: A case study of vermiremediation of PAHs contaminated soil. *The Environmentalist* [Internet]. 2008 [Cited 2020 March 23]; *28*, 466–475. https://doi.org/10.1007/s10669008-9171-7

Sinha, R. K., Herat, S., Karmegam, N., Chauhan, K., & Chandran, V. (2010). Vermitechnology-The emerging 21st century bioengineering technology for sustainable development and protection of human health and environment: A review. *Dynamic Soil, Dynamic Plant, 4*(Special Issue 1), 22–47.

Sönmez, İ, Sönmez, S., & Kaplan, M. (2002). Çöp kompostunun bitki besim maddesi içerikleri ve bazı organik gübrelerle karşılaştırılması. *Selçuk Üniversitesi Ziraat Fakültesi Dergisi, 16*(29), 31–38.

Swati, A., & Hait, S. (2018). Greenhouse gas emission during composting and vermicomposting of organic wastes–A review. *Clean—Soil Air Water, 46*(6), 1700042.

Tomati, U., & Gali, E. (1995). Earthworms, soil fertility and plant productivity. *Acta Zoologica Fennica, 196*, 11–14.

Wang, J., et al. (2014). Emissions of ammonia and greenhouse gases during combined precomposting and vermicomposting of duck manure. *Waste Management, 34*(8), 1546–1552.

Zhao, S., Ma, X., Fang, S., & Zhu, L. (2016). Behaviors of N-ethyl perfluorooctane sulfonamide ethanol (N-EtFOSE) in a soil-earthworm system: Transformation and bioaccumulation. *Science of the Total Environment, 554–555*, 186–191.

Zhao, S., & Zhu, L. (2017). Uptake and metabolism of 10:2 fluorotelomer alcohol in soil earthworm (Eisenia fetida) and soil-wheat (Triticum aestivum L.) systems. *Environmental Pollution, 220*, 124–131.

Correction to: Vermiwash: A Vermicompost By-Product for Sustainable Agriculture

Pawan Kumar Rose, Sivaraman Balaji, Surojit Bera, Sandip Mondal, Manjeet Bansal, Mithilesh Kumar Jha, and Sagnik Chakraborty

Correction to:
Chapter 17 in: S. A. Aransiola and N. R. Maddela (eds.), *Vermitechnology: Economic, Environmental and Agricultural Sustainability*, https://doi.org/10.1007/978-3-031-96212-7_17

The original version of the chapter "Vermiwash: A Vermicompost By-Product for Sustainable Agriculture" was inadvertently published with the co-author name incorrectly mentioned as "Surojit Das" instead of "Surojit Bera". The correct author name has been updated in the chapter.

The updated version of this chapter can be found at
https://doi.org/10.1007/978-3-031-96212-7_17

S. A. Aransiola and N. R. Maddela (eds.), *Vermitechnology: Economic, Environmental and Agricultural Sustainability*, https://doi.org/10.1007/978-3-031-96212-7_19

MIX
Papier aus verantwortungsvollen Quellen
Paper from responsible sources
FSC® C105338

If you have any concerns about our products,
you can contact us on
ProductSafety@springernature.com

In case Publisher is established outside the EU,
the EU authorized representative is:
Springer Nature Customer Service Center GmbH
Europaplatz 3, 69115 Heidelberg, Germany

Printed by Libri Plureos GmbH
in Hamburg, Germany